Microwave Filters Using Parallel Coupled Lines

Leo Young

ARTECH HOUSE, INC.
Dedham, Massachusetts

affiliated with
Horizon House-Microwave Inc.

MICROWAVE FILTERS
USING
PARALLEL COUPLED LINES

Printed and bound in the United States of America.

Library of Congress catalog card number: 72-168945

INTRODUCTION

This book of reprints is a companion volume to another on PARALLEL COUPLED LINES AND DIRECTIONAL COUPLERS. Sections I, II, and III of that volume are equally applicable to this volume in that they form the link between the theoretical design of a filter or coupler in terms of its electrical parameters, on the one hand, and the physical realization of various structures (rectangular bars, round rods, stripline, micro-strip), on the other hand. Similar remarks can be made about Section IX of that volume, which covers slot line.

It was originally our intention to compile a book of reprints on all types of microwave filters, but the selection from so many good papers proved an insuperable task, if the book were to remain useful in connection with any given problem. There were too many related papers that tended to complement each other so that one could not be left out without including the other. It would have been possible to make a collection of **representative** papers, but the reader would not have found the volume self-contained. More important papers would have been left out than could have been included. So the idea evolved of making more than one collection of reprints, each on a more or less self-contained subject. The first two subjects selected were (1) filters using coupled lines (including interdigital, combline, and parallel coupled filters) and (2) TEM-mode parallel-line directional couplers (as a companion volume). As a result these two volumes are quite slim. We hope no one will have to pay for information of no interest to him. An attempt was made to select only those papers that seemed of lasting value, or that have been and still are frequently referenced, or that present generally useful design information. It is hoped to follow up these two volumes with other slim volumes, each based on a well-defined filter topic.

The papers reprinted here are grouped together under eight headings. The oldest paper is given first under each heading, followed by the newer papers in chronological order.

We would like to reference some books exclusively on the subject of microwave filters. The most widely used is probably Matthaei, Young and Jones.[1] A more recent excellent book with unique contributions is edited by Matsumoto[2] with several eminent contributors from Japan (K. Kuroda, J. Ishii, K. Hatori, R. Sato, N. Saito, T. Ikeda; as well as A. Matsumoto). Another unique book from Japan (also in English) is Uchida's[3] on multiwire coupled lines. One section of the present book of reprints is on diplexers and multiplexers; in this context we wish to reference the chapter by Matthaei and Cristal[4] in the **ADVANCES IN MICROWAVES**. The interested reader is also referred to three review papers on microwave filters, by Young[5,6] and Wenzel.[7] Professor H.J. Carlin has given lectures and written some most interesting tutorial reports with an original point of view, but these are not generally available. There are of course numerous books on lumped-constant filters, which underlie or provide prototypes for many distributed-constant filters.

There are many more filter types than covered in the papers reprinted here. We have confined ourselves to only one class—but an extremely important class—of filters, namely those using parallel coupled lines. For a review of many kinds of microwave filters, see books by Matthaei, Young and Jones,[1] and Matsumoto;[2] and for the latest developments in microwave filters, see especially the **IEEE Transactions on Microwave Theory and Techniques.**

It is hoped to publish other small volumes of reprints covering certain important aspects of microwave filters sometime in the future.

ACKNOWLEDGMENTS

We wish to express our appreciation to all the authors and publishers for permission to reprint the papers in this volume. The great majority of these papers originated in publications of the IEEE.

We have also received particularly helpful correspondence or comments from H.J. Carlin, S.B. Cohn, M. Dishal, T. Fjällbrant, W.J. Getsinger, M.C. Horton, J. Ishii, K. Kuroda, R.M. Kurzrok, R. Levy, A. Matsumoto, W. W. Mumford, J. Neirynck, R. Pregla, J.D. Rhodes, H.J. Riblet, B.M. Schiffman, C.P. Tresselt, H. Uchida, R.J. Wenzel, and H.A. Wheeler.

It is always a pleasure to acknowledge the assistance made available by Stanford Research Institute. In particular, I am grateful to Mrs. Dianna Tseu for lightening some of my editorial tasks.

Leo Young

TABLE OF CONTENTS

SECTION VII MULTIPLEXERS

SECTION VIII LINEAR PHASE FILTERS

REFERENCES

1. G.L. Matthaei, L. Young, and E.M.T. Jones, **Microwave Filters, Impedance Matching Networks, and Coupling Structures** (McGraw-Hill, New York, 1964).
2. A. Matsumoto, editor, **Microwave Filters**, Supplement 1 to **Advances in Microwaves** (general editor, L. Young) (Academic Press, New York, 1970).
3. H. Uchida, **Fundamentals of Coupler Lines and Multi-wire Antennas** (Sasaki Printing and Publishing Co., Ltd., Sendai, Japan, 1967).
4. G.L. Matthaei and E.G. Cristal, "Theory and Design of Diplexers and Multiplexers," **Advances in Microwaves**, Vol. 2, L. Young, editor, (Academic Press, New York, 1967).
5. L. Young, "Microwave Filters – 1965," **IEEE Trans. on Microwave Theory and Techniques**, Vol. MTT-13, No. 5, pp. 488-508 (September 1965).
6. L. Young, "Reflections on Microwave Filters and Couplers," **The Microwave Journal**, Vol. 11, No. 4, pp. 54-63 (April 1968).
7. R.J. Wenzel, "The Modern Network Theory Approach to Microwave Filter Design," **IEEE Trans. on Electromagnetic Compatibility**, Vol. EMC-10, pp. 196-209 (June 1968).
8. J.T. Bolljahn and G.L. Matthaei, "A Study of the Phase and Filter Properties of Arrays of Parallel Conductors Between Ground Planes," **Proc. IRE**, Vol. 50, pp. 299-311 (March 1962).
9. L.A. Robinson, "Wideband Interdigital Filters with Capacitively Loaded Resonators," **1965 G-MTT Symposium Digest**, pp. 33-37, Clearwater, Florida, May 1965.
10. E.G. Cristal, "Coupled Circular Cylindrical Rods Between Parallel Ground Planes," **IEEE Trans. on Microwave Theory and Techniques**, Vol. MTT-12, pp. 428-439 (July 1964).
11. H.J. Riblet, "An Explicit Derivation of the Relationship Between the Parameters of an Interdigital Structure and the Equivalent Transmission-Line Cascade," **IEEE Trans. on Microwave Theory and Techniques**, Vol. MTT-15, pp. 161-166 (March 1967) and Correction in **IEEE Trans. on Microwave Theory and Techniques**, Vol. MTT-16, p. 55 (January 1968).
12. J.D. Rhodes, "The Theory of Generalized Interdigital Networks," **IEEE Trans. on Circuit Theory**, Vol. CT-16, pp. 280-288 (August 1969).
13. R. Sato and E.G. Cristal, "Simplified Analysis of Coupled Transmission-Line Networks, **IEEE Trans. on Microwave Theory and Techniques**, Vol. MTT-18, pp. 122-131 (March 1970).
14. M. Dishal, Chapter 8 in **Reference Data for Radio Engineers**, Fifth Edition (International Telephone and Telegraph Corporation, 1968).
15. E.G. Cristal, "Data for Partially Decoupled Round Rods Between Parallel Ground Planes," **IEEE Trans. on Microwave Theory and Techniques**, Vol. MTT-16, pp. 311-314 (May 1968).
16. B.M. Schiffman, "Two Nomograms for Coupled-Line Sections for Band-Stop Filters," **IEEE Trans. on Microwave Theory and Techniques**, Vol. MTT-14, pp. 297-299 (June 1966).
17. M.C. Horton and R.J. Wenzel, "Realization of Microwave Filters with Equal Ripple Response in Both Pass and Stop Bands," **Proceedings of the Symposium on Generalized Networks**, pp. 257-288 (Polytechnic Institute of Brooklyn, New York, 1966).
18. B.M. Schiffman and L. Young, "Design Tables for an Elliptic-Function Band-Stop Filter (N=5)," **IEEE Trans. on Microwave Theory and Techniques**, Vol. MTT-14, pp. 474-482 (October 1966).
19. I. Rubinstein and R.L. Sleven, "Elliptic-Function Filter," **Microwaves**, Vol. 5, No. 12, pp. 50-55 (December 1966).
20. W. Steenart, "The Synthesis of Coupled Transmission Line All-Pass Networks in Cascades of 1 to n," **IEEE Trans. on Microwave Theory and Techniques**, Vol. MTT-11, pp. 23-29 (January 1963).
21. E.G. Cristal, "Theory and Design of Transmission Line All-Pass Equalizers," **IEEE Trans. on Microwave Theory and Techniques**, Vol. MTT-17, pp. 28-38 (January 1969) and "Analysis and Exact Synthesis of Cascaded Commensurate Transmission-Line C-Section All-Pass Networks," **IEEE Trans. on Microwave Theory and Techniques**, Vol. MTT-14, pp. 285-291 (June 1966).
22. B.M. Schiffman, "Multisection Microwave Phase Shift Network," **IEEE Trans. on Microwave Theory and Techniques**, Vol. MTT-16, p. 209 (April 1966).
23. B.M. Schiffman, "A New Class of Broadband Microwave 90-Degree Phase Shifters," **IEEE Trans. on Microwave Theory and Techniques**, Vol. MTT-6, pp. 232-237 (April 1958).
24. Leo Young, "Group Delay and Dissipation Loss in Transmission-Line Filters," **IRE Trans. on Microwave Theory and Techniques**, Vol. MTT-11, pp. 215-217 (May 1963).
25. S.B. Cohn, "Dissipation Loss in Multi-Coupled-Resonator Filters," **Proc. IRE**, Vol. 47, pp. 1342-1348 (August 1959).
26. H.A. Wheeler, "Formulas for the Skin Effect," **Proc. IRE**, Vol. 30, pp. 412-424 (September 1942).
27. Leo Young, "Peak Internal Fields in Direct-Coupled-Cavity Filters," **IRE Trans. on Microwave Theory and Techniques**, Vol. MTT-8, pp. 612-616 (November 1960).

17. R.J. Wenzel, "Exact Design of TEM Microwave Networks Using Quarter Wave Lines." Mr. D. D. Tang of the Massachusetts Institute of Technology, Lincoln Laboratory, has pointed out an error in the comparison between lumped and distributed bandpass filter networks given in Section IV-B. The slope of the lumped element filter in Equation (3) is incorrect and should be quite similar, with the lumped element bandpass prototype being slightly *superior* to the TEM highpass prototype when used as a bandpass filter. The correct slope ratio is found to be independent of the number of filter sections (n), and is equal to unity at zero bandwidth (B=0) and is equal to $\pi/4$ at 100% bandwidth (B=1).

LIST OF CORRECTIONS

$$\left.\frac{d\,|t|_L^2}{dX}\right|_{X=1-\frac{B}{2}} = \frac{n}{B(1-\frac{B}{2})} \qquad (1)$$

This leads to the corrected Equation (31),

$$\frac{\text{Slope}_{TEM}}{\text{Slope}_L} = \frac{\pi B}{2}\,\frac{(1-\frac{B}{2})}{\sin\frac{\pi B}{2}} \qquad (2)$$

and significantly changes the plot in Figure 9. A plot of the correct relationship is shown in Figure 1, and leads to the conclusion that the slopes of the two filters are

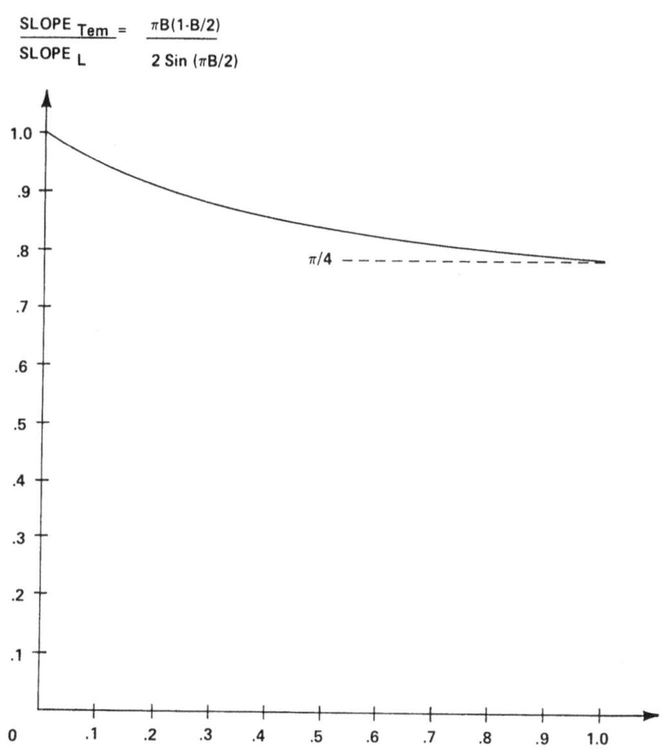

Fig. 1 Ratio of low band-edge slopes for TEM and lumped element bandpass filters vs. fractional bandwidth (B) for Butterworth response.

From the above results, the choice between a lumped element bandpass filter and a commensurate linelength highpass prototype used as a bandpass filter can be based on the practicality of the resulting structures rather than on any substantial theoretical selectivity advantage. It is emphasized that a further disadvantage of the commensurate linelength filter is the existence of repeating passbands which have not been considered in the above comparison.

List of Corrections (and One Addition) to "Microwave Filters, Impedance Matching Networks, and Coupling Structures" by G. L. Matthaei, Leo Young, and E.M.T. Jones (McGraw-Hill Book Co., New York, 1964)

On p. 134, below Eq. (4.10-16), it should read:

$$\theta = \pi/2n$$

On p. 164, Eq. (5.02-8), the tan δ should be in the numerator:

$$d_d = \frac{Bt^{\tan\delta}}{2}$$

On pp. 174 and 193 opposite conventions are followed: on p. 174 the odd mode refers to equal and opposite currents, but on p. 193, equal and opposite voltages are implied

On p. 174, Eq. (5.05-2), for $\left(\frac{\pi}{2}\cdot\frac{w+s}{2}\right)$ substitute:

$$\left(\frac{\pi}{2}\cdot\frac{w+s}{b}\right)$$

On p. 181, the source of Fig. (5.05-4) is Ref. 6 (and not Ref. 5, as stated).

On p. 220, the equivalent circuit on the top right of the page requires the addition of an ideal 1:-1 transformer (i.e., a phase reversal).

On p. 226, make same correction as on p. 220.

On p. 227, the two equivalent circuits on the right apply only when the source admittance is Y_A (impedance is Z_A). In general, a unit element of admittance Y_A (impedance Z_A) should be inserted immediately to the left of the ideal transformer in both Fig. 5.09-3(a) and (b).

On p. 240, the first expression should be:

$$\left| S_{13} \right|^2 = \left| S_{23} \right|^2 = \frac{4}{9 + (B_a + 2B_b)^2}$$

On p. 307, in the middle of the page, 172° should be 162°.

On p. 306, line 5, the referenced equation should be Eq. (6.09-2), and not Eq. (6.09-1).

On p. 354, Ref. 24, should refer to Vol. 8, and not Vol. 9.

On p. 392, line 10 from bottom, the last word on the line should be *wider*, and not narrower.

On p. 432, Fig. 8.02-3, expressions (6) and (8) (which are both correct) may be coordinated with Fig. 9.02-1(a), p. 523, and Fig. 9.04-1, p. 532, respectively, by adding $(Q_e)_A = \pi V_1/2$ in (6) and adding

$$k_{j,\,j+1}\Big/_{j=1 \text{ to } n-1} = 2\Big/\left(\pi\sqrt{V_{j,\,j+1}}\right) \quad \text{in} \qquad (8)$$

On. p. 677, Eq. (11.07-2), there is no need for g_0 or g_{n+1} or w'_1 to be set equal to unity. Then a factor w'_1 should be added on the right hand side of Eq. (11.07-3).

On p. 739, Eq. (12.05-7), the denominator of the last term should not be π, but $(1\text{-}\cos 2\delta)$.

On p. 752, Eq. (12.08-5), the left-hand side should be $Y_2 \tan\phi_0$, and not $Y_0 \tan\phi_0$.

On p. 761, expression for Z_4, the second denominator should be

$$Ag_4\,(1 + g_5\, g_6)^2$$

i.e., g_4 there should be g_6.

On p. 849, Eq. (4), the first equation should be

$$w = \frac{\omega_2 - \omega_1}{\omega_0}$$

GENERAL THEORY

I

Synthesis of a Class of Strip-Line Filters*

H. OZAKI† AND J. ISHII††

I. Introduction

RECENTLY, microwave filters and directional couplers using strip-lines have been greatly acclaimed because of their numerous advantages as a substitute for waveguide and coaxial lines. In this article the authors present a design theory for a class of strip-line filters on an insertion loss basis.

The terms "networks" or "filters" will be used throughout the paper to indicate that the networks under consideration are composed of lumped resistors and lossless transmission-lines with equal electrical length, unless the other meaning is definitely specified.

Some fundamental properties pertaining to the filters are described in this introductory section.

In general, all the parameters of the networks are expressed by algebraic functions of the tanh KL, where K is the propagation constant and L is the length of component lines. From this fact, Richards[1] proved that the networks may be treated exactly the same way as conventional lumped constant networks when they are discussed with respect to the new variable defined by

$$s = j \tan \pi f/2f_0, \qquad (1)$$

where f_0 is the real constant frequency (of which L is just equal to a quarter wavelength), and f is a real frequency variable. It is seen from the frequency transformation (1) that the frequency range $f(-f_0, f_0)$ corresponds to $s(-j\infty, j\infty)$ and the networks have a periodic characteristic with the period $2f_0$, as shown in Fig. 1. Therefore, to consider

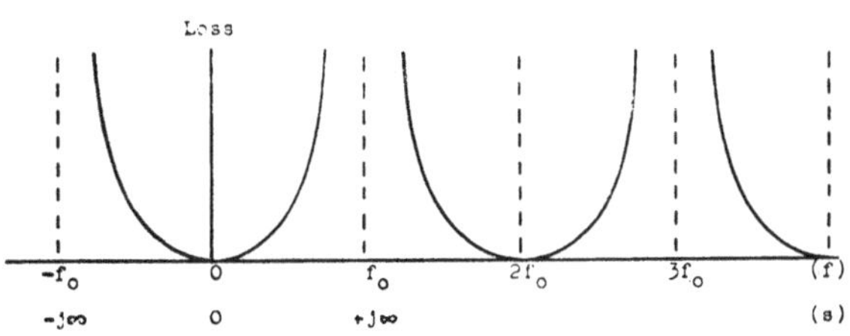

Fig. 1—Periodic characteristic of the filters with maximally flat low-pass character.

the network function over the whole range $s(-j\infty, j\infty)$ is equivalent to considering it over the real frequency range $f(-f_0, f_0)$. In design work the determination of f_0 is a basic matter in filter synthesis problems. Hereafter let f_0 denote the "standard frequency" of the filters. Close correspondence may be rigorously established between lumped and distributed parameter circuits by making the frequency transformation (1).

For practical operation and fabrication, the distributed-line ladder type filter with simple modes is suitable. Two kinds of ladders may exist. One consists of the chains of quarter-wave couplers (line type,[2] bar type,[3] or circuit type A[4]), and the other is made up of the alternative chains of lumped reactance elements and quarter-wave couplers.

Define[5] the "unit element" with the characteristic impedance Z_0 as a distributed element which has the transfer matrix

$$\frac{1}{\sqrt{1-s^2}}\begin{bmatrix} 1 & Z_0 s \\ s/Z_0 & 1 \end{bmatrix}.$$

In the case of a single line, a tandem unit element is a quarter-wave coupler and an output-short-circuited or output-open-circuited unit element is analogous to an inductance or a capacitance, respectively. In the case of a coupled line a unit element may be a rather abstract element, as described in Section III.

With the exception of the line type, it is desirable that all or some of the tandem unit elements contribute to the filter characteristic. But in this case synthesizing procedures require somewhat different considerations[4,6-8] which take the distributed nature into account. When all the tandem unit elements do not contribute to filter characteristic but serve only as connectors between reactive elements, whole operations[9] are quite analogous to those of lumped constant networks. The filters considered here belong to this class. However, it should be noted in both cases that there are two difficulties related to such realization through the existence of tandem unit elements. 1) It is impossible to apply directly ordinary frequency transformation by which low-pass character can be transformed into high-pass, band-pass, or band-elimination character. 2) Frequency normalization is

* Manuscript received by the PGCT, May 2, 1957; revised manuscript received, December 16, 1957.

† Dept. of Elec. Commun., Osaka University, Osaka, Japan.
†† Dept. of Elec., Kinki University, Osaka, Japan.

[1] P. I. Richards, "Resistor transmission-line circuits," Proc. IRE, vol. 36, pp. 217–220; February, 1948.

[2] H. Seidel, "Synthesis of a class of microwave filters," IRE Trans. on Microwave Theory and Techniques, vol. MTT-5, pp. 107–114; April, 1957.

[3] K. Kuroda, "Design of transmission-line filters," to be published.

[4] H. Ozaki and J. Ishii, "Synthesis of transmission-line networks and the design of uhf filters," IRE Trans. on Circuit Theory, vol. CT-2, pp. 325–336; December, 1955.

[5] This definition is slightly modified from that used in reference 4; the one presented here will be more adequate.

[6] Y. Kasahara, H. Ozaki, and T. Fujisawa, "Design of distributed-constant filter," *Tech. Rep. Osaka Univ.*, vol. 4, pp. 227–236.

[7] Y. Kasahara and T. Fujisawa, "Design of distributed-constant filter," *J. Inst. Elec. Commun. Engrs. Japan*, vol. 37, pp. 21–26; January, 1954. (In Japanese.)

[8] N. Ikeno, "Design of distributed constant networks," reviewed by H. Ozaki and J. Ishii, IRE Trans. on Circuit Theory, vol. CT-4, pp. 22–23; March, 1957.

[9] K. Kuroda, "Derivation methods of distributed constant filters from lumped constant filters," text for lectures at Joint Meeting of Kansai Branch, of Inst. of Elec. Commun., of Elec., and of Illumin. Engrs. of Japan, p. 32; October, 1952. (In Japanese.)

impossible with respect to s and this must be done by changing the ratio f/f_0. Therefore the reference frequency of normalization is uniquely determined and is equal to $f_0/2$. This means that the standard frequency f_0 is not independently determined.

The canonical section to be considered here is coupled strip-transmission-lines with equal electrical length. For the filters composed of this canonical section, a design method was published by Jones and Bolljahn[10] on an image parameter basis. The method adopted for design in this paper may be an extension of their paper; the filters synthesized here are over the band of interest $(-f_0, f_0)$. By using the frequency transformation (1), equivalent circuits are obtained for ten kinds of canonical sections, and these form the important part of this paper. Employing the equivalent circuits, it is possible to design line type, low-pass, high-pass, and band-pass ladder type filters in coupled strip-lines on an insertion loss basis. Synthesis procedures are those which were formerly developed for the realization of other TEM transmission-line filters. Respective transformations and physical configurations are depicted. Finally, a development is given of characteristic immittance formulas which will serve as a basis for practical fabrication of some filter constructions.

II. Immittance Matrices of Canonical Sections

The canonical section to be considered has shielded lossless coupled-strip-lines with equal electrical length, Fig. 2, p. 106. It is supposed that the center strip-conductors are supported midway and parallel to each other.

In design work of strips, we define the characteristic impedances and admittances of coupled lines corresponding to even and odd excitations.[10] Let these be represented by

$Z^a_{oe}(Y^a_{oe})$ = characteristic impedance (admittance) of a line with even excitation,

$Z^a_{oo}(Y^a_{oo})$ = characteristic impedance (admittance) of a line with odd excitation,

$Z^b_{oe}(Y^b_{oe})$ and $Z^b_{oo}(Y^b_{oo})$ = those of b line.

Following the method of Jones and Bolljahn,[9] postulating four current sources, we can get the impedance matrix of coupled lines. As stated in the previous section, the impedance matrix of a canonical section is expressed in terms of the new frequency sensitive variable s defined by (1) and it becomes

$$\begin{bmatrix} Z_{11} & Z_{12} & Z_{13} & Z_{14} \\ Z_{21} & Z_{22} & Z_{23} & Z_{24} \\ Z_{31} & Z_{32} & Z_{33} & Z_{34} \\ Z_{41} & Z_{42} & Z_{43} & Z_{44} \end{bmatrix} = \begin{bmatrix} \frac{Z^a_{oe}+Z^a_{oo}}{2}\cdot\frac{1}{s}, & \frac{Z^a_{oe}-Z^a_{oo}}{2}\cdot\frac{1}{s}, & \frac{Z^a_{oe}-Z^a_{oo}}{2}\cdot\frac{\sqrt{1-s^2}}{s}, & \frac{Z^a_{oe}+Z^a_{oo}}{2}\cdot\frac{\sqrt{1-s^2}}{s} \\ \frac{Z^b_{oe}-Z^b_{oo}}{2}\cdot\frac{1}{s}, & \frac{Z^b_{oe}+Z^b_{oo}}{2}\cdot\frac{1}{s}, & \frac{Z^b_{oe}+Z^b_{oo}}{2}\cdot\frac{\sqrt{1-s^2}}{s}, & \frac{Z^b_{oe}-Z^b_{oo}}{2}\cdot\frac{\sqrt{1-s^2}}{s} \\ \frac{Z^b_{oe}-Z^b_{oo}}{2}\cdot\frac{\sqrt{1-s^2}}{s}, & \frac{Z^b_{oe}+Z^b_{oo}}{2}\cdot\frac{\sqrt{1-s^2}}{s}, & \frac{Z^b_{oe}+Z^b_{oo}}{2}\cdot\frac{1}{s}, & \frac{Z^b_{oe}-Z^b_{oo}}{2}\cdot\frac{1}{s} \\ \frac{Z^a_{oe}+Z^a_{oo}}{2}\cdot\frac{\sqrt{1-s^2}}{s}, & \frac{Z^a_{oe}-Z^a_{oo}}{2}\cdot\frac{\sqrt{1-s^2}}{s}, & \frac{Z^a_{oe}-Z^a_{oo}}{2}\cdot\frac{1}{s}, & \frac{Z^a_{oe}+Z^a_{oo}}{2}\cdot\frac{1}{s} \end{bmatrix}.$$

Postulating voltage sources instead of current sources, the resulting admittance matrix is as follows:

$$\begin{bmatrix} Y_{11} & Y_{12} & Y_{13} & Y_{14} \\ Y_{21} & Y_{22} & Y_{23} & Y_{24} \\ Y_{31} & Y_{32} & Y_{33} & Y_{34} \\ Y_{41} & Y_{42} & Y_{43} & Y_{44} \end{bmatrix}$$

$$= \begin{bmatrix} \frac{Y^a_{oo}+Y^a_{oe}}{2}\cdot\frac{1}{s}, & \frac{Y^a_{oo}-Y^a_{oe}}{2}\cdot\frac{1}{s}, & \frac{Y^a_{oo}-Y^a_{oe}}{2}\cdot\frac{\sqrt{1-s^2}}{s}, & \frac{Y^a_{oo}+Y^a_{oe}}{2}\cdot\frac{\sqrt{1-s^2}}{s} \\ \frac{Y^b_{oo}-Y^b_{oe}}{2}\cdot\frac{1}{s}, & \frac{Y^b_{oo}+Y^b_{oe}}{2}\cdot\frac{1}{s}, & \frac{Y^b_{oo}+Y^b_{oe}}{2}\cdot\frac{\sqrt{1-s^2}}{s}, & \frac{Y^b_{oo}-Y^b_{oe}}{2}\cdot\frac{\sqrt{1-s^2}}{s} \\ \frac{Y^b_{oo}-Y^b_{oe}}{2}\cdot\frac{\sqrt{1-s^2}}{s}, & \frac{Y^b_{oo}+Y^b_{oe}}{2}\cdot\frac{\sqrt{1-s^2}}{s}, & \frac{Y^b_{oo}+Y^b_{oe}}{2}\cdot\frac{1}{s}, & \frac{Y^b_{oo}-Y^b_{oe}}{2}\cdot\frac{1}{s} \\ \frac{Y^a_{oo}+Y^a_{oe}}{2}\cdot\frac{\sqrt{1-s^2}}{s}, & \frac{Y^a_{oo}-Y^a_{oe}}{2}\cdot\frac{\sqrt{1-s^2}}{s}, & \frac{Y^a_{oo}-Y^a_{oe}}{2}\cdot\frac{1}{s}, & \frac{Y^a_{oo}+Y^a_{oe}}{2}\cdot\frac{1}{s} \end{bmatrix}.$$

Of course, these are the symmetrical matrices: $Z^a_{oe} - Z^a_{oo} = Z^b_{oe} - Z^b_{oo}$ and $Y^a_{oo} - Y^a_{oe} = Y^b_{oo} - Y^b_{oe}$.

[10] E. M. T. Jones and J. T. Bolljahn, "Coupled-strip-transmission line filters and directional couplers," IRE Trans. on Microwave Theory and Techniques, vol. MTT-4, pp. 75–81; April, 1956.

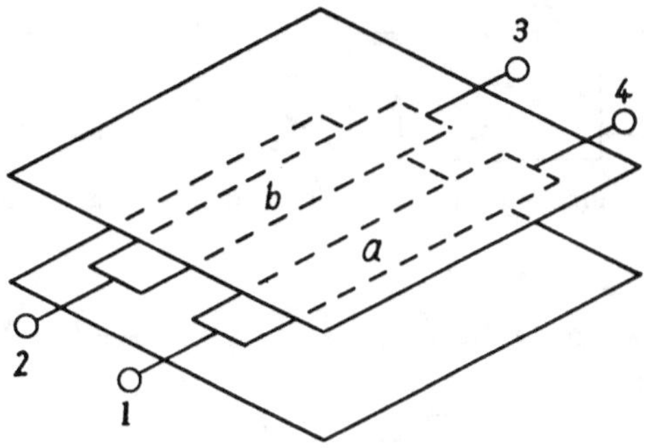

Fig. 2—Shielded coupled strip-lines: canonical section.

III. Equivalent Transformations

Since it is possible to open-circuit or short-circuit any two terminals of a canonical section, it can be regarded as an unbalanced two-terminal pair network. Tables I and II show equivalent transformations[11] for the ten kinds of canonical sections. They are obtained from the immittance matrices for various terminal connections and show the close correspondence between lumped and coupled-line distributed parameter circuits. Although the tables also include cases of rare availability, these are presented for the reader's information.

Table I shows the relationships between element values of original and transformed circuits in the case of equal widths of center conductors, where $Z^a_{oe} = Z^b_{oe} = Z_{oe}$ and $Z^a_{oo} = Z^b_{oo} = Z_{oo}$. We can determine the dimensions of infinitely thin strip-lines through Cohn's impedance monograms[12] for given element values of equivalent circuits.

In Table II, the relationships of element values are presented for the case of different widths of center conductors; the canonical sections with equal widths are more or less impractical to utilize in synthesis procedures.

The term "ue" means unit element, and Z_o and Y_o are the characteristic impedance and admittance of a unit element, respectively. Subsidiary conditions in equivalent transformations 4 and 7 must be satisfied, otherwise one of the characteristic impedances may have a complex value. The element values in equivalent circuits 5 and 9 give the admittances $(Y^a_{oe} + Y^a_{oo})/2$, etc., and all the others give the impedances $(Z^a_{oe} + Z^a_{oo})/2$, etc. Circuits 7 and 8 contain an ideal transformer.

TABLE I

Original circuits	Equivalent circuits	Element values (equal widths)
1)	Z_o, u.e.	$Z_o = \dfrac{Z_{oe} + Z_{oo}}{2}$
2)	Y_o, u.e.	$Y_o = \dfrac{1}{2}\cdot\left(\dfrac{1}{Z_{oe}} + \dfrac{1}{Z_{oo}}\right)$
3)	Y_o, L, u.e.	$Z_{oe} = \dfrac{Y_oL+1+\sqrt{Y_oL(Y_oL+1)}}{Y_o}$, $Z_{oo} = \dfrac{Y_oL+1+\sqrt{Y_oL(Y_oL+1)}}{Y_o\left[2Y_oL+1+2\sqrt{Y_oL(Y_oL+1)}\right]}$
4)	C, L	$Z_{oe} = \dfrac{L}{C}\cdot\dfrac{1}{Z_{oo}}$, $Z_{oo} = L - \sqrt{L(L-\frac{1}{C})}$
		Subsidiary condition: $LC > 1$

TABLE II

Original circuits	Equivalent circuits	Element values (different widths) $(Z^a_{oe}+Z^a_{oo})/2$, $(Y^a_{oe}+Y^a_{oo})/2$	$(Z^a_{oo}+Z^b_{oo})$, $(Y^a_{oe}+Y^b_{oe})$	$(Z^b_{oe}+Z^b_{oo})/2$, $(Y^b_{oe}+Y^b_{oo})/2$
(5)	L_1, Y_o u.e., L_2	$Y_o + \dfrac{1}{L_1}$	$\dfrac{1}{L_1} + \dfrac{1}{L_2}$	$\dfrac{1}{L_2} + Y_o$
(6)	C_1, Y_o u.e., C_2	$Z_o + \dfrac{1}{C_1}$	$\dfrac{1}{C_1} + \dfrac{1}{C_2}$	$\dfrac{1}{C_2} + Z_o$
(7)	C, 1:n, L $(1<n)$	$\dfrac{1}{C} + \dfrac{L}{n^2}$	$\dfrac{1}{C} + L(1-\dfrac{1}{n})^2$	L
		Subsidiary condition: $n^2 > LC\ n-1)$		
(8)	1:n $(n<0)$, C, L	$\dfrac{1}{C} + \dfrac{L}{n^2}$	$L(1-\dfrac{1}{n})^2$	$\dfrac{1}{C} + L$
(9)	L_1, L_3, L_2	$\dfrac{1}{L_3} + \dfrac{1}{L_1}$	$\dfrac{1}{L_1} + \dfrac{1}{L_2}$	$\dfrac{1}{L_2} + \dfrac{1}{L_3}$
(10)	C_1, C_2, C_3	$\dfrac{1}{C_3} + \dfrac{1}{C_1}$	$\dfrac{1}{C_1} + \dfrac{1}{C_2}$	$\dfrac{1}{C_2} + \dfrac{1}{C_3}$

IV. Synthesis Procedures

The equivalent circuits presented in the previous section offer the possibility of designing strip-line filters by using the synthesis techniques of other TEM transmission-line filters which have already been evolved.

This section is devoted to the synthesis procedures for four types of circuit: line type, low-pass ladder, high-pass ladder, and band-pass ladder.

[11] These transformations may be obtained in more dexterous fashion by using K. Kuroda's method which was included in a dissertation submitted in partial fulfillment of the degree of Doctor of Engineering, 1955. Since his thesis was published in Japanese, the calculations made here are by the method in Jones and Bolljahn, *op. cit.* We shall review Kuroda's method in the near future.

[12] S. B. Cohn, "Shielded coupled-strip transmission line," IRE Trans. on Microwave Theory and Techniques, vol. MTT-3, pp. 29–38; October, 1955.

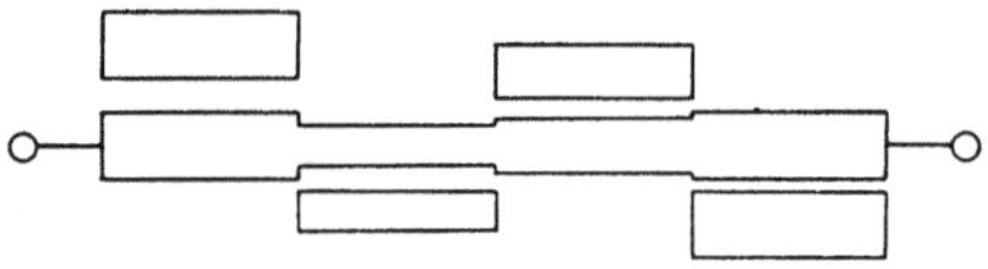

Fig. 3—Physical configuration of line type.

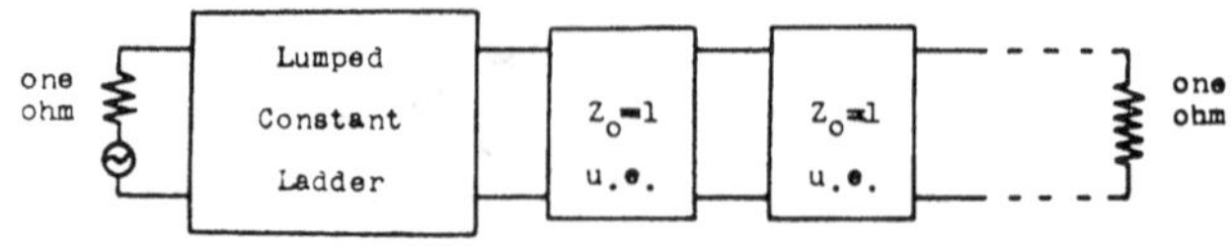

Fig. 4—A lumped constant ladder containing degenerate line type.

TABLE III

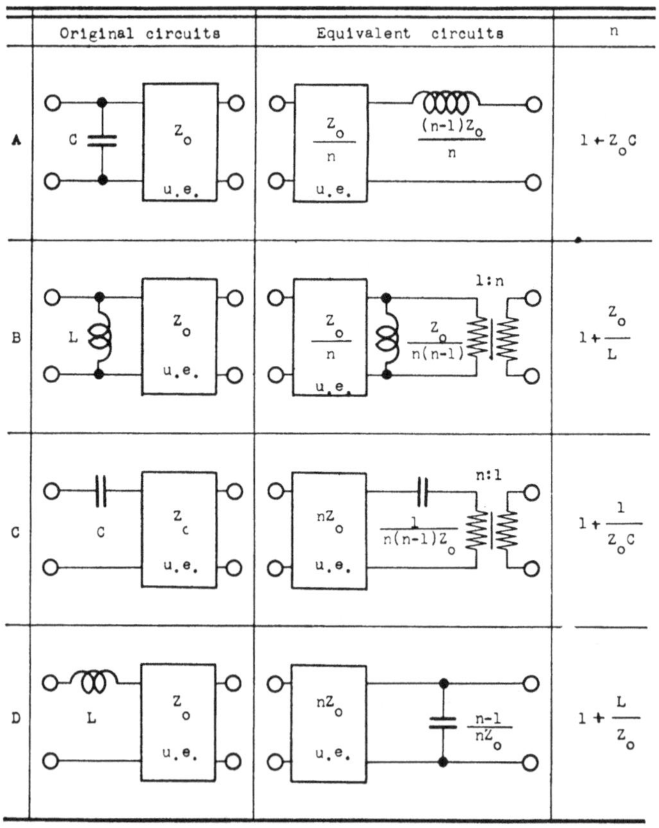

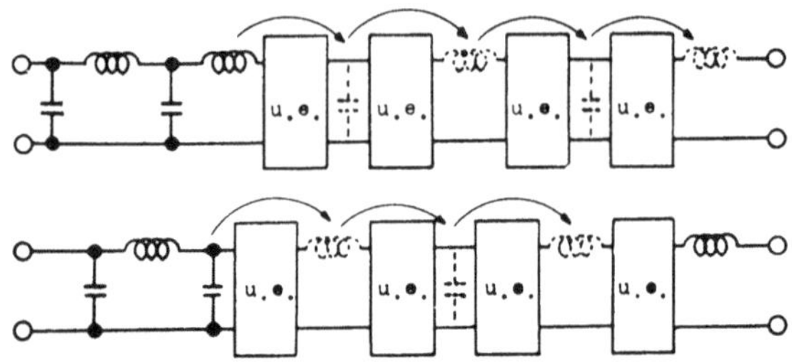

Fig. 5—Transformation procedures of low-pass ladder.

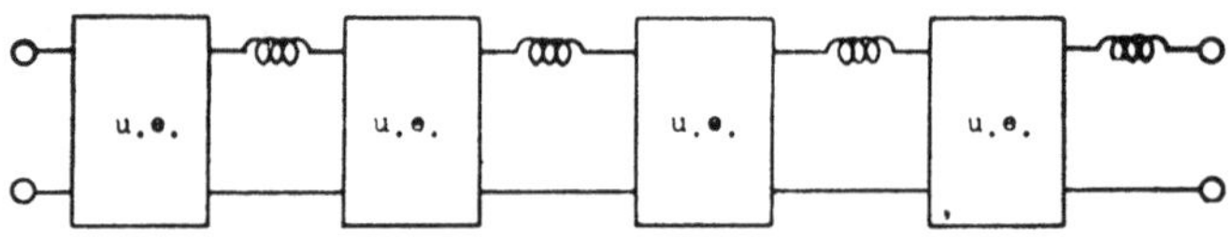

Fig. 6—A network with series inductances equivalent to low-pass ladder

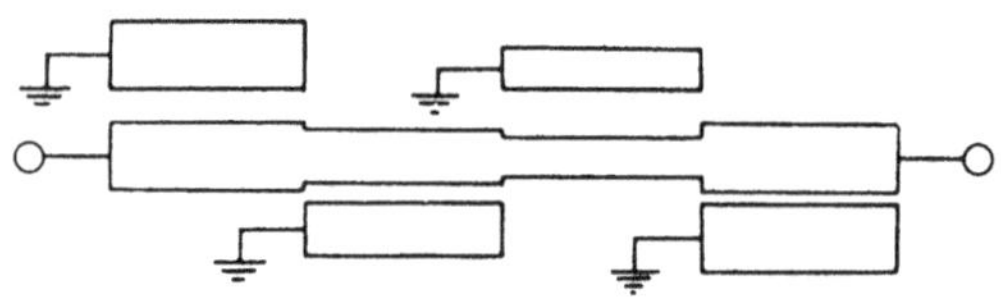

Fig. 7—Physical configuration of low-pass ladder.

A. Line Type

Circuits 1 and 2 in Table I may be used. Fig. 3 shows an example of the physical configuration. The details on realization techniques are omitted since they are relatively well known.[3,4,6-8]

The method used for synthesis in the succeeding three subsections is based on the equivalent transformations which were presented by Kuroda.[3,9] Necessary equivalent circuits for present applications are shown in Table III. A short description of this method will be given below for convenience.

We can generally suppose that any lumped constant ladder has a one-ohm terminal end resistance. Every insertion loss characteristic may be left unchanged by inserting a certain number of unit elements with a 1-ohm wave impedance between the lumped constant ladder and load resistances, as shown in Fig. 4. This inserted network is in effect an all-pass circuit or a "degenerate line type." In this case connect the degenerate line type at the output terminals of the ladder. Applying the equivalent transformations shown in Table III, the original lumped constant ladder is identically transformed into the network in which appropriate lumped reactive elements and unit elements are chained alternatively. The transformed circuit thus obtained leads to a realizable distributed parameter filter representation.

B. Low-Pass Ladder with No Finite Transmission Zeros

The equivalent transformations *A* and *D* in Table III may be used and the schematic diagram shown in Fig. 5 may help to make the argument clear. After one obtains the transformed network shown in Fig. 6, it may be realized by using the equivalent circuit 3 in Table I. Fig. 7 shows the physical configuration.

C. High-Pass Ladder with Manifold Transmission Zeros at the Origin

For the high-pass prototype shown in Fig. 8(a), the transformation procedures are entirely similar to those of the low-pass ladder. Circuits 5–7 in Table II and the transformations *B* and *C* in Table III may be utilized. In this case, however, the following operations are required.

1) An ideal transformer combined with a shunt inductance and a series capacitance is equivalent to circuit 7 in Table II. As pointed out by Jones and Bolljahn,[9] it is necessary to realize this circuit at accessible terminals in order to prevent the production of stray couplings in the

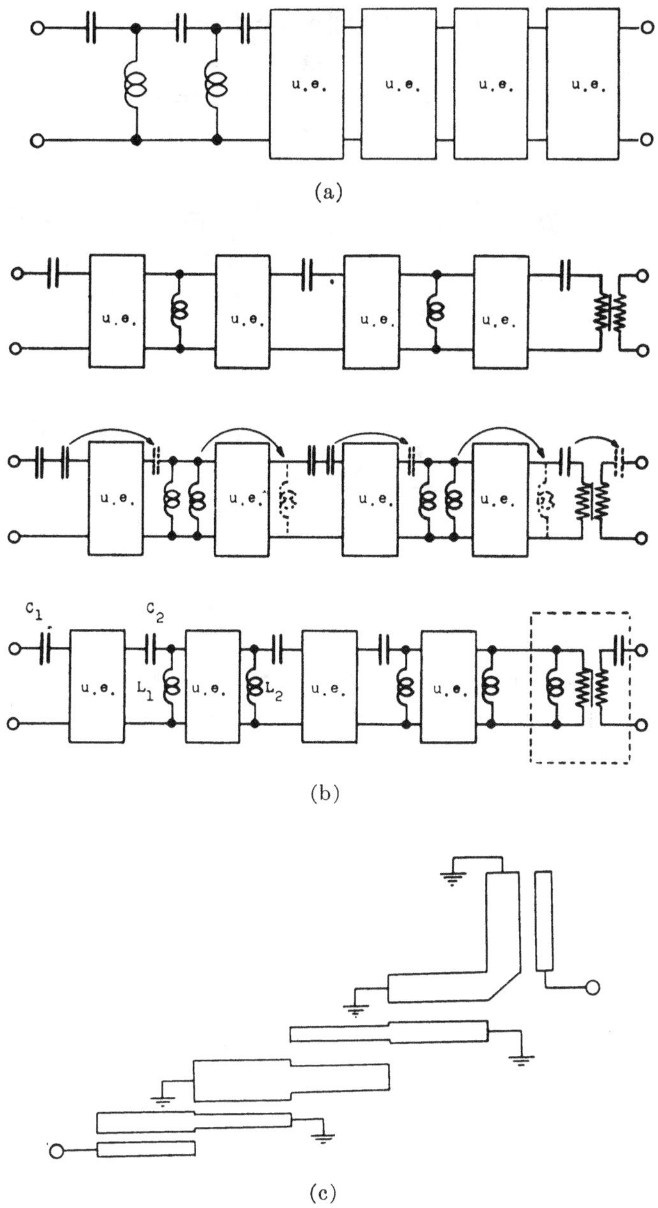

Fig. 8—(a) High-pass ladder containing degenerate line type. (b) Illustrative transformation procedures and transformed circuits of high-pass ladder. (c) Physical configuration of high-pass ladder.

filter section array. Every ideal transformer appearing in each step of the equivalent transformations must be transposed into accessible terminals. This operation is always possible since an ideal transformer has a diagonal transfer matrix.

2) A reactive element in any shunt or series arm is appropriately divided into two parts and one of these elements is transposed to another side of a tandem unit element as shown in Fig. 8(b). The "appropriate" separations of a reactive element must be done so that the quivalent circuit 7 is realizable. Therefore the widths of strip-lines must generally be made unequal since symmetrical conditions, *e.g.*, $C_1 = C_2$, $L_1 = L_2$, $\cdots$, etc. in Fig. 8(b), may not always be retained. After this operation, the equivalent circuits 5–7 in Table II are applied.

The illustrative transformation procedures and transformed circuits, and the physical layout of the high-pass prototype are furnished by Fig. 8(b) and 8(c), respectively.

D. Band-Pass Ladder

The band-pass prototype derived from the low-pass ladder shown in Fig. 9 is to be constructed through the same procedures used for the two former types, but in many cases it will not be feasible to realize this prototype because of the difficulty noted in Section IV-C, 1).

The series capacitive coupled and shunt inductive coupled band-pass ladder, shown in Fig. 10(a) and Fig. 11(a), respectively, may be realized. Their transformed circuits are illustrated in Fig. 10(b) and Fig. 11(b); physical layouts are drawn in Fig. 10(c) and Fig. 11(c).

V. Characteristic Immittances

Characteristic-impedance formulas for coupled strip-lines when two center conductors have equal dimensions have been discussed in detail.[12] In this section, as an approach to design work for some filter construction, a development is given of characteristic-immittance formulas for determining static capacities of coupled thin strip-lines with different widths.

Since the two lines have mutual coupling and respective capacitances to ground, they are analogous to a capacitive π network as shown in Fig. 12, where $I_a = I_b$ ($V_a = V_b$) and $I_a = -I_b$ ($V_a = -V_b$) mean even and odd current (voltage) excitations, respectively, and give the characteristic impedances (admittances) in each case. The resulting characteristic immittances are as follows:

$$
\begin{aligned}
Z^a_{oe} &= \frac{P}{Q}\cdot(3C_1 + C_2 - C_{12}) \text{ ohms},\\
Z^a_{oo} &= \frac{P}{Q}\cdot(C_1 - C_2 + C_{12}) \text{ ohms},\\
Z^b_{oe} &= \frac{P}{Q}\cdot(C_1 + 3C_2 - C_{12}) \text{ ohms},\\
Z^b_{oo} &= \frac{P}{Q}\cdot(-C_1 + C_2 + C_{12}) \text{ ohms},\\
Y^a_{oe} &= \frac{1}{P}\cdot(-C_1 + C_2 + C_{12}) \text{ mhos},\\
Y^a_{oo} &= \frac{1}{P}\cdot(C_1 + 3C_2 - C_{12}) \text{ mhos},\\
Y^b_{oe} &= \frac{1}{P}\cdot(C_1 - C_2 + C_{12}) \text{ mhos},\\
Y^b_{oo} &= \frac{1}{P}\cdot(3C_1 + C_2 - C_{12}) \text{ mhos},
\end{aligned} \tag{2}
$$

where $P = 200\sqrt{\epsilon_r}/3$, ϵ_r = specific dielectric constant of surrounding medium of lines, $Q = 4(C_1 \cdot C_{12} + C_{12} \cdot C_2 + C_2 \cdot C_1) - (C_1 + C_{12} + C_2)^2$, C_1 = the self capacity ($\mu\mu$f/cm) of b line to ground when a line is grounded, C_2 = the self capacity ($\mu\mu$f/cm) of a line to ground when b line is grounded, and C_{12} = the capacity ($\mu\mu$f/cm) between the parallel combination of the two center lines and

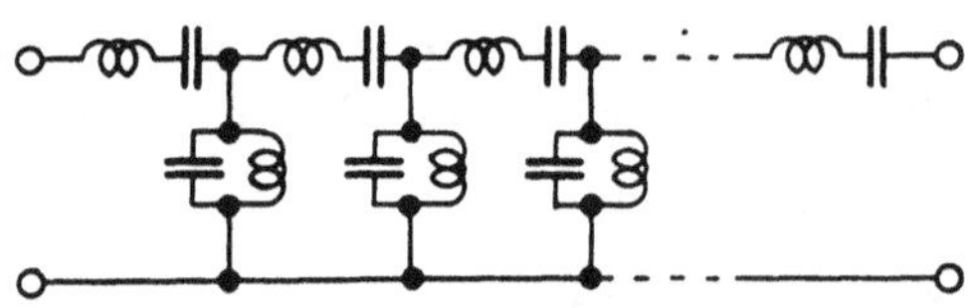

Fig. 9—Band-pass ladder derived from low-pass prototype.

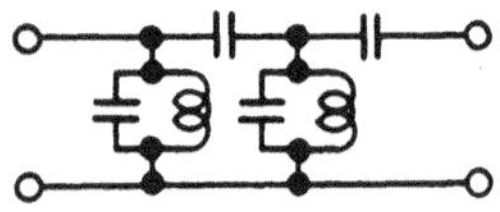

(a)

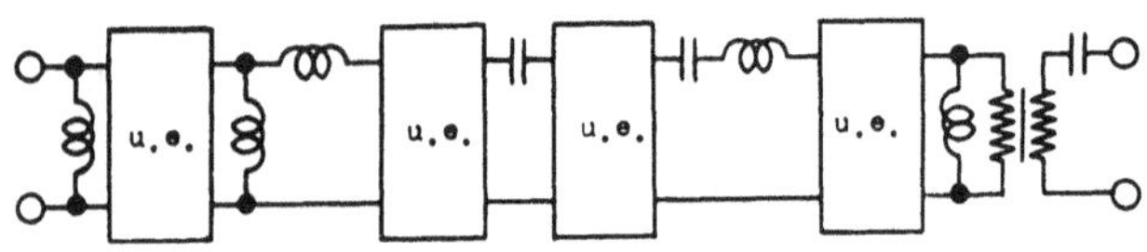

(b)

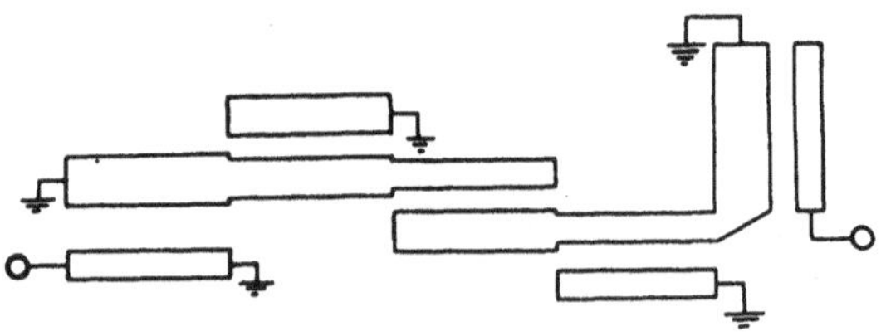

(c)

Fig. 10—(a) Series capacitive coupling band-pass filter. (b) Transformed circuit equivalent to series capacitive coupling band-pass filter. (c) Physical configuration of series capacitive coupling band-pass filter.

ground. By solving the simultaneous equations (2) with respect to C_1, C_{12}, and C_2, the following relations result:

$$
\begin{aligned}
C_1 &= \frac{4P}{R}\cdot(Z^a_{oe} + Z^a_{oo}) = \frac{P}{4}\cdot(Y^b_{oe} + Y^b_{oo}),\\
C_{12} &= \frac{8P}{R}\cdot(Z^a_{oo} + Z^b_{oo}) = \frac{P}{2}\cdot(Y^a_{oe} + Y^b_{oe}), \qquad (3)\\
C_2 &= \frac{4P}{R}\cdot(Z^b_{oe} + Z^b_{oo}) = \frac{P}{4}\cdot(Y^a_{oe} + Y^a_{oo}),
\end{aligned}
$$

where

$$
\begin{aligned}
R = 2[(Z^a_{oe} + Z^a_{oo})\cdot 2(Z^a_{oo} + Z^b_{oo}) &+ 2(Z^a_{oo} + Z^b_{oo})\\
\cdot(Z^b_{oe} + Z^b_{oo}) &+ (Z^a_{oe} + Z^a_{oo})\cdot(Z^b_{oe} + Z^b_{oo})]\\
- [(Z^a_{oe} + Z^a_{oo})^2 &+ 4(Z^a_{oo} + Z^b_{oo})^2 + (Z^b_{oe} + Z^b_{oo})^2].
\end{aligned}
$$

We can determine the static capacities of strips by (3) for given impedances $(Z^a_{oe} + Z^a_{oo})/2$, etc., and admittances $(Y^a_{oe} + Y^a_{oo})/2$, etc., in Table II.

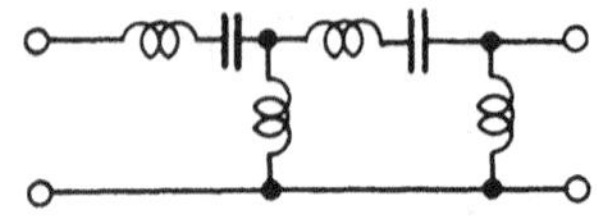

(a)

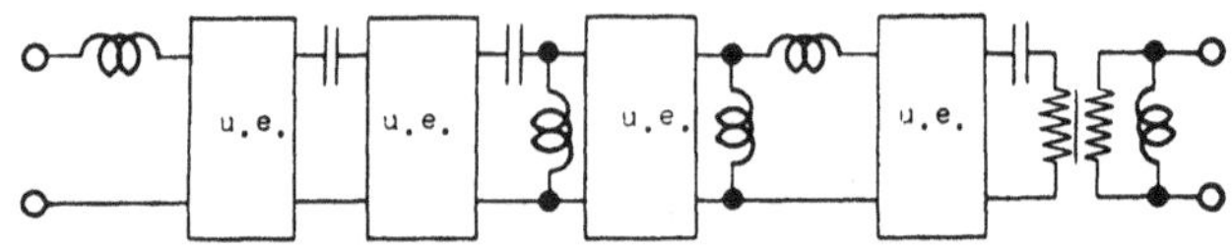

(b)

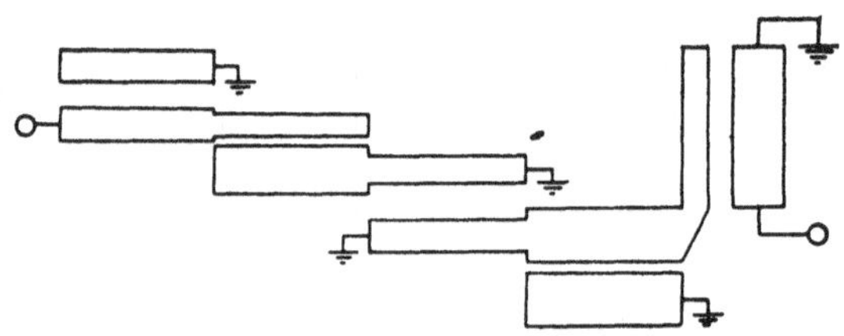

(c)

Fig. 11—(a) Shunt inductive coupling band-pass filter. (b) Transformed circuit equivalent to shunt inductive coupling band-pass filter. (c) Physical configuration of shunt inductive coupling band-pass filter.

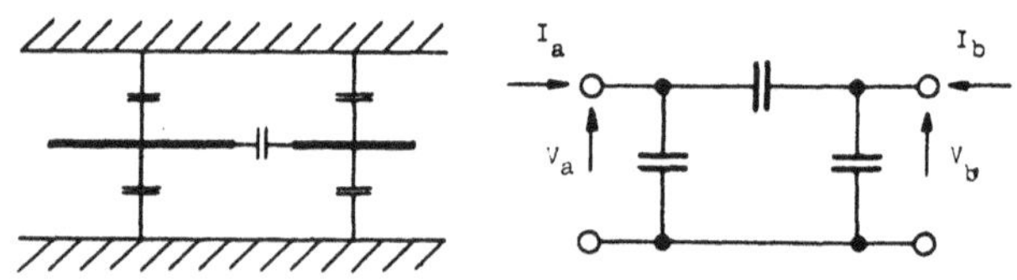

Fig. 12—Coupled strip-lines analogous to capacitive π network.

VI. Conclusion

All filters in this paper utilize distributed coupling effects. The designing methods adopted here are naturally suitable for synthesis in both broad and narrow band. Line type, low-pass and high-pass ladder networks may be particularly fitting for an impedance matching device and the direct coupled resonator filters presented in the Section IV may be compatible for narrow band-pass filters with clear performance.

Tables I and II are applicable for any kind of TEM transmission-line and lead to new network configurations with rigorous correspondence to lumped parameter filters.

VII. Acknowledgment

The authors are indebted to Prof. K. Kuroda of the Department of Electronics, University of Shizuoka, for his valuable suggestions and kindly discussions.

Reprinted from the IRE TRANSACTIONS OF THE PROFESSIONAL GROUP ON *CIRCUIT THEORY*
Volume CT-5, Number 2, June, 1958
PRINTED IN THE U.S.A.

Richards' Theorem Extended to a Two-Terminal-Pair Network*

9

By

Nobuji SAITO

(Received December 20, 1961)

Synopsis

Richards' theorem is well known as the key theorem that Bott and Duffin applied to the present method of impedance synthesis not requiring a transformer. When an n-terminal-pair network is synthesized, a positive real matrix $[Y(p)]$, instead of a admittance $Y(p)$, will be used to represent a network. Dealing with Richards' theorem as a matrix, we can obtain Richards' theorem extending into an n-terminal-pair network. The extended Richard's theorem has proposed an important method for synthesizing an n-terminal-pair network. And in the case of a distributed constant network, the extraction of a coupled transmission line element has been proposed as a consequential means of a distributed constant network synthesis.

1. Richards' Theorem

P.I. Richards reports in his paper[1] the following theorem.

If $Y(p)$ is a positive real function and p_1 is any positive real number,

$$Y'(p) = Y(p_1)\frac{pY(p) - p_1 Y(p_1)}{pY(p_1) - p_1 Y(p)} \tag{1}$$

is also a positive real function. This is Richards' theorem, which is famous as the key theorem that Bott and Duffin applied to the present method of impedance synthesis[2] not requiring a transformer.

2. Extension of Richards' Theorem to an n-Terminal-Pair Network.

When an n-terminal-pair network is synthesized, a positive real admittance matirx $[Y(p)]$ instead of an admittance $Y(p)$ will be used to represent a network. On an assumption that Richards' theorem could be extended to an n-terminal-pair network, it might be rewritten as follows:

If the matrix $[Y(p)]$ is a real possitive one, p_1 being any positive real number,

$$[Y'(p)] = [[1]p_1 - [Y(p)][Y(p_1)]^{-1}p]^{-1}\cdot[[Y(p)]p_1 - [Y(p_1)]p] \tag{2}$$

* This paper was published in Japanese by the auther on the Convention Record of Tohoku Section of I.E.C.J. (The Insitute of Electrical Communication Engineers of Japan) Octover, 1959, and on the Journal of I.E.C.J., July, 1961 (44, 7, p. 1033)

is also a positive real matrix. Eq. (2) may be a general form of Richards' theorem extended to an n-terminal-pair network. In fact, it is possible to prove that if $[Y(p)]$ is a positive real matrix, then $[Y'(p)]$ is also positive real. The process of verification will be explained in detail in the following section.

3. $[Y'(p)]$ of a Two-Terminal-Pair Network.

A demonstration of the above theorem will be given for a two-terminals-pair network which is the simplest one of n-terminal-pair networks. In this case, elements of $[Y'(p)]$ from eq. (2) are as follows:

$$\left.\begin{aligned}
Y_{11}' &= \frac{1}{\nabla}\Big[Y_{11}(p_1^2-p^2) - \frac{z_{22}\,p}{z_{11}z_{22}-z_{12}^2}\cdot \\
&\quad \cdot\{p_1 - p(z_{11}Y_{11}+z_{22}Y_{22}+2z_{12}Y_{12})\} - z_{22}\,pp_1(Y_{11}Y_{22}-Y_{12}^2)\Big] \\
Y_{22}' &= \frac{1}{\nabla}\Big[Y_{22}(p_1^2-p^2) - \frac{z_{11}\,p}{z_{11}z_{22}-z_{12}^2}\cdot \\
&\quad \cdot\{p_1 - p(z_{11}Y_{11}+z_{22}Y_{22}+2z_{12}Y_{12})\} - z_{11}\,pp_1(Y_{11}Y_{22}-Y_{22}^2)\Big] \\
Y_{12}' &= \frac{1}{\nabla}\Big[Y_{12}(p_1^2-p^2) + \frac{z_{12}\,p}{z_{11}z_{22}-z_{12}^2}\cdot \\
&\quad \cdot\{p_1 - p(z_{11}Y_{11}+z_{22}Y_{22}+2z_{12}Y_{12})\} + z_{12}\,pp_1(Y_{11}Y_{22}-Y_{12}^2)\Big] \\
\nabla &= p_1^2 - pp_1(z_{11}Y_{11}+z_{22}Y_{22}+2z_{12}Y_{12}) \\
&\quad + p^2(z_{11}z_{22}-z_{12}^2)(Y_{11}Y_{22}-Y_{12}^2)\,,
\end{aligned}\right\} \qquad (3)$$

where Y_{11}, Y_{22} and Y_{12} are the elements of $[Y(p)]$ and z_{11}, z_{22} and z_{12} are those of $[Y(p_1)]^{-1}$ respectively.

4. Division of $[Z(p)]$

Each element of $[Z(p)]$, which is the inverse matrix of $[Y(p)]$, will be reduced to $Z^{(a)}$ and $Z^{(b)}$ like the following equations:

$$\left.\begin{aligned}
Z &= Z^{(a)} + Z^{(b)} \\
Z^{(a)} &= \frac{p(pZ - p_1 z)}{p^2 - p_1^2} \\
Z^{(b)} &= \frac{p_1(pz - p_1 Z)}{p^2 - p_1^2}
\end{aligned}\right\} \qquad (4)$$

Then, $[Z(p)]$ will be as follows:

$$\begin{aligned}
[Z] &= \begin{bmatrix} Z_{11}, & Z_{12} \\ Z_{12}, & Z_{22} \end{bmatrix} \\
&= \begin{bmatrix} Z_{11}^{(a)}, & Z_{12}^{(a)} \\ Z_{12}^{(a)}, & Z_{22}^{(a)} \end{bmatrix} + \begin{bmatrix} Z_{11}^{(b)}, & Z_{12}^{(b)} \\ Z_{12}^{(b)}, & Z_{22}^{(b)} \end{bmatrix}
\end{aligned}$$

$$= \begin{pmatrix} \dfrac{p(pZ_{11}-p_1z_{11})}{p^2-p_1^2} & \dfrac{p(pZ_{12}-p_1z_{12})}{p^2-p_1^2} \\ \dfrac{p(pZ_{12}-p_1z_{12})}{p^2-p_1^2} & \dfrac{p(pZ_{22}-p_1z_{22})}{p^2-p_1^2} \end{pmatrix} + \begin{pmatrix} \dfrac{p_1(pz_{11}-p_1Z_{11})}{p^2-p_1^2} & \dfrac{p_1(pz_{12}-p_1Z_{12})}{p^2-p_1^2} \\ \dfrac{p_1(pz_{12}-p_1Z_{12})}{p^2-p_1^2} & \dfrac{p_1(pz_{22}-p_1Z_{22})}{p^2-p_1^2} \end{pmatrix} .$$

If both $[Z^{(a)}]$ and $[Z^{(b)}]$ are realizable impedance matrices, $[Z]$ can be given by the circuit $[Z^{(a)}]$ connected with $[Z^{(b)}]$ in series. Hence, it is worth-while first to exmaine whether they are positive real or not. As $[Z^{(a)}]$ is a symmetrical matrix, the necessary and sufficient condition for it to be a realizable impedance matrix is that the following quadratic form $Z_{\text{in}}^{(a)}$ is a positive real function, where t_1 and t_2 are arbitrary real numbers:

$$Z_{\text{in}}^{(a)} = t_1^2 Z_{11}^{(a)} + 2t_1t_2Z_{12}^{(a)} + t_2^2 Z_{22}^{(a)} .$$

To verify $Z_{\text{in}}^{(a)}$ being positive real, let us exmaine first that its resistans is always positive on the imaginary axis of the p plan. Its resistance is given by

$$\begin{aligned} R_e Z_{\text{in}}^{(a)} &= R_e \{t_1^2 Z_{11}^{(a)} + 2t_1 t_2 Z_{12}^{(a)} + t_2^2 Z_{22}^{(a)}\} \\ &= t_1^2 R_e Z_{11}^{(a)} + 2t_1 t_2 R_e Z_{12}^{(a)} + t_2^2 R_e Z_{22}^{(a)} \\ &= \frac{t_2^2}{R_e Z_{11}^{(a)}} \left[\left\{ \frac{t_1}{t_2} R_e Z_{11}^{(a)} + R_e Z_{12}^{(a)} \right\}^2 \right. \\ &\qquad \left. - \{(R_e Z_{12}^{(a)})^2 - R_e Z_{11}^{(a)} \cdot R_e Z_{22}^{(a)}\} \right] , \end{aligned}$$

and at a point $j\omega_1$ on the imaginary axis the first term of the above equation must be positive, because $[Z]$ is positive real, $t_2^2 \{R_eZ_{11}^{(a)} t_1/t_2 + R_eZ_{12}^{(a)}\}^2 \geqq 0$, and $R_e Z_{11}^{(a)}(j\omega_1) = R_eZ_{11}(j\omega_1) \cdot \omega_1^2/(\omega_1^2+p_1^2) \geqq 0$. The second term of it can be rewritten as follows:

$$R_e \frac{p\{pZ_{11}-p_1z_{11}\}}{p^2-p_1^2} \cdot R_e \frac{p\{pZ_{22}-p_1z_{22}\}}{p^2-p_1^2} - \left[R_e \frac{p\{pZ_{12}-p_1z_{12}\}}{p^2-p_1^2} \right]^2$$

$$= \frac{\omega_1^4}{(\omega_1^2+p_1^2)^2} [R_e Z_{11}(j\omega_1) \cdot R_e Z_{22}(j\omega_1) - \{R_e Z_{12}(j\omega_1)\}^2] ,$$

and this is positive, too, because the inside of [ ], det $[R_eZ(j\omega_1)]$, is positive. Accordingly,

$$R_e Z_{\text{in}}^{(a)}(j\omega_1) \geqq 0 ,$$

which shows that the real part of $Z_{\text{in}}^{(a)}$ is always positive at $j\omega_1$.

Next, it seems that poles of $Z^{(a)}$ occur at the points $p=\pm p_1$ and $Z=\infty$, but since the factor p-p_1 is cancelled from the numerator and the denominator of $Z^{(a)}$, there is no pole at $p=p_1$. Moreover, Z is positive real and has no any poles on the right-half-plane of p; hence, $Z^{(a)}$ and $Z_{in}{}^{(a)}$ are also analytic there.

Finally, if $Z^{(a)}$ has some poles on j-axis, they must occur only at the same points with poles of Z. And the residue of the pole of Z_{jj} on j-axis, a_{jj}, is positive; hence the residues of $Z_{jj}{}^{(a)}$ at the point $p=j\omega_i$,

$$\frac{-\omega}{-(\omega_i^2+p_1^2)}\,a_{jj}$$

are also positive. If Z_{11}, Z_{22} and Z_{12}, the elements of $[Z]$, have poles at the same point, the following relation exists among their residues,

$$a_{11}\,a_{22}-a_{12}^2 \geqq 0\ ,$$

therefore the following inequality always holds for $p=j\omega_i$.

$$R_e Z_{11}{}^{(a)}(j\,\omega_i)\cdot R_e Z_{22}{}^{(a)}(j\,\omega_i)-\{R_e Z_{12}{}^{(a)}(j\,\omega_i)\}^2 \geqq 0\ .$$

Accordingly, the residues of the poles of $Z_{in}{}^{(a)}$ on j-axis are always positive.

As has been explained, $Z_{in}{}^{(a)}$ must satisfy all the necessary and sufficient conditions in order to be a positive real function. It has been proved that $[Z^{(a)}]$ is also a positive real matrix. We can proceed similarly with $[Z^{(b)}]$.

5. Verification of $[Y'(p)]$ to be Positive Real.

As has been proved in the previous section, $[Z^{(b)}]$ is a positive real matrix, accordingly, its inverse matrix $[Z^{(b)}]^{-1}$ must also be a positive real matrix, and can be written as follows:

$$\left.\begin{aligned}
[Z^{(b)}]^{-1} &= \begin{pmatrix} \dfrac{p_1(pz_{11}-p_1Z_{11})}{p^2-p_1^2} & \dfrac{p_1(pz_{12}-p_1Z_{12})}{p^2-p_1^2} \\[2ex] \dfrac{p_1(pz_{12}-p_1Z_{12})}{p^2-p_1^2} & \dfrac{p_1(pz_{22}-p_1Z_{22})}{p^2-p_1^2} \end{pmatrix}^{-1} \\
&= \frac{(p^2-p_1^2)/p_1}{(pz_{11}-p_1Z_{11})(pz_{22}-p_1Z_{22})-(pz_{12}-p_1Z_{12})^2} \\
&\quad \cdot \begin{bmatrix} pz_{22}-p_1Z_{22} & -(pz_{12}-p_1Z_{12}) \\ -(pz_{12}-p_1Z_{12}) & pz_{11}-p_1Z_{11} \end{bmatrix} \\
&= \frac{(p^2-p_1^2)/p_1}{\nabla} \begin{bmatrix} pz_{22}(Y_{11}Y_{22}-Y_{12}^2)-p_1Y_{11}\,, & -\{pz_{12}(Y_{11}Y_{22}-Y_{12}^2)+p_1Y_{12}\} \\ -\{pz_{12}(Y_{11}Y_{22}-Y_{12})+p_1Y_{12}\}\,, & pz_{11}(Y_{11}Y_{22}-Y_{12})-p_1Y_{22} \end{bmatrix} \\
&= \frac{1}{z_{11}z_{22}-z_{12}^2}\begin{bmatrix} z_{22} & -z_{12} \\ -z_{12} & z_{11} \end{bmatrix}\frac{p}{p_1}+[Y'(p)]
\end{aligned}\right\} \quad (5)$$

On the other hand, each element of $[Z^{(b)}]$ becomes zero at $p=\infty$, so each element of $[Z^{(b)}]^{-1}$ inversely has a pole at $p=\infty$. If the residue of its pole are designated by $[g_\infty]$, the following inequalities must hold for them:

$$\left.\begin{aligned} g_{11\infty} &= \frac{(z_{22}-p_1 a_{22\infty})/p_1}{(z_{11}-p_1 a_{11\infty})(z_{22}-p_1 a_{22\infty})-(z_{12}-p_1 a_{12\infty})^2} \geqq 0 \\ g_{22\infty} &= \frac{(z_{11}-p_1 a_{11\infty})/p_1}{(z_{11}-p_1 a_{11\infty})(z_{22}-p_1 a_{22\infty})-(z_{12}-p_1 a_{12\infty})^2} \geqq 0 \\ g_{11\infty}\cdot g_{22\infty}-g_{12\infty}^2 &= \frac{1/p_1^2}{(z_{11}-p_1 a_{11\infty})(z_{22}-p_1 a_{22\infty})-(z_{12}-p_1 a_{12\infty})^2} \geqq 0\,, \end{aligned}\right\} \quad (6)$$

where $[a_\infty]$ is the residue of the pole of $[Z]$ at $p=\infty$. From the above relations the following inequalities result:

$$\left.\begin{aligned} z_{22}-p_1 a_{22\infty} &\geqq 0 \\ z_{11}-p_1 a_{11\infty} &\geqq 0 \end{aligned}\right\} \quad (7)$$

Moreover, $g_{11\infty}$ can be rewritten as follows:

$$g_{11\infty} = \frac{1/p_1}{\dfrac{z_{11}z_{22}-z_{12}^2}{z_{22}} - p_1\dfrac{a_{11\infty}a_{22\infty}-a_{12\infty}^2}{a_{22\infty}} - \dfrac{p_1 z_{22}(a_{12\infty}-z_{12}a_{22\infty}/z_{22})^2}{a_{22\infty}(z_{22}-p_1 a_{22\infty})}}\,,$$

where the second term of the denominator is negative, because $a_{11\infty}\cdot a_{22\infty}-a_{12\infty}$ is the determinat of $[a_\infty]$ and positive. The third is also negative from eq. (7); hence, it follows

$$g_{11\infty} > \frac{1}{p_1\dfrac{z_{11}z_{22}-z_{12}^2}{z_{22}}} = \frac{1}{p_1}Y_{11}(p_1) > 0\,. \quad (8\cdot a)$$

Similarly in the relation with $g_{22\infty}$, the following inequalities can exist:

$$g_{22\infty} > \frac{z_{11}}{p_1(z_{11}z_{22}-z_{12}^2)} = \frac{1}{p_1}Y_{22}(p_1) > 0\,. \quad (8\cdot b)$$

Extracting from $[Z^{(b)}]^{-1}$ its pole at $p=\infty$, the reduced matrix $[Z_\infty^{(b)}]^{-1}$ must also be positive real. And the relation between them is represented by

$$[Z^{(b)}]^{-1} = [g_\infty]p + [Z_\infty^{(b)}]^{-1}\,.$$

Therefore, $[Y'(p)]$, the subject of verification in this section, is given from the above equation and eq. (5) as follows:

$$[Y'(p)] = [Z_\infty^{(b)}]^{-1} + \left\{[g_\infty] - \frac{1}{p_1}[Y(p_1)]\right\}p$$

$$= [Z_\infty^{(b)}]^{-1} + \begin{pmatrix} g_{11\infty} - \dfrac{1}{p_1} Y_{11}(p_1) & g_{12\infty} - \dfrac{1}{p_1} Y_{12}(p_1) \\ g_{12\infty} - \dfrac{1}{p_1} Y_{12}(p_1) & g_{22\infty} - \dfrac{1}{p_1} Y_{22}(p_1) \end{pmatrix} p.$$

Here $\{[g_\infty]-[Y(p_1)]/p_1\}\, p$ is also positive real, because the following relations yields from eq. (6) and eq. (8).

$$g_{11\infty} - \frac{1}{p_1} Y_{11}(p_1) \geqq 0$$

$$g_{22\infty} - \frac{1}{p_1} Y_{22}(p_1) \geqq 0$$

$$\left\{g_{11\infty} - \frac{1}{p_1} Y_{11}(p_1)\right\}\left\{g_{22\infty} - \frac{1}{p_1} Y_{22}(p_1)\right\} - \left\{g_{12\infty} - \frac{1}{p_1} Y_{12}(p_1)\right\}^2$$

$$= \frac{a_{11\infty}\, a_{22\infty} - a_{12\infty}{}^2}{(z_{11} z_{22} - z_{12}{}^2)\{(z_{11} - p_1 a_{11\infty})(z_{22} - p_1 a_{22\infty}) - (z_{12} - p_1 a_{12\infty})^2\}} \geqq 0.$$

Accordingly $[Y'(p)]$ which is given by the sum of two real positive matirces $[Z_\infty^{(b)}]^{-1}$ and $\{[g_\infty]-[Y(p_1)]/p_1\}\, p$ must be positive real.

Thus, it has been shown that $[Y'(p)]$ is a positive real matrix. In the same way, it can be done also for $[Z^{(a)}]$. Not only for $[Y'(p)]$ of two-terminal-pair network, but also for that of any n-terminal-pair networks to be positive real, it can be proved by the identical procedure of verification although it is laborious work because of the matrix having elements of n-lines and n-columns.

6. Physical Meaning

Eq. (2) is rewritten as follows:

$$[Y(p)] = [p[Y(p_1)] + p_1[Y'(p)]] \cdot [p_1[1] + p[Y(p_1)]^{-1}[Y'(p)]]^{-1}$$

$$= \left[[Y(p_1)]^{-1}\frac{p_1}{p} + [Y(p_1)]^{-1}[Y'(p)][Y(p_1)]^{-1}\right]^{-1}$$

$$+ \left[[Y'(p)]^{-1} + [Y(p_1)]^{-1}\frac{p}{p_1}\right]^{-1}.$$

In the above equation $[Y(p_1)]^{-1}[Y'(p)][Y(p_1)]^{-1}$ is also positive real, for $[Y(p_1)]$ and $[Y'(p)]$ are positive real. Accordingly, $[Y(p)]$ is given as the sum of positive real matrices, and obtained by the circuite shown in fig. (1. a). Similarly, $[Z(p)]$, the inverse matrix of $[Y(p)]$, is represented by the following equation:

$$[Z(p)] = \left[[Y(p_1)]\frac{p}{p_1} + [Y'(p)]\right]^{-1}$$

$$+\left[[Y(p_1)][Y'(p)]^{-1}[Y(p_1)]+[Y(p_1)]\frac{p_1}{p}\right]^{-1}.$$

It may be realized by the circuit like that in fig. (1. b).

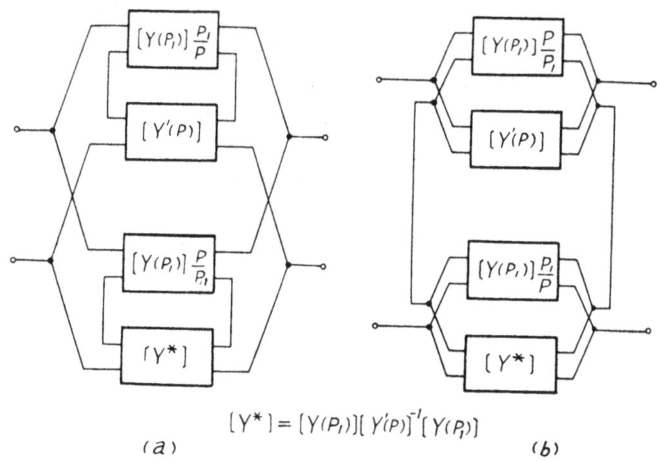

Fig. 1. Analyzed $[Y(p)]$ or $[Z(p)]$ into (a) parallel and (b) siries connection type netweek.

The degree of $[Y'(p)]$ in relate with p given by eq. (3), seems to be apparently larger than the degree of $[Y(p)]$, but the factor $(p-p_1)^2$ is cancelled from the numerator and the denominator of $[Y'(p)]$, so $[Y'(p)]$ and $[Y(p)]$ may be of the same degree. When $[Y(p)]$ is a pure reactance, the other factor $(p+p_1)^2$ is also cancelled from them, and finally $[Y'(p)]$ will be two degrees less than $[Y(p)]$ is.

Another important application of extended Richards' theorem is to extract a coupled transmission line element from the prescribed impedance $[Z(p)]$. This corresponds to the Richards' method by which a coaxial line element can be extracted in cascade from impedance $Z(p)$[3]. $[Z(1)]$ in the following case of $p_1=1$ and $p=1$:

$$[Z(1)]=\begin{bmatrix} Z_{11}(1), & Z_{12}(1) \\ Z_{12}(1), & Z_{22}(1) \end{bmatrix}$$

represents a characteristic impedance matrix of a coupled transmission line to be extracted in the manner shown in fig. (2). And parameters of the reduced circuit is just equal to $[Y'(p)]$ when $p_1=1$.

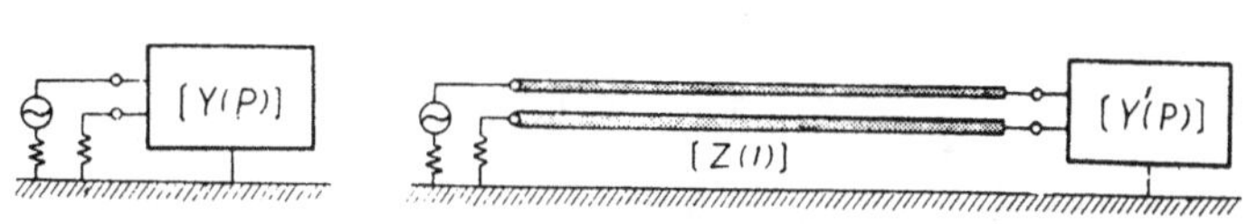

Fig. 2. Extraction of one set of coupled transmission line from $[Y(\mathrm{p})]$.

7. Conclusion

Dealing with Richards' theorem as a matrix, we have obtained a method for analyzing an n-terminal-pair network. We have not yet succeeded in obtaining a method which will not require a transformer, corresponding to Bott & Duffin method on a one-terminal-pair network, but it is one of the important methods for synthesizing an n-terminal-pair network. And the extraction of a coupled transmission line element proposed previously may also be a consequential means of network synthesis in the case of a distributed constant network.

8. Acknowledgement

The author wishes to thank Profs. K. Nagai and R. Sato who supervised this work and gave many helpful suggestions.

Bibliography

1. P.I. Richards: "Special class of functions with positive real part in a half-plane", Duke Math. J., 14, (1947)
2. R. Bott and J. Duffin: "Impedance synthesis without use of transformer", J.A. Phys., 20, 8, p 816, (1949)
3. P.I. Richards: "Resistor transmission-line circuit", I.R.E., 36, 2, p 217, (1948)

Exact Design of TEM Microwave Networks Using Quarter-Wave Lines*

R. J. WENZEL†, MEMBER, IEEE

Summary—**Modern network theory procedures, based on Ozaki-Ishii[1] synthesis techniques, are reviewed for application in the design of TEM mode microwave networks using parallel coupled bars and/or series and shunt stubs. The circuit equivalences and identities obtained are theoretically valid over the entire frequency spectrum and lead to several physical configurations having identical response functions. These equivalent circuits often allow simplification of the physical circuitry and realization of both broad and narrow bandwidths. The problem of practical circuit configurations is discussed from the viewpoint of bandwidth and circuit element values. Neglecting multiple responses, TEM low-pass, high-pass and band-pass Butterworth filters are shown to offer steeper bandedge characteristics than those of corresponding lumped element filters. The use of complementary filters to match a source and load over a wide frequency range is outlined and TEM realizations of these complements are obtained. A simple procedure for obtaining element values of Butterworth complements is described. An analysis of parallel coupled filters is made and a simplified equivalent circuit is obtained. An exact synthesis procedure for parallel coupled bar filters and their equivalent forms is given. Construction details and experimental results are described for two filters which use series stubs.**

I. Introduction

THE PURPOSE of this paper is to present a unified account of an exact modern network synthesis applicable to the design of distributed TEM mode networks. Emphasis is placed on microwave circuit aspects rather than network theory wherever possible. In the past, most TEM microwave networks have been designed by use of image parameter methods. The depth of understanding of lumped element circuits afforded by modern network theory has not been completely extended to distributed networks. A discussion of the relative merits of modern network and image parameter network theories can be found in literature.[2,3] The details of modern network synthesis are beyond the scope of this paper and may be reviewed in existing publications.[2,4,5]

This paper is presented as follows: Section II introduces the transformation that allows distributed TEM networks to be treated in a manner analogous to lumped element networks. Equivalent circuits for various configurations are derived and tabulated. The "unit element" is introduced as a necessary circuit element and Kuroda's identities are discussed. The section concludes with a design example and a discussion of impedance and frequency scaling.

Section III describes a number of equivalent physical structures that realize the same circuit response. Achievement of practical bandwidths and element values is related to the circuit equivalences.

Section IV contains a comparison between lumped element and distributed filter characteristics. The repetitive nature of the microwave circuits is not considered. Under this condition an analysis of the ability of each type of network to approximate the ideal filter characteristic is given.

Section V applies the theory of the preceding sections to some interesting network problems. Parallel coupled line filters are analyzed and equivalent circuits are obtained. Complementary microwave filters are introduced and network realizations are obtained. Construction details and experimental results are given for two filters containing series stubs.

Appendix I demonstrates the methods for exact filter synthesis of uniform parallel coupled bars, lines with shunt shorted stubs and lines with series open stubs. Appendix II outlines a procedure for eliminating ideal transformers in networks having parallel coupled bars.

II. Modern Network Filter Theory

The basis for modern network theory was established in the mid-1920's with the work of Foster and Cauer and was culminated with the conditions for realizability of driving-point impedances stated by Brune in 1930.[6] From a consideration of the energy functions associated with a LLFPB (lumped, linear, finite, passive and bilateral) network, Brune showed that, for a function of a complex variable to be a realizable driving point impedance, it is necessary and sufficient that it be a "positive real" function—where a positive real (p.r.) function is one that

1) is real for real values of the complex frequency variable $s=\sigma+j\omega$ and
2) has a positive real part for values of s having a positive real part.

* Received May 17, 1963; revised manuscript received September 18, 1963.

† The Bendix Corporation, Research Laboratories Division, Southfield, Mich.

[1] H. Ozaki and J. Ishii, "Synthesis of a class of strip-line filters," IRE Trans. on Circuit Theory, vol. CT-5, pp. 104–109; June, June, 1958.

[2] L. Weinberg, "Network Analysis and Synthesis," McGraw-Hill Book Co., Inc., New York, N. Y.; 1962

[3] "Reference Data for Radio Engineers," published by International Telephone and Telegraph Corp., 4th ed., pp. 187–188; 1957.

[4] E. A. Guillemin, "Synthesis of Passive Networks," John Wiley and Sons, Inc., New York, N. Y.; 1957.

[5] N. Balabanian, "Network Synthesis," Prentice-Hall, Inc., Englewood Cliffs, N.J.; 1958.

[6] O. Brune, "Synthesis of a finite 2-terminal network whose driving point impedance is a prescribed function of frequency," *J. Math. and Phys.*, vol. 10, pp. 191–236; 1930.

From these two conditions, it is possible to determine if a given impedance function of s can be realized.

The extension of lumped network synthesis techniques to microwave networks was demonstrated by Richards[7] in 1948. In microwave networks, the impedances of sections of transmission line are expressed as functions of $j \tan \beta l$ where β is the phase constant along the transmission line and l is the line length. Richards showed that microwave networks composed
18 of lumped resistors and lossless equal length transmission lines can be treated as lumped networks using a transformation to the complex frequency variable

$$S = j \tan \frac{\pi f}{2f_0} \tag{1}$$

where f_0 is the real constant frequency at which l is a quarter wavelength and f is the real frequency variable. As can be seen, S is periodic in $2f_0$ and the response of a microwave network will repeat at this interval. The mapping properties of functions of the variable S are illustrated in Fig. 1 for the case of a high pass S-plane filter.

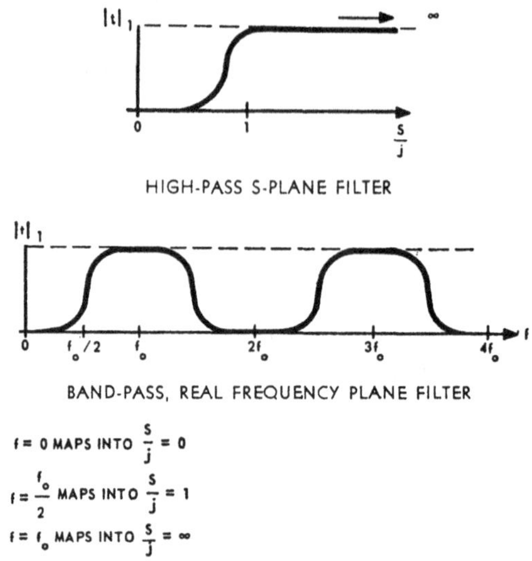

Fig. 1—Mapping properties of a function of the variable $S/j = \tan \pi f/2f_0$.

The synthesis procedure to be used becomes apparent from Fig. 1.[8] The required response is synthesized using lumped elements in the S plane and is then converted to a microwave structure in the f plane. The only remaining problem is to express the microwave structures as functions of S. One TEM mode structure to be considered for application in modern network design is shown in Fig. 2.

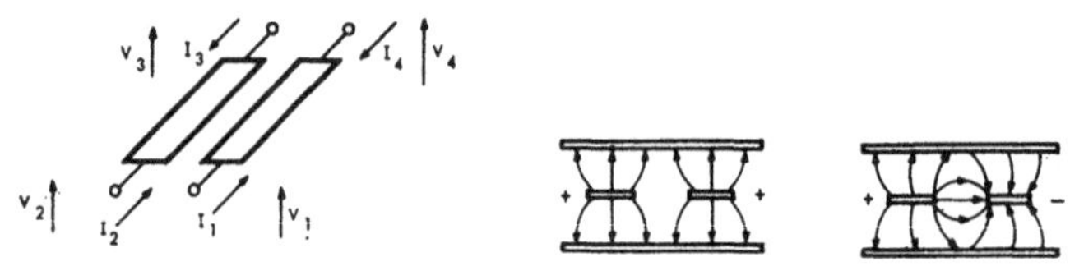

(a) (b) (c)

Fig. 2—Parallel coupled stripline four-port. (a) Four-port parameters. (b) Even mode field configuration. (c) Odd mode field configuration.

Jones and Bolljahn[9] have derived the impedance matrix for the configuration of Fig. 2(a).

$$\begin{bmatrix} V_1 \\ V_2 \\ V_3 \\ V_4 \end{bmatrix} = \begin{bmatrix} Z_{11} & Z_{12} & Z_{13} & Z_{14} \\ Z_{21} & Z_{22} & Z_{23} & Z_{24} \\ Z_{31} & Z_{32} & Z_{33} & Z_{34} \\ Z_{41} & Z_{42} & Z_{43} & Z_{44} \end{bmatrix} \begin{bmatrix} I_1 \\ I_2 \\ I_3 \\ I_4 \end{bmatrix}$$

where

$$Z_{11} = Z_{22} = Z_{33} = Z_{44} = -j(Z_{0e} + Z_{0o}) \frac{\cot \theta}{2}$$

$$Z_{12} = Z_{21} = Z_{34} = Z_{43} = -j(Z_{0e} - Z_{0o}) \frac{\cot \theta}{2}$$

$$Z_{13} = Z_{31} = Z_{24} = Z_{42} = -j(Z_{0e} - Z_{0o}) \frac{\csc \theta}{2}$$

$$Z_{14} = Z_{41} = Z_{23} = Z_{32} = -j(Z_{0e} + Z_{0o}) \frac{\csc \theta}{2}$$

where

θ = electrical length of the line,
Z_{0e} = even mode impedance,
Z_{0o} = odd mode impedance.

The above relations hold for equal width strips. The matrix for unequal strip widths, as well as methods for calculating even and odd mode impedances, are given by Ozaki and Ishii.[1] By using unequal strip widths, an additional degree of freedom is obtained as required in many practical cases. The analysis of unequal width strips follows the same procedure as that used for equal width strips. For simplicity, the derivation will be carried out for the equal width case. Substituting

$$S = j \tan \frac{\pi f}{2f_0} = j \tan \theta \text{ gives}$$

$$Z_{11} = Z_{22} = Z_{33} = Z_{44} = \frac{Z_{0e} + Z_{0o}}{2S}$$

$$Z_{12} = Z_{21} = Z_{34} = Z_{43} = \frac{Z_{0e} - Z_{0o}}{2S}$$

$$Z_{13} = Z_{31} = Z_{24} = Z_{42} = \frac{Z_{0e} - Z_{0o}}{2S} \sqrt{1 - S^2}$$

$$Z_{14} = Z_{41} = Z_{23} = Z_{32} = \frac{Z_{0e} + Z_{0o}}{2S} \sqrt{1 - S^2}. \tag{2}$$

[7] P. I. Richards, "Resistor transmission-line circuits," PROC. IRE, vol. 36, pp. 217–220; February, 1948.

[8] The basic synthesis procedure used follows that of Ozaki and Ishii.[1]

[9] E. M. T. Jones and J. T. Bolljahn, "Coupled-strip-transmission line filters and directional couplers," IRE TRANS. ON MICROWAVE THEORY AND TECHNIQUES, vol. 19, pp. 75–81; April, 1956.

To determine the impedance parameters when the two-strip configuration is used as a twoport, it is only necessary to apply the pertinent port conditions. For example, assume that terminal (2) is grounded, (3) is open, (1) is the input port and (4) is the output port. The port conditions are then $V_2=0$ and $I_3=0$. It follows that

$$\begin{bmatrix} V_1 \\ 0 \\ V_3 \\ V_4 \end{bmatrix} = \begin{bmatrix} Z_{11} & Z_{12} & Z_{13} & Z_{14} \\ Z_{21} & Z_{22} & Z_{23} & Z_{24} \\ Z_{31} & Z_{32} & Z_{33} & Z_{34} \\ Z_{41} & Z_{42} & Z_{43} & Z_{44} \end{bmatrix} \begin{bmatrix} I_1 \\ I_2 \\ 0 \\ I_4 \end{bmatrix} \tag{3}$$

and, solving for V_1 and V_4 in terms of I_1 and I_4, gives

$$V_1 = \left(Z_{11} - \frac{Z_{12}Z_{21}}{Z_{22}}\right) I_1 + \left(Z_{14} - \frac{Z_{12}Z_{24}}{Z_{22}}\right) I_4$$

$$V_4 = \left(Z_{41} - \frac{Z_{21}Z_{42}}{Z_{22}}\right) I_1 + \left(Z_{44} - \frac{Z_{24}Z_{42}}{Z_{22}}\right) I_4. \tag{4}$$

Substituting (2) and arranging them in matrix form gives

$$\begin{bmatrix} V_1 \\ V_4 \end{bmatrix} = \begin{bmatrix} \dfrac{2Z_{0e}Z_{0o}}{(Z_{0e}+Z_{0o})S} & \dfrac{2Z_{0e}Z_{0o}\sqrt{1-S^2}}{(Z_{0e}+Z_{0o})S} \\ \dfrac{2Z_{0e}Z_{0o}\sqrt{1-S^2}}{(Z_{0e}+Z_{0o})S} & \dfrac{S^2(Z_{0e}-Z_{0o})^2+4Z_{0e}Z_{0o}}{2S(Z_{0e}+Z_{0o})} \end{bmatrix} \begin{bmatrix} I_4 \\ I_4 \end{bmatrix}. \tag{5}$$

Before proceeding further, it is necessary to introduce one additional circuit element, the "unit element" (u.e.), defined by the $ABCD$ matrix as

$$\frac{1}{\sqrt{1-S^2}} \begin{bmatrix} 1 & Z_0 S \\ \dfrac{S}{Z_0} & 1 \end{bmatrix}. \tag{6}$$

The matrix of the unit element is the same as that of a transmission line of electrical length $\theta=\pi f/(2f_0)$ and characteristic impedance Z_0. A series of unit elements has the following interesting properties when inserted between a lumped network and a termination of impedance Z_0.

1) The driving point impedance remains exactly the same as without the u.e.'s,
2) the magnitude of the transfer impedance remains unchanged,
3) the phase of the transfer impedance may be changed.

If unit elements are inserted before the lumped network instead of between the network and its termination,

1) the input and transfer impedances are changed,
2) the magnitude of the reflection coefficient remains unchanged,
3) the phase of the reflection coefficient is changed.

These properties are readily apparent if one thinks of the u.e.'s as lengths of line inserted between a network and its load or between the network and its driving source.

Now consider the cascade of a unit element and an S-plane inductor[10] [Fig. 3(a)]. The $ABCD$ matrix is

$$\begin{bmatrix} A & B \\ C & D \end{bmatrix} = \frac{1}{\sqrt{1-S^2}} \begin{bmatrix} 1 & Z_0 S \\ \dfrac{S}{Z_0} & 1 \end{bmatrix} \begin{bmatrix} 1 & LS \\ 0 & 1 \end{bmatrix}$$

$$= \frac{1}{\sqrt{1-S^2}} \begin{bmatrix} 1 & (L+Z_0)S \\ \dfrac{S}{Z_0} & \dfrac{S^2L}{Z_0}+1 \end{bmatrix}. \tag{7}$$

The corresponding impedance matrix is

$$[Z] = \begin{bmatrix} \dfrac{Z_0}{S} & \dfrac{Z_0\sqrt{1-S^2}}{S} \\ \dfrac{Z_0\sqrt{1-S^2}}{S} & \dfrac{S^2L+Z_0}{S} \end{bmatrix}. \tag{8}$$

Comparing this with (5) shows the two to be equivalent, if

$$\text{(a)}\quad Z_0 = \frac{2Z_{0e}Z_{0o}}{Z_{0e}+Z_{0o}} \qquad \text{(b)}\quad L = \frac{(Z_{0e}-Z_{0o})^2}{2(Z_{0e}+Z_{0o})}. \tag{9}$$

Solving for Z_{0e} and Z_{0o} gives

$$Z_{0e} = Z_0 + L + \sqrt{L(Z_0+L)},$$

$$Z_{0o} = \frac{Z_0 Z_{0e}}{2Z_{0e} - Z_0} = \frac{Z_0 + L + \sqrt{L(Z_0+L)}}{1 + \dfrac{2L}{Z_0} + \dfrac{2}{Z_0}\sqrt{L(Z_0+L)}}. \tag{10}$$

Thus the equivalence of the networks shown in Fig. 3(b) has been established where Z_{0e}, Z_{0o}, Z_0 and L are related by (10).

A similar procedure can be applied for other port conditions and equivalent circuits may be developed. A list of all possible circuit configurations and their equivalent S-plane circuits is given in Table I for the case of unequal strip widths. The equal width case is obtained by setting $Z_{0e}^{a}=Z_{0e}^{b}$ and $Z_{0o}^{a}=Z_{0o}^{b}$.

The synthesis procedure to be used is based upon a series of equivalent circuits known as Kuroda's identities.[11] Kuroda showed the equivalence of the circuits listed in Table II (p. 98). As a sample proof, the equivalence of the networks in Fig. 4 will be demonstrated by

[10] An S-plane inductor L represents the characteristic impedance of a transmission line and has units of ohms.[7]

[11] K. Kuroda, "Derivation Methods of Distributed Constant Filters from Lumped Constant Filters," test for lectures at Joint Meeting of Konsoi Branch of Institute of Elec. Commun., of Elec., and of Illumin. Engrs. of Japan, p. 32; October, 1952. (In Japanese.)

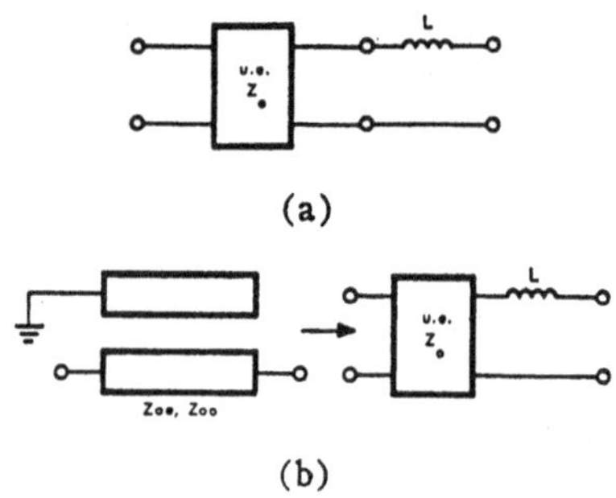

Fig. 3—(a) Cascade connection of a unit element and an S-plane inductor. (b) Equivalence of stripline and S-plane networks.

TABLE I

EQUIVALENT CIRCUITS FOR UNEQUAL STRIP WIDTHS

TEM Circuit	Equivalent Circuit	Element Values		
(a) 	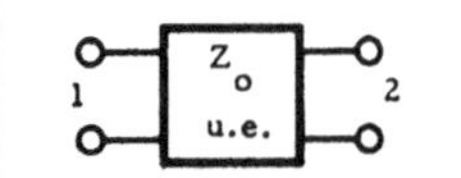	$Z_o = \frac{Z_{oe} + Z_{oo}}{2}$		
(b) 	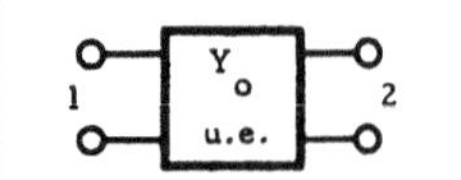	$Y_o = \frac{1}{2}\left(\frac{1}{Z_{oe}} + \frac{1}{Z_{oo}}\right)$		
(c) 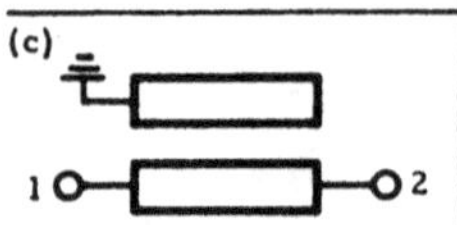	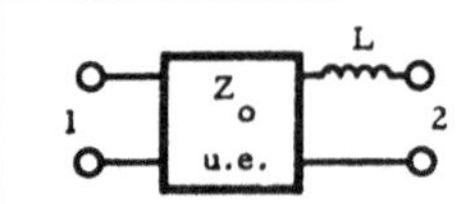	$Z_{oe} = Z_o + L + \sqrt{L(Z_o + L)}$ $Z_{oo} = \frac{Z_o + L + \sqrt{L(Z_o + L)}}{1 + \frac{2L}{Z_o} + \frac{2}{Z_o}\sqrt{L(Z_o + L)}}$		
(d) 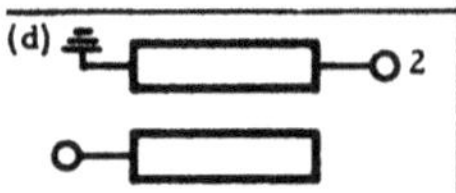	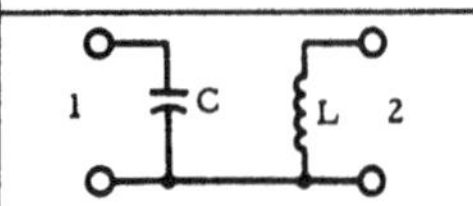	$Z_{oe} = \frac{L}{C Z_{oo}}$ $Z_{oo} = L - \sqrt{L\left(L - \frac{1}{C}\right)}$ $LC > 1$		
(e) 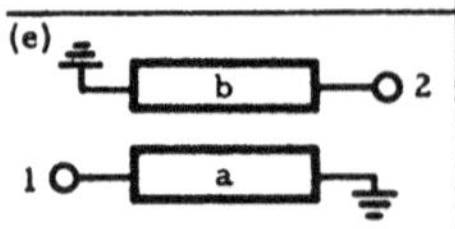	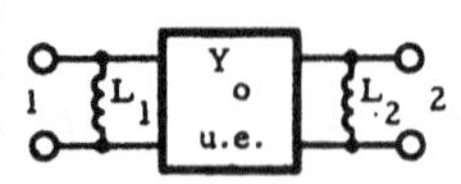	$\frac{Y_{oe}^{a} + Y_{oo}^{a}}{2} = Y_o + \frac{1}{L_1}$	$Y_{oe}^{a} + Y_{oe}^{b} = \frac{1}{L_1} + \frac{1}{L_2}$	$\frac{Y_{oe}^{b} + Y_{oo}^{b}}{2} = Y_o + \frac{1}{L_2}$
(f) 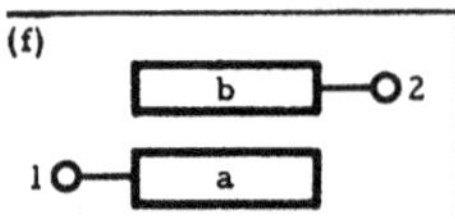	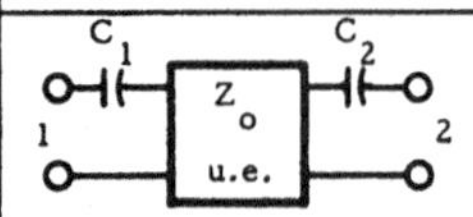	$\frac{Z_{oe}^{a} + Z_{oo}^{a}}{2} = Z_o + \frac{1}{C_1}$	$Z_{oo}^{a} + Z_{oo}^{b} = \frac{1}{C_1} + \frac{1}{C_2}$	$\frac{Z_{oe}^{b} + Z_{oo}^{b}}{2} = Z_o + \frac{1}{C_2}$
(g) 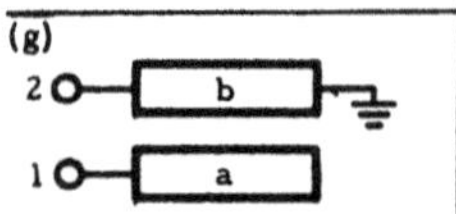	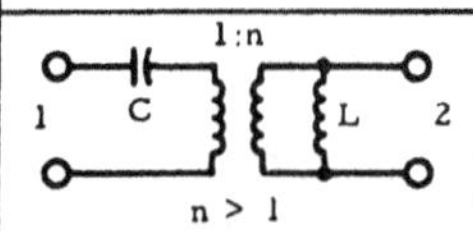	$\frac{Z_{oe}^{a} + Z_{oo}^{a}}{2} = \frac{1}{C} + \frac{L}{n^2}$	$Z_{oo}^{a} + Z_{oo}^{b} = \frac{1}{C} + L\left(1 - \frac{1}{n}\right)^2$	$\frac{Z_{oe}^{b} + Z_{oo}^{b}}{2} = L$ $n^2 > LC(n-1)$
(h) 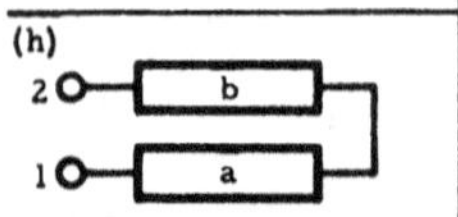	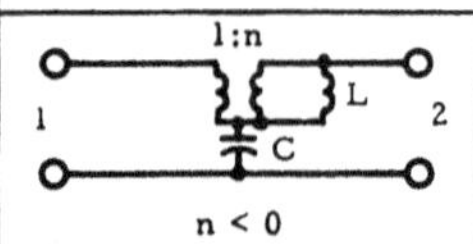	$\frac{Z_{oe}^{a} + Z_{oo}^{a}}{2} = \frac{1}{C} + \frac{L}{n^2}$	$Z_{oo}^{a} + Z_{oo}^{b} = L\left(1 - \frac{1}{n}\right)^2$	$Z_{oe}^{b} + Z_{oo}^{b} = \frac{1}{C} + L$
(i) 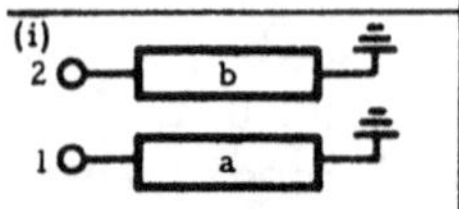	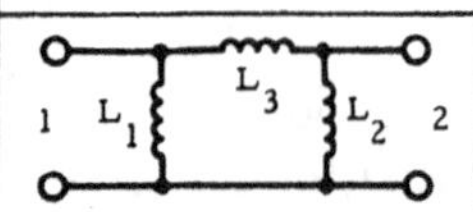	$\frac{Y_{oe}^{a} + Y_{oo}^{a}}{2} = \frac{1}{L_1} + \frac{1}{L_3}$	$Y_{oe}^{a} + Y_{oe}^{b} = \frac{1}{L_1} + \frac{1}{L_2}$	$\frac{Y_{oe}^{b} + Y_{oo}^{b}}{2} = \frac{1}{L_2} + \frac{1}{L_3}$
(j) 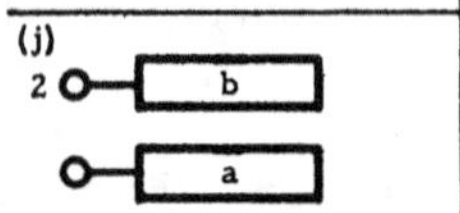	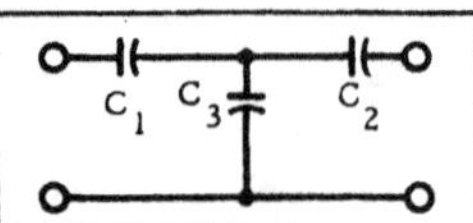	$\frac{Z_{oe}^{a} + Z_{oo}^{a}}{2} = \frac{1}{C_1} + \frac{1}{C_3}$	$Z_{oo}^{a} + Z_{oo}^{b} = \frac{1}{C_1} + \frac{1}{C_2}$	$\frac{Z_{oe}^{b} + Z_{oo}^{b}}{2} = \frac{1}{C_2} + \frac{1}{C_3}$

TABLE II
KURODA'S IDENTITIES

Original Circuit	Equivalent Circuit	n
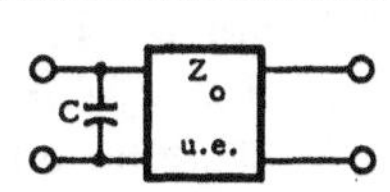	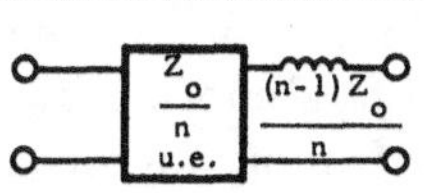	$1+Z_oC$
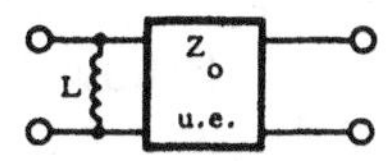	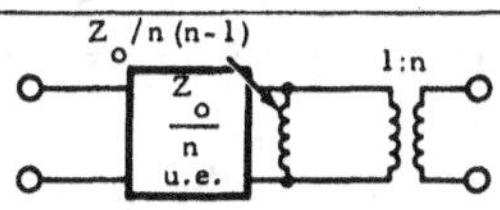	$1+\frac{Z_o}{L}$
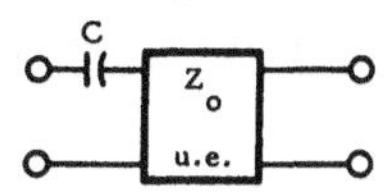	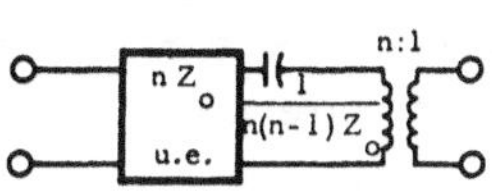	$1+\frac{1}{Z_oC}$
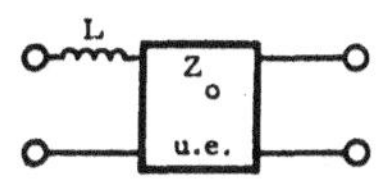	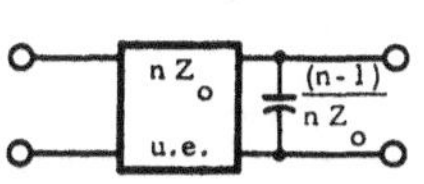	$1+\frac{L}{Z_o}$

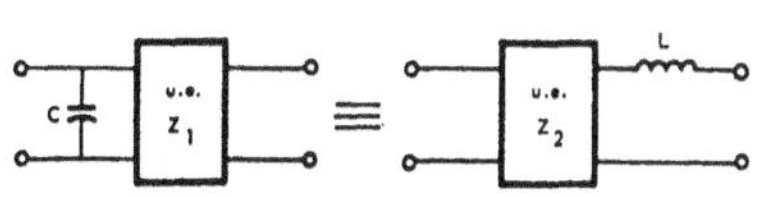

Fig. 4—Kuroda identity.

writing the $ABCD$ matrix. For the network on the left of Fig. 4,

$$\frac{1}{\sqrt{1-S^2}}\begin{bmatrix}1 & 0\\ CS & 1\end{bmatrix}\begin{bmatrix}1 & Z_1S\\ \dfrac{S}{Z_1} & 1\end{bmatrix}$$

$$=\frac{1}{\sqrt{1-S^2}}\begin{bmatrix}1 & Z_1S\\ \left(C+\dfrac{1}{Z_1}\right)S & Z_1CS^2+1\end{bmatrix}\quad(11)$$

and, for the network on the right of Fig. 4,

$$\frac{1}{\sqrt{1-S^2}}\begin{bmatrix}1 & Z_2S\\ \dfrac{S}{Z_2} & 1\end{bmatrix}\begin{bmatrix}1 & SL\\ 0 & 1\end{bmatrix}$$

$$=\frac{1}{\sqrt{1-S^2}}\begin{bmatrix}1 & (Z_2+L)S\\ \dfrac{S}{Z^2} & \dfrac{S^2L}{Z_2}+1\end{bmatrix}.\quad(12)$$

The two networks are equivalent if

$$\text{(a) } Z_1=Z_2+L,\quad \text{(b) } C+\frac{1}{Z_1}=\frac{1}{Z_2},\quad \text{(c) } \frac{L}{Z_2}=Z_1C.\quad(13)$$

If a parameter n is defined by $n=1+Z_1C$, then

$$Z_2=\frac{Z_1}{n};\qquad L=\frac{n-1}{n}Z_1.\quad(14)$$

All techniques necessary for design have now been presented. The procedure to be followed is:

1) Choose the function of S to be synthesized that provides the desired response.
2)[12] Perform the synthesis by modern network techniques to yield a lumped network.
3) Insert a series of unit elements of characteristic impedance Z_0 between the lumped network and its load (of impedance Z_0), if the input impedance is to be preserved. If the quantity of interest is a reflection coefficient, unit elements of characteristic impedance Z_0 can also be inserted in front of the network.
4) Use Kuroda's identities to obtain a configuration of unit elements and lumped elements that can be separated into a cascade of circuits similar to those in Table I.
5) From the L, C and Z_0 values determine Z_{0e} and Z_{0o} for each section.

Example—It is desired to design a resistively terminated, Butterworth filter having a reflection coefficient characteristic of order three. The normalized element values for this network, obtained by consulting tables,[2,13] are shown in Fig. 5(a). Since the quantity of interest is a reflection coefficient, unit elements can be added both before the load and after the source. Adding three unit elements of characteristic impedance $Z_0=1$, two after the source and one before the load, gives the network of Fig. 5(b). The response of this network differs from that of Fig. 5(a) only in the phase of the reflection coefficient.

Kuroda's identities (Table II) are now applied to the shunt capacitors to obtain a cascade of realizable TEM elements. The transformed circuit is shown in Fig. 5(c).

Consulting the equivalent circuits in Table I shows the above circuit can be realized as shown in Fig. 5(d).

Z_{0e} and Z_{0o} for each section are obtained from the corresponding L and Z_0 in Fig. 5(c). Once Z_{0e} and Z_{0o} are known, the dimensions of the elements can be found from available tables[14–20] or by the method given by

[12] Lumped circuit element values have been computed and tabulated for Butterworth and Chebyshev filter characteristics.[2,13] The tabulated element values must be chosen for the proper characteristic. For example, element values for a Butterworth transfer impedance which assumes an ideal current source are not the same as those for a network having a Butterworth transmission coeficient which assumes a matched source. In general, steps 1) and 2) can be omitted by consulting the many excellent tables available.

[13] L. Weinberg, "Network design by use of modern synthesis techniques and tables," *Proc. Natl. Electronics Conf.*, vol. 12, pp. 704–817; 1956.

[14] S. B. Cohn, "Shielded coupled-strip transmission line," IRE TRANS ON MICROWAVE THEORY AND TECHNIQUES, vol. MTT-3, pp. 29–38; October, 1955.

[15] S. B. Cohn, "Parallel-coupled transmission-line resonator filters," IRE TRANS. ON MICROWAVE THEORY AND TECHNIQUES, vol. MTT-6, pp. 223–231; April, 1958.

[16] W. J. Getsinger, "Coupled rectangular bars between parallel plates," IRE TRANS. ON MICROWAVE THEORY AND TECHNIQUES, pp. 65–72; January, 1962.

[17] W. J Getsinger, "A coupled strip-line configuration using printed-circuit construction that allows very close coupling," IRE TRANS. ON MICROWAVE THEORY AND TECHNIQUES, vol. 9, pp. 535–544; November, 1961.

[18] S. B. Cohn, "Characteristic impedances of broadside-coupled strip transmission lines," IRE TRANS ON MICROWAVE THEORY AND TECHNIQUES, vol. 8, pp. 633–637; November, 1960.

[19] S. B. Cohn," Thickness correction for capacitive obstacles and strip conductors," IRE TRANS. ON MICROWAVE THEORY AND TECHNIQUES, vol. 8, pp. 638–644; Novermber, 1960.

[20] J. D. Horgan, "Coupled strip transmission lines with rectangular inner conductors," IRE TRANS. ON MICROWAVE THEORY AND TECHNIQUES, vol. 5, pp. 92–99; April, 1957.

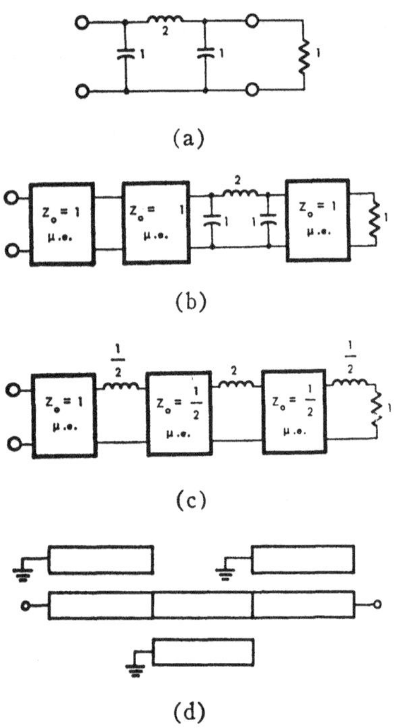

Fig. 5—(a) Third-order Butterworth filter. (b) Butterworth filter with unit elements added. (c) Equivalent circuit after application of Kuroda's identities. (d) Microwave network that realizes a third-order Butterworth response.

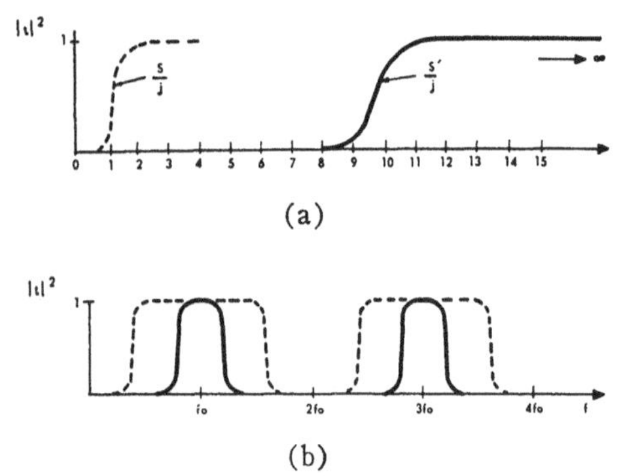

Fig. 6—(a) Renormalized S'-plane high-pass filter. (b) Corresponding f-plane filter.

Ozaki and Ishii for the case of unequal stripwidths.

To make filter computations as simple as possible, as in the above example, most networks are designed with a load impedance level of 1 Ω and a normalized band-edge frequency of $\omega=1$.

Changing the impedance level of a normalized S-plane network to Z_0 is accomplished by multiplying all normalized resistance and inductance values by Z_0 and dividing all capacitance values by Z_0.

Frequency normalization of S-plane networks cannot be accomplished in the same manner as that associated with lumped elements. The impedance of an S-plane inductor is $Z=jL \tan \pi f/2f_0$. Frequency normalization in the S plane, therefore, can be accomplished only by a change in f_0 which is tantamount to a change in the line lengths used in the TEM structure. For lumped elements, a frequency change is accomplished by multiplying the frequency variable ω by a constant. It is obvious that the analogous S-plane variable S is not multiplied by a constant in the above normalization procedure.

The effect of a renormalization in the S-plane variable $j \tan \pi f/2f_0$ is to change the location of the band-edge response of the network in the real frequency interval between 0 and f_0. Renormalizations of this type are the bandwidth controlling factors in TEM distributed line filters. As an example, consider a normalized S-plane high-pass filter which transforms to a band-pass filter in the f plane (see Fig. 1). This filter has a bandwidth of 100 per cent about the point f_0 and has a response which is periodic in $2f_0$. Assume that a change is made from S to the new variable $S'=10S$. The pertinent mapping points are now

$$\frac{S}{} = \frac{S'}{10j} = 0 \quad \text{maps into } f = 0,$$

$$\frac{S}{j} = \frac{S'}{10j} = 1 \quad \text{maps into } f = 0.938f_0,$$

$$\frac{S}{j} = \frac{S'}{10j} = \infty \quad \text{maps into } f = f_0.$$

The renormalized mapping properties in the S' plane and in the f plane are shown in Fig. 6, together with the responses of the original filter. Thus it is seen that the bandwidth of the distributed TEM filter is controlled by a renormalization of the S-plane variable and the effect of this renormalization is to change the characteristic impedance of the component elements in the TEM structure.

III. Equivalent TEM Filter Structures

When a filter has been designed by the methods of the preceding section and the pertinent element values have been obtained, the problem of constructing the network remains. The design techniques presented do not directly take into account the practical limitations on element values and circuit arrangement. However, judicious use of the techniques can often greatly simplify the construction of the device.

A. Element Values

When working with S-plane capacitors and inductors, it is very important to keep in mind the range of element values that can be realized. With lumped circuit capacitances, element value ranges of 10^6:1 can be achieved easily. In microwave circuits, the available range is much more limited. For example, consider the real frequency TEM circuit realization of an S-plane capacitor which is represented by a length of open circuited transmission line [see Fig. 7(a)].

The constant $1/C$ is associated with the characteristic impedance of the line Z_0. By consulting graphs of Z_0 as a function of line width,[21,22] it becomes apparent that

[21] "The Microwave Engineers Handbook," Horizon House-Microwave, Inc., Brookline, Mass.; 1962.

[22] R. H. T. Bates, "The characteristic impedance of the shielded slab line," IRE Trans. on Microwave Theory and Techniques, vol. 4, pp. 28–33; January, 1956.

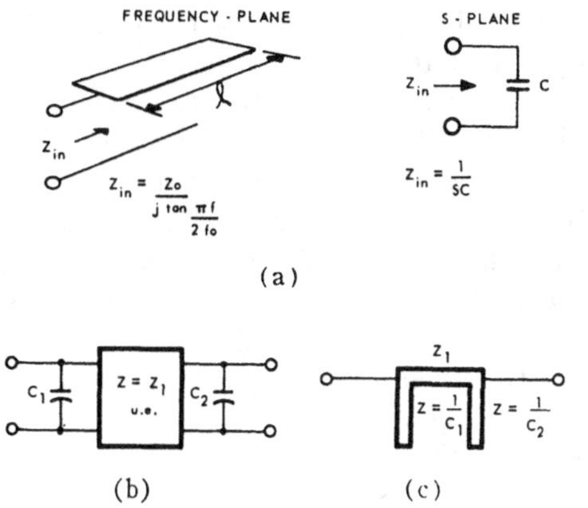

Fig. 7—(a) Stripline realization of S-plane capacitor. (b) S-plane network. (c) Stripline realization.

for $Z_0<10$ the lines become very wide and for $Z_0>300$ they become extremely narrow. Thus, the realizable range of Z_0, and consequently C, is only on the order of 30:1. As might be expected, this places a severe restriction on the responses that can be obtained with a given circuit configuration. Because of this restriction, the practical realization of a filter with a specified response may appear to be difficult if not impossible. Fortunately, in many cases, it is possible to realize the same response with more than one physical configuration. Before proceeding with the realization problem, it is useful to investigate the behavior of some TEM line sections and to obtain, where possible, different physical forms that realize similar responses.

B. Equivalent TEM Elements

Assume that it is desired to obtain a TEM line realization of the S-plane network shown in Fig. 7(b). There is no simple circuit, as found by consulting Table I, that realizes this network. However, the circuit can be realized in a TEM network as a combination of two shunt capacitors and a series unit element as shown in Fig. 7(c).

Table III lists a number of S-plane equivalent circuits for elements with series and shunt open and short circuited lines. The series stubs are constructed by using either a double coaxial structure or a single coaxial structure between flat ground planes. The impedance Z_2 of the unit element is the impedance of the inner coaxial structure, with respect to the outer conductor, and the impedance of the inner coaxial structure Z_1 is made numerically equal to L for the case of an inductor or $1/C$ for a capacitor. The inner coax is filled with a dielectric such that the series stub will fit within the unit element length l_0. Series stubs are useful in realizing large series capacitors or small series inductors. Construction of the shunt stubs is self-explanatory.

The importance of the equivalent circuits of Table III can be seen if one considers the realization of a unit element followed by an S-plane inductor (see Table I). Application of Kuroda's first identity (Table II) shows this to be the same as a shunt capacitor followed by a

TABLE III

EQUIVALENT CIRCUITS FOR ELEMENTS WITH SERIES AND SHUNT OPEN AND SHORT CIRCUITED LINES

TEM Circuit	Equivalent S-Plane Circuit

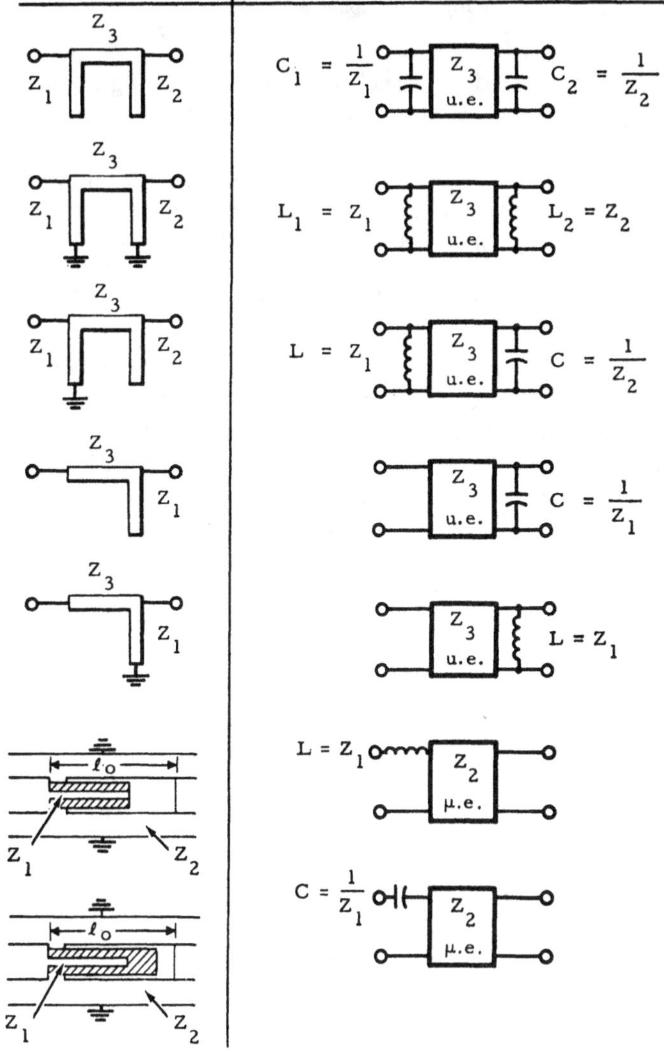

unit element. This configuration can be realized as shown in Fig. 8.

For a filter element of this type to have a broad stop band, a large value of L or C is required. Assume values of $Z_1=1$ and $L=2$. Then it follows that $Z_2=3$ and $C=\frac{2}{3}$. Normalizing to an impedance level of 50 Ω gives, for the series line with shunt C, $Z_2=150\ \Omega$ and $Z_3=75\ \Omega$. Impedances of this magnitude are within the realizable range and present no problem.

For the parallel coupled line realization, the even and odd mode impedances are

$$Z_{0e}=Z_0+L+\sqrt{L(Z_0+L)}=5.45$$

$$Z_{0o}=\frac{Z_{0e}}{1+\dfrac{2L}{Z_0}+\dfrac{2}{Z_0}L(Z_0+L)}=0.55.$$

Normalizing to 50 Ω gives $Z_{0e}=272\ \Omega$ and $Z_{0o}=27.5\ \Omega$. Attempts at realizing impedances of this order in

23

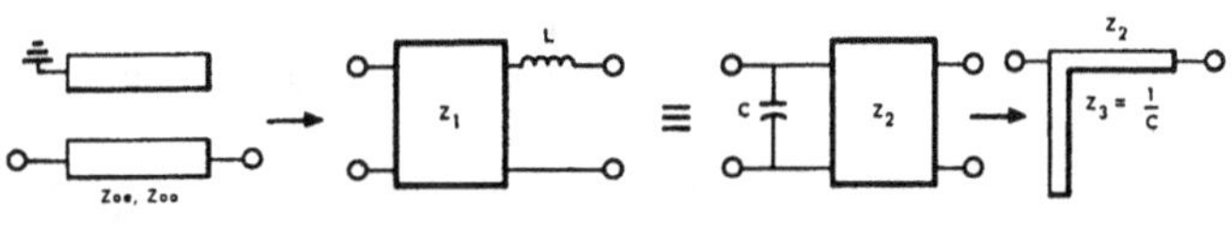

Fig. 8—Stripline equivalents.

parallel coupled lines lead to impractical configurations requiring very small strips having extremely close coupling. Conversely, if it is desired to achieve a narrow
24 stop-band, the parallel coupled element may well prove easier to construct. This is a direct consequence of the limited range of TEM impedance values obtainable.

C. Classification of Elements from Bandwidth Considerations

In designing a filter, bandwidth is a major consideration. For the case of microwave TEM filters, because of the limited range of practical impedance values, it is very important that the designer choose the physical form applicable for a given bandwidth.

The starting point for the bandwidth consideration is the synthesis of the S-plane filter. Filters that are synthesized on a normalized basis will have their cutoff at $S/j=1$ and the corresponding TEM line realization will have a band-edge cutoff at $f_0/2$. In general, the normalized filters will have bandwidths[23] of 100 per cent centered about f_0 and will be periodic in $2f_0$. For changes in bandwidth, the variable S is changed to $S'=S/K$, $(K>0)$ by substituting S' for S in the desired filter function. This changes the band-edge cutoff f_c from $f_0/2$ to

$$\tan \frac{\pi f_c}{2f_0} = K.$$

Then

$$f_c = \frac{2f_0}{\pi} \tan^{-1} K. \tag{19}$$

The fractional bandwidth then becomes

$$\frac{2(f_0 - f_c)}{f_0} = 2\left(1 - \frac{2}{\pi} \tan^{-1} K\right). \tag{20}$$

For $K>1$, the bandwidth decreases and for $K<1$ it increases.

Substituting the variable S' for the variable S changes the original filter elements L and C to the new elements $L'=L/K$ and $C'=C/K$. This means that for narrow bandwidth $(K>1)$, small L's and small C's are required, whereas large L's and large C's are required for wide bandwidths $(K<1)$.

These considerations give a criterion for establishing the bandwidth properties of TEM line configurations. For the case of parallel coupled lines, practical realization requires that Z_{0e} and Z_{0o} be reasonably close to Z_0 (a practical range being $Z_{0e}<150\ \Omega$ and $Z_{0o}>20\ \Omega$). The relations in Table I provide the information necessary to evaluate the bandwidth capabilities of these sections. For example, consider the strip-line element in Table I(c). This section has a stop-band centered about f_0, the bandwidth of which is controlled by the value of L. Inspection of the equations for Z_{0e} and Z_{0o} show that these values are close to Z_0 only when L is small. Thus this section is practically realizable only for $L<1$ and is a good element for use in realizing narrow, stop-band filters. On the other hand, its shunt stub-line equivalent (Table III) can be realized for a wide stop band without encountering difficulty in achieving the element values.

For the case of the element in Table I(f), the equations for Z_{0e} and Z_{0o} show that this section will be difficult to realize for small C's the bandwidth capabilities of all the sections shown in Tables I and III can be determined in a similar manner.

IV. Comparison of TEM and Lumped Element Filters[24]

When a filter is synthesized in the S plane, its response in the real frequency plane is not identical to its S-plane response. An S-plane Butterworth filter, for example, does not have a Butterworth response with respect to the frequency variable ω in its TEM realization.

A. High-Pass and Low-Pass Filters

Theorem: The frequency plane response of a normalized S-plane Butterworth filter whose frequency variable is $\tan \pi\omega/2\omega_0$ provides a better approximation to the ideal low-pass or high-pass filter characteristic than does a frequency plane Butterworth filter whose variable is ω for $0<\omega<\omega_0$.

Proof I: Low-pass filter with band-edge cutoff at $\omega=\omega_0/2$. The Butterworth expression that approximates the ideal characteristic is

$$|t|^2 = \frac{1}{1 + (\text{frequency variable})^{2n}}. \tag{21}$$

The ideal characteristic has a value of 1 in the range $0<\omega<\omega_0/2$. Further, in this frequency range,

$$\tan \frac{\pi\omega}{2\omega_0} < \omega$$

and therefore

$$\frac{1}{1 + \left(\tan \dfrac{\pi\omega}{2\omega_0}\right)^{2n}} > \frac{1}{1 + \omega^{2n}}. \tag{22}$$

Thus the variable $\tan \pi\omega/2\omega_0$ provides a better approximation to the ideal characteristic than does the variable ω. In the frequency range $\omega_0/2<\omega<\omega_0$, the

[23] The term "bandwidth" will be taken as synonymous with "bandwidth about f_0" and can be either a stop band or a pass band

[24] This comparison is made only over the interval $0<\omega\leq\omega_0$ for the case of low-pass and high-pass filters and over the range $0\leq\omega\leq 2\omega_0$ for band-pass filters. Thus, the repetititve nature of the distributed filter is not considered in the comparison.

ideal characteristic has a value of zero. In this range, $\tan \pi\omega/2\omega_0 > \omega$ and thus

$$\frac{1}{1+\left(\tan\frac{\pi\omega}{2\omega_0}\right)^{2n}} < \frac{1}{1+\omega^{2n}}. \tag{23}$$

Therefore, once again, the variable $\tan \pi\omega/2\omega_0$ provides a better approximation to the characteristic than does the variable ω.

Proof II: The proof for the case of a high-pass filter with band-edge cutoff at $\omega_c = \omega_0/2$ follows in an analogous manner.

B. Band-Pass Filters

The low-pass normalized lumped element Butterworth filter of order n has the transmission characteristic

$$|t(j\Omega)|^2 = \frac{1}{1+\Omega^{2n}}.$$

An unnormalized low-pass filter of bandwidth ω_c then has the characteristic

$$\left|t\left(\frac{j\Omega}{\omega_c}\right)\right|^2 = \frac{1}{1+\left(\frac{\Omega}{\omega_c}\right)^{2n}}.$$

The low-pass to band-pass transformation[4] is

$$\Omega = \frac{\omega^2 - \omega_L^2}{\omega}$$

where

$\omega_2 - \omega_1 = \omega_c$ the filter bandwidth,
$\omega_1\omega_2 = \omega_L^2$ the geometric center of the lumped element pass-band,
$\omega_2 =$ upper band-edge frequency,
$\omega_1 =$ lower band-edge frequency.

Then

$$|t(j\omega)|^2 = |t|_L^2 = \frac{1}{1+\left[\frac{\omega^2-\omega_L^2}{\omega\omega_c}\right]^{2n}}. \tag{24}$$

The TEM line realization of a high-pass S-plane Butterworth filter can be used as a band-pass filter symmetric about $\omega = \omega_0$. For the normalized S-plane high-pass filter,

$$\left|t\left(j\tan\frac{\pi\omega}{2\omega_0}\right)\right|^2 = \frac{1}{1+\left[\frac{1}{\tan\frac{\pi\omega}{2\omega_0}}\right]^{2n}}. \tag{25}$$

For a bandwidth of ω_c about $\omega = \omega_0$,

$$\text{fractional bandwidth} = \frac{\omega_c}{\omega_0} = 2\left(1 - \frac{2}{\pi}\tan^{-1} K\right)$$

[see (20)], then

$$\left|t\left(\frac{j\tan\frac{\pi\omega}{2\omega_0}}{K}\right)\right|^2 = |t|_{\text{TEM}}^2 = \frac{1}{1+\left[\frac{K}{\tan\frac{\pi\omega}{2\omega_0}}\right]^{2n}}.$$

Solving for K gives

$$K = \tan\frac{\pi}{2}\left(1-\frac{\omega_c}{2\omega_0}\right)$$

or

$$|t|_{\text{TEM}}^2 = \frac{1}{1+\left[\frac{\tan\frac{\pi}{2}\left(1-\frac{\omega_c}{2\omega_0}\right)}{\tan\frac{\pi\omega}{2\omega_0}}\right]^{2n}}. \tag{26}$$

The response of the TEM realization of the S-plane filter is symmetric about $\omega = \omega_0$. The band-edges of both the TEM and the lumped element band-pass filters can be made identical provided that

$$\omega_2 = \omega_0 + \frac{\omega_c}{2}$$

$$\omega_1 = \omega_0 - \frac{\omega_c}{2}$$

$$\omega_1\omega_2 = \omega_0^2 - \frac{\omega_c^2}{4} = \omega_L^2.$$

Defining $B = \omega_c/\omega_0$, the filter fractional bandwidth restricted to the range $0 < B < 1$, and $X = \omega/\omega_0$, in the range $0 < \omega < 2\omega_0$, allows the lumped element characteristic to be written as

$$|t|_L^2 = \frac{1}{1+\left[\frac{1-X^2-\frac{B^2}{4}}{BX}\right]^{2n}} \tag{27}$$

and the TEM characteristic as

$$|t|_{\text{TEM}}^2 = \frac{1}{1+\left[\cot\frac{\pi B}{4}\cot\frac{\pi X}{2}\right]^{2n}}. \tag{28}$$

A comparison of the two filters is made by evaluating the characteristic slopes at the low band-edge frequency for fractional bandwidths in the range $0 < B < 1$.

25

The slope of the TEM filter response at band-edge $X=1-B/2$ is given by

$$\left.\frac{d\,|t|^2_{\text{TEM}}}{dX}\right|_{X=1-B/2} = \frac{\pi n}{2\sin\pi\left(1-\dfrac{B}{2}\right)}. \tag{29}$$

26

The slope of the lumped element filter is

$$\left.\frac{d\,|t|^2_L}{dX}\right|_{X=1-B/2} = \frac{2n\left(1-\dfrac{B}{2}\right)^{2n-1}\left[1+\dfrac{2}{B}\left(1-\dfrac{B}{2}\right)^2\right]}{\left[1+\left(1-\dfrac{B}{2}\right)^{2n}\right]^2}. \tag{30}$$

Defining $A=1-B/2$, the ratio of slopes is given by

$$\frac{\text{Slope}_{\text{TEM}}}{\text{Slope}_L} = \frac{\pi(1-A)[1+A^{2n}]^2}{4A^{2n-1}[A^2-A+1]\sin\pi A}. \tag{31}$$

The slope ratio is plotted in Fig. 9. From the graph, it is apparent that the slope of the TEM filter characteristic at band-edge is always greater than the corresponding slope of the lumped element filter, the factor becoming greater as the bandwidth is increased. Both have the same limiting slope for narrow bandwidths ($B\rightarrow 0$). The TEM filter has a sharper cutoff slope than the lumped element filter, especially for moderate to wide bandwidths and for a large number of sections.

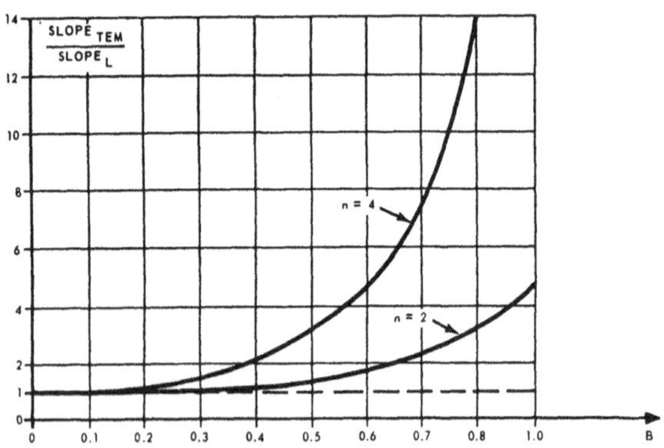

Fig. 9—Ratio of low band-edge slopes for the TEM and lumped elements band-pass filters vs fractional bandwidth for two- and four-section Butterworth network.

V. Applications

A. Equivalence of Stub and Parallel Coupled Line Filters

A common TEM line configuration used in many filters is shown in Fig. 10(a) along with its equivalent circuit. Cohn[15] obtained a design procedure for this type of filter by use of image parameter theory. The methods of the preceding sections can be used to obtain an exact synthesis procedure. Application of Kuroda's identities provide some insight into the synthesis problem. The basic identity to be used is shown in Fig. 10(b). Application of this identity to the end filter section gives the circuit of Fig. 10(c). The transformer is brought out to accessible terminals by transforming each component from left to right sides of the transformer, Fig. 10(d). The network between the transformer and the last unit element has the same general form as the original circuit. Continuation of the above process leads to the circuit of Fig. 10(e). The cascade of transformers can be replaced by a single transformer and the capacitor can be transformed to the input terminals as shown in Fig. 10(f).

The dual network along with its equivalent circuit is shown in Fig. 11(a). Application of a procedure analogous to the one in Fig. 10 gives the dual equivalent circuit shown in Fig. 11(b). The following discussion pertinent to an exact synthesis procedure applies equally well to both the circuit of Fig. 10(a) and its dual Fig. 11(a).

One apparent result of the preceding transformation is that the equivalent circuit, Fig. 10(f), has only $N+2$ degrees of freedom: the N unit elements, the transformer with turns ratio n; and the capacitor C. For the case of symmetric networks, the transformation can be performed in a manner such that $n=1$. This allows the practical realization of these networks in the form of stepped impedance lines with one series or shunt element.

On the other hand, the circuit of Fig. 10(a) contains $2N+1$ parameters but has the same order characteristic polynominal as Fig. 10(f) and is thus a redundant structure. In fact, for many practical filter structures, the N unit element values can be set equal to the characteristic impedance of the system and the $N+1$ capacitors still provide enough degrees of freedom to allow realization of the desired network function. Also, the parallel-coupled filter has no S plane, L-C ladder equivalent, valid over the entire frequency spectrum. Thus, any attempt to obtain an exact synthesis procedure for this type of network by using an L-C ladder prototype will fail.

An exact synthesis of both the circuits of Fig. 10(a) and (f) is possible, but the procedure is not the same as for networks that have an L-C ladder, S-plane prototype. An example of this method is given in Appendix I.

Two important points should be emphasized as a result of the above discussion:

1) From the viewpoint of exactly synthesizing parallel coupled bar filters [of the type in Fig. 10(a) and Fig. 11(a)], the only S-plane elements required are C's, L's, ideal transformers and unit elements. Furthermore, all filters of this type have been shown to require only one L or one C and

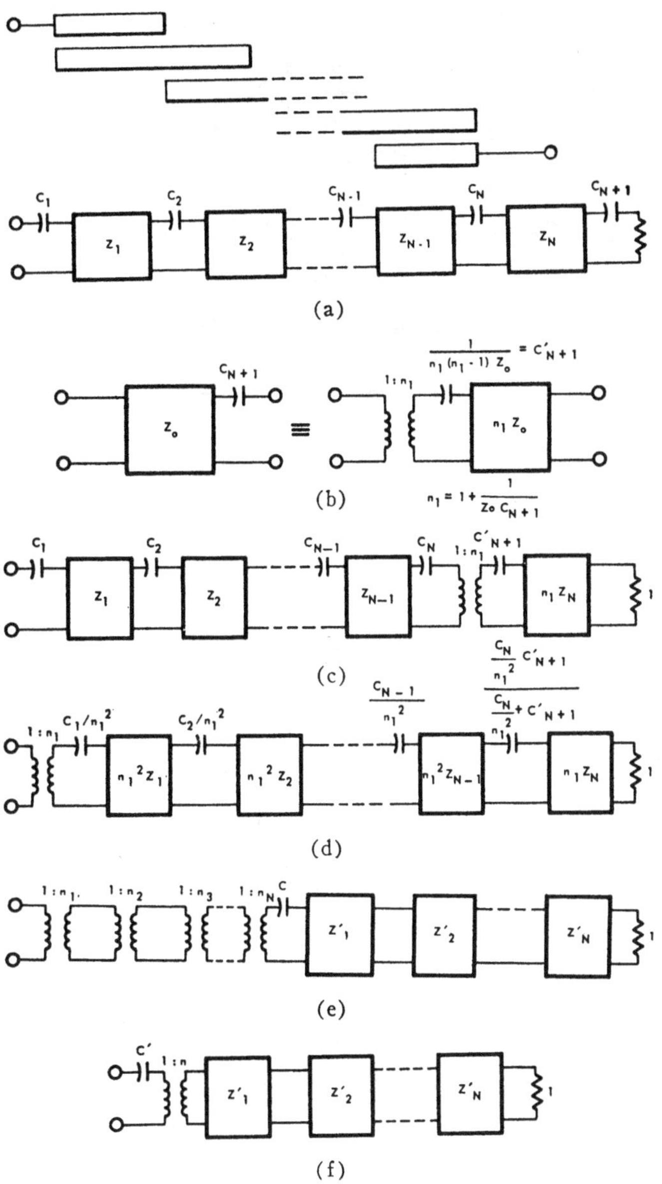

Fig. 10—(a) Stripline circuit and equivalent circuit. (b) Kuroda identity. (c) Application of Kuroda identity to end section of filter. (d) Network obtained by bringing the transformer to accessible terminals. (e) Equivalent circuit after n cycles of transformation. (f) Equivalent circuit with single transformer.

can always be realized theoretically in the alternate forms which do not require parallel coupled bars.

2) By carrying out the synthesis procedure outlined in Appendix I for a large number of cases, tables[25] of element values can be obtained which enable the microwave engineer to design filters of the type shown in Figs. 10 and 11 with known responses

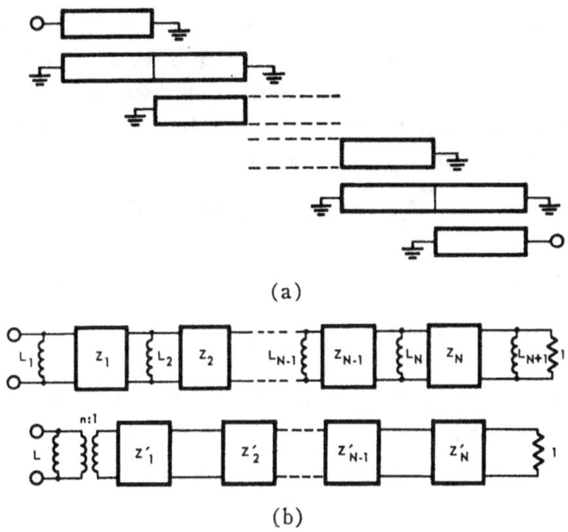

Fig. 11—(a) Stripline circuit and equivalent circuit. (b) Equivalent circuit without parallel coupled bars.

over the entire frequency range. There is no bandwidth limitation in the synthesis procedure. The limitations that arise will be due to physical realizability of the required element values.

An appreciation of the problem of physical realizability is important. The parallel coupled bar filter, Fig. 10(a); the uniform line with series open stubs; or the stepped impedance filter with one series element, Fig. 10(f), realize the same response. However, for a given bandwidth, one configuration may be more practical from the standpoint of realizable element values.

B. Complementary Filters

In many situations, it is desirable to design a lossless network to couple two resistances over a specified band of frequencies in such a way that the driving source will be properly matched outside the band. Assume that the input impedance of the load and coupling network is Z_1. Then, if another impedance Z_c is added in series such that Z_1+Z_c is constant, the over-all input impedance is resistive and a perfect match can be provided over the entire frequency range. The dual situation is the addition of admittances in parallel such that Y_1+Y_c is constant. Impedances or admittances that add to give a constant independent of frequency are said to be complementary.[26]

Theorem: A lossless ladder network, terminated in a 1 Ω resistor that realizes a Butterworth transfer impedance $Z_{12}(S)$, has a driving point impedance $Z_1(S)$ which always has a complementary impedance $Z_c(S)$ that can be realized in a lossless ladder terminated in a 1 Ω resistor. The network for $Z_c(S)$ can be obtained by replacing the L's and C's of $Z_1(S)$ with $\mathbf{C}$'s and $\mathbf{L}$'s, respectively, where $\mathbf{C}=1/L$, and $\mathbf{L}=1/C$.

[25] The many tables now available do not contain the element values obtained by the above synthesis procedure. This is a consequence of the fact that these networks do not have an S-plane, L-C ladder equivalent. The insertion loss function of these networks is of a form different than those with L-C equivalents and thus leads to different element values. See Appendix I (41) and (42) and the following discussion.

[26] For a thorough discussion of complementary filters, see Guillemin,[4] p. 476.

As an example, if $Z_1(S)$ is realized in the ladder form of Fig. 12(b), then $Z_c(S)$ is realized as shown in Fig. 12(c). The network definitions are shown in Fig. 12(a).

Proof: For the network of Fig. 12(a), the input power is dissipated in the 1 Ω resistor, so that

$$|I_1|^2 \text{ Re } [Z_1(j\omega)] = |V_2|^2.$$

Thus,

28

$$\text{Re } [Z_1(j\omega)] = |Z_{12}(j\omega)|^2 = \frac{1}{1+\omega^{2n}} \tag{32}$$

for a Butterworth response.[27] From the definition of complementary filters, it follows that

$$\begin{aligned} \text{Re } [Z_c(j\omega)] &= 1 - \text{Re } [Z_1(j\omega)] && (33) \\ &= 1 - |Z_{12}(j\omega)|^2 \\ &= 1 - \frac{1}{1+\omega^{2n}} \\ &= \frac{1}{1+\left(\frac{1}{\omega}\right)^{2n}} \cdot && (34) \end{aligned}$$

Substituting $\omega' = 1/\omega$ gives

$$\text{Re } [Z_c(j\omega)] = \frac{1}{1+(\omega')^{2n}} \cdot \tag{35}$$

Thus, the network of $Z_c(S)$ has the same geometry and the change in variable results in converting every L and C of $Z(S)$ into a $\mathbf{C}$ and $\mathbf{L}$ where

$$\mathbf{C} = \frac{1}{L} \qquad \mathbf{L} = \frac{1}{C} \cdot$$

The prototype elements of Butterworth filters have been tabulated in several references[13,28] and, as a result of the above theorem, the complementary network prototypes can be obtained by inspection.

By using the methods described in Section III it is possible to design complementary microwave filters. For practical microwave realization, the series connection required for complementary impedances is in most cases not as desirable as the parallel connection of admittance complements. The complementary admittance lumped element prototypes for a third-order Butterworth response are shown in Fig. 13(a).

For the admittance Y_1, the procedure for obtaining the TEM line realization follows that of the example in Section II. Since the input admittance is to be preserved, unit elements can only be inserted before the

[27] Note that this filter is designed on a transfer impedance basis and assumes a current source. When the two complementary filters are connected, their common input is purely resistive (R) and provides a matched load to a generator with internal resistance equal to R.

[28] P. R. Geffe, "Computer prepared tables enable design of ultra-flat networks," *Electronic Design*, vol. 8, pp. 48–51; August 31, 1960.

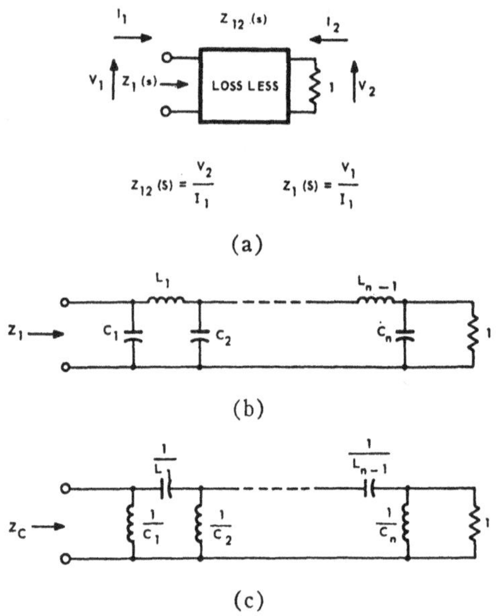

Fig. 12—(a) Network parameters. (b) Butterworth ladder network. (c) Complementary ladder network.

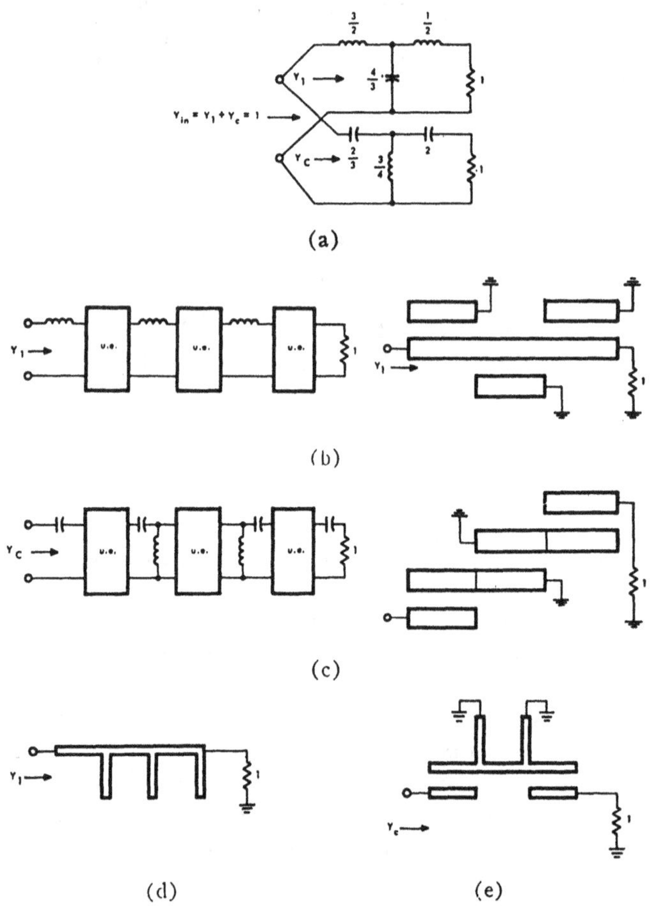

Fig. 13—(a) Admittance complements. (b) Realization of Y_1. (c) Realization of Y_c. (d) Realization of Y_1 without parallel coupled lines. (e) Realization of Y_c with a reduced number of parallel coupled bars.

load. The final equivalent circuit along with a TEM line realization is shown in Fig. 13(b). A realization of Y_c can be obtained in a similar manner, but will, in general, involve ideal transformers. These transformers can often be eliminated by using a technique illustrated in Appendix II. The final transformerless equivalent circuit for Y_c, along with a TEM realization, is shown in Fig. 13(c).

There are two major drawbacks to the TEM circuits shown in Figs. 13(b) and (c): 1) the parallel coupled sections with shorted ends are difficult to construct and 2) the filter of Fig. 13(b) can provide only relatively narrow stop bands due to the restriction previously discussed. These difficulties can be reduced if the circuits are realized in a different form by using the identities in Table III. Thus, the circuit of Fig. 13(b) can be realized as shown in (d)[29] and that of (c) can be realized as shown in (e).[30] By using the identities in Table III, it is possible to reduce filters with parallel coupled elements to an alternate form.

C. Experimental Results

A number of filters employing most of the network configurations discussed have been built and tested. Excellent agreement with theory was obtained in all cases. For construction details of the stop-band configurations of open shunt stubs [Fig. 13(d)] or parallel coupled bars [Fig. 13(b)], the reader is referred to the paper by Matthaei and Schiffman.[31] Construction details of parallel coupled filters are available in the literature.[32] To the author's knowledge, no construction details of filters employing series stubs have been reported. In view of this, two filters employing these elements, one a band pass about f_0, the other a band stop about f_0, will be described.

The S-plane prototype for the band-pass filter is shown in Fig. 14(a). The element values are those corresponding to a Butterworth reflection coefficient characteristic of order 3 and a bandwidth of 100 per cent. Note that the usual 1, 2, 1 element distribution is not used but rather the reciprocals of these values. This arises from the fact that the 1, 2, 1 values provide a low-pass S-plane response. A band pass about f_0 corresponds to a high-pass S-plane filter. The low-pass to

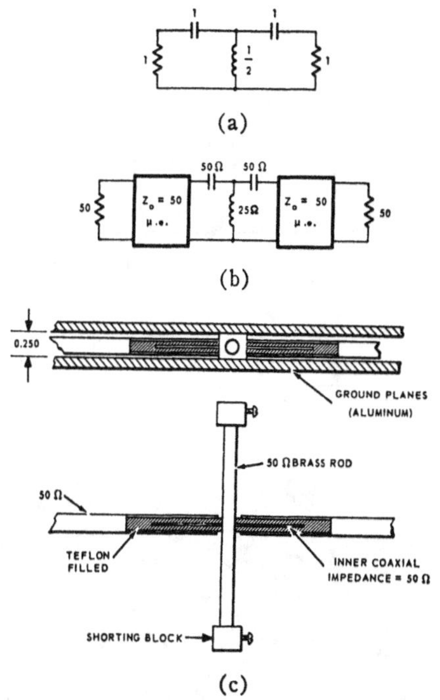

Fig. 14—(a) S-plane band-pass prototype. (b) Prototype after addition of two unit elements and renormalization to $Z_0=50\ \Omega$. (c) Construction details of three-section Butterworth filter.

high-pass transformation of a normalized network requires replacing each shunt element by a dual series element with a reciprocal value and each series element by a dual shunt element with reciprocal value. (See Guillemin,[4] p. 602.)

Since the filter was to provide a specified reflection coefficient, a unit element was added before and after the lossless part of the network. Consulting Table III, the network is seen to be realizable as the cascade of a series capacitor stub, a shunt inductor stub and another series capacitor stub. No transformations were required. Normalization to $Z_0=50\ \Omega$ gives the final equivalent circuit [Fig. 14(b)].

The filter was constructed between ground planes having a 0.250-inch spacing with a 50 Ω round center conductor. The design center frequency was 2 Gc. The shunt 25 Ω inductor was obtained by using two 50 Ω stubs, one on either side of the center conductor. Construction details are shown in Fig. 14(c). The response and VSWR of the filter is shown in Fig. 15(a) along with the theoretical values. A photograph of the filter is shown in Fig. 15(b). The primary purpose in constructing this filter was to establish the degree to which junction effects would alter the multiple response at 6 Gc. As is evident from Fig. 15(a), these effects were small. Insertion loss in the pass bands was less than 0.4 db and could be further improved by using silver plated conductors rather than brass and aluminum.

The S-plane prototype for the stop-band filter is shown in Fig. 16(a). The element values are those for a five-section, 0.1-db ripple Chebyshev reflection coeffi-

[29] The equivalence of the networks in Fig. 13(b) and (d) was discussed by Matthaei and Schiffman at the 1963 PTGMTT National Symposium in Santa Monica, Calif. Their paper, to be published in the IEEE Transactions under the title "Exact Design of Band-Stop Microwave Filters," provides information relative to constructing these networks.

[30] This network can be constructed entirely without parallel coupled bars by using series stubs. See Section V-C.

[31] G. L. Matthaei and B. M. Schiffman, "Exact Design of Band-Stop Microwave Filters," presented at 1963 PTGMTT National Symposium, Santa Monica, Calif., May 20–22.

[32] G. L. Matthaei, "Design of wide-band (and narrow-band) band-pass microwave filters on the insertion loss basis," IRE Trans. on Microwave Theory and Techniques, vol. 8, pp. 580–593; November, 1960.

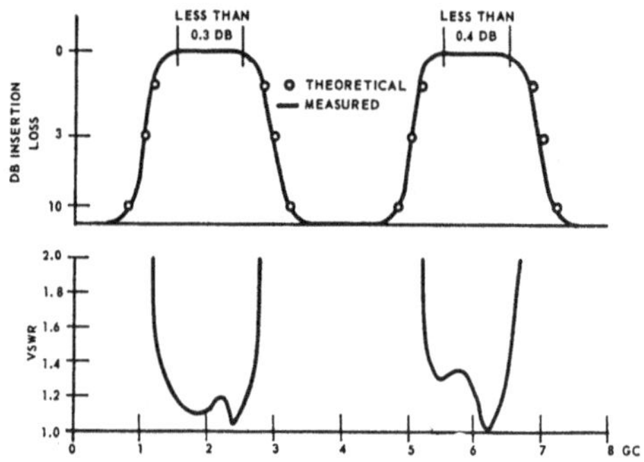

(a)

(b)

Fig. 15—(a) Insertion loss and VSWR of third-order Butterworth filter. (b) three-section Butterworth filter.

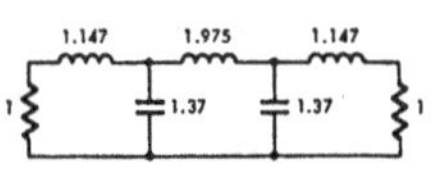

(a)

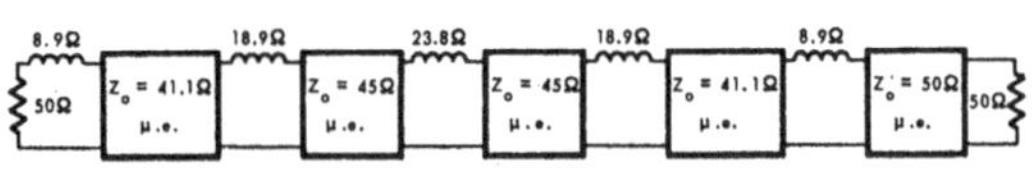

(b)

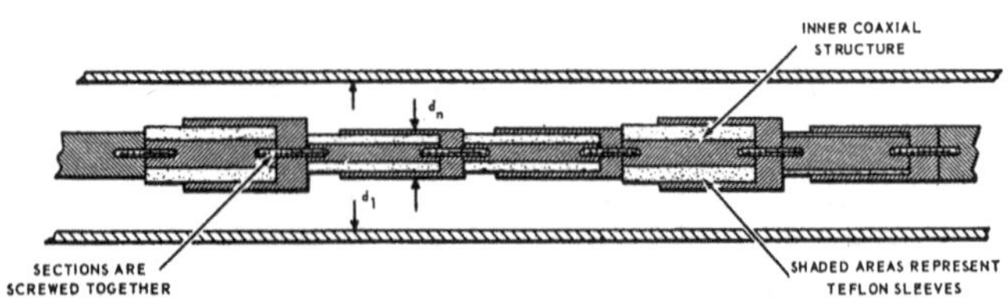

(c)

Fig. 16—(a) 5-section 0.1-db Chebyshev filter prototype. (b) Filter circuit after application of Kuroda's Identities and renormalization to 50 Ω. (c) Construction of stop-band filter with series shorted stubs. The impedance of the unit element is realized by using the correct d_1/d_n ratio and that of the inductors by using the correct inner coaxial impedance.

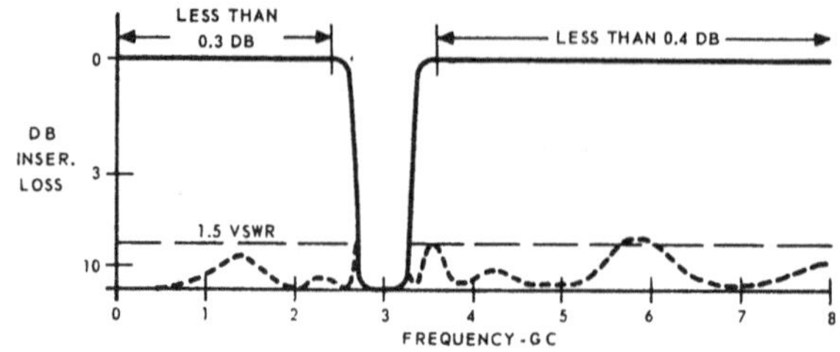

(a)

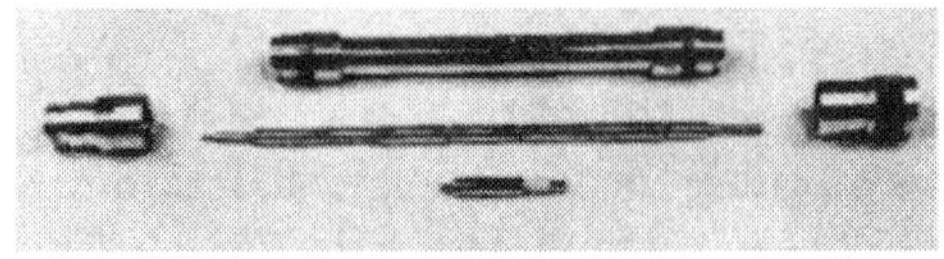

(b)

Fig. 17—(a) Insertion loss and VSWR of five-section Chebyshev band-stop filter with series stubs. (b) Band-stop filter.

cient characteristic. The filter was designed to have a 30 per cent band stop about f_0. For a normalized band-edge frequency of $\Omega = 1$, the 0.1-db bandwidth[33] is 100 per cent. Dividing Ω by a constant K, where K is greater than one, reduces the bandwidth in accordance with the discussion in Section III-C. The required constant for a 30 per cent stop band is

$$K = \tan \frac{\pi 0.85 f_0}{2 f_0} = 4.16.$$

Application of Kuroda's identities and renormalization to $Z_0 = 50\ \Omega$ gives the final equivalent circuit of Fig. 16(b).

The filter was constructed using a double coaxial structure. Construction details are shown in Fig. 16(c). The response and VSWR of the filter is shown in Fig. 17(a) and a photograph of the filter is shown in Fig. 17(b).

Little difficulty was encountered in constructing the filters with series stubs. In fact, junction effects for the series stubs were found to be less significant than for shunt stubs, especially when low impedance values were required.

VI. Conclusions

TEM networks designed by modern network theory techniques have been shown to exhibit the following advantages over those based on approximate methods:

1) The equivalent circuits used are theoretically valid over the entire frequency range.
2) Different physical structures that realize identical responses can often be obtained.
3) Limitations are readily found on the response obtainable with a given physical configuration.

[33] A Chebyschev filter characteristics as a 3-db point whose locations depend on the number of elements. For $\Omega = 1$, the characteristic differs from the ideal by a value equal to the specified ripple and thus serves as a convenient bandwidth reference point.

4) The networks, in general, employ a minimum number of elements.
5) It is possible to construct networks that realize exactly a desired input impedance, reflection coefficient or phase response.

The small range of realizable impedance values has been shown to be a limiting factor in the response obtainable in several TEM filter configurations. Equivalent network elements have been demonstrated to allow simplification in physical circuitry as well as improvement in response.

The real frequency response of TEM realizations of S-plane high-pass or low-pass Butterworth filters was shown to be better than that of corresponding lumped element filters. An S-plane high-pass filter, used as a band-pass filter in the frequency plane, was shown to have steeper cutoff at the band-edge than the corresponding lumped element filter. In this comparison, multiple responses of the microwave filter were not considered.

Complementary microwave filters are shown to be useful in providing an exact match of source to load over a wide frequency band. Following the procedure given, the elements of Butterworth complements are easily obtained. Techniques for eliminating ideal transformers in networks with series capacitors and shunt inductors are shown to be useful in obtaining practical circuit configurations. The exact synthesis procedure for parallel coupled line filters can be used to obtain tables that will enable these filters to be designed in a simple manner.

The basic circuit elements required for synthesizing network configurations discussed in this paper are S-plane inductors and capacitors, ideal transformers and unit elements. The parallel coupled bar is not a necessary element; however, it allows a different physical realization for networks involving basic elements.

Application of the basic network theory described in Section III is not restricted to the structures described. Couplers, transformers and many other common microwave components can be analyzed and designed by the use of these methods.

Appendix I

The following example demonstrates an exact synthesis procedure applicable to

1) parallel coupled bars with open ends,
2) parallel coupled bars with shorted ends,
3) stepped quarter-wave sections with one series open stub,
4) stepped quarter-wave sections with one shunt shorted stub,
5) uniform line with quarter-wave spaced series open stubs,
6) uniform line with quarter-wave spaced shunt shorted stubs.

Types 1), 3) and 5) are equivalent and 2), 4) and 6) are their respective duals. The basic prototype to be considered is shown in Fig. 18(a).

The transmission function for a network of this type is[34]

$$|t|^2 = \frac{-S^2(1-S^2)^N}{P_{N+1}(-S^2)} = \frac{\Omega^2(1+\Omega^2)^N}{P_{N+1}(\Omega^2)} \tag{36}$$

where

$$S = j\Omega = j\tan\frac{\pi f}{2f_0}$$

N = number of unit elements

P_{N+1} is a polynominal of degree $N+1$ in Ω^2.

In accordance with the discussion in Section V-A, the N unit element impedance values can be arbitrarily assigned. A convenient choice is to set $Z_1 = Z_2 = \cdots Z_N = 1$. The first step in the synthesis procedure is to obtain a suitable form for $|t|^2$.

Letting $\Omega = \tan, \theta$ gives

$$|t|^2 = \frac{\tan^2\theta(1+\tan^2\theta)_N}{P_{N+1}(\tan^2\theta)} = \frac{\sin^2\theta}{Q_{N+1}(\cos^2\theta)} \tag{37}$$

where

Q_{N+1} is a polynomial of degree $N+1$ having different coefficients than those of P_{N+1}.

Now, letting $x = \cos\theta$, (37) becomes

$$|t|^2 = \frac{1-x^2}{Q_{N+1}(x^2)} = \frac{x^2-1}{(x^2-1) - G_{N+1}(x^2)}$$

where

$$Q_{N+1}(x^2) = G_{N+1}(x^2) + (1-x^2)$$

or

$$|t|^2 = \frac{1}{1 - F_{N+1}(x^2)} \tag{38}$$

where

$$F_{N+1}(x^2) = \frac{G_{N+1}(x^2)}{x^2-1}.$$

The next step in the synthesis procedure is the approximation problem; that is, a criterion must be established that defines the manner in which the function $F_{N+1}(x^2)$ is to be chosen to approximate the desired $|t|^2$ characteristic. This is, in general, a difficult problem and the reader is referred to literature[2,4,5] for detailed discussions.

A common approximation is the maximally flat response which is obtained by setting all but the $(N+1)$st

[34] This can be obtained by mutiplying the $ABCD$ matrices for the network.

31

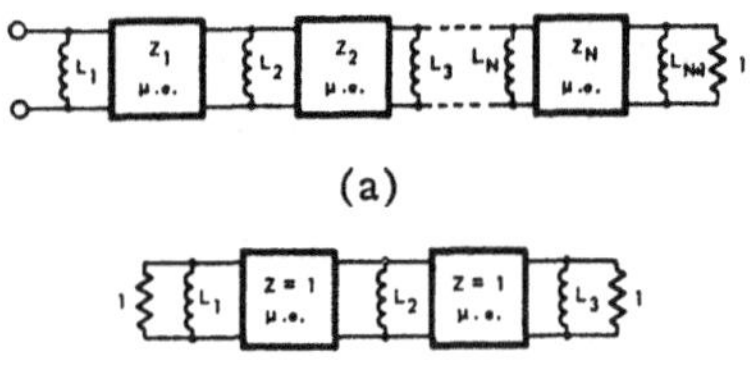

Fig. 18—(a) Prototype for 1) parallel coupled bars with shorted ends, 2) Line with quarter-wave short circuited stubs. (b) Prototype for third-order Butterworth filter.

32

coefficient of $F_{N+1}(x^2)$ equal to zero. Applying this criterion gives

$$|t|^2 = \frac{1}{1 - \dfrac{Gx^{2(N+1)}}{x^2 - 1}}$$

where

G is the $(N+1)$st coefficient of the polynomial $G_{N+1}(x^2)$.

The prototype for a third-order maximally flat filter is shown in Fig. 18(b).

There are two unit elements; therefore, $N=2$. Then

$$|t|^2 = \frac{1}{1 - \dfrac{Gx^6}{x^2 - 1}}.$$

Since

$$x^2 = \frac{1}{1 - S^2},$$

$$|t|^2 = \frac{1}{1 - \dfrac{G}{S^2(1 - S^2)^2}}$$

or

$$|\rho|^2 = 1 - |t|^2 = \frac{G}{G - S^2(1 - S^2)^2}. \quad (39)$$

The bandwidth of the filter is determined by the constant G; using $S = j \tan \pi f_c/2f_0$ and setting $|\rho|^2 = 0.5$ allows (39) to be solved for G. The fractional bandwidth of the filter is

$$\frac{2(f_0 - f_c)}{f_0}.$$

The general network configuration of the filter is known and it is sufficient for synthesis purposes to obtain the input impedance that corresponds to the given $|\rho|^2$. For the details of this procedure, see Guillemin,[4] p. 458.

For $G=36$, the bandwidth is 70 per cent and

$$Z_{in}(S) = \frac{S^3 + 4S^2 + 7S}{S^3 + 4S^2 + 7S + 12}. \quad (40)$$

The synthesis procedure is different depending upon which of the possible configurations is desired. Richard's theorem,[7] basically common to each procedure, is used to remove unit elements.[35] This theorem states that if $m_1m_2 - n_1n_2|_{S^7} = 0$, then a unit element of impedance $Z_{in}(1)$ can be removed with the remaining impedance being

$$Z'_{in}(S) = Z_{in}(1)\frac{SZ_{in}(1) - Z_{in}(S)}{SZ_{in}(S) - Z_{in}(1)}.$$

Under these conditions, both the numerator and denominator of $Z'_{in}(S)$ contain the factor (S^2-1) and $Z'_{in}(S)$ is thus one degree lower than $Z_{in}(S)$. The product of the even parts of numerator and denominator, respectively, of $Z_{in}(S)$ is given by m_1m_2 and the product of the odd parts is given by n_1n_2. A one stub network can be obtained by applying Richard's theorem to (40).

$$m_1m_2 - n_1n_2|_{S=1} = 4S^2(4S^2 + 12) - (S^3 + 7S)^2|_{S=1} = 0.$$

A unit element of impedance $Z_{in}(1) = \frac{1}{2}$ can be removed. The input impedance of the remaining network is

$$Z'_{in}(S) = Z_{in}(1)\frac{SZ_{in}(1) - Z_{in}(S)}{SZ_{in}(S) - Z_{in}(1)} = \frac{S^2 + 2S}{4S^2 + 14S + 24}.$$

The common S^2-1 factor has been cancelled. $Y'_{in}(S) = 1/Z'_{in}(S)$ has a pole at zero that can be removed by taking out a shunt inductor,

$$L = \frac{1}{12}; \qquad 2S + S^2 \overline{\big)\, 24 + 14S + 4S^2} \quad \text{quotient } \frac{12}{S}$$

$$\begin{array}{r} 24 + 12S \\ \hline 2S + 4S^2. \end{array}$$

The remaining impedance is $Z''_{in}(S) = (S+2)/(4S+2)$ after cancellation of the common factor. Applying Richard's theorem to $Z''_{in}(S)$ shows that the remaining impedance consists of a unit element of impedance $\frac{1}{2}$ and a 1-Ω resistance. The final circuit is shown in Fig. 19(a).

To obtain the equivalent three-stub network with unit elements of impedance $Z_0 = 1$, partially remove a pole (shunt inductor) of $Y_{in}(S) = 1/Z_{in}(S)$ so that the remaining input impedance $Z'_{in}(S)$ satisfies $Z'_{in}(1) = 1$. This insures that the unit element with $Z_0 = 1$ can be removed and thus reduces the order of $Z_{in}(S)$ in accordance with Richard's theorem.

For example, using (40),

$$7S + 4S^2 + S^3 \overline{\big)\, 12 + 7S + 4S^2 + S^3} \quad \text{quotient } \frac{a}{s}$$

$$\begin{array}{r} 7a + 4aS + aS^2 \\ \hline (12 - 7a) + (7 - 4a)S + (4 - a)S^2 + S^3. \end{array}$$

[35] This theorem is discussed further by Grayzel, "A synthesis procedure for transmission line networks," IRE Trans. on Circuit Theory, Vol. CT-5, pp. 172–181; September, 1958.

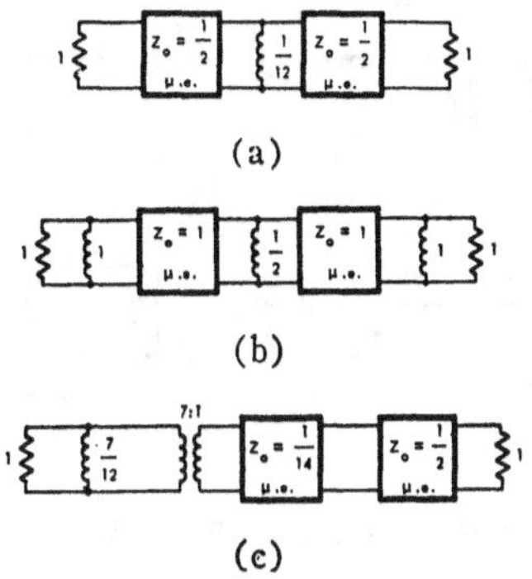

Fig. 19—(a) One-stub third-order Butterworth filter. (b) Three-stub third-order. Butterworth filter. (c) Third-order Butterworth filter synthesized in the equivalent form of Fig. 11(b).

Then

$$Z'_{\text{in}}(S) = \frac{S^3 + 4S^2 + 7S}{S^3 + (4 - a)S^2 + (7 - 4a)S + (12 - 7a)}.$$

For $Z'_{\text{in}}(1) = 1$, $a = 1$; therefore, $L = 1/a = 1$.

The unit element of $Z_0 = 1$ can now be removed from $Z'_{\text{in}}(S)$. The above process is now repeated using the remaining impedance. The final network is shown in Fig. 19(b).

A third form of the network can be obtained by noting that the network must have an equivalent circuit with a shunt inductor, a transformer and a cascade of two unit elements as demonstrated in Section V-A. For

$$Z_{\text{in}}(S) = \frac{S^3 + 4S^2 + 7S}{S^3 + 4S^2 + 7S + 12},$$

removal of the shunt inductor (pole of Y_{in} at zero) gives

$$\begin{array}{r|l} & \dfrac{12}{7S} \\ \hline 7S + 4S^2 + S^3 & 12 + 7S + 4S^2 + S^3 \\ & 12 + \dfrac{48S}{7} + \dfrac{12S^2}{7} \\ \hline & \dfrac{S}{7} + \dfrac{16S^2}{7} + S^3 \end{array}$$

or

$$L = \frac{7}{12}.$$

Then

$$Z'_{\text{in}}(S) = \frac{S^2 + 4S + 7}{S^2 + \dfrac{16S}{7} + \dfrac{1}{7}} = \frac{7S^2 + 28S + 49}{7S^2 + 16S + 1}.$$

The turns ratio of the transformer can be obtained by noting that

$$\lim_{S \to 0} Z'_{\text{in}}(S) = n^2$$

[see Fig. 19(c)]. This gives $n^2 = 49$ or $n = 7$. Removal of the transformer yields

$$Z''_{\text{in}}(S) = \frac{\dfrac{S^2}{7} + \dfrac{4S}{7} + 1}{7S^2 + 16S + 1}.$$

Application of Richard's theorem shows the remaining network consists of a cascade of two unit elements of impedance $Z = 1/14$ and $Z = \frac{1}{2}$ terminated in a 1-Ω resistor. The final equivalent circuit is shown in Fig. 19(c).

The networks of Fig. 19 are exact equivalents. The network of Fig. 19(b) can be realized as a line with shunt stubs or with parallel coupled bars by using the relations in Table I. Elements for filters with a Chebyshev characteristic can be obtained in a similar manner by applying a suitable approximation criterion to (38).

The filters discussed above are said to have a "maximally flat" or "Chebyshev" characteristic, but these responses are not the same as those of networks that have an S-plane, L-C ladder equivalent. The basic equation for a maximally flat filter with an L-C prototype is

$$|\rho|^2 = \frac{1}{1 + \Omega^{2n}} \tag{41}$$

whereas, for the filters discussed above, the basic equation is

$$|\rho|^2 = \frac{1}{1 + \Omega^2(1 + \Omega^2)^{n-1}}. \tag{42}$$

It can be shown that filters of the type described by (41) have steeper band-edge slopes than those whose characteristics are given by (42). In the pass band of the filters ($\Omega \gg 1$), (41) becomes nearly equal to (42). Thus an L-C ladder can serve as the prototype for networks with a characteristic given by (42), but the response realized follows that of the prototype only for f in the vicinity of f_0. Whereas this approximation may lead to a suitable filter, it cannot be used to realize a response that is specified over the entire frequency range.

Appendix II

S-plane ladder networks with series capacitors and shunt inductors generally will have ideal transformers in their TEM realizations. These transformers can be consolidated and transposed to the input or output terminals where they can be realized in a section such as that shown in Table I(g). In many cases, however, it is possible to eliminate the ideal transformers. As an example, consider the lower half circuit in Fig. 13(a) that realizes Y_e. Adding three unit elements of characteristic impedance $Z_0 = 1$ to Fig. 13(a) gives the circuit of Fig. 20(a). Application of Kuroda's identities gives Fig. 20(b). Transforming the 4/3 capacitor to the right hand side of the second transformer, consolidating the trans-

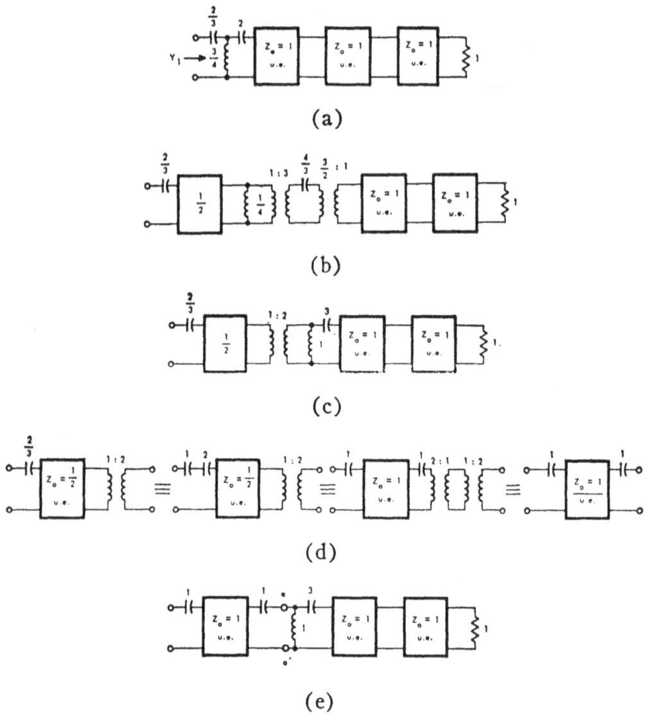

Fig. 20—Elimination of ideal transformers.

formers and transforming the 1/4 inductor to the right hand side of the resultant transformer gives Fig. 20(c). The transformer can be eliminated, if the series 2/3 capacitor is divided into a series combination, so that the capacitor closest to the unit element, when transformed by Kuroda's third identity, yields a transformer with a 2:1 turns ratio or

$$n = 2 = 1 + \frac{1}{Z_0C} = 1 + \frac{2}{C},$$

$$\therefore\ C = 2.$$

The procedure is illustrated in Fig. 20(d) and the resulting circuit is shown in Fig. 20(e). A similar procedure can be applied to the circuit to the right of terminals aa' to obtain a realization without transformers. For the case of a shunt inductor, the inductance is divided into a parallel combination of inductors so that the required turns ratio results.

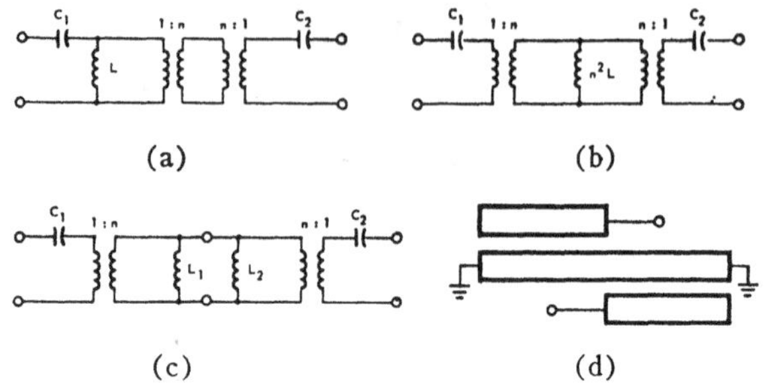

Fig. 21—Realization using sections with ideal transformers.

The above admittance function can also be realized in another way. The circuit of Fig. 20 is left unchanged, if a pair of back to back ideal transformers of turns ratio n is introduced as shown in Fig. 21(a). Transforming the inductor to the region between transformers gives Fig. 21(b). Dividing the inductor, so that

$$\frac{L_1L_2}{L_1 + L_2} = n^2L,$$

gives the circuit of Fig. 21(c) which can be realized as the cascade of two elements of the form shown in Table I(g) where n is chosen to make the realization practical. The TEM realization is shown in Fig. 21(d).

Acknowledgment

The author wishes to thank M. C. Horton of Bendix Research Laboratories for his many helpful suggestions and his assistance in the preparation of this paper. Thanks are also due V. R. Schoep and W. P. Harokopus of Bendix Research Laboratories for their constructive comments.

TABLE I

WAVE CASCADE MATRIX, R, FOR DISTRIBUTED LC LADDER AND UNIT ELEMENTS. THE CONSTANT MATRIX, A, IS DEFINED IN (3), AND I IS THE IDENTITY MATRIX

FILTER ELEMENTS		SCHEMATIC	R-MATRIX
LOW PASS	L	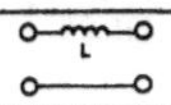	$LS\left(\frac{1}{LS} I + \tilde{A}\right)$
	C	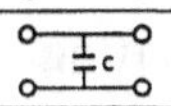	$CS\left(\frac{1}{CS} I + A\right)$
	u.e.	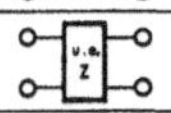	$\frac{S}{\sqrt{1-S^2}}\left[\frac{1}{S} I + (Z\tilde{A} + Z^{-1}A)\right]$
HIGH PASS	u.e.	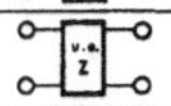	$\frac{1}{\sqrt{1-S^2}}\left\|(Z\tilde{A} + Z^{-1}A)S + I\right\|$
	C	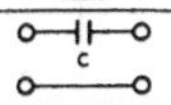	$\frac{1}{CS}(CS\,I + \tilde{A})$
	L	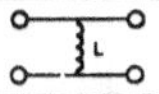	$\frac{1}{LS}(LS\,I + A)$

first. Such a filter may be comprised of distributed series C's, shunt L's, u.e.'s, and a unit terminating load. The C's, L's, and u.e.'s may occur in random sequence; however, in order to be a nonredundant filter, no two C's nor two L's may occur adjacent to each other even if separated by one or more u.e.'s.[2] It is noted in Table I that each of the lossless high-pass elements (C's, L's, u.e.'s) has an R-matrix which comprises a scalar denominator factor multiplied by a matrix which is linear in the frequency parameter S. The scalar denominator factor of each R-matrix contains S to the first order. Thus, an optimum high-pass filter, having a mixed cascade of m high-pass ladder elements and n unit elements terminated in a unit load, will have an overall R-matrix of the form

$$R = \left(\frac{1}{S}\right)^m \left(\frac{1}{\sqrt{1-S^2}}\right)^n B_{m+n}(S) \tag{4}$$

where $B_{m+n}(S)$ is an $(m+n)$th degree 2×2 matrix polynomial in S,

$$B_{m+n}(S) = \begin{bmatrix} b_{11}(S) & b_{12}(S) \\ b_{21}(S) & b_{22}(S) \end{bmatrix}_{m+n}. \tag{5}$$

The R-matrix element of interest in (2) is $r_{12} = s_{11}/s_{21}$ representing the ratio of input reflected wave to that transmitted into the load.

$$r_{12} = s_{11}/s_{21} = \left(\frac{1}{S}\right)^m \left(\frac{1}{\sqrt{1-S^2}}\right)^n b_{12m+n}(S). \tag{6}$$

The overall cascade input reflection coefficient $\rho = s_{11}$ and the transmission coefficient into the unit load $t = s_{21}$ are now renamed for ease in recognition. Total power into the filter is conserved, thus

$$|\rho|^2 + |t|^2 = 1 \tag{7}$$

which can be rearranged to show the dependence of the power transmission response on $r_{12} = \rho/t$, i.e.,

[2] Otherwise the elements may be combined, reducing the total number, by simple parallel or series combinations possibly in conjunction with use of one of Kuroda's Identities.

$$|t|^2 = \frac{1}{1 + |\rho|^2/|t|^2} = \frac{1}{1 + |r_{12}|^2}. \tag{8}$$

Equation (6) for $r_{12}(S)$, when multiplied by $r_{12}(-S)$ to give $|r_{12}|^2$, results in a ratio of $(m+n)$th degree polynomials in $(-S^2)$. The general form of the resultant numerator polynomial, which has real coefficients, will not change if each term is multiplied by a real constant involving $S_c^2 = (j \tan \theta_c)^2$, where $\theta_c = \pi\omega_c/2\omega_0$ and ω_c is designated to be the filter cutoff frequency. Then 37

$$\frac{|\rho|^2}{|t|^2} = \left(\frac{-S_c^2}{-S^2}\right)^m \left(\frac{1-S_c^2}{1-S^2}\right)^n P_{m+n}\left(\frac{-S^2}{-S_c^2}\right) \tag{9}$$

or from (1),

$$\frac{|\rho|^2}{|t|^2} = \left(\frac{\tan\theta_c}{\tan\theta}\right)^{2m} \left(\frac{\cos\theta}{\cos\theta_c}\right)^{2n} P_{m+n}\left(\frac{\tan^2\theta}{\tan^2\theta_c}\right) \tag{10}$$

where P_{m+n} is an $(m+n)$th degree polynomial in $-S^2/-S_c^2$.

2) *Low-pass prototype:* The individual R-matrices of an optimum low-pass (L-P) filter having m low-pass ladder elements, n unit elements, and a unit termination, may be multiplied to obtain the overall R-matrix of the cascade. The low-pass optimum filter may be comprised of series L's, shunt C's and u.e.'s in random sequence; however, to be nonredundant, L's must be adjacent to C's if not separated by a u.e., or L's must be adjacent to each other (and likewise C's) if separated by a u.e.

By applying a procedure similar to that used above for the high-pass filter, the low-pass prototype response ratio of reflected to transmitted power is given by:

$$\frac{|\rho|^2}{|t|^2} = \left(\frac{-S^2}{-S_c^2}\right)^m \left(\frac{-S^2(1-S_c^2)}{-S_c^2(1-S^2)}\right) Q_{m+n}\left(\frac{-S_c^2}{-S^2}\right) \tag{11}$$

or by

$$\frac{|\rho|^2}{|t|^2} = \left(\frac{\tan\theta}{\tan\theta_c}\right)^{2m} \left(\frac{\sin\theta}{\sin\theta_c}\right)^{2n} Q_{m+n}\left(\frac{\tan^2\theta_c}{\tan^2\theta}\right) \tag{12}$$

where Q_{m+n} is an $(m+n)$th degree polynomial in $(-S_c^2/-S^2)$.

B. Step 2—Approximating Functions

The ideal microwave prototype filter has a transmission power-amplitude response, (8), which is rectangular; i.e., as frequency increases through cutoff, it is desired that transmission $|t|^2$ change from zero to unity for a high-pass prototype and from unity to zero for a low-pass prototype, as shown in Fig. 2.

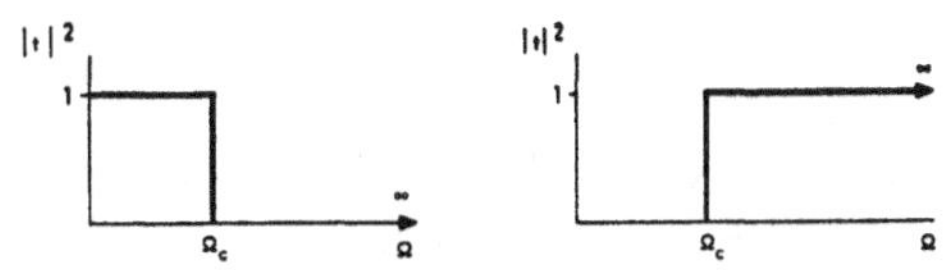

(a) (b)

Fig. 2. Ideal rectangular transmission amplitude power response. (a) L-P prototpye. (b) H-P prototype.

GENERAL THEORY

This rectangular response cannot be realized exactly and, therefore, must be approximated in some manner. The two most common approximations, which are considered here for application to the optimum multipole filter are the maximally flat (Butterworth) and the equal ripple (Chebyshev).

1) *Butterworth:* The maximally flat approximation for the high-pass prototype results from choosing coefficients of the $(m+n)$th degree polynomial P_{m+n} (9) in a manner such that all but the highest order derivative of $|\rho|^2/|t|^2$, taken with respect to S^{-1}, are zero at $S^{-1}=0$ $(S=\infty)$. This criterion requires that all coefficients in P_{m+n} be zero except the constant term. A constant term of unity defines the cutoff frequency $\Omega_c = S_c/j$ to occur at half power (-3 dB) as verified by inspection of (8) and (9).

Similarly, all coefficients of Q_{m+n} (11) are set equal to zero, except the constant coefficient of unity, to obtain the low-pass maximally flat approximation. In the low-pass case, all but the highest order derivative of $|\rho|^2/|t|^2$, taken with respect to S, are zero at $S=0$.

The Butterworth approximating functions are then given by:

$$\text{H-P:}\quad \frac{|\rho|^2}{|t|^2} = \left(\frac{S_c}{S}\right)^{2m}\left(\frac{\sqrt{1-S_c^2}}{\sqrt{1-S^2}}\right)^{2n} = \left(\frac{\tan\theta_c}{\tan\theta}\right)^{2m}\left(\frac{\cos\theta}{\cos\theta_c}\right)^{2n}$$

$$\text{L-P:}\quad \frac{|\rho|^2}{|t|^2} = \left(\frac{S}{S_c}\right)^{2m}\left(\frac{S\sqrt{1-S_c^2}}{S_c\sqrt{1-S^2}}\right)^{2n} = \left(\frac{\tan\theta}{\tan\theta_c}\right)^{2m}\left(\frac{\sin\theta}{\sin\theta_c}\right)^{2n}. \qquad (13)$$

Note that the low-pass approximation can be obtained directly from the high-pass by replacing each frequency parameter (i.e., S, S_c) by its respective inverse (S^{-1}, S_c^{-1}), and vice versa.

2) *Chebyshev:* An equal ripple approximation for the high-pass prototype results from choosing coefficients of the polynomial, P_{m+n} (9), in such a manner that the pass band response ripples between the values of unity and $(1+\epsilon^2)^{-1}$. The polynomial form which exhibits this response is developed (see Appendix) from a constant amplitude rational fraction all-pass function.

This same development, with each frequency parameter replaced by its inverse, gives the proper coefficients of Q_{m+n} (11) for equal ripple pass band response of the low-pass prototype.

The Chebyshev approximating functions are given by:

38

$$\text{H-P:}\quad \frac{|\rho|^2}{|t|^2} = \epsilon^2\left[T_m\left(\frac{S_c}{S}\right)T_n\left(\frac{\sqrt{1-S_c^2}}{\sqrt{1-S^2}}\right) - U_m\left(\frac{S_c}{S}\right)U_n\left(\frac{\sqrt{1-S_c^2}}{\sqrt{1-S^2}}\right)\right]^2$$
$$= \epsilon^2\left[T_m\left(\frac{\tan\theta_c}{\tan\theta}\right)T_n\left(\frac{\cos\theta}{\cos\theta_c}\right) - U_m\left(\frac{\tan\theta_c}{\tan\theta}\right)U_n\left(\frac{\cos\theta}{\cos\theta_c}\right)\right]^2$$

$$\text{L-P:}\quad \frac{|\rho|^2}{|t|^2} = \epsilon^2\left[T_m\left(\frac{S}{S_c}\right)T_n\left(\frac{S\sqrt{1-S_c^2}}{S_c\sqrt{1-S^2}}\right) - U_m\left(\frac{S}{S_c}\right)U_n\left(\frac{S\sqrt{1-S_c^2}}{S_c\sqrt{1-S^2}}\right)\right]^2$$
$$= \epsilon^2\left[T_m\left(\frac{\tan\theta}{\tan\theta_c}\right)T_n\left(\frac{\sin\theta}{\sin\theta_c}\right) - U_m\left(\frac{\tan\theta}{\tan\theta_c}\right)U_n\left(\frac{\sin\theta}{\sin\theta_c}\right)\right]^2 \qquad (14)$$

where

$$T_m(x) = \cos(m \arc\cos x) \quad \text{and} \quad U_m(x) = \sin(m \arc\cos x)$$

are unnormalized mth degree Chebyshev polynomials of the first and second kinds, respectively [19].

C. Step 3—Network Synthesis

The final step in obtaining an optimum filter is solving the realization problem; i.e., it is desired to synthesize a distributed network with a physical response corresponding to the admissible approximating function of (13) or (14). The general form of these functions was obtained by considering a cascade of unit elements and distributed L's and C's. If the input impedance of this cascade is determined from the specified power function, Richards' Theorem[3] can be applied to determine the unit element values, and pole removing techniques [13]–[15] can be used to determine the LC values. The power reflection coefficient $|\rho|^2 = 1-|t|^2 = \rho\bar{\rho}$ where the bar denotes complex conjugate, can be obtained from the approximating function for $|\rho|^2/|t|^2$. Because the network is to be physically realizable, the reflection coefficient ρ must have no poles in the right half plane. Since the network functions under consideration do not have j-axis poles, the desired reflection coefficient can be determined from the squared approximating function $|\rho|^2$ by finding roots of the numerator and denominator polynomials and associating the left half-plane poles with ρ and the right half-plane poles with $\bar{\rho}$. The only restriction on the zeros of ρ is that they be chosen in conjugate pairs such that the numerator of $\rho\bar{\rho}$ is the numerator of $|\rho|^2$.[4] For many practical microwave filters, all zeros lie on the imaginary axis and no alternate choices are possible.

[3] For a statement and discussion of Richard's Theorem see [4], [5], [6], [8], and [16].

[4] For a discussion of the effects of different zero distributions, see Weinberg [14], pp 592–595.

Once a reflection coefficient has been obtained by the above procedure, the transformation

$$Z_{\text{in}} = \frac{1 + \rho}{1 - \rho} \tag{15}$$

yields the input impedance of one possible network and the opposite choice for the algebraic sign of ρ gives the input impedance of the dual network. The choice of either network is dictated by the physical configuration of the desired realization and the element values obtained from the synthesis procedure.

Many variations in network form, each having a minimum number of elements, are possible. Specifically, there exists a number of configurations equal to the number of unique networks that can be obtained by applying Kuroda's Identities. Furthermore, redundancy (for example, replacing a series inductor by two inductors or adding u.e.'s whose impedance is the same as that of the load) allows the possibility of additional network forms. Any one of the minimum element networks is a permissible mathematical solution; however, a practical realization may require that various network forms be investigated.

The schematic forms of two symmetric, doubly terminated, optimum filters and their respective duals are shown in Table II. The forms shown utilize the maximum number of elements that can usually be obtained in a practical network.

TABLE II

EXAMPLES OF PRACTICAL OPTIMUM MULTIPOLE FILTER NETWORKS AND THEIR RESPECTIVE DUALS

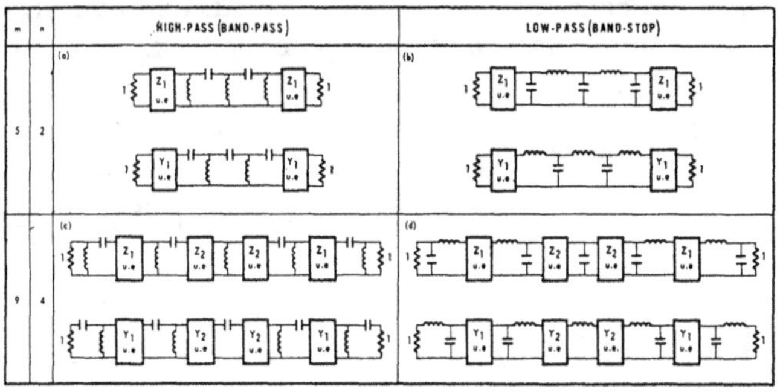

TABLE III

SOME COMMON FILTER FORMS CONTAINING FEWER THAN MAXIMUM ELEMENTS OR REDUNDANT ELEMENTS THAT CAN BE EXACTLY DESIGNED TO GIVE AN OPTIMUM MULTIPOLE RESPONSE BY SUITABLE CHOICE OF m AND n

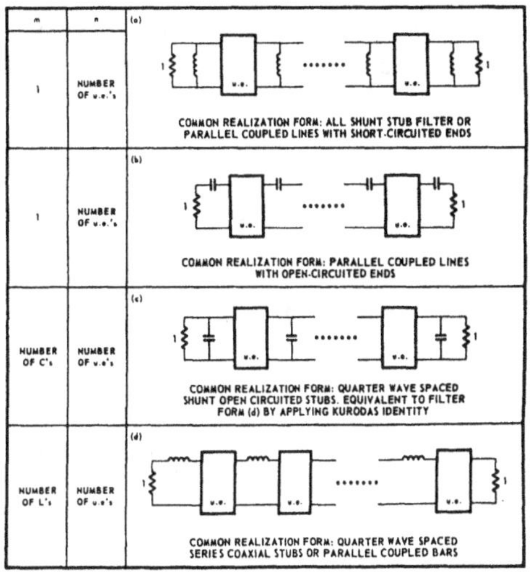

In some specific cases, series elements can be contained within shunt stubs, and multiple stubs can be contained within u.e.'s, thereby increasing filter complexity for a given number of unit elements. In general, this can be accomplished only in relatively wide-band designs and at lower microwave frequencies where element value restrictions and complicated junctions do not present serious limitations. Symmetry in the assumed forms requires equal terminating impedances; however, for the band-pass case, Kuroda's Identities often can be applied to provide for unequal terminations. In general, the filters need not be symmetric and all combinations of m and n may be used. The networks shown in Table II can be used as a guide in applying Richards' Theorem and pole removing techniques to Z_{in}, together with application of Kuroda's Identities, to obtain equivalent forms. Because of the assumed symmetry, only part of each network need be synthesized. For example, in the symmetrical case of five ladder elements ($m=5$) and two unit elements ($n=2$), the network is synthesized by applying Richards' Theorem once to obtain the unit element value followed by three pole removals.

It is evident that an optimum filter can contain many sections in a relatively short overall length. In some applications, use of the full number of sections possible in a given length may not be necessary or practical. For example, networks having two u.e.'s can contain up to five LC elements (not including the possibility of series stubs within shunt elements and multiple stubs in u.e.'s). The use of fewer LC elements or redundancy may in some cases simplify the construction of the filter and may also ease the element value requirements. Several common filter forms that contain fewer than a maximum number of elements, or redundant elements, are listed in Table III. Networks of this type with optimum response can be exactly synthesized by using the appropriate m and n in (13) or (14).

Bandwidth scaling of optimum filter network element values is not as easily accomplished as it is in networks incorporating only ladder elements. For networks that have a ladder prototype, a set of normalized element values can be determined and bandwidth renormalization may be accomplished by multiplying these values by a constant. Variation of the bandwidth determining constant (S_c) for the general optimum filter changes the element values in a manner which may not permit a simple calculation to yield the new values. A digital computer, however, can be used to obtain tables of element values for a wide range of bandwidths.

The manner in which prototype element values are converted to physical dimensions and a general discussion of equivalent network forms can be found in the literature [4], [5], [6], [9], [12].

V. Experimental Evaluation

A seven-section equal ripple band-pass filter of 3 to 1 bandwidth[5] will be considered in detail as a design example. The filter, consisting of two unit elements and five LC elements, will be designed to have 0.1 dB ripple in the pass band. The normalized cutoff frequency for a 3 to 1 band is $S_c = j$. From (8) and (14) the transmission function is given by:

$$|t|^2 = \frac{1}{1 + 0.0233\left[T_5\left(\frac{S_c}{S}\right)T_2\left(\frac{\sqrt{1-S_c^2}}{\sqrt{1-S^2}}\right) - U_5\left(\frac{S_c}{S}\right)U_2\left(\frac{\sqrt{1-S_c^2}}{\sqrt{1-S^2}}\right)\right]^2}. \tag{16}$$

Substitution of $S_c = j$, simplification of the resultant ratio of polynomials, and use of the identity $|\rho|^2 + |t|^2 = 1$ gives:

$$|\rho|^2 = \frac{[0.98S^6 + 8.15S^4 + 17.64S^2 + 10.78]^2}{-S^{14} + 2.96S^{12} + 14.96S^{10} + 100.96S^8 + 308.69S^6 + 487.06S^4 + 380.46S^2 + 116.23}. \tag{17}$$

Roots of the denominator are obtained with the use of a computer and the left-hand plane poles are associated with ρ. Since the numerator of $|\rho|^2$ is a perfect square, the numerator of ρ is obtainable by inspection:

$$\rho = -\frac{0.98S^6 + 8.15S^4 + 17.64S^2 + 10.78}{S^7 + 6.16S^6 + 17.47S^5 + 32.36S^4 + 39.13S^3 + 37.57S^2 + 20.73S + 10.78}. \tag{18}$$

Use of the linear transformation $Z_{in} = (1+\rho)/(1-\rho)$ yields

$$Z_{in} = \frac{S^7 + 5.18S^6 + 17.47S^5 + 24.21S^4 + 39.13S^3 + 19.93S^2 + 20.73S}{S^7 + 7.13S^6 + 17.47S^5 + 40.51S^4 + 39.13S^3 + 55.21S^2 + 20.73S + 21.56}. \tag{19}$$

One possible network form is shown in Table II(a). Determination of element values begins with the removal of a unit element by application of Richards' Theorem to give $Z_{u.e.} = Z_{in}(1) = 0.63$. The impedance of the remaining network after removal of the u.e. is:

$$Z_{in}'(S) = Z_{in}(1)\frac{SZ_{in}(1) - Z_{in}(S)}{SZ_{in}(S) - Z_{in}(1)}$$

or

$$Z_{in}'(S) = \frac{1.84S^6 + 10.20S^5 + 18.84S^4 + 33.65S^3 + 20.08S^2 + 20.98S}{4.64S^6 + 21.09S^5 + 64.83S^4 + 82.37S^3 + 128.05S^2 + 60.52S + 62.96} \tag{20}$$

after cancellation of the common $S^2 - 1$ factor.

Next, a shunt inductor of value $L_1 = 0.33$ is removed by dividing the numerator of $Y_{in}'(S) = 1/Z_{in}'(S)$ by its denominator.[6] A series capacitor of value $C = 1.27$ and a shunt inductor of value $L_2 = 0.26$ are obtained from the impedance of the remaining network,

$$Z_{in}''(S) = \frac{5.54S^5 + 30.73S^4 + 56.77S^3 + 101.42S^2 + 60.52S + 62.96}{13.98S^5 + 46.87S^4 + 102.76S^3 + 77.15S^2 + 80.25S} \tag{21}$$

[5] As with conventional Chebyshev-type characteristics, the bandwidth is measured between the outer ripple points.

[6] This step removes a simple pole of Y_{in} at $S = 0$ and is the same technique that is used to synthesize lumped element ladder networks [13]–[15].

in a similar manner. A diagram of the synthesis steps, together with the completely synthesized network, is shown in Fig. 3.

In order to obtain a network configuration that is convenient to construct, the following Kuroda Identity transformations are performed. Referring to Fig. 4, the 0.33 ohm distributed inductors and 1.27 mho distributed capacitors are transformed to the outside of the unit elements. Because the network is symmetric, the transformers can be moved to the center of the network and eliminated as in Fig. 4(c) [4]. The network of Fig. 4(c) contains a minimum number of elements and is still in optimum multipole form.[7] However, the center inductor value is low, requiring a 5.15 ohm shunt stub in a 50 ohm terminated filter realization. Two parallel 10.3 ohm stubs could be used, one on either side of the center line, but these values are still low to realize easily. To obtain more convenient center element values, redundant elements are introduced by splitting the center inductor into three inductors. The three element values are chosen such that application of Kuroda's Identity and elimination of the resulting transformers gives element values that do not differ greatly from the characteristic impedance of the terminations. The final network is shown in Fig. 4(f). The introduction of redundant elements in the manner described has yielded a network with realizable element values, but has not resulted in increasing the filter length. The final filter is not in optimum multipole form, but it does realize the optimum response for $m=5$ and $n=2$.

A trial filter was constructed based on the parameters used in the design example. A ground plane spacing of 0.200 inch was chosen and center conductor rod dimensions were obtained with the aid of standard graphs [20]. The unit elements were made one-quarter wavelength long at 2.175 Gc. Stub lengths were determined experimentally by adjusting each to produce a zero of transmission at 4.350 Gc by temporarily shortening all other stubs. A line drawing of the entire filter is shown in Fig. 5 and a photograph is shown in Fig. 6. To reduce undesirable junction effects, the two outer junctions were mitered as shown in Fig. 5. When first constructed, the bandwidth of the filter was about one per cent too narrow and the center frequency was about two per cent low. Slight adjustment of the shunt stub lengths resulted in a measured response that agrees very closely with the theoretical characteristics. Both theoretical and experimental curves are shown in Fig. 7. Note especially the measured VSWR vs. frequency which contains seven ripples commensurate with the five ladder elements and two unit elements incorporated in the design.

41

[7] This form can be obtained directly from (19) without the aid of Kuroda's Identities by removing a shunt inductor, series capacitor, unit element, and shunt inductor using the synthesis techniques described.

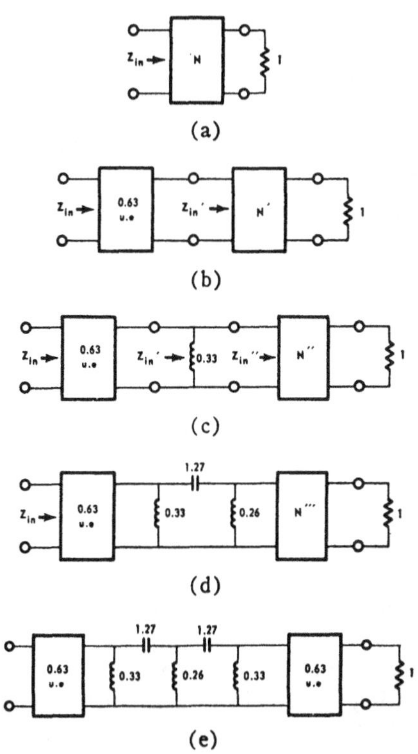

Fig. 3. Synthesis steps in seven-section filter design example. (a) Unsynthesized network. (b) Removal of a unit element by use of Richards' Theorem. (c) Removal of a shunt stub by removing a pole of $Y_{in}'(S)$ at $S=0$. (d) Removal of two more stubs in a manner similar to (c). (e) Final network obtained by symmetry.

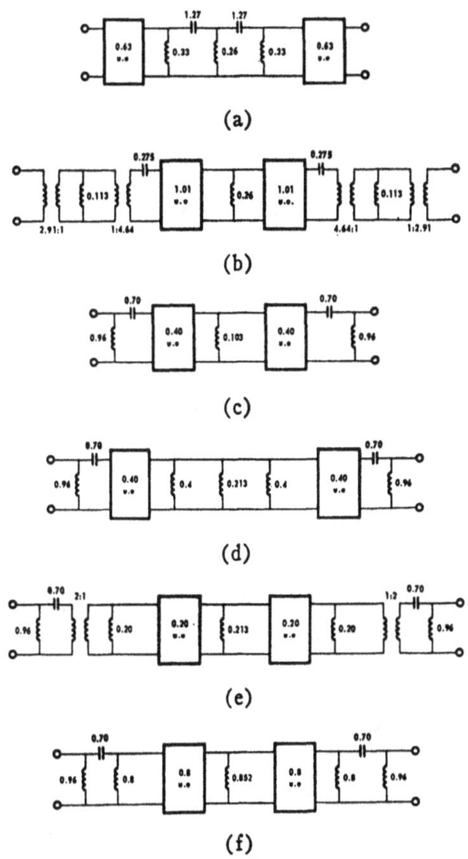

Fig. 4. Application of Kuroda's Identities to obtain convenient element values. (a) Initial prototype network. (b) Transformation of 0.33 ohm distributed inductors and 1.27 mho distributed capacitors. (c) Elimination of transformers. (d) Introduction of redundant elements. (e) Application of Kuroda's Identity. (f) Final network after elimination of transformers.

42

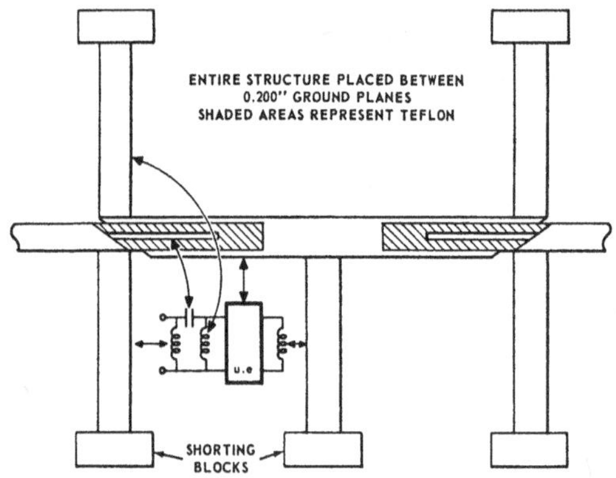

Fig. 5. Detailed view of an experimental filter that realizes a seven-section filter ($m=5$, $n=2$).

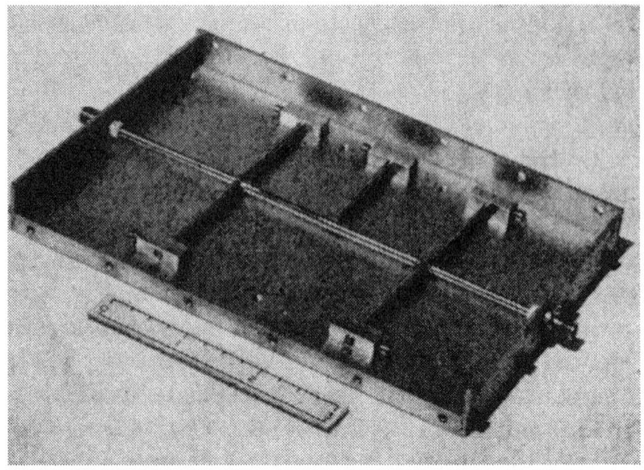

Fig. 6. Experimental seven-section filter ($m=5$, $n=2$).

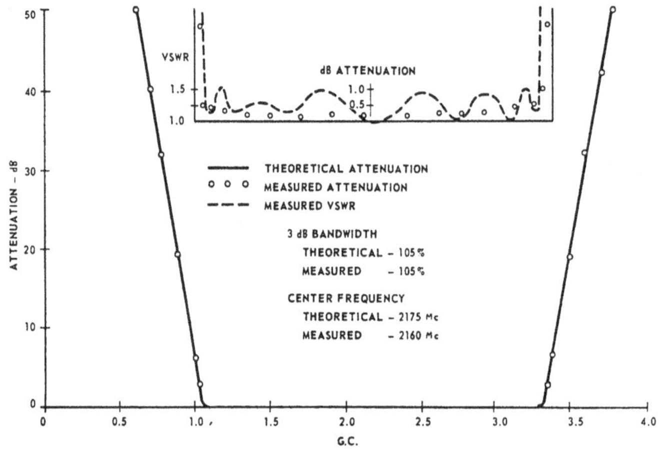

Fig. 7. Theoretical and experimental response characteristics of a seven-section filter ($m=5$, $n=2$).

VI. Comparison of Optimum Filter Networks

The optimum response of a filter network with a given number of nonredundant elements is obtained by use of (8) and (13) or (14). Given k nonredundant elements, m and n can be chosen in any manner such that $m+n=k$, and each choice will result in an optimum multipole filter of order k. The response of each network will be different, and the relation of one response to another will change with filter bandwidth. In narrow-band cases, practical considerations dictate the manner in which m and n should be chosen whereas for wide-band filters size considerations are of most importance.

Optimum multipole filter responses can be compared by investigating the filtering properties of unit elements in comparison with LC-type elements. The following comparison will be carried out for the Chebyshev high-pass prototype (band-pass filter) with the results for the low-pass case being obtained by the high-pass to low-pass transformation. Similar results apply for the Butterworth case. For convenient calculation of stop-band attenuation, (14) is written as:

$$\frac{|\rho|^2}{|t|^2} = \epsilon^2 \cosh^2 \left[m \cosh^{-1} \frac{\Omega_c}{\Omega} + n \cosh^{-1} \frac{\sqrt{1+\Omega_c^2}}{\sqrt{1+\Omega^2}} \right] \quad \text{when } \Omega < \Omega_c. \tag{22}$$

This form follows from (35) and (36) of the Appendix, together with (1), using the identities $\cos^{-1} x = j \cosh^{-1} x$ and $\cos jx = \cosh x$. If n is zero, the function corresponds to the response of an LC ladder prototype and available nomographs [21] can be used to obtain the network response. The general optimum multipole function (22) can be adapted for use with these nomographs in the following manner. Inspection of the functions $\cosh^{-1} \Omega_c/\Omega$ and $\cosh^{-1} (\sqrt{1+\Omega_c^2}/\sqrt{1+\Omega^2})$ shows that the ratio

$$R = \frac{\cosh^{-1} \dfrac{\Omega_c}{\Omega}}{\cosh^{-1} \dfrac{\sqrt{1+\Omega_c^2}}{\sqrt{1+\Omega^2}}} \tag{23}$$

is always greater than one, for $\Omega < \Omega_c$. Thus, in effect, it takes more unit elements than LC elements to give a specified attenuation slope. A plot of R vs. Ω for several Ω_c values is given in Fig. 8. For increasing Ω_c, and for $(\Omega_c/2) < \Omega < \Omega_c$, the ratio approaches unity. This indicates that in the region immediately beyond cutoff for narrow to moderate bandwidth filters, a unit element gives almost the same attenuation characteristic as a stub-type element. However, as the filter bandwidth increases, the unit element becomes less effective and more u.e.'s must be used to obtain the same attenuation as that of an LC element. The graph of Fig. 8 can be used to obtain a comparison of different m and n choices by letting

$$\frac{|\rho|^2}{|t|^2} = \epsilon^2 \cosh^2 \left[m' \cosh^{-1} \frac{\Omega_c}{\Omega} \right] \tag{24}$$

and using the referenced nomographs, where

$$m' = m + \frac{n}{R} \cdot \tag{25}$$

In practice, one estimates an average value of R from Fig. 8, obtains m' from (25), and finds the approximate network response from the appropriate nomograph in Kawakami [21]. A more accurate approximation can be obtained by using a new value of R for each Ω/Ω_c value investigated.

As an example consider the number of nonredundant elements $k=m+n=7$. Several choices of m and n are listed in Table IV along with an average m' for both a narrow-band and wide-band filter. For the narrow-band case, R is so close to unity that all forms give almost the same response. Thus, the choice of m and n is dictated by the practicality of a given network form. Stub network forms usually become unrealizable for narrow bandwidths, and coupled line structures, normally having a high value for n, are most practical. For example, the familiar parallel coupled filter employs $m=1$ and $n=k-1$, and is very practical to construct for narrow bandwidths.

For the wide-band case, the responses differ substantially for different choices of m and n and the networks containing more LC-type elements are observed to be superior. Referring to Table IV, the $m=3$ and $n=4$ case has a response corresponding to approximately 5.6 LC-type elements while the $m=0$, and $n=7$ case provides a characteristic similar to that of 4.5 LC elements. These approximations are only estimates for $(\Omega/\Omega_c > 0.6$. When Ω/Ω_c is less than 0.6, R becomes larger, resulting in a more degraded performance for those filters containing unit elements. However, most practical applications of wide-band microwave filters require steep skirt characteristics in which the behavior near cutoff is of greatest importance, and the above method of estimation is sufficient. The exact attenuation curve can be obtained from (8) and (22), if desired.

For wide-band filters, most choices of m and n result in prototypes that can be more easily realized with stub-type networks than with coupled line structures. The designer has the freedom of choosing m and n to meet specific size and performance requirements. Some unit elements are usually required, since only a limited number of stubs can be placed at a single junction. Thus,

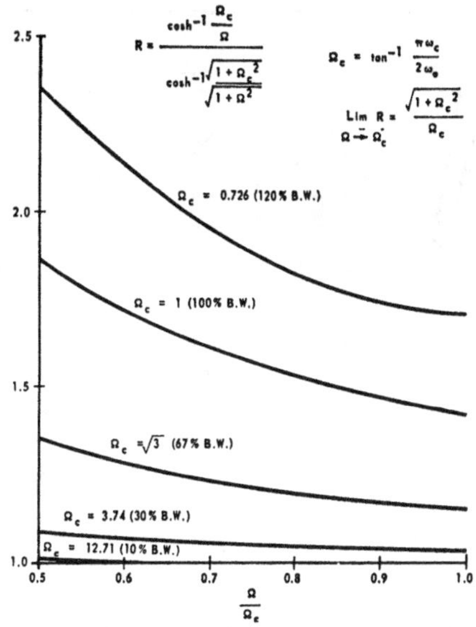

Fig. 8. Plot of R vs. Ω/Ω_c. Comparison of unit element and stub-type element attenuation properties for high-pass prototypes.

TABLE IV

PARAMETERS FOR COMPARISON OF DIFFERENT OPTIMUM MULTIPOLE FILTERS WITH $k=m+n=7$

10% BANDWIDTH			100% BANDWIDTH (3:1)		
m	n	$m' = m + \frac{n}{R}$	m	n	$m' = m + \frac{n}{R}$ *
7	0	7	7	0	7
3	4	≈7	3	4	≈5.6
0	7	≈7	0	7	≈4.5

*R WAS CHOSEN TO BE 1.55

the $m=7$, and $n=0$ filter has the best response for all cases listed in Table IV, but a practical network would require the introduction of at least three redundant unit elements. The use of $m=7$ and $n=3$ would result in an improved filter by allowing the unit elements to contribute to the response. In the design example, a choice of $m=5$ and $n=2$ resulted in a response which is but slightly degraded from that of the case of $m=7$, $n=0$ in a structure only two thirds as long.

To obtain the best response from a filter of minimum size, the designer should try to incorporate as many LC elements as is practical. For wide-band designs the use of a high n/m ratio leads to a large network with a relatively poor response and should be avoided. In fact, in many cases, a low value of k with a low n/m ratio provides approximately the same attenuation as a high k value with a high n/m ratio. As an example, the response characteristic of a conventional filter constructed of shunt-shorted stubs placed a quarter wavelength apart is compared in Fig. 9 with the characteristic of the design example of Section V. The all-shunt-stub network form was designed using $m=1$ and $n=8$, and is four times longer than the $m=5$ and $n=2$ case, the latter of which exhibits a better response.

Redundant filter elements often must be incorporated to obtain practical element values. The redundancy can

43

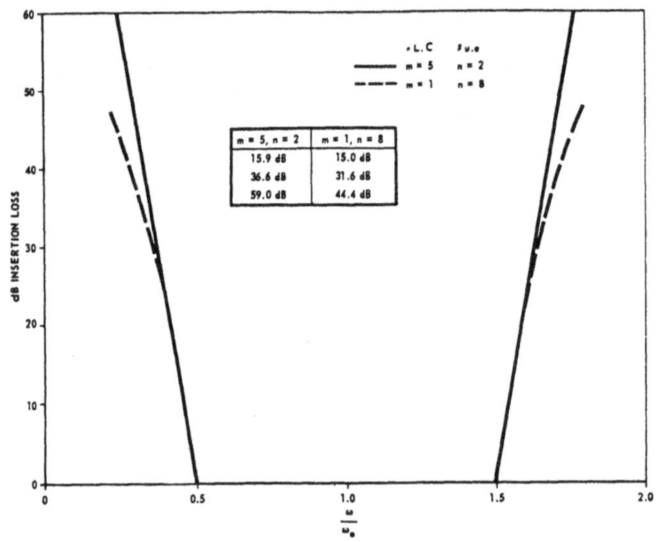

Fig. 9. Comparison of filter characteristics that demonstrates the size advantage obtained by using a low n/m ratio for wide-band designs.

either be added in the LC part of the network as in the design example, or in the unit element section [4], [12]. Redundant unit elements add length to the filter and thus it is more desirable to add the redundancy into the LC part of the network if possible. In wide-band designs it is usually most convenient to obtain the optimum multipole network and add redundant elements with the aid of Kuroda's Identities. For narrow-band designs, where redundancy is often high, it is more convenient to introduce redundant elements through the synthesis procedure.[8]

VII. Conclusions

The synthesis and design of a broad class of TEM mode filters has been presented in a three-step procedure. General transmission response functions were developed that give the "best" response (in a Butterworth or Chebyshev sense) for a given number of quarter-wave elements. A synthesis and design procedure was described that allows the realization of practical distributed networks in the form of quarter-wave lines. Direct application of the methods presented permits the exact design of most conventional TEM mode filter forms. By utilizing every quarter-wave element to augment the filter response, an optimum multipole filter can be obtained. A comparison of the filtering properties of stub-type elements with unit elements showed the stub types to be always superior and increasingly better for wider bandwidths. The unit element was shown to give effective band-pass filtering for bandwidths up to two octaves. The method of comparison presented can be used to determine the response characteristics of different optimum multipole network forms using bandwidth as a parameter. Comparisons of this type are useful in determining an optimum network form of minimum size. Use of the low-pass to high-pass transformation permits the band-pass results described to be applied to band-stop filters.

The procedures presented are also applicable to the design of certain types of waveguide filters for bandwidths of 20 per cent or less. These include those filter forms in which the response of the waveguide elements closely approximates that of TEM mode stub-type elements.

The basic approximating functions developed can also be used to design multiplexing filter networks by choosing the functions to represent squared magnitudes of the transfer admittances ($|Y_{12}|^2$) or transfer impedances ($|Z_{12}|^2$) and properly positioning the 3-dB cutoff frequencies [11]. Synthesis procedures applicable to obtaining a practical realization can be found in the literature [11], [13]–[15].

To verify the optimum multipole theory, a seven-section band-pass filter of 3 to 1 bandwidth was designed and constructed. The experimental results followed very closely those predicted by the theory.

VIII. Appendix

The equal ripple approximation to the ideal filter characteristic will be obtained by direct development from the definition of a constant amplitude rational fraction all-pass function on the complex W-plane.[9]

A unit amplitude all-pass transfer function on the imaginary axis in the complex W-plane is defined by:

$$e^{j2\phi_i(W)} = \frac{W_i + W}{W_i - W}. \tag{26}$$

The root parameter W_i must be real in order that (26) be a rational unit amplitude function describing the transfer characteristic of a realizable filter. However, if a product of such functions is taken, the resulting unit amplitude function will be rational if the W_i's are either real or occur in conjugate pairs. Thus a rational unit amplitude transfer function on the imaginary axis in W can be represented by the product,

$$e^{j2\phi(W)} = \Pi e^{j2\phi_i(W)} = \frac{\Pi(W_i + W)}{\Pi(W_i - W)} \tag{27}$$

[8] As described, for example, in Wenzel [4], pp 109–110.

[9] A similar approach has been used by Bennett [22], [23] for application to RC lumped element networks and leads to the desired equal ripple approximation via relatively simple logic. The rational fraction functions used by Sharpe [24] and Helman [25], [26] for lumped RC networks are also applicable but no justification is given for their choice. Kohler and Carlin [27] go through a lengthy justification after having chosen the proper approximation for the distributed optimum low-pass filter. Perhaps the earliest discovery of the proper equal ripple approximating function for the distributed optimum filter was recorded in a personal letter from Prof. Kuroda to Prof. Ozaki in 1954, reference to which was found in unpublished notes written by Prof. Fujisawa while a visiting lecturer at the University of Michigan in 1963.

The high-pass equal ripple function $F_H(S)=f(W)$ is obtained by substituting transformation (31) into (30) to give:

$$F_H(S) = \frac{\Pi(\sqrt{S_i^2 - S_c^2} + \sqrt{S^2 - S_c^2}) + \Pi(\sqrt{S_i^2 - S_c^2} - \sqrt{S^2 - S_c^2})}{2\Pi\sqrt{S_i^2 - S^2}}. \tag{32}$$

where the symbol Π indicates the product taken over all roots W_i which must occur in conjugate pairs if not real. The magnitude of $e^{j2\phi(W)}$ is unity on the imaginary axis in W and the phase variation is monotonic. Similarly, the positive square root of (27) is of constant unit amplitude,

$$e^{j\phi} = \Pi e^{j\phi_i} = e^{j\Sigma\phi_i} = \frac{\Pi\sqrt{W_i + W}}{\Pi\sqrt{W_i - W}} \tag{28}$$

with the symbol Σ indicating summation on all angles ϕ_i associated with roots W_i. A real function $f(W)$ is defined by

$$f(W) = \tfrac{1}{2}(e^{j\phi} + e^{-j\phi}) = \cos\phi = \cos\Sigma\phi_i. \tag{29}$$

This function has a maximum amplitude of unity and is the cosine of a monotonic varying angle ϕ along the imaginary axis in W. Using (28) in (29),

$$f(W) = \frac{\Pi(W_i + W) + \Pi(W_i - W)}{2\Pi\sqrt{W_i^2 - W^2}} \tag{30}$$

which is equal ripple along the entire imaginary axis in the W-plane.

A. High Pass

Equal ripple variation may be restricted to the high-pass range on the imaginary S-plane axis $S>S_c$ by applying to (30) the transformation:

$$W = \sqrt{1 - S^2/S_c^2}; \qquad W_i = \sqrt{1 - S_i^2/S_c^2}. \tag{31}$$

This transform maps that portion of the positive imaginary S-plane axis for which $S>S_c$ onto the entire positive imaginary W-plane axis, as shown in Fig. 10. The remainder of the positive imaginary S-plane axis $S<S_c$ maps onto the positive real axis in the W-plane between unity and zero.

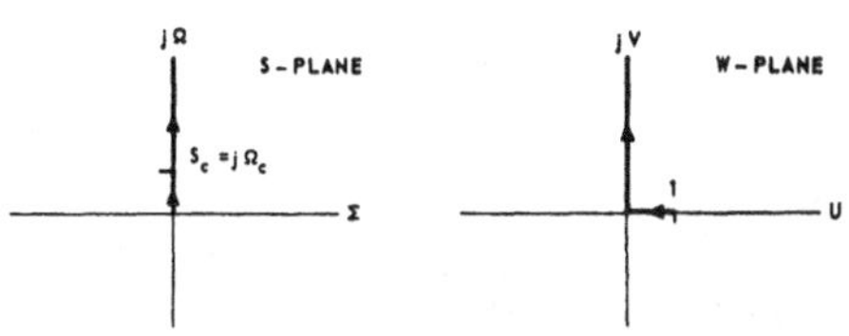

Fig. 10. Mapping of the positive imaginary S-plane axis onto the W-plane by the high-pass transformation $W=\sqrt{1-S^2/S_c^2}$.

The roots S_i in (32) may be chosen arbitrarily as long as they occur in conjugate pairs if not real. It is noted that the real roots $S_i=0$ and $S_i=-1$ permit the square of F_H in (32) to have the same pole locations as those of $|\rho|^2/|t|^2$ in (9). If the pole locations of F_H are now restricted such that m poles occur at $S_i=0$, corresponding to m high-pass LC ladder elements, and n poles occur at $S_i=-1$, corresponding to n. u.e.'s, then (32) for the optimum multipole equal ripple filter becomes

$$F_H(S) = \frac{(jS_c + \sqrt{S^2 - S_c^2})^m(\sqrt{1 - S_c^2} + \sqrt{S^2 - S_c^2})^n + (jS_c - \sqrt{S^2 - S_c^2})^m(\sqrt{1 - S_c^2} - \sqrt{S^2 - S_c^2})^n}{2(jS)^m(\sqrt{1 - S^2})^n}. \tag{33}$$

This algebraic fraction, representing a unit amplitude optimum equal ripple approximation, can be expressed alternately in Chebyshev polynomials by noting that $F(S)=f(W)=\cos\phi=\cos\Sigma\phi_i$ in (29). Thus the ith component phase angle ϕ_i is obtained by considering only the ith term from each of the products in (32) for F_H:

$$\cos\phi_i = \frac{\sqrt{S_i^2 - S_c^2}}{\sqrt{S_i^2 - S^2}}. \tag{34}$$

The high-pass LC ladder element phase angle ϕ_{LC} is established by setting $S_i=0$; similarly, the u.e. phase angle $\phi_{\text{u.e.}}$ is established by setting $S_i=-1$, yielding

$$\cos\phi_{LC} = \frac{jS_c}{jS} = \frac{\tan\theta_c}{\tan\theta};$$

$$\cos\phi_{\text{u.e.}} = \frac{\sqrt{1 - S_c^2}}{\sqrt{1 - S^2}} = \frac{\cos\theta}{\cos\theta_c}. \tag{35}$$

From (29) the high-pass approximating function F_H may be written alternately as

$$F_H = \cos(m\phi_{LC} + n\phi_{\text{u.e.}})$$

$$F_H = T_m\left(\frac{\tan\theta_c}{\tan\theta}\right)T_n\left(\frac{\cos\theta}{\cos\theta_c}\right) - U_m\left(\frac{\tan\theta_c}{\tan\theta}\right)U_n\left(\frac{\cos\theta}{\cos\theta_c}\right) \tag{36}$$

where $T_m(x)=\cos(m \text{ arc } \cos x)$ and $U_m(x)=\sin(m \text{ arc } \cos x)$ are un-normalized mth degree Chebyshev polynominals of the first and second kinds, respectively [19].

B. Low Pass

A low-pass approximating function F_L can be obtained from the high-pass function F_H by replacing each frequency parameter by its inverse. Thus, (35), (33), and (36) become now, for the low-pass case,

$$\cos\phi_{LC} = \frac{S}{S_c} = \frac{\tan\theta}{\tan\theta_c}; \quad \cos\phi_{\text{u.e.}} = \frac{S\sqrt{1-S_c^2}}{S_c\sqrt{1-S^2}} = \frac{\sin\theta}{\sin\theta_c} \tag{37}$$

$$F_L = \frac{(S+\sqrt{S^2-S_c^2})^m(S\sqrt{1-S_c^2}+\sqrt{S^2-S_c^2})^n + (S-\sqrt{S^2-S_c^2})^m(S\sqrt{1-S_c^2}-\sqrt{S^2-S_c^2})^n}{2(S_c)^{m+n}(\sqrt{1-S^2})^n} \tag{38}$$

$$F_L = \cos(m\phi_{LC} + n\phi_{\text{u.e.}})$$

$$F_L = T_m\left(\frac{\tan\theta}{\tan\theta_c}\right)T_n\left(\frac{\sin\theta}{\sin\theta_c}\right) - U_m\left(\frac{\tan\theta}{\tan\theta_c}\right)U_n\left(\frac{\sin\theta}{\sin\theta_c}\right). \tag{39}$$

The functions F_H and F_L exhibit equal ripple characteristics and monotonic phase variation in their respective pass bands and have pole locations corresponding to the R-matrix multiplication of unit elements and LC ladder elements. The unit amplitude equal ripple functions, F_H of (33) or (36) and F_L of (38) or (39), are each multiplied by an arbitrary voltage ripple factor ϵ and then squared in order to form the Chebyshev approximation to the power reflection-to-transmission ratio.

$$\text{H-P:} \quad \frac{|\rho|^2}{|t|^2} = \epsilon^2 F_H^2$$

$$\text{L-P:} \quad \frac{|\rho|^2}{|t|^2} = \epsilon^2 F_L^2. \tag{40}$$

Acknowledgment

The authors wish to thank Dr. W. G. Jaeckle and P. C. Goodman for their discussions and valuable suggestions during the course of this work. Acknowledgment is also due T. E. Gordon, who constructed the experimental filter and carried out the measurements.

References

[1] Cohn, S. B., Parallel-coupled transmission-line-resonator filters, *IRE Trans. on Microwave Theory and Techniques*, vol MTT-6, Apr 1958, pp 223–231.
[2] Matthaei, G. L., et al., Design criteria for microwave filters and coupling structures, Final Rept, SRI Project 2326, Contract DA36-039SC7-4862, Stanford Research Inst,. Menlo Park, Calif., Jan 1961.
[3] Matthaei, G. L., Design of wide-band (and narrow-band) band-pass microwave filters on the insertion loss basis, *IRE Trans. on Microwave Theory and Techniques*, vol MTT-8, Nov 1960, pp 580–593.
[4] Wenzel, R. J., Exact design of TEM microwave networks using quarter-wave lines, *IEEE Trans. on Microwave Theory and Techniques*, vol MTT-12, Jan 1964, pp 94–111.
[5] Ozaki, H., and J. Ishii, Synthesis of transmission-line networks and the design of UHF filters, *IRE Trans. on Circuit Theory*, vol CT-2, Dec 1955, pp 325–336.
[6] ——, Synthesis of a class of strip-line filters, *IRE Trans. on Circuit Theory*, vol CT-5, Jun 1958, pp 104–109.
[7] Jones, E. M. T., Synthesis of wide-band microwave filters to have prescribed insertion loss, *1956 IRE Conv. Rec.*, pt 5, p 119–128.
[8] Grayzel, A. I., A synthesis procedure for transmission line networks, *IRE Trans. on Circuit Theory*, vol CT-5, Sep 1958, pp 172–181.
[9] Horton, M. C., and R. J. Wenzel, Exact design of filters by network techniques, *Microwaves*, vol 3, Apr 1964, pp 16–21.
[10] Wenzel, R. J., Summary of methods for practical synthesis of distributed filter networks, in *Symp. Digest, 7th Midwest Symp. on Circuit Theory*, University of Michigan, Ann Arbor, Mich., May 1964.
[11] Wenzel, R. J., Application of exact synthesis methods to multichannel filter design, *IEEE Trans. on Microwave Theory and Techniques*, vol MTT-13, Jan 1965, pp 5–15.
[12] Schiffman, B. M., and G. L. Matthaei, Exact design of band-stop microwave filters, *IEEE Trans. on Microwave Theory and Techniques*, vol MTT-12, Jan 1964, pp 6–15.
[13] Guillemin, E. A., *Synthesis of Passive Networks*. New York: Wiley, 1957.
[14] Weinberg, L., *Network Analysis and Synthesis*. New York: McGraw-Hill, 1962.
[15] Balabanian, N., *Network Synthesis*. Englewood Cliffs, N. J.: Prentice-Hall, 1958.
[16] Richards, P. I., Resistor-transmission-line circuits, *Proc. IRE*, vol 36, Feb 1948, pp 217–220.
[17] Kuroda, K., Derivation methods of distributed constant filters from lumped constant filters, text for lectures at Joint Meeting of Konsoi Branch of Institute for Elec. Commun., of Elec., and of Illumin. Engrs. of Japan, Oct 1952, p 32 (in Japanese).
[18] Beatty, R. W., and D. M. Kerns, Relationships between different kinds of network parameters, not assuming reciprocity or equality of the waveguide or transmission line characteristic impedances, *Proc. IEEE (Correspondence)*, vol 52, Jan 1964, p 84.
[19] Clement, P. R., The Chebyshev approximation method, *Quart. appl. Math.*, vol 11, Jul 1953, pp 167–183.
[20] *The Microwave Engineers Handbook*. Brookline, Mass.: Horizon House-Microwave, Inc., 1963.
[21] Kawakami, M., Nomographs for Butterworth and Chebyshev filters, *IEEE Trans. on Circuit Theory*, vol CT-10, Jun 1963, pp 288–289.
[22] Bennett, B. J., Synthesis of electric filters with arbitrary phase characteristics, *1953 IRE Conv. Rec.*, pt 5, pp 19–26.
[23] ——, Rebuttal, *IRE Trans. on Circuit Theory (Correspondence)*, vol CT-2, Jun 1955, p 218.
[24] Sharpe, C. B., A general Chebyshev rational function, *Proc. IRE*, vol 42, Feb 1954, pp 454–457.
[25] Helman, D., Chebyshev approximations for amplitude and delay with rational functions, *Proc. of the Symp. on Modern Network Synthesis*, Apr 1955, pp 385–402.
[26] ——, Synthesis of electric filters with arbitrary phase characteristics, *IRE Trans. on Circuit Theory (Correspondence)*, vol CT-2, Jun 1955, pp 217–218.
[27] Kohler, W., and H. J. Carlin, Equiripple transmission line networks, Rept PIB MRI-1089-62, Microwave Research Instit., Polytechnic Institute of Brooklyn, New York, Jan 1963.

Reprinted from IEEE TRANSACTIONS
ON *MICROWAVE THEORY AND TECHNIQUES*
Volume MTT-14, Number 12, December, 1966
pp. 635-647

Theoretical and Practical Applications of Capacitance Matrix Transformations to TEM Network Design

R. J. WENZEL, MEMBER, IEEE

Abstract—TEM propagation on an array of parallel conductors is described in terms of the normalized static capacitance matrix. Important properties of capacitance matrices are discussed and a physical and network interpretation is given to a useful linear transformation of the static capacitance matrix. Several practical applications of capacitance matrix transformations are given. These include 1) equivalent circuits for directional couplers with equal terminations, 2) design procedures for directional couplers with unequal terminations, and 3) element value tables and design details for compact coaxial filter-transformers. Construction details and experimental results are presented for a 3:1 bandwidth filter-transformer constructed with multiple re-entrant coaxial lines.

Manuscript received June 17, 1966; revised August 2, 1966. This work was supported by the U. S. Army Electronics Laboratory, Fort Monmouth, N. J., under Contracts DA 28-043 AMC-00399(E) and DA 28-043 AMC-01869(E).
The author is with The Bendix Corporation, Research Laboratories Division, Southfield, Mich.

I. INTRODUCTION

AN IMPORTANT ASPECT of TEM quarter-wave network synthesis and design is the multiplicity of physical configurations that yield identical response characteristics. Different network configurations are often required to realize a desired response for moderate changes in design parameters because of the small range of realizable impedance values. In most design procedures (either exact or approximate), practical circuit element values are obtained by application of suitable equivalent circuit relationships.

48

In a previous paper [1], the concept of obtaining invariant transmission characteristics using transformations of the static capacitance matrix of a parallel-coupled line array was introduced and was applied to the exact design of interdigital bandpass filters. The purpose of this paper is to demonstrate in more detail the importance of the static capacitance matrix, to show how numerous equivalent circuits can be derived by its transformation, and to give a simple network and physical interpretation to the transformation.

The body of this paper is presented in Sections II and III. In Section II-A, the description of TEM transmission on an array of parallel conductors is given in terms of the static capacitance matrix and some basic properties of capacitance matrices are described and related to specific network geometries. In Section II-B, the capacitance matrix transformation is defined and a network and physical interpretation of this transformation is presented.

The application of the capacitance matrix transformation to the design of directional couplers with equal or unequal terminations is presented in Sections III-A and III-B. Element value tables and design information relative to the construction of compact filter-transformers based on the interdigital filter prototype are presented in Section III-C. Measured performance characteristics for an experimental 3:1 bandwidth filter-transformer are also given.

II. TEM Mode Coupled Line Theory and the Capacitance Matrix Transformation

A. The Static Capacitance Matrix C

TEM propagation on an array of $N+1$ parallel conductors can be described in terms of an $N\times N$ static capacitance matrix by consideration of the $2N$ coupled partial differential equations that apply to such a system. This problem has been considered by several authors [2]–[4] and it is readily shown that the $2N$-port voltages and currents defined in Fig. 1 are related by the admittance matrix $\boldsymbol{Y}$ as follows:

$$\begin{bmatrix} I_A \\ \hline I_B \end{bmatrix} = \frac{1}{\eta_0 S}\left[\begin{array}{c|c} \mathbf{C} & -\mathbf{C}\sqrt{1-S^2} \\ \hline -\mathbf{C}\sqrt{1-S^2} & \mathbf{C} \end{array}\right] \cdot \begin{bmatrix} V_A \\ \hline V_B \end{bmatrix} \tag{1}$$

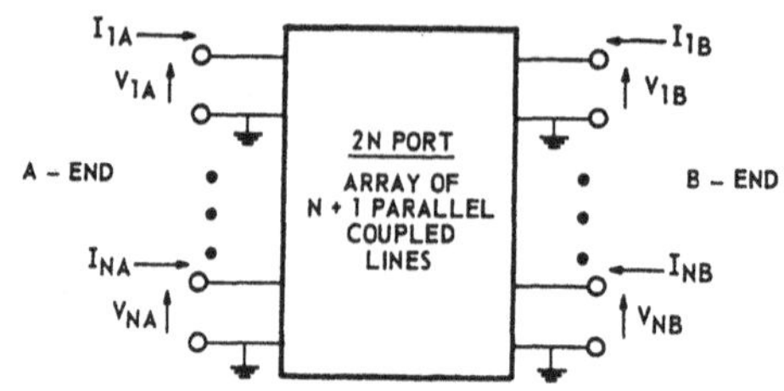

Fig. 1. $2N$-port network with admittance matrix $\boldsymbol{Y}$.

where

$\mathbf{C}=N\times N$ normalized static capacitance matrix with elements c,

$c=c'/\epsilon=$ ratio of the static capacitance between conductors per unit length to the permittivity of the medium (this ratio is independent of the dielectric medium and depends only on cross-section geometry). This dimensionless ratio is convenient for practical use with available design data [5], [6].

$\eta_0=376.6$ ohms, the impedance of free space,

$S=j\tan\theta$

$\theta=\pi\omega/2\omega_0=$ electrical length of the lines in radians,

and

$\omega_0=$ the frequency for which the lines are a quarter wavelength.

Other network matrices ($\boldsymbol{Z}$, $\boldsymbol{ABCD}$, $\boldsymbol{S}$, etc.) can be obtained by well-known transformations of (1) (see Reference [7]).

Some general properties of capacitance matrices will be discussed by considering the arbitrary array of four conductors shown in Fig. 2(a). Conductor 4 is taken as reference (ground). The 3 by 3 capacitance matrix has main diagonal terms given by the sum of all capacitors connected to each of the three ungrounded conductors and off-diagonal terms given by the negative of the capacitors between conductors. Therefore

$$\mathbf{C} = \left[\begin{array}{c|c|c} c_1+c_{12}+c_{13} & -c_{12} & -c_{13} \\ \hline -c_{12} & c_2+c_{12}+c_{23} & -c_{23} \\ \hline -c_{13} & -c_{23} & c_3+c_{13}+c_{23} \end{array}\right]. \tag{2}$$

The elements of matrix (2) satisfy the following relationships [8], [9]:

$$\begin{aligned} &\text{a)}\ \sum_{k=1}^{N} c_{ik} \geq 0 \qquad i=1,2,\cdots,N \\ &\text{b)}\ c_{ik}=c_{ki} \qquad i\neq k \\ &\text{c)}\ c_{ii}\geq 0 \\ &\text{d)}\ c_{ik}\leq 0 \qquad i\neq k. \end{aligned} \tag{3}$$

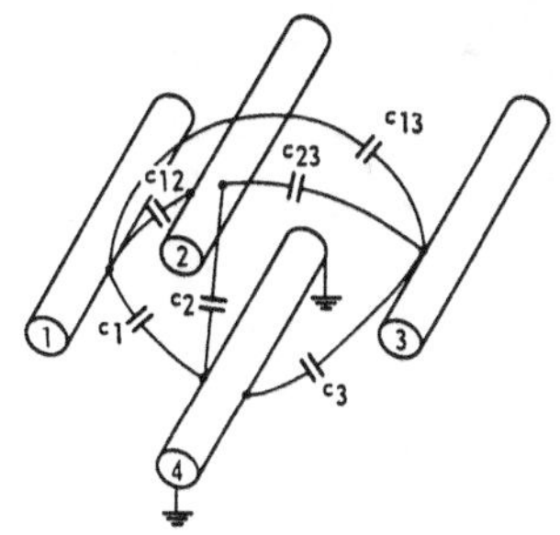

(a) General Array of Parallel Coupled Lines

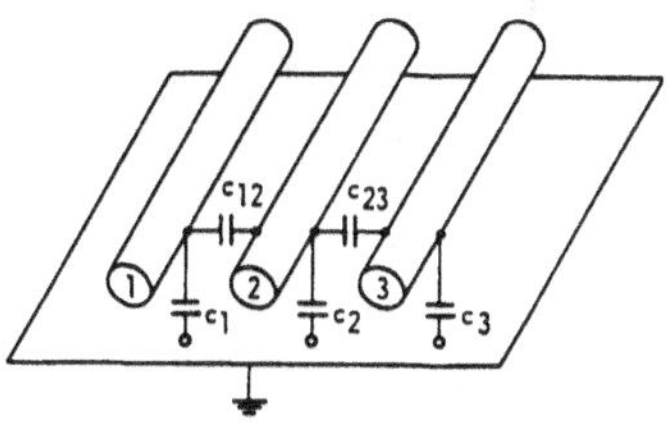

(b) Parallel Coupled Array With Coupling to Nearest Neighbors Only

Fig. 2. Static capacitances of parallel coupled line arrays.

To obtain a physical realization for a network of the type shown in Fig. 2(a), starting from a given matrix C, requires that physical dimensions be determined either experimentally or analytically by solution of Laplace's equation. Unfortunately, this is an extremely difficult problem, for most geometric configurations and solutions are known only for a few special cases. One case of particular interest is obtained by assuming that each conductor is coupled directly only to its nearest neighbors. This configuration is shown in Fig. 2(b), where conductor 4 is assumed to be a ground plane. The capacitance matrix for this network is given by

$$C = \left[\begin{array}{c|c|c} c_1 + c_{12} & -c_{12} & 0 \\ \hline -c_{12} & c_{12} + c_2 + c_{23} & -c_{23} \\ \hline 0 & -c_{23} & c_3 + c_{23} \end{array}\right]. \qquad (4)$$

The C matrix for a like array with any number of conductors is obtained in a similar manner. Note that the assumption of coupling only to nearest neighbors results in a matrix with nonzero elements on the main diagonal and adjacent to the main diagonal only. Given a C matrix of the above type, data exist from which physical dimensions can be determined [5], [6]. It is emphasized that the assumption of coupling to nearest neighbors only is not necessarily desirable from the viewpoint of obtaining optimum device performance from a given number of lines. Many useful devices might result if simple solutions to Laplace's equation were known for a larger number of geometries. On the other hand, networks with lines coupled only to nearest neighbors can realize a wide class of network functions and allow considerable versatility of cross-section geometry, as will be shown in succeeding sections. Throughout the remainder of the paper, only network geometries with lines directly coupled to nearest neighbors will be considered.

B. The Capacitance Matrix Transformation[1]

Different network configurations that leave some specified network response invariant are said to be "equivalent." A method of investigating possible equivalent circuits is to use a linear transformation of network variables (voltage and/or current, for example). Linear transformations of network variables can be performed in a manner such that network realizability is assured (at least from a theoretical viewpoint). However, any such transformation will be of practical utility only if it is easily interpreted and/or leads to a desirable physical structure.

Guillemin [8] has shown that a useful linear transformation is obtained by multiplying the rows and columns of a parameter matrix by suitable constants. In carrying out this transformation on a static capacitance matrix, symmetry requires that each operation be applied to both a row and its corresponding column.

Mathematically, the transformation is given by

$$C' = n_N C n_N \rightarrow \begin{array}{c} \\ n_1 \rightarrow \\ \\ n_2 \rightarrow \\ \vdots \\ \\ n_N \rightarrow \end{array} \begin{array}{c} \begin{array}{cccc} n_1 \downarrow & n_2 \downarrow & \cdots & n_N \downarrow \end{array} \\ \left[\begin{array}{c|c|c|c} c_{11} & -c_{12} & 0 & \\ \hline -c_{12} & c_{22} & & \\ \hline 0 & & & \\ \hline & & & c_{NN} \end{array}\right] \end{array} \qquad (5)$$

where n_N is a diagonal matrix with elements $n_1, n_2, \cdots, n_N$. The transformed capacitance matrix C' is obtained by pre- and post-multiplication of a given matrix C by the diagonal matrix n_N. The right-hand portion of (5) indicates a convenient notation for use in applying the transformation to practical problems and serves to illustrate how each ith row and its corresponding column is multiplied by the same constant n_i.

To give a network interpretation to the capacitance matrix transformation of (5), consider the addition of ideal transformers of turns ratio n_i:1 to the $2N$-ports of the N-line network shown in Fig. 3(a). The resultant network is shown in Fig. 3(b). Physically, a new network has been formed with voltage and current variables related to those of the original network by

[1] This section is directed toward the definition and interpretation of capacitance matrix transformations. A more general discussion of equivalent network concepts can be found in Guillemin [8], pp. 141–176.

$$\begin{bmatrix} V_A' \\ \hline V_B' \end{bmatrix} = \left[\begin{array}{c|c} n_N^{-1} & 0 \\ \hline 0 & n_N^{-1} \end{array}\right] \begin{bmatrix} V_A \\ \hline V_B \end{bmatrix} \quad \text{and} \quad \begin{bmatrix} I_A' \\ \hline I_B' \end{bmatrix} = \left[\begin{array}{c|c} n_N & 0 \\ \hline 0 & n_N \end{array}\right] \begin{bmatrix} I_A \\ \hline I_B \end{bmatrix} \tag{6}$$

where n_N is a diagonal $N \times N$ matrix with elements $n_1, n_2, \cdots, n_N$, and n_N^{-1} is a diagonal $N \times N$ matrix with elements $1/n_1, 1/n_2, \cdots, 1/n_N$. Then from (1)
50 and (6)

$$\begin{bmatrix} I_A' \\ \hline I_B' \end{bmatrix} = \frac{1}{\eta_0 S} \left[\begin{array}{c|c} n_N & 0 \\ \hline 0 & n_N \end{array}\right] \left[\begin{array}{c|c} C & -C\sqrt{1-S^2} \\ \hline -C\sqrt{1-S^2} & C \end{array}\right] \left[\begin{array}{c|c} n_N & 0 \\ \hline 0 & n_N \end{array}\right] \begin{bmatrix} V_A' \\ \hline V_B' \end{bmatrix} \tag{7}$$

where

$$\left[\begin{array}{c|c} n_N^{-1} & 0 \\ \hline 0 & n_N^{-1} \end{array}\right]^{-1} = \left[\begin{array}{c|c} n_N & 0 \\ \hline 0 & n_N \end{array}\right].$$

Thus

$$\begin{bmatrix} I_A' \\ \hline I_B' \end{bmatrix} = \frac{1}{\eta_0 S} \left[\begin{array}{c|c} n_N C n_N & -n_N C n_N \sqrt{1-S^2} \\ \hline -n_N C n_N \sqrt{1-S^2} & n_N C n_N \end{array}\right] \begin{bmatrix} V_A' \\ \hline V_B' \end{bmatrix}$$

$$\frac{1}{\eta_0 S} \left[\begin{array}{c|c} C' & -C'\sqrt{1-S^2} \\ \hline -C'\sqrt{1-S^2} & C' \end{array}\right] \begin{bmatrix} V_A' \\ \hline V_B' \end{bmatrix}. \tag{8}$$

Referring to (8), the new matrix $C' = n_N C n_N$ is exactly that given by the capacitance matrix transformation of (5). However, the new matrix C' is also a matrix corresponding to a parallel line array and can be realized by simply changing the cross-sectional geometry from that of the original network as depicted in Fig. 3(c). The following statement summarizes the network and physical implications of the capacitance matrix transformation:

> given a parallel line array, transformation of the corresponding matrix C results in a matrix C' that corresponds to a new network identical to the original with ideal transformers of turns ratio n_i:1 added at both ends of each line. Physically, the new network can be realized by a change in cross-sectional geometry.

The factors n_i used in performing a capacitance matrix transformation can be chosen arbitrarily as long as they lead to a transformed capacitance matrix whose elements obey the restrictions given by (3).

The remaining question to be investigated is the following: under what conditions can the cross-sectional geometry be altered while a specified network response is held invariant? The answer becomes clear upon studying the diagrams of Fig. 3. If each of the $2N$-ports of the parallel line array of Fig. 3(a) is terminated in a passive admittance Y_i $(i = 1 \cdots 2N)$ then addition of ideal transformers [as in Fig. 3(b)] will affect only the network admittance level at any driven port. However, each terminating admittance Y_i can be transformed through the ideal transformers to yield primed terminating admittance $Y_i' = n_i^2 Y_i$. The original network and ideal transformers can then be replaced by a new network with a different cross-sectional geometry as described. The above statements are summarized as follows:

> the transmission response of a parallel line network with capacitance matrix C and line terminations Y_i is identical to that of a parallel line network with capacitance matrix C' and terminations $Y_i' = n_i^2 Y_i$, where the n_i are the constants utilized in transforming C into C'.

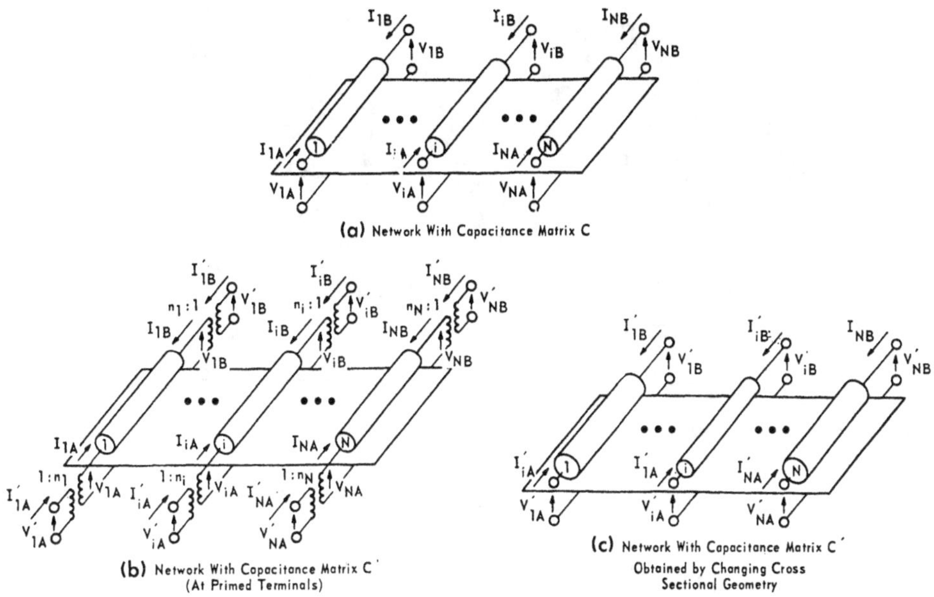

Fig. 3. Network and physical interpretation of the capacitance matrix transformation.

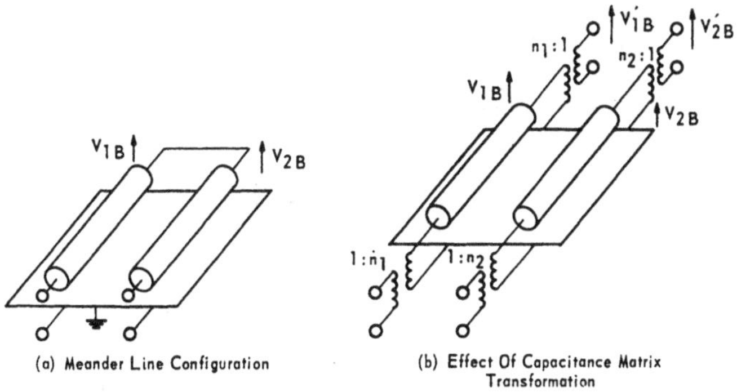

Fig. 4. Example illustrating inadmissible terminating conditions.

From the above statement, all complex terminating admittances, including opens and shorts, can be accommodated. It is emphasized that only terminations connected at the ports defined between the end of each line and the ground plane be considered in obtaining equivalent circuits by a change in cross-sectional geometry. An example of a network configuration for which an equivalent circuit cannot be realized by a change in cross-sectional geometry only is the familiar meander line shown in Fig. 4(a). Only two lines are shown for simplicity. Transformations of the capacitance matrix of the basic array introduce ideal transformers as shown in Fig. 4(b). As is readily seen, the original boundary condition that $V_{1B}=V_{2B}$ cannot be satisfied by making $V'_{1B}=V'_{2B}$ except when $n_1=n_2$. However, choosing $n_1=n_2$ is a trivial transformation that merely changes the impedance level of the entire network. The above example does not invalidate the use of the capacitance matrix transformation, but emphasizes that only under the specified terminating conditions can the transformation be interpreted as a simple change in cross-sectional geometry.

III. TEM Network Applications of Capacitance Matrix Transformations

A. Symmetric Directional Coupler Equivalent Circuits

As an example of some familiar equivalent circuits obtainable by application of the capacitance matrix transformation, consider the matched single-section four-port parallel line directional coupler shown in Fig. 5(a). The matched constraint requires that $Z_o=\sqrt{Z_{oe}Z_{oo}}$, where Z_{oe} and Z_{oo} are the even- and odd-mode impedances and Z_o is the characteristic impedance of the terminating lines [10]. The normalized [1] even- and odd-mode static capacitance values (c) are related to the even- and odd-mode impedances [10] by

$$Z_{oe}=\frac{\eta_o}{\sqrt{\epsilon_r}\cdot C_{oe}}\quad\text{and}\quad Z_{oo}=\frac{\eta_o}{\sqrt{\epsilon_r}\cdot C_{oo}}. \tag{9}$$

The corresponding self and mutual static capacitance values are then those given in Fig. 5(b). The circuit of Fig. 5(b) can be considered as a degenerate three-line network above a grounded plane in which the center line is coupled to both outer lines, but is decoupled from

52

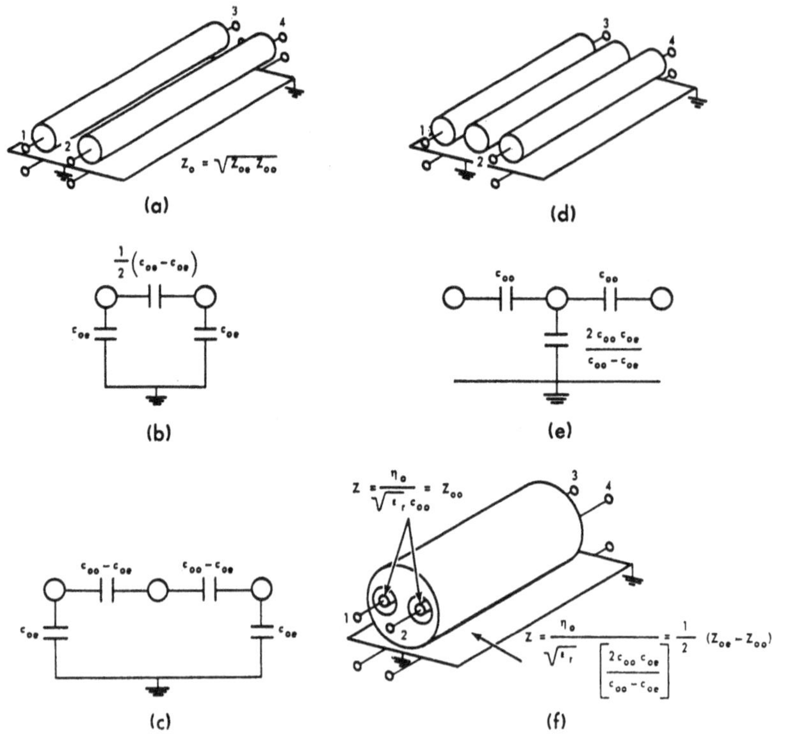

Fig. 5. Directional coupler equivalent circuits obtained by capacitance matrix transformations.

ground as shown in Fig. 5(c). The electrical performance of this network at ports one through four is identical to that of the two-line network of Fig. 5(b). The capacitance matrix for the network of Fig. 5(c) is

$$\left[\begin{array}{c|c|c} C_{oe} & -(C_{oo}-C_{oe}) & 0 \\ \hline -(C_{oo}-C_{oe}) & 2(C_{oo}-C_{oe}) & -(C_{oo}-C_{oe}) \\ \hline 0 & -(C_{oo}-C_{oe}) & C_{oo} \end{array}\right] \begin{array}{c} \\ \leftarrow n. \\ \\ \end{array} \quad (10)$$
$$\uparrow n$$

If the center row and column of this matrix is multiplied by an admissible constant n, the electrical performance at ports one through four is unchanged. The effect of a partial capacitance matrix transformation on the corresponding physical realization is shown in Fig. 5(d). A complete transformation results when the multiplying constant is chosen such that the outer lines are decoupled from the ground plane. This requires that the sum of elements in the first row or column of the capacitance matrix be zero. Therefore,

$$C_{oo} - n'(C_{oo} - C_{oe}) = 0 \quad \text{or} \quad n' = \frac{C_{oo}}{C_{oo} - C_{oe}}. \quad (11)$$

The completely transformed network capacitance values are shown in Fig. 5(e) and a physical realization with appropriate impedance values is shown in Fig. 5(f). The transformed network is seen to be the re-entrant section described by Cohn [11]. The two-line directional coupler and the re-entrant coupled section are limiting cases of a more general three-line network. The two-line network results from a transformation that decouples the center line from the ground plane and the re-entrant section from a transformation that decouples the outer lines from the ground plane.

B. Directional Couplers with Unequal Terminations

The capacitance matrix transformation allows the simple design of directional couplers with unequal terminations. To demonstrate the technique, consider again the matched four-port directional coupler with coupling coefficient k shown in Fig. 6(a). The corresponding capacitance diagram is shown in Fig. 6(b) where the capacitance values are those given in Section III-A. The coupling coefficient is given by

$$k = \frac{\frac{1}{2}(C_{oo} - C_{oe})}{\frac{1}{2}(C_{oo} - C_{oe}) + C_{oe}} = \frac{C_{oo} - C_{oe}}{C_{oo} + C_{oe}}. \quad (12)$$

The appropriate capacitance matrix is

$$\frac{1}{2}\left[\begin{array}{c|c} C_{oo} + C_{oe} & C_{oe} - C_{oo} \\ \hline C_{oe} - C_{oo} & C_{oo} + C_{oe} \end{array}\right] \begin{array}{c} \\ \leftarrow n \end{array}. \quad (13)$$
$$\uparrow n$$

If an asymmetric transformation is performed by

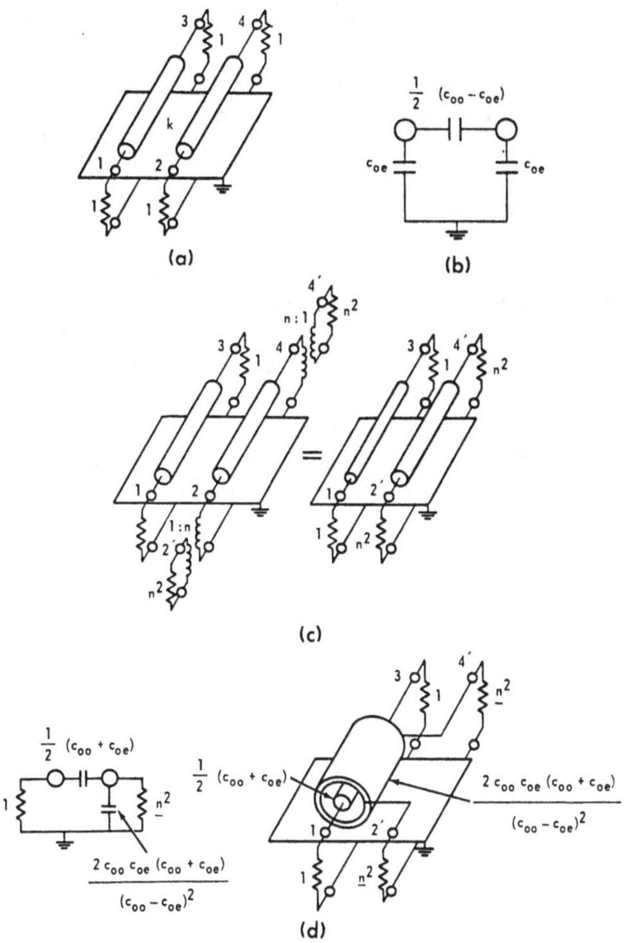

Fig. 6. Directional couplers with unequal terminations.

multiplying the second row and column by a factor n, a new capacitance matrix is obtained whose network realization is the original network with ideal transformers of turns ratio $n:1$ added to both ends of the second line. As described in Section II-B, if the unit terminating conductances are changed to a value of n^2, the original network with the ideal transformers added can be realized by a change in cross-sectional geometry while maintaining the transmission properties invariant. The network interpretation and a physical realization of a coupler corresponding to a partial transformation are shown in Fig. 6(c).

A complete transformation results if the self-capacitance from the first node is eliminated. From (13) this requires

$$n(c_{oo} - c_{oe}) = c_{oo} + c_{oe}$$

or

$$n = \frac{c_{oo} + c_{oe}}{c_{oo} - c_{oe}} = \frac{1}{k} \cdot \tag{14}$$

A coaxial physical realization of a coupler corresponding to a complete capacitance matrix transformation is shown in Fig. 6(d) along with the appropriate capacitance network. From (14), the maximum termination ratio is seen to be dependent on coupling level, the ratio being given by

$$n_{\max}^2 = \frac{1}{k^2} \cdot \tag{15}$$

As an example, a 3-dB coupler can yield at most a

$$\frac{1}{k^2} = \frac{1}{(0.707)^2} = 2:1$$

termination ratio. Also, as is evident from (15), the lower the coupling coefficient, the higher the possible termination ratio. The above procedure yields a transformation from a unity admittance value to a larger admittance value. Transformations from unity to smaller admittance values are obtained by changing the admittance level of the entire network and using the transformed line as reference.

C. *Compact Filter-Transformers*

In a previous paper [1], symmetric capacitance matrix transformations were utilized to obtain partial or entire coaxial realizations of interdigital filters. By applying an asymmetric transformation, compact coaxial interdigital filters that impedance transform in the passband, can be designed. These designs use the element value tables developed for the exact design of interdigital filters [12]. As with all impedance transforming devices, there are constraints imposed on the impedance ratio that can be achieved over a given bandwidth with a given tolerance and number of sections. These constraints have been investigated by straightforward application of the capacitance matrix transformation. The new network forms were found to be comparable in impedance transforming properties to quarter-wave stepped impedance designs using a like number of sections.

The normalized self and mutual static capacitance element values for interdigital filters with open-circuited terminating lines (OCTL) are given in Tables I–III for four, five, and six section designs, respectively.[2] Two ripple values, 0.01 dB (1.10 VSWR maximum) and 0.10 dB (1.36:1 VSWR maximum) are given. The tabulated element values can be used directly with available data, [1], [5], [6], and [13] to obtain physical dimensions for interdigital filters.

[2] More extensive element value tables including networks with short-circuited terminating lines (SCTL) are given in Reference [12]. The OCTL networks give a higher termination ratio for a specified number of sections and passband ripple than do the SCTL networks. The three tables given should be adequate for most filter-transformer requirements.

The following symbols are used in the tables presented:

- N_f = number of filter sections (4, 5, or 6)
- m = 3, number of L-C elements in basic prototype[3]
- n = number of unit elements in basic prototype (1, 2, or 3)[3]
- r = ripple value in dB (0.01 dB and 0.10 dB)
- $VSWR_M$ = maximum theoretical standing-wave ratio in the filter passband
- % BW = percentage bandwidth
- F_H/F_L = (200+% BW)/(200−% BW) = ratio of high passband edge frequency to low passband edge frequency
- R_S/R_L = maximum termination ratio determined by analysis. General formulas are given in each table for the specific configuration shown.

Concerning the filtering properties of the filter-transformers, the stop-band attenuation at transformed frequency Ω_A is given by [12], [14];

$$\frac{1}{\sqrt{\epsilon_r}}\begin{bmatrix} 11.0 & -11.0 & 0 & 0 \\ -11.0 & 23.58 & -7.19n_1 & 0 \\ 0 & -7.19n_1 & 23.58n_1^2 & -11.0n_1n_2 \\ 0 & 0 & -11.0n_1n_2 & 11.0n_2^2 \end{bmatrix} \begin{matrix} \\ \\ \leftarrow n_1 \\ \leftarrow n_2 \end{matrix} . \qquad \begin{matrix} \uparrow & \uparrow \\ n_1 & n_2 \end{matrix} \tag{18}$$

$$A = 10 \log_{10}\left[1 + \epsilon^2 \cosh^2\left(m \cosh^{-1}\frac{\Omega_c}{\Omega_A} + n \cosh^{-1}\frac{\sqrt{1+\Omega_c^2}}{\sqrt{1+\Omega_A^2}}\right)\right] \tag{16}$$

where

- A = attenuation in dB
- $\epsilon^2 = (10^{r/10} - 1)$
- $\Omega_c = \tan \pi f_c/2f_0$
- $\Omega_A = \tan \pi f_A/2f_0$
- f_0 = frequency at which the lines are a quarter wavelength
- f_c = cutoff frequency $= f_0[1 - \%BW/200]$
- f_A = frequency at which the attenuation is A dB

To demonstrate the use of the tables and the design techniques involved, a 3:1 bandwidth four-section filter transformer with 0.01 dB passband ripple will be described. The prototype self and mutual static capacitance values are obtained from Table I-A and are shown in Fig. 7(a). The capacitance matrix is:

$$\frac{1}{\sqrt{\epsilon_r}}\begin{bmatrix} 11.0 & -11.0 & 0 & 0 \\ -11.0 & 23.58 & -7.19 & 0 \\ 0 & -7.19 & 23.58 & -11.0 \\ 0 & 0 & -11.0 & 11.0 \end{bmatrix}. \tag{17}$$

The network of Fig. 7(a) is symmetric and is designed to work between equal terminating impedances. Unequal terminating impedances can be obtained by an asymmetrical transformation of the capacitance matrix, as discussed in Section II.

A desirable network configuration is one that contains only one shunt c in the corresponding capacitance network. This allows all other elements to be incorporated coaxially in the element represented by the single shunt c.

Referring to matrix (18), a single shunt c can be achieved by choosing n_1 and n_2 such that the desired network configuration, as shown in Fig. 7(b), is obtained.[4]

This requires that:

$$-11.0 + 23.58 - 7.19n_1 = 0$$
$$11.0n_1n_2 = 11.0n_2^2$$
$$\therefore\ n_1 = n_2 = 1.75. \tag{19}$$

Then R_S/R_L = termination ratio $= n_2^2 = 3.06$.

Multiplication of the outer row and column of matrix (17) by $n_2 = 1.75$, as indicated in matrix (18), lowers the impedance level by $n_2^2 = 3.06$, and requires the use of a $1/n_2^2 = 0.327$ ohm load[5] to maintain a 0.01 dB ripple response in the passband. The transformed network thus acts as a 3.06:1 impedance transformer. The impedance ratio that can be achieved is limited by the choice of n_1 and n_2, which are constrained, as described in Section II. A higher impedance ratio can be achieved by applying another transformation such that the single shunt capacitor is connected to the fourth node. Referring to matrix (18), this requires that:

[3] The significance of m and n is described in detail in References [12] and [14].

[4] Note that the formulas for R_S/R_L given in each table allow the calculation of the achievable termination ratio for the configurations shown. Actual capacitance values are obtained as described in the design example.

[5] On a normalized 1-ohm basis.

$$-11.0 + 23.58 - 7.19n_1 = 0$$
$$-7.19n_1 + 23.58n_1^2 - 11.0n_1n_2 = 0. \quad (20)$$

Solving gives

$$n_1 = 1.75$$
$$n_2 = 3.09, \quad (21)$$

and results in a termination ratio of $n_2^2 = 9.6{:}1$. The corresponding capacitance network is shown in Fig. 7(c). Whereas the network of Fig. 7(c) is capable of providing a higher impedance ratio over a given band than the network of Fig. 7(b), all elements must be incorporated coaxially in the end element. This requires the end element to be large and results in a difficult connector problem. For this reason a network form such as that of Fig. 7(b) is often preferable. Furthermore, in some applications it is desirable to have not only a low impedance but also to have a region of high current density. This can be achieved by using a configuration such as that of Fig. 7(b).

Before describing experimental results, it is interesting to compare the transformer properties of the coaxial filter-transformer with those of quarter-wave stepped impedance lines. From Young's Tables [15], for 3:1 bandwidth, four quarter-wave sections give a maximum VSWR of 1.07:1 for a 3:1 impedance ratio and a maximum VSWR of 1.18:1 for a 10:1 impedance ratio. The partially transformed filter network [Fig. 7(b)] is comparable to the four-section quarter-wave stepped impedance design and the fully transformed network is superior to a stepped impedance design.

A trial transformer was constructed based on the capacitance network shown in Fig. 7(b). The corresponding *S*-plane equivalent circuit on a 50-ohm basis is shown in Fig. 8(a). The element values (C_1, Z, L, C_2) are directly related to the static capacitor values, as shown in (22). The corresponding impedance values on a 50-ohm basis for all-coaxial realization are:

$$\frac{1}{C_1} = Z_1 = \frac{\eta_0}{\sqrt{\epsilon_r}\, c_0} = \frac{376.7}{11.0} = 35.2 \text{ ohms}$$
$$Z = Z_2 = \frac{\eta_0}{\sqrt{\epsilon_r}\, c_2} = \frac{376.7}{12.58} = 30.0 \text{ ohms}$$
$$L = Z_3 = \frac{\eta_0}{\sqrt{\epsilon_r}\, c_3} = \frac{376.7}{35.8} = 14.6 \text{ ohms} \quad (22)$$
$$\frac{1}{C_2} = Z_4 = \frac{\eta_0}{\sqrt{\epsilon_r}\, c_4} = \frac{376.7}{33.7} = 11.2 \text{ ohms}$$

where

$c_0 - c_4$ = static capacitance values shown in Fig. 7(b).

The corresponding diameter ratios are [13] (for teflon, $\sqrt{\epsilon_r} = 1.44$)

$$d_{Z_1} = 2.28_{\text{in teflon}}$$
$$d_{Z_2} = 1.65_{\text{in air}}$$
$$d_{Z_3} = 1.27_{\text{in air}}$$
$$d_{Z_4} = 1.34_{\text{in teflon}}. \quad (23)$$

A schematic cross-sectional drawing showing the location of the various elements and appropriate impedances is shown in Fig. 8(b). Symbols have been added to important junction points in Figs. 8(a) and 8(b) to aid in visualizing the relationship between the prototype network and its physical realization. A detailed line drawing of the filter assembly, including dimensions, is given in Fig. 9.

The design center frequency chosen was 2.0 GHz. The line elements were arbitrarily shortened by 0.100 inch to compensate for end effects. Since a 50/3.06 = 16.3 ohm load was not available, a 50-ohm connector was attached to the low impedance end and tests were conducted by measuring the VSWR and insertion loss on a 50-ohm basis. The input impedance of a doubly terminated Chebyshev filter with small passband ripple is essentially real in its passband; therefore, a good indication of trial design performance can be obtained by measuring the input VSWR at the low impedance end. The VSWR value should be very nearly the transformer impedance ratio of 3.06:1. Again, because of the essentially real character of the input impedance, the performance that would have been achieved if a 16.3 ohm load had been available can be computed to good approximation by

$$\text{VSWR}_{16.4\text{ ohm load}} = \frac{\text{VSWR}_{50\text{ ohm load}}}{3.06}, \quad \text{when } \text{VSWR}_{50\text{ ohm load}} \geq 3.06 \quad (24)$$
$$\text{VSWR}_{16.4\text{ ohm load}} = \frac{3.06}{\text{VSWR}_{50\text{ ohm load}}}, \quad \text{when } \text{VSWR}_{50\text{ ohm load}} \leq 3.06.$$

The measured and computed performance characteristics, together with theoretical values, are shown in Fig. 10. The experimental characteristics shown were achieved with the initial design—no alterations in dimensions were made. The performance is observed to be excellent in the initial design and indicates that transformers of the type described offer the possibility of achieving wideband impedance transforming networks in a very compact configuration. A photograph of the disassembled filter-transformer is shown in Fig. 11.

The filter-transformers described need not use an all-coaxial realization. Partial capacitance matrix transformations can be performed that allow coupled bar realizations to be employed, as described in [1]. Various geometric configurations can be investigated using the capacitance matrix transformation to achieve a desired configuration and/or more convenient line impedance values.

55

TABLE I

SELF AND MUTUAL STATIC CAPACITORS FOR CHEBYSHEV INTERDIGITAL FILTERS WITH OPEN-CIRCUITED TERMINATING LINES $m=3$, $n=1$, $N_f=4$ SECTIONS

A

$r = 0.01$ DB, $VSWR_M = 1.10:1$

% BW	$\sqrt{\epsilon_r}\, c_0$	$\sqrt{\epsilon_r}\, c_1$	$\sqrt{\epsilon_r}\, c_2$
30.0	.255E+01	.316E+02	.719E+01
35.0	.300E+01	.262E+02	.719E+01
40.0	.346E+01	.221E+02	.719E+01
45.0	.393E+01	.190E+02	.719E+01
50.0	.443E+01	.165E+02	.719E+01
55.0	.494E+01	.144E+02	.719E+01
60.0	.547E+01	.127E+02	.719E+01
65.0	.602E+01	.113E+02	.719E+01
70.0	.661E+01	.101E+02	.719E+01
75.0	.722E+01	.902E+01	.719E+01
80.0	.788E+01	.811E+01	.719E+01
85.0	.857E+01	.730E+01	.719E+01
90.0	.932E+01	.659E+01	.719E+01
95.0	.101E+02	.596E+01	.719E+01
100.0	.110E+02	.539E+01	.719E+01
110.0	.129E+02	.441E+01	.719E+01
120.0	.153E+02	.360E+01	.719E+01
130.0	.183E+02	.292E+01	.719E+01
140.0	.222E+02	.233E+01	.719E+01
150.0	.276E+02	.183E+01	.719E+01

B

$r = 0.10$ DB, $VSWR_M = 1.36:1$

% BW	$\sqrt{\epsilon_r}\, c_0$	$\sqrt{\epsilon_r}\, c_1$	$\sqrt{\epsilon_r}\, c_2$
30.0	.164E+01	.354E+02	.648E+01
35.0	.192E+01	.295E+02	.648E+01
40.0	.222E+01	.250E+02	.648E+01
45.0	.252E+01	.215E+02	.648E+01
50.0	.283E+01	.188E+02	.648E+01
55.0	.315E+01	.165E+02	.648E+01
60.0	.349E+01	.146E+02	.648E+01
65.0	.384E+01	.130E+02	.648E+01
70.0	.421E+01	.116E+02	.648E+01
75.0	.459E+01	.105E+02	.648E+01
80.0	.500E+01	.942E+01	.648E+01
85.0	.544E+01	.850E+01	.648E+01
90.0	.590E+01	.769E+01	.648E+01
95.0	.640E+01	.697E+01	.648E+01
100.0	.693E+01	.631E+01	.648E+01
110.0	.815E+01	.519E+01	.648E+01
120.0	.962E+01	.425E+01	.648E+01
130.0	.115E+02	.345E+01	.648E+01
140.0	.138E+02	.276E+01	.648E+01
150.0	.171E+02	.216E+01	.648E+01

$$\frac{F_H}{F_L} = \frac{200 + \%\,BW}{200 - \%\,BW}$$

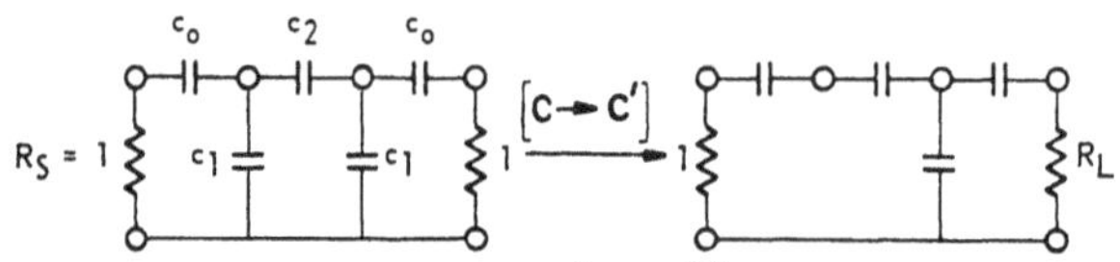

$$\frac{R_S}{R_L} = \left[1 + \frac{c_1}{c_2}\right]^2$$

TABLE II

SELF AND MUTUAL STATIC CAPACITORS FOR CHEBYSHEV INTERDIGITAL FILTERS WITH OPEN-CIRCUITED TERMINATING LINES $m=3$, $n=2$, $N_f=5$ SECTIONS

A

$r = 0.01$ DB, $VSWR_M = 1.10:1$

% BW	$\sqrt{\epsilon_r}\, c_0$	$\sqrt{\epsilon_r}\, c_1$	$\sqrt{\epsilon_r}\, c_2$	$\sqrt{\epsilon_r}\, c_3$
30.0	.240E+01	.350E+02	.688E+01	.293E+02
35.0	.282E+01	.291E+02	.688E+01	.235E+02
40.0	.326E+01	.246E+02	.689E+01	.193E+02
45.0	.371E+01	.212E+02	.690E+01	.160E+02
50.0	.417E+01	.184E+02	.691E+01	.134E+02
55.0	.465E+01	.162E+02	.692E+01	.113E+02
60.0	.515E+01	.143E+02	.693E+01	.960E+01
65.0	.567E+01	.127E+02	.694E+01	.819E+01
70.0	.622E+01	.113E+02	.695E+01	.701E+01
75.0	.681E+01	.101E+02	.697E+01	.601E+01
80.0	.742E+01	.910E+01	.698E+01	.515E+01
85.0	.808E+01	.820E+01	.699E+01	.442E+01
90.0	.879E+01	.739E+01	.701E+01	.379E+01
95.0	.955E+01	.667E+01	.702E+01	.325E+01
100.0	.104E+02	.602E+01	.703E+01	.278E+01
110.0	.123E+02	.491E+01	.706E+01	.201E+01
120.0	.145E+02	.398E+01	.708E+01	.143E+01
130.0	.174E+02	.320E+01	.710E+01	.994E+00
140.0	.212E+02	.254E+01	.713E+01	.665E+00
150.0	.265E+02	.196E+01	.714E+01	.423E+00

B

$r = 0.10$ DB, $VSWR_M = 1.36:1$

% BW	$\sqrt{\epsilon_r}\, c_0$	$\sqrt{\epsilon_r}\, c_1$	$\sqrt{\epsilon_r}\, c_2$	$\sqrt{\epsilon_r}\, c_3$
30.0	.158E+01	.376E+02	.629E+01	.322E+02
35.0	.186E+01	.313E+02	.630E+01	.261E+02
40.0	.214E+01	.266E+02	.630E+01	.215E+02
45.0	.243E+01	.229E+02	.630E+01	.180E+02
50.0	.274E+01	.200E+02	.631E+01	.152E+02
55.0	.305E+01	.176E+02	.631E+01	.129E+02
60.0	.337E+01	.156E+02	.632E+01	.111E+02
65.0	.371E+01	.139E+02	.632E+01	.955E+01
70.0	.407E+01	.124E+02	.633E+01	.825E+01
75.0	.444E+01	.112E+02	.634E+01	.714E+01
80.0	.484E+01	.101E+02	.634E+01	.618E+01
85.0	.526E+01	.910E+01	.635E+01	.536E+01
90.0	.571E+01	.823E+01	.636E+01	.464E+01
95.0	.619E+01	.745E+01	.636E+01	.402E+01
100.0	.671E+01	.675E+01	.637E+01	.347E+01
110.0	.789E+01	.553E+01	.639E+01	.257E+01
120.0	.933E+01	.451E+01	.640E+01	.187E+01
130.0	.111E+02	.365E+01	.642E+01	.132E+01
140.0	.135E+02	.291E+01	.643E+01	.907E+00
150.0	.167E+02	.227E+01	.644E+01	.589E+00

$$\frac{F_H}{F_L} = \frac{200 + \%\,BW}{200 - \%\,BW}$$

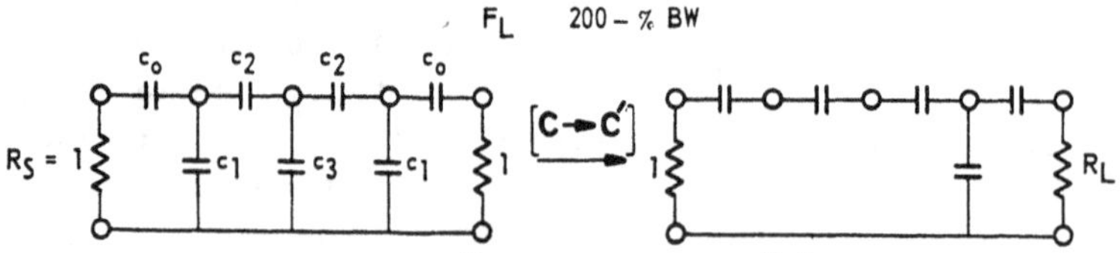

$$\frac{R_S}{R_L} = \left[\left(1 + \frac{c_1}{c_2}\right)\left(2 + \frac{c_3}{c_2}\right) - 1\right]^2$$

TABLE III

SELF AND MUTUAL STATIC CAPACITORS FOR CHEBYSHEV INTERDIGITAL FILTERS WITH OPEN-CIRCUITED TERMINATING LINES $m=3$, $n=3$, $N_f=6$ SECTIONS

A

r = 0.01 DB VSWR$_M$ = 1.10:1

% BW	$\sqrt{\epsilon_r}\,c_0$	$\sqrt{\epsilon_r}\,c_1$	$\sqrt{\epsilon_r}\,c_2$	$\sqrt{\epsilon_r}\,c_3$	$\sqrt{\epsilon_r}\,c_4$
30.0	.232E+01	.368E+02	.678E+01	.314E+02	.640E+01
35.0	.273E+01	.306E+02	.679E+01	.253E+02	.641E+01
40.0	.315E+01	.260E+02	.679E+01	.208E+02	.642E+01
45.0	.358E+01	.224E+02	.680E+01	.174E+02	.644E+01
50.0	.403E+01	.195E+02	.681E+01	.146E+02	.645E+01
55.0	.449E+01	.171E+02	.682E+01	.124E+02	.647E+01
60.0	.497E+01	.151E+02	.683E+01	.106E+02	.649E+01
65.0	.548E+01	.135E+02	.684E+01	.906E+01	.651E+01
70.0	.601E+01	.120E+02	.685E+01	.778E+01	.654E+01
75.0	.656E+01	.108E+02	.687E+01	.670E+01	.656E+01
80.0	.716E+01	.971E+01	.688E+01	.577E+01	.659E+01
85.0	.779E+01	.875E+01	.690E+01	.497E+01	.662E+01
90.0	.847E+01	.789E+01	.691E+01	.427E+01	.664E+01
95.0	.920E+01	.713E+01	.693E+01	.367E+01	.667E+01
100.0	.999E+01	.644E+01	.694E+01	.314E+01	.671E+01
110.0	.118E+02	.525E+01	.698E+01	.229E+01	.677E+01
120.0	.140E+02	.426E+01	.701E+01	.163E+01	.684E+01
130.0	.168E+02	.342E+01	.704E+01	.113E+01	.690E+01
140.0	.205E+02	.270E+01	.708E+01	.751E+00	.697E+01
150.0	.257E+02	.208E+01	.711E+01	.473E+00	.703E+01

B

r = 0.10 DB VSWR$_M$ = 1.36:1

% BW	$\sqrt{\epsilon_r}\,c_0$	$\sqrt{\epsilon_r}\,c_1$	$\sqrt{\epsilon_r}\,c_2$	$\sqrt{\epsilon_r}\,c_3$	$\sqrt{\epsilon_r}\,c_4$
30.0	.155E+01	.386E+02	.624E+01	.334E+02	.601E+01
35.0	.182E+01	.322E+02	.624E+01	.271E+02	.601E+01
40.0	.210E+01	.274E+02	.624E+01	.225E+02	.602E+01
45.0	.239E+01	.236E+02	.625E+01	.188E+02	.603E+01
50.0	.268E+01	.206E+02	.625E+01	.160E+02	.604E+01
55.0	.299E+01	.182E+02	.626E+01	.136E+02	.605E+01
60.0	.331E+01	.161E+02	.626E+01	.117E+02	.606E+01
65.0	.364E+01	.144E+02	.627E+01	.101E+02	.607E+01
70.0	.399E+01	.129E+02	.627E+01	.874E+01	.608E+01
75.0	.435E+01	.116E+02	.628E+01	.758E+01	.609E+01
80.0	.474E+01	.104E+02	.629E+01	.658E+01	.610E+01
85.0	.515E+01	.943E+01	.630E+01	.571E+01	.612E+01
90.0	.559E+01	.854E+01	.630E+01	.496E+01	.613E+01
95.0	.606E+01	.773E+01	.631E+01	.430E+01	.615E+01
100.0	.657E+01	.701E+01	.632E+01	.372E+01	.617E+01
110.0	.773E+01	.575E+01	.634E+01	.276E+01	.620E+01
120.0	.914E+01	.469E+01	.636E+01	.201E+01	.624E+01
130.0	.109E+02	.380E+01	.638E+01	.143E+01	.628E+01
140.0	.132E+02	.302E+01	.640E+01	.975E+00	.632E+01
150.0	.164E+02	.235E+01	.642E+01	.631E+00	.636E+01

$$\frac{F_H}{F_L} = \frac{200 + \%\,BW}{200 - \%\,BW}$$

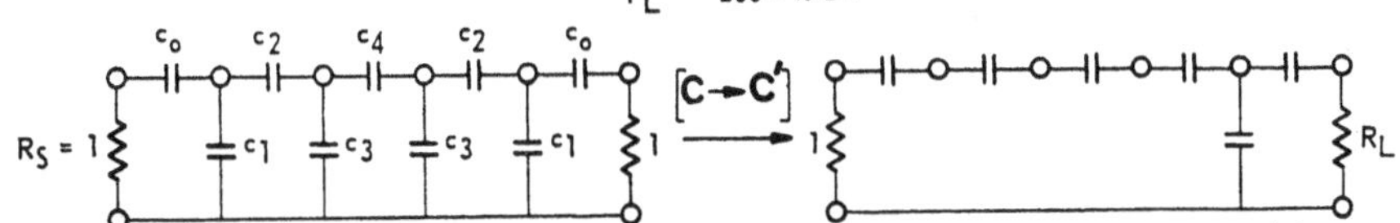

$$\frac{R_S}{R_L} = n_1^2 n_2^2 n_3^2 \text{ WHERE}$$

$$n_1 = 1 + \frac{c_1}{c_2} \qquad n_2 = 1 + \frac{c_2 + c_3}{c_4} - \frac{c_2}{c_4 n_1} \qquad n_3 = 1 + \frac{c_3 + c_4}{c_2} - \frac{c_4}{c_2 n_2}$$

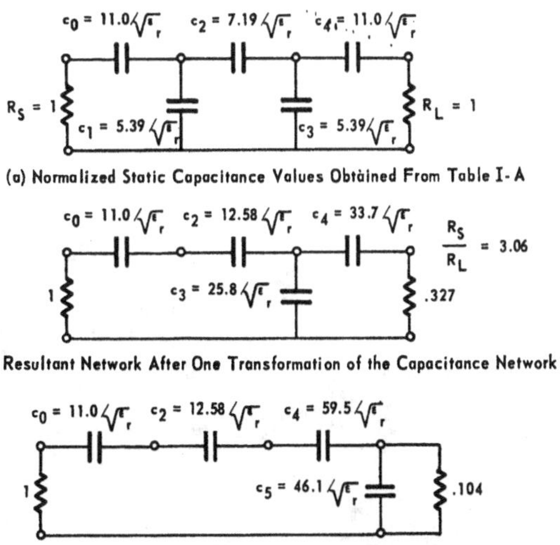

(a) Normalized Static Capacitance Values Obtained From Table I-A

(b) Resultant Network After One Transformation of the Capacitance Network

(c) Network with Largest Possible Transformer Ratio

Fig. 7. Coaxial filter-transformer design.

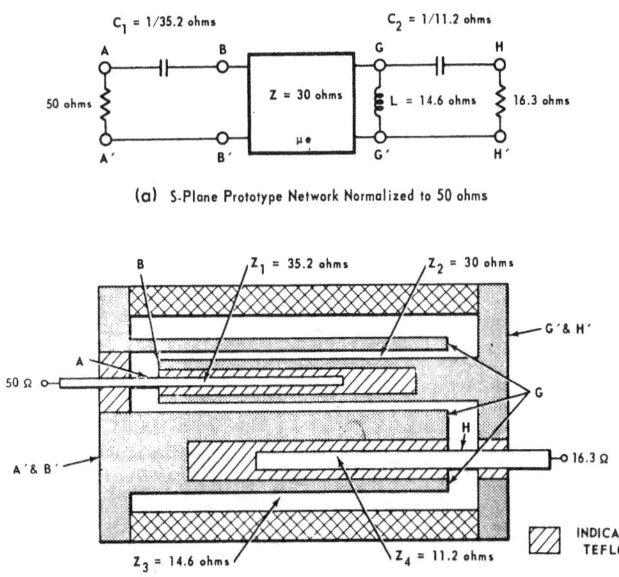

(a) S-Plane Prototype Network Normalized to 50 ohms

(b) Schematic Cross-Sectional Drawing

Fig. 8. Coaxial filter-transformer schematic diagram.

58

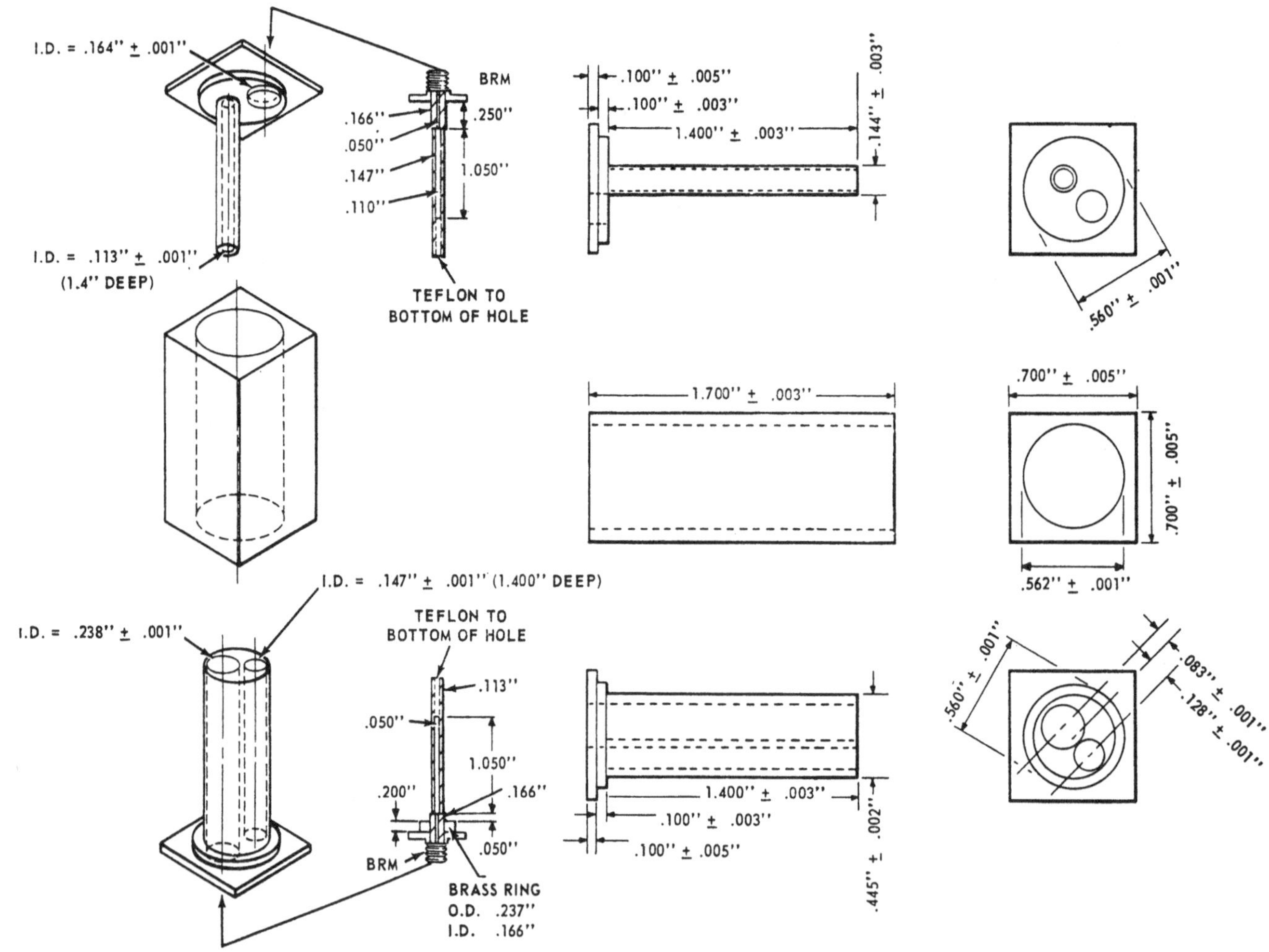

Fig. 9. Assembly drawing of 3:1 bandwidth filter-transformer.

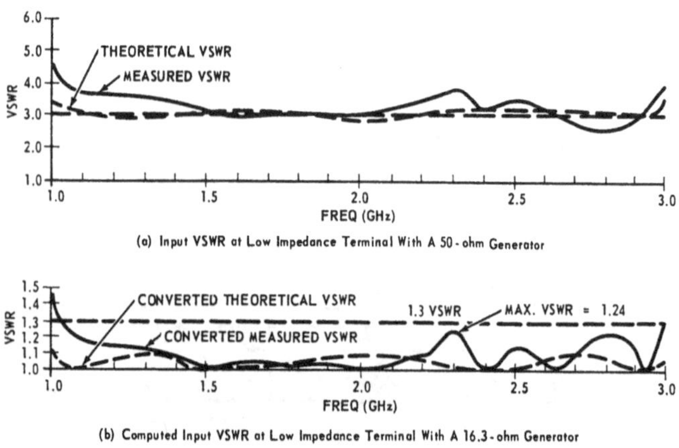

Fig. 10. Theoretical and measured performance of 3:1 bandwidth filter-transformer.

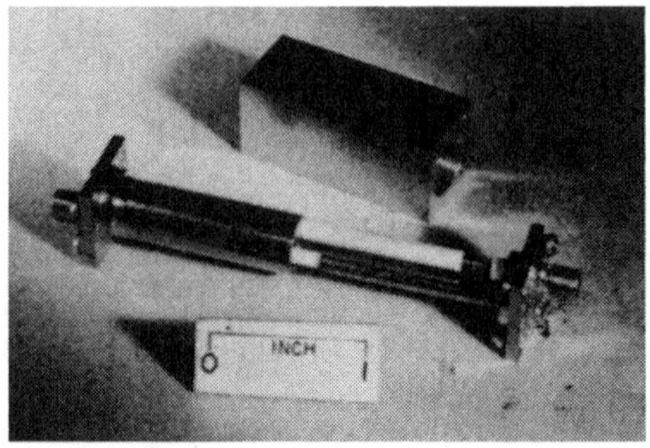

Fig. 11. Four-section coaxial filter-transformer.

Conclusions

The description of TEM propagation on an array of parallel coupled lines in terms of the static capacitance matrix allows a unified treatment of many heretofore seemingly unrelated network configurations. The use of the capacitance matrix transformation provides a simple method, devoid of complicated mathematics, for investigating possible equivalent circuit structures and obtaining new network forms. Furthermore, the techniques described provide considerable physical insight into the meaning of equivalent TEM networks and should be a very useful tool, both computationally and conceptually, to the network designer.

References

[1] R. J. Wenzel, "Exact theory of interdigital band-pass filters and related coupled structures," *IEEE Trans. on Microwave Theory and Techniques*, vol. MTT-13, pp. 559–575, September 1965.

[2] S. O. Rice, "Steady-state solutions of transmission line equations," *Bell Sys. Tech. J.*, vol. 20, pp. 131–178, April 1941.

[3] D. C. Youla, "An introduction to coupled-line network theory," MRI, Polytechnic Institute of Brooklyn, Brooklyn, N. Y., Rept. 960–961, 1961.

[4] E. Ott, "A network approach to the design of multiline $2N$-port directional couplers," Polytechnic Institute of Brooklyn, Brooklyn, N. Y., Rept. PIBMRI-1236-64; Air Research Development Command, Rome Air Development Center, Griffiss AFB, Rome, N. Y., Contract AF 30(602)-2868, Tech. Rept. RADC-TR-65-41, April 1965.

[5] W. J. Getsinger, "Coupled rectangular bars between parallel plates," *IRE Trans. on Microwave Theory and Techniques*, vol. MTT-10, pp. 65–72, January 1962.

[6] E. G. Cristal, "Coupled circulator cylindrical rods between parallel ground planes," *IEEE Trans. on Microwave Theory and Techniques*, vol. MTT-12, pp. 428–439, July 1964.

[7] R. W. Beatty and D. M. Kerns, "Relationships between different kinds of network parameters, not assuming reciprocity or equality of the waveguide or transmission line characteristic impedances," *Proc. IEEE* (*Correspondence*), vol. 52, p. 84, January 1964.

[8] E. A. Guillemin, *Synthesis of Passive Networks*. New York: Wiley, 1957.

[9] R. M. Fano, L. J. Chu, and R. B. Adler, *Electromagnetic Fields, Energy and Forces*. New York: Wiley, 1960.

[10] E. M. T. Jones and J. T. Bolljahn, "Coupled-strip-transmission-line filters and directional couplers," *IRE Trans. on Microwave Theory and Techniques*, vol. MTT-4, pp. 75–81, April 1956.

[11] S. B. Cohn, "The re-entrant cross section and wide-band 3-dB hybrid couplers," *IEEE Trans. on Microwave Theory and Techniques*, vol. MTT-11, pp. 254–258, July 1963.

[12] R. J. Wenzel and M. C. Horton, "Exact design techniques for microwave TEM filters," Final Rept., U. S. Army Electronics Lab., Fort Monmouth, N. J., 00399(E), Bendix Research Labs., Southfield, Mich., April 1965.

[13] *The Microwave Engineers Handbook and Buyer's Guide*. Brookline, Mass.: Horizon House-Microwave, Inc., 1966, p. 91.

[14] M. C. Horton and R. J. Wenzel, "General theory and design of optimum quarter-wave TEM filters," *IEEE Trans. on Microwave Theory and Techniques*, vol. MTT-13, pp. 316–327, May 1965.

[15] L. Young, "Tables for cascaded homogeneous quarter-wave transformers," *IRE Trans. on Microwave Theory and Techniques*, vol. MTT-9, pp. 233–237, April 1959.

PARALLEL COUPLED FILTERS

II

This relatively early paper proved to be very popular. Theoretically, it may be considered the fore-runner of the interdigital filter, to which it is closely related in performance and design.[1] It has an important advantage for printed-circuits, as in stripline and microstrip (the most suitable line for microwave integrated circuits), namely that it does not require grounding. A good ground is difficult to achieve in microstrip without causing radiation. The main drawback of the parallel-coupled filter in some situations is that it becomes rather long and thin, where interdigital and combline filters are more compact.

It may be worth remarking that one can also construct parallel coupled filters in slot line, which have the same shape factor as these filters in strip line or microstrip. Design information for coupled slots is not as readily available as for coupled strips, but could be generated with the aid of the last two papers (by Cohn and by Wen) reprinted and discussed in the companion volume on **Parallel Coupled Lines and Directional Couplers.**

Reprinted from IRE TRANSACTIONS
ON MICROWAVE THEORY AND TECHNIQUES
Volume MTT-6, Number 2, April, 1958

PRINTED IN THE U.S.A.

Parallel-Coupled Transmission-Line-Resonator Filters*

SEYMOUR B. COHN†

***Summary*—This paper describes the synthesis of band-pass transmission-line filters consisting of series of half-wavelength resonant conductors such as strips. The design differs from the usual end-coupled strip configuration in that successive strips are parallel coupled along a distance of a quarter-wavelength. The resulting coupling between resonators is partly electric and partly magnetic. Several important advantages are gained by this arrangement: 1) the length of the filter is approximately half that of the end-coupled type; 2) the gaps are larger and therefore less critical; and 3) the insertion-loss curve is symmetrical on a frequency scale with the first spurious response occurring at three times the center frequency of the pass band.**

Formulas are derived for the parallel-coupled-resonator transmission-line filter that permit accurate design for Tchebycheff, maximally flat, or any other physically realizable response. The formulas are theoretically exact in the limit of zero bandwidth, but frequency-response calculations show them to give good results for bandwidths up to about 30 per cent. An experimental strip-line filter of this type has been constructed, and the data given in this paper show that excellent performance has been obtained.

Introduction

As shown in Fig. 1(a),[1-3] multiple-coupled-resonator band-pass filters in strip line (or other TEM transmission line) have been most commonly composed of half-wavelength strips coupled end to end. In this paper, an alternative arrangement is treated in which the half-wavelength strips are parallel-coupled, as in Fig. 1(b) and 1(c). Parallel coupling offers a number of important advantages over end coupling: 1) the length of the filter is reduced approximately by half; 2) a symmetrical insertion-loss-vs-frequency response is obtained with the first spurious response occurring at three times the center frequency, and 3) a much larger gap between adjacent strips is permitted. The last advantage is of particular importance, since it eases the tolerance on the gaps for a given bandwidth, or permits a broader bandwidth for a given tolerance. Furthermore, the larger gap permits a higher power rating of the filter.

Formulas have been derived that allow the parallel-coupled-resonator filter to be designed in a straightforward manner, and to have any desired physically realizable response, such as maximally flat or equal-ripple. The accuracy of the design formulas has been checked by exact computation, and they have been found to give excellent results for bandwidths up to about 20 per cent in the case of maximally flat response and 30 per cent in the case of equal ripple response. The method of analysis is basically that used in an earlier paper[3] for a number of other types of direct-coupled-resonator filters. In this method, the lumped-constant low-pass prototype filter having the desired response is made equivalent to a set of either series- or parallel-resonant LC arms interconnected by broad-band quarter-wavelength transformers. An approximate equivalence then is established between the latter circuit and the actual coupled-resonator structure.

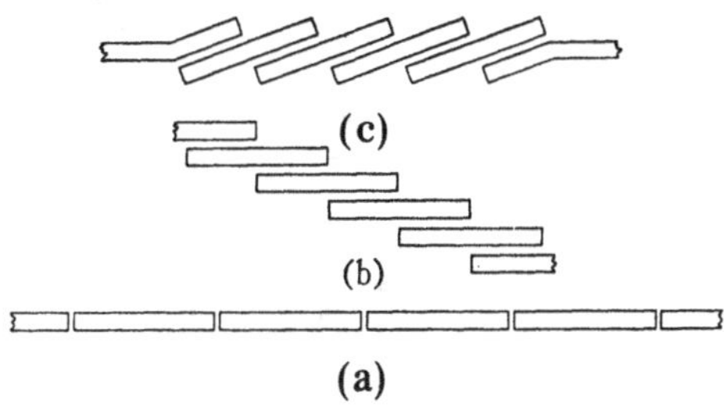

Fig. 1—Coupled-resonator strip-line filters: (a) end coupled; (b) and (c) parallel coupled.

Design Procedure

The design formulas for an n-resonator parallel-coupled filter are given in Table I. They utilize the element values g_1, g_2, $\cdots$, g_n of the prototype low-pass filter, which may be computed for either maximally flat or equal-ripple response by means of formulas given in Table II.[3] The schematic diagram in Fig. 2 shows that the filter is assembled from $n+1$ sections, resulting in a structure containing n resonators. The sections are of equal length (one-quarter wavelength at the center frequency), and their electrical design is completely specified by two characteristic impedances—Z_{oe} of the even-mode wave on the parallel conductors, and Z_{oo} of the odd-mode wave. These characteristic impedances have previously been defined, and formulas and graphs relating them to the dimensions of the cross section are available.[4] The total filter structure always will be symmetrical for maximally flat or equal-ripple response.

In the design of a parallel-coupled-resonator filter one should first select the type of response function and the number of resonators that will yield the desired in-

* Manuscript received by the PGMTT, October 21, 1957. The work described in this paper was supported by the U. S. Army Signal Eng. Labs. under Contract No. DA 36-039-sc-64625.

† Stanford Research Institute, Menlo Park, Calif.

[1] E. G. Fubini, W. E. Fromm, and H. S. Keen, "Microwave applications of high-Q strip components," IRE Convention Record, Pt. 8, pp. 98–103; March, 1954.

[2] E. A. Bradley, "Design and development of strip-line filters," IRE Trans., vol. MTT-4, pp. 86–93; April, 1956.

[3] S. B. Cohn, "Direct-coupled-resonator filters," Proc. IRE, vol. 45, pp. 187–196; February, 1957.

[4] S. B. Cohn, "Shielded coupled-strip transmission line," IRE Trans., vol. MTT-3, pp. 29–38, October, 1955.

TABLE I

FORMULAS FOR PARALLEL-COUPLED TRANSMISSION-LINE-RESONATOR FILTER

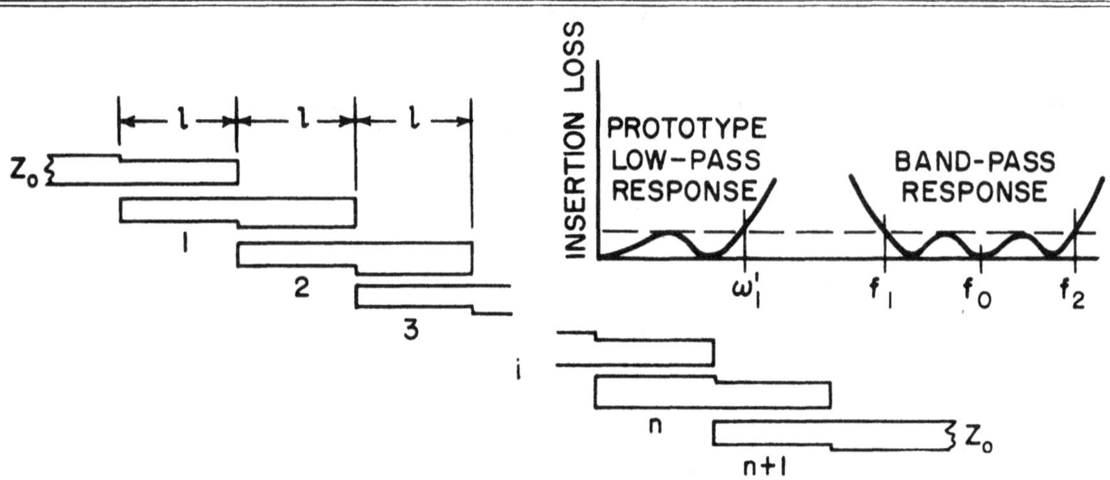

(For either maximally-flat or equal-ripple response, the structure is symmetrical.)

$l = \lambda_0/4$, where λ_0 = wavelength in transmission line at f_0

$f_0 = (f_1 + f_2)/2$

$$Z_{oe_i} = Z_o\left\{1 + \frac{Z_o}{K_{i-1,i}} + \left(\frac{Z_o}{K_{i-1,i}}\right)^2\right\}, \qquad i = 1 \text{ to } n+1$$

$$Z_{oo_i} = Z_o\left\{1 - \frac{Z_o}{K_{i-1,i}} + \left(\frac{Z_o}{K_{i-1,i}}\right)^2\right\}, \qquad i = 1 \text{ to } n+1$$

$$\frac{Z_o}{K_{i-1,i}} = \frac{\pi}{\omega_1'}\left(\frac{f_2 - f_1}{f_2 + f_1}\right)\left(\frac{1}{g_{i-1}g_i}\right)^{1/2}$$

$g_1, g_2, \cdots, g_n$ = prototype elements in farads and henries from Table II.

$$g_0 = \frac{\pi}{\omega_1'}\left(\frac{f_2 - f_1}{f_2 + f_1}\right)$$

$$g_{n+1} = \frac{\pi}{r\omega_1'}\left(\frac{f_2 - f_1}{f_2 + f_1}\right)$$

r = right-hand load resistance in schematic of Table II.
ω_1' = pass band edge of prototype filter.
f_1, f_2 = corresponding pass band edges of transmission-line filter.
Z_{oe_i} = even-mode characteristic impedance with respect to ground of each conductor in ith section.
Z_{oo_i} = odd-mode characteristic impedance with respect to ground of each conductor in ith section.

(The dimensions for strip-line construction may be obtained as function of Z_{oe_i} and Z_{oo_i} from Cohn.[4])

sertion-loss function in the pass and stop bands. This may be done with the aid of the insertion-loss formulas in Table II and the following approximate relationship between the frequency scales of the prototype filter and the parallel-coupled-resonator filter:

$$\omega' = 2\omega_1'\left(\frac{f - f_0}{f_2 - f_1}\right). \tag{1}$$

The exact response curves and the bandwidth-error curve (which appear later) show how response errors are introduced when the relative bandwidth exceeds 0.1. The bandwidth error may be reduced by preadjusting the design with the aid of these curves.

Next, the element values $g_1, g_2, \cdots, g_n$ may be computed, and in terms of these the Z_{oe} and Z_{oo} values of each of the $n+1$ sections. Then, the transmission-line dimensions in each section should be designed to yield these characteristic impedances. This may be done easily in the case of thin strip conductors, and with somewhat more difficulty in the case of thick strips, by means of the available graphs and formulas.[4] It is noted that the strip widths and spacings will, in general, differ from section to section, and hence the width of the resonators will not be constant, as shown in Fig. 2. This variation in width is necessary to compensate for differences in the coupling gap. The amount of variation decreases as the bandwidth is reduced and, except perhaps for the first and last section, the variation is negligible in the case of bandwidths less than a few per cent.

One further important step in the filter design is to alter the length of the resonators to compensate for fringing capacitance at their ends. This may be done

65

TABLE II

PROTOTYPE LOW-PASS FILTER AND ITS DESIGN EQUATIONS FOR MAXIMALLY-FLAT AND TCHEBYCHEFF RESPONSE

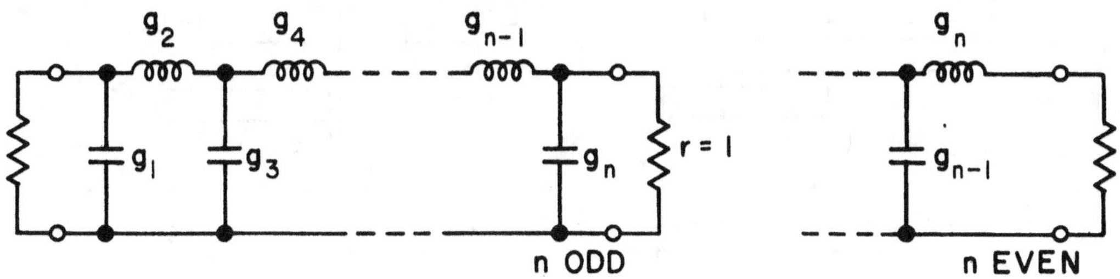

Maximally-Flat Response

$r = 1$ for all n

$$g_k = 2 \sin\left[\frac{(2k-1)\pi}{2n}\right], \qquad k = 1, 2, \cdots, n$$

$$A = 10 \log_{10} (1 + \omega'^{2n})\text{db}$$

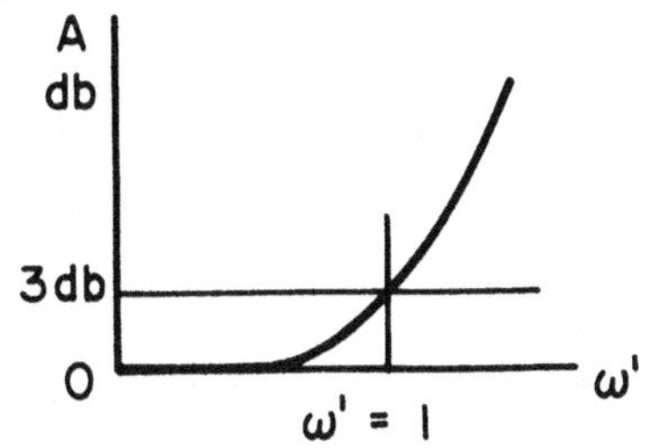

Tchebycheff Response

$r = 1$ for n odd, $\qquad r = \tanh^2 (\beta/4)$ for n even

$$g_1 = 2a_1/\gamma$$

$$g_k = \frac{4a_{k-1}a_k}{b_{k-1}g_{k-1}}, \qquad k = 2, 3, \cdots, n$$

check: $g_n = g_1 r$

$$a_k = \sin\left[\frac{(2k-1)\pi}{2n}\right], \qquad k = 1, 2, \cdots, n$$

$$b_k = \gamma^2 + \sin^2\left(\frac{k\pi}{n}\right), \qquad k = 1, 2, \cdots, n$$

$$\beta = \log_e\left(\coth\frac{A_m}{17.37}\right), \qquad A_m \text{ in db}$$

$$\gamma = \sinh\left(\frac{\beta}{2n}\right)$$

$$A = 10 \log_{10} [1 + (10^{A_m/10} - 1) \cos^2 (n \cos^{-1} \omega')]\text{db}, \qquad \omega' \leq 1$$

$$A = 10 \log_{10} [1 + (10^{A_m/10} - 1) \cosh^2 (n \cosh^{-1} \omega')]\text{db}, \qquad \omega' \geq 1$$

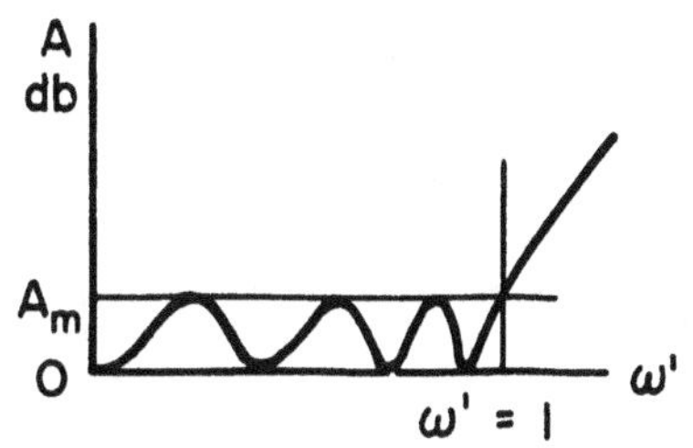

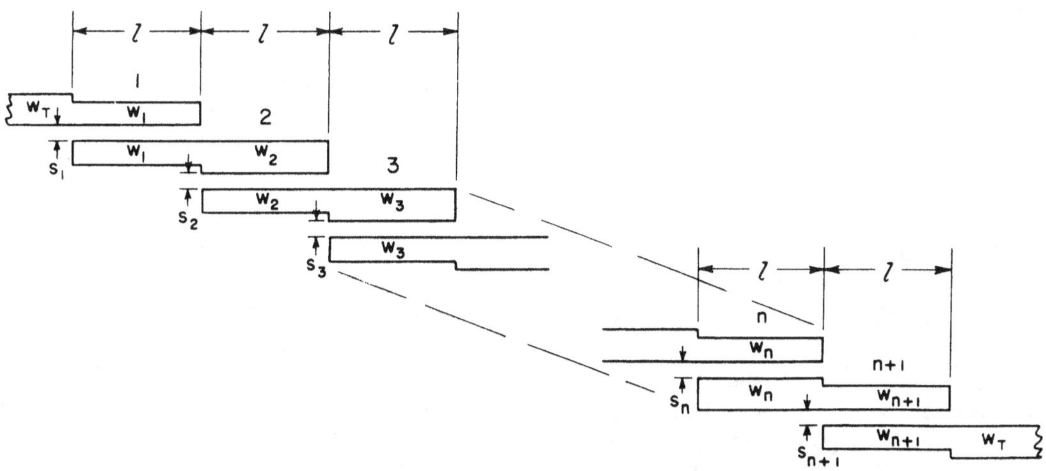

Fig. 2—Basic dimensions of the strip-line parallel-coupled-resonator filter.

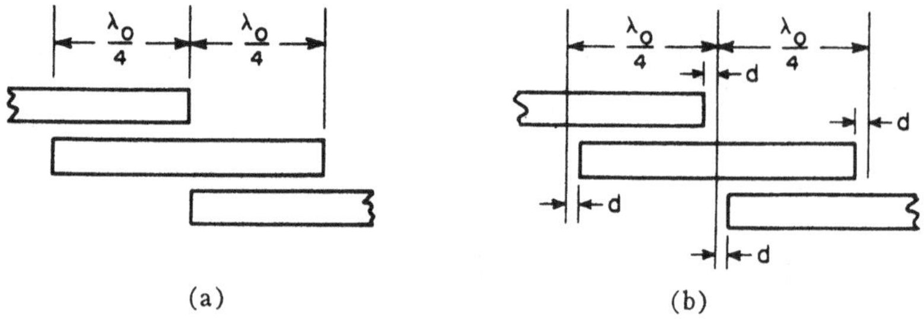

Fig. 3—(a) Strip-resonator design neglecting fringing capacitance at ends. (b) Suggested compensation for fringing capacitance at ends of resonators.

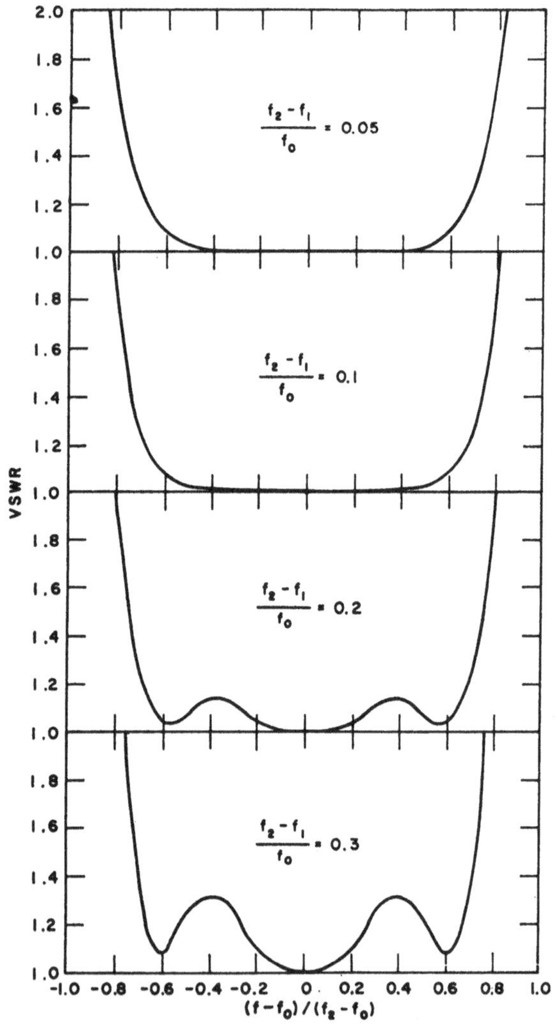

Fig. 4—VSWR curves of six-resonator, parallel-coupled filter designed for maximally-flat response.

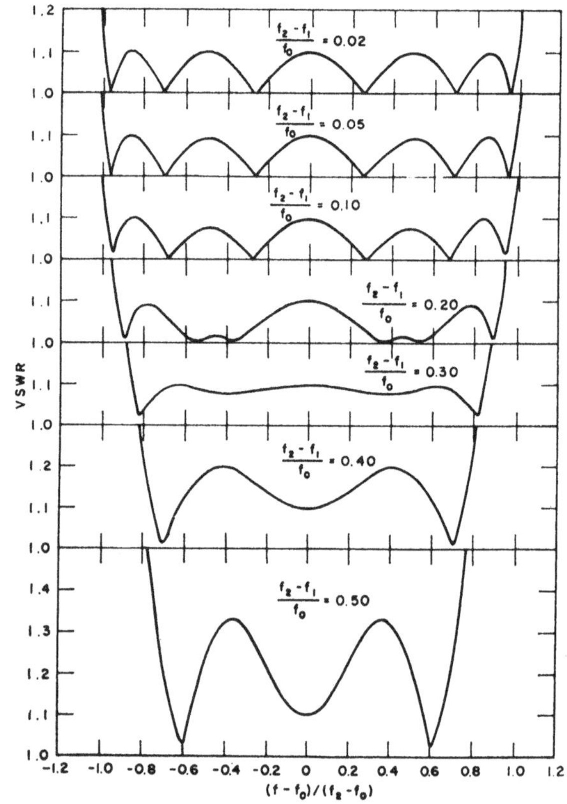

Fig. 5—VSWR curves of six-resonator, parallel-coupled filter designed for equal-ripple response—vswr = 1.10.

as shown in Fig. 3(b), where each end is cut back by the average length, d. The dimension d would be expected to be somewhat less than $(C_f'/\epsilon)\cdot(b/2)$, where C_f'/ϵ is a quantity plotted elsewhere as a function of the strip-thickness-to-plate-spacing ratio, t/b.[5] Measurements on the experimental filter described below indicate that d should be approximately 0.75 $(C_f'/\epsilon)\cdot(b/2)$. For very thin strips, this would be 0.75 $(0.220b) = 0.165b$. In wide-bandwidth filters an error in d would be unimportant, while in narrow-bandwidth filters the error would be cancelled by the tuning adjustments that are already needed to overcome the effects of constructional tolerances. This tuning may be done, for example, by means of screws or dielectric slugs in high-electric-field regions of the resonators.

Verification of Design Accuracy

Because of the approximations necessary in the derivation of the design formulas, their accuracy was checked theoretically. This was done by exact computation of the insertion-loss and vswr responses of the actual transmission-line filter networks specified by the formulas for various bandwidths and for either maximally flat or Tchebycheff response. The computations were performed on an electronic computer by a matrix-

[5] S. B. Cohn, "Problems in strip transmission line," IRE Trans., Vol. MTT-3, pp. 119–126; March, 1955.

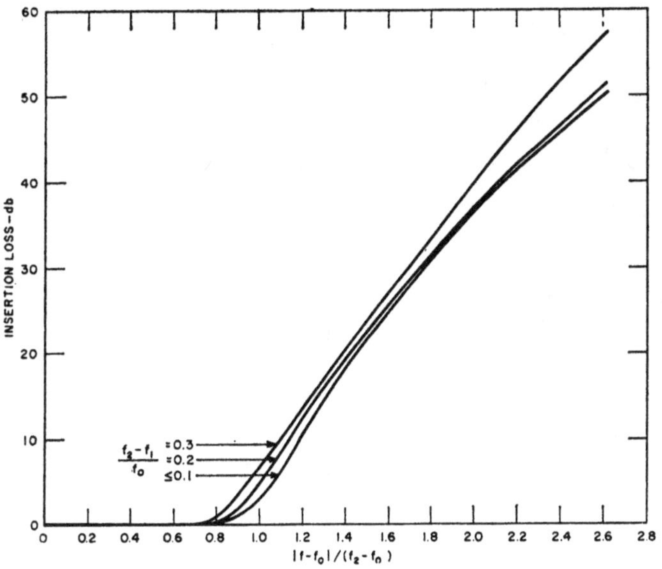

Fig. 6—Insertion-loss curves of six-resonator, parallel-coupled filter designed for maximally-flat response.

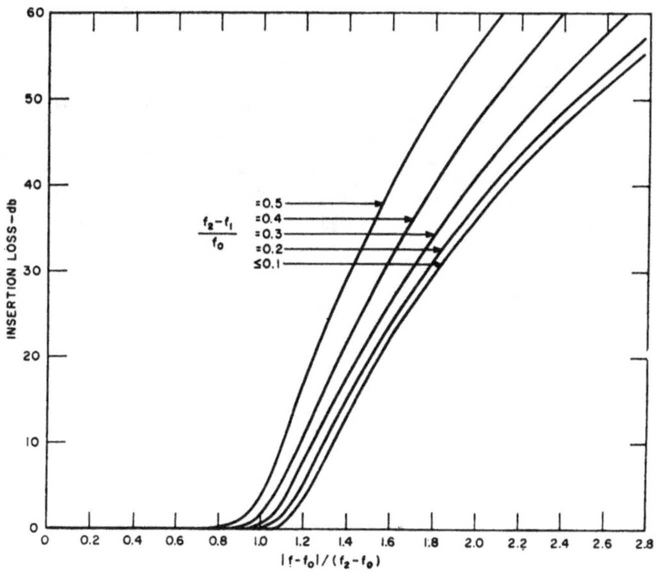

Fig. 7—Insertion-loss curves of six-resonator, parallel-coupled filter designed for equal-ripple response—vswr = 1.10.

multiplication method.[6] Because the filter is symmetrical, it is sufficient to compute the matrix of half the filter. The insertion loss and vswr then were computed as functions of the elements of this matrix.

The pass band vswr curves for six-resonator maximally flat filters of various bandwidths are shown in Fig. 4. The relative bandwidth is defined as

$$W = \frac{f_2 - f_1}{f_0}, \tag{2}$$

where $f_0 = (f_1 + f_2)/2$, and f_1 and f_2 are the 3-db points of the response curve. The response is seen to be truly maximally flat for relative bandwidths up to 0.1, while at 0.2 the deviation from maximally flat is slight. At 0.3, however, the deviation would be serious in the more critical applications. The corresponding curves are shown in Fig. 5 for six-resonator filters designed to have an equal ripple vswr of 1.10 in the pass band. In this case, f_1 and f_2 are the pass band limits for the equal ripple-level vswr of 1.10. The desired response is obtained very accurately for relative bandwidths up to 0.05, and gradually deteriorates as the bandwidth is increased further. In spite of this deterioration, the pass band vswr limit of 1.10 is not exceeded for relative bandwidths up to 0.3, and even at 0.4 or 0.5 the response is adequate for many applications. But, it is important to note that these conclusions are drawn from specific cases considered, and may vary somewhat with other numbers of resonators or with other equal ripple levels.

The insertion-loss curves for the above cases are plotted against a normalized frequency scale in Fig. 6 and Fig. 7. Because the response is symmetrical with

[6] P. I. Richards, "Applications of matrix algebra to filter theory," Proc. IRE, vol. 34, pp. 145P–150P; March, 1946.

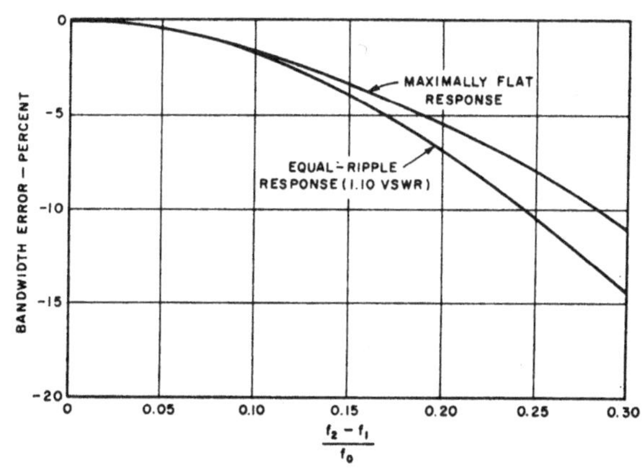

Fig. 8—Bandwidth error vs relative bandwidth for six-resonator parallel-coupled filter. Error is expressed as a percentage of the *bandwidth;* e.g., at $(f_2-f_1)/f_0=0.10$, corrected bandwidth $=0.098$.

frequency, only half of each pass band is shown. In both figures, the insertion loss is virtually identical with that of the prototype function for relative bandwidths up to 0.1, and even for greater bandwidths, the deviation is quite small.

In addition to the errors in response-curve shape that develop as the bandwidth is increased, the predicted bandwidth is slightly in error. In Fig. 8 the bandwidth error in per cent is plotted vs relative bandwidth for the cases considered above. It is seen that the discrepancy does not exceed 2 per cent of the bandwidth for relative bandwidths up to 0.1, and is only about 6 per cent at 0.2. The actual bandwidth of the filter appears to be always less than the value assumed in the design, and therefore, in the case of the wider bandwidths, it would be desirable in calculating the parameters of a given filter to use a somewhat larger bandwidth than actually is required. Fig. 8 may be used as an approximate guide in selecting the bandwidth design value.

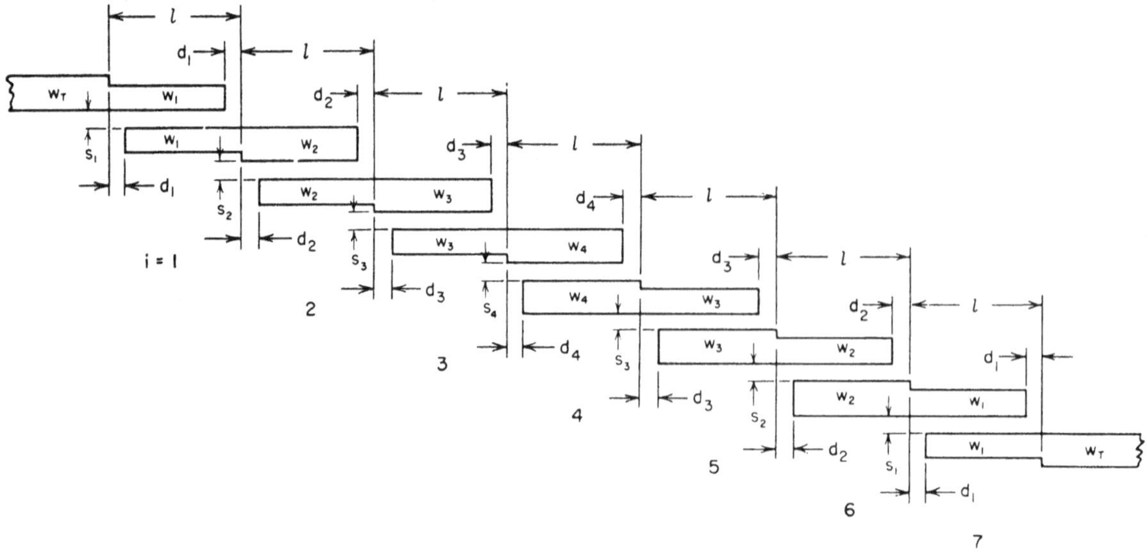

Fig. 9—Layout of parallel-coupled-resonator filter.

TABLE III

DESIGN PARAMETERS FOR EXPERIMENTAL PARALLEL-COUPLED STRIP-LINE-RESONATOR FILTER

i	$Z_o / K_{i-1,i}$	Z_{oe_i}	Z_{oo_i}	w_i	s_i	d_i
1	0.449	82.5 ohms	37.6 ohms	0.236 inch	0.021 inch	0.073 inch
2	0.1529	58.8 ohms	43.5 ohms	0.346 inch	0.110 inch	0.084 inch
3	0.1038	55.7 ohms	45.3 ohms	0.360 inch	0.158 inch	0.085 inch
4	0.0976	55.4 ohms	45.6 ohms	0.361 inch	0.163 inch	0.085 inch

DESIGN OF EXPERIMENTAL FILTER

A strip-line parallel-coupled-resonator filter has been constructed and tested in order to demonstrate the feasibility of this type of design. As shown in Fig. 9, the filter has six resonators. The two ground planes were spaced 0.5 inch apart by a polystyrene dielectric. The strips are of copper foil, 0.0017 inch thick. The filter was designed for 10-per cent bandwidth, centered at 1200 mc, and an equal ripple pass band vswr of 1.10. The exact theoretical vswr and the insertion-loss curves of the actual transmission-line filter network, as calculated by an electronic computer, are shown in Fig. 5 and Fig. 7. In each figure the appropriate curve is labeled $(f_2-f_1)/f_0=0.1$. The details of the design procedure and a discussion of the experimental results are given below.

Fig. 9 shows that the filter has seven sections, but because of the symmetry of the filter, it is necessary to compute the parameters of only the first four sections. The first step in the design procedure is to compute the element values g_1, g_2, g_3, and g_4 of the prototype seven-element, low-pass filter. This is done by means of the formulas for Tchebycheff response in Table II, with $n=7$ and $A_m=0.00986$ db (which corresponds to an input vswr of 1.10). The resulting values are

$$g_1 = 0.77968, \quad g_2 = 1.35921, \quad g_3 = 1.68800, \quad g_4 = 1.53454.$$

Next, the quantity $Z_o/K_{i-1,i}$ and the even- and odd-mode characteristic impedances Z_{oe_i} and Z_{oo_i} are computed from the formulas in Table I. In this calculation, $(f_2-f_1)/f_0$ is set equal to 0.1, and Z_o to 50 ohms. Then the strip widths w_i and separations s_i of the various sections are obtained from the nomograms of a previous paper,[4] with ϵ_r equal to 2.55, b equal to 0.5 inch, and strip thickness assumed to be zero. Finally, the resonators are shortened at their ends by d_i to compensate for fringing capacitance. The various quantities referred to above are contained in Table III.[7]

A further quantity required is the width w_T of the terminating 50-ohm strips. This may be obtained from a graph of characteristic impedance vs strip width.[5,8] The result is $w_T/b=0.744$ or $w_T=0.372$ inch. The section length l is a quarter-wavelength in the dielectric, and hence equals 1.540 inch.

Photographs of the completed filter are shown in Fig. 10. The structure is a sandwich of copper foil separated by a pair of polystyrene plates each machined to have flat surfaces and a thickness of 0.250 inch. Thus, the total ground-plane spacing is one-half inch. The copper foil was cemented to the polystyrene plates with Dow

[7] The values of d_2 in Table III were computed from an approximate formula, and are quite close to the value $0.165b$ mentioned above. Because the formula is rather complex, and its assumptions are as yet unproved, it is not given here.

[8] S. B. Cohn, "Characteristic impedance of the shielded-strip transmission line," IRE TRANS., vol. MTT-2, pp. 52–57; July, 1954.

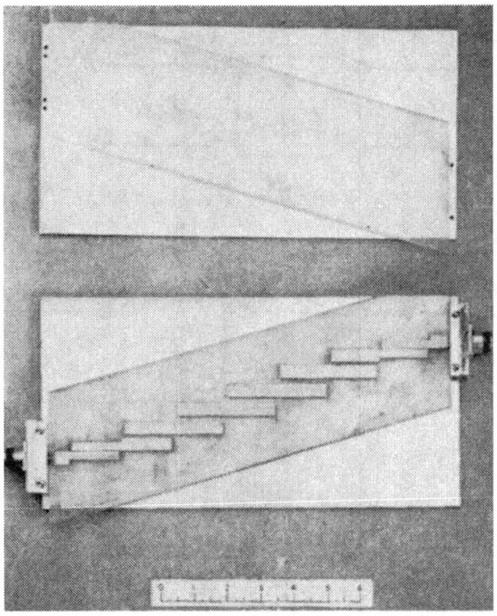

(a)

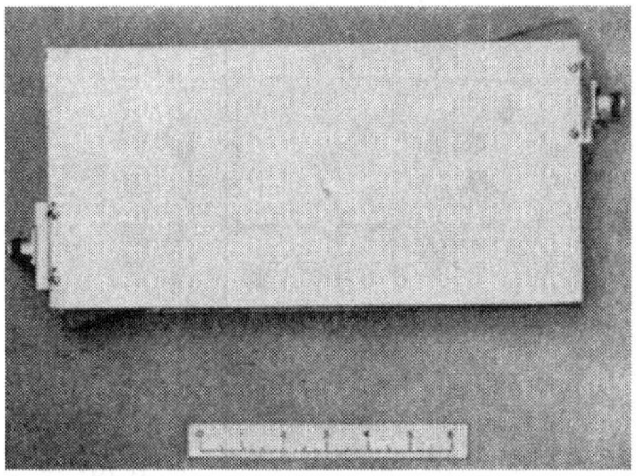

(b)

Fig. 10—Photographic views of parallel-coupled resonator filter: (a) with upper plate removed; (b) completely assembled.

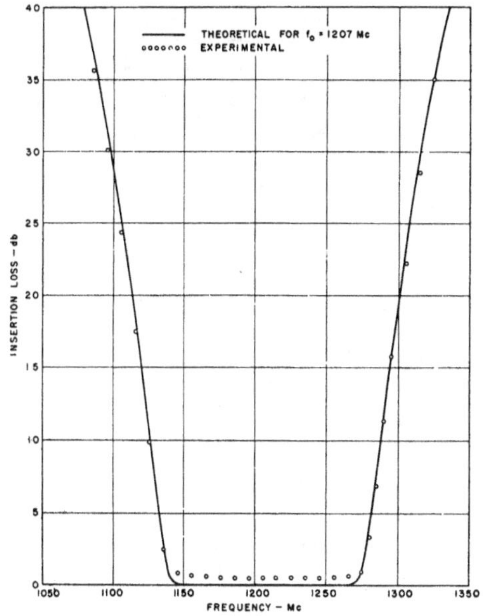

Fig. 11—Insertion loss of parallel-coupled-resonator filter.

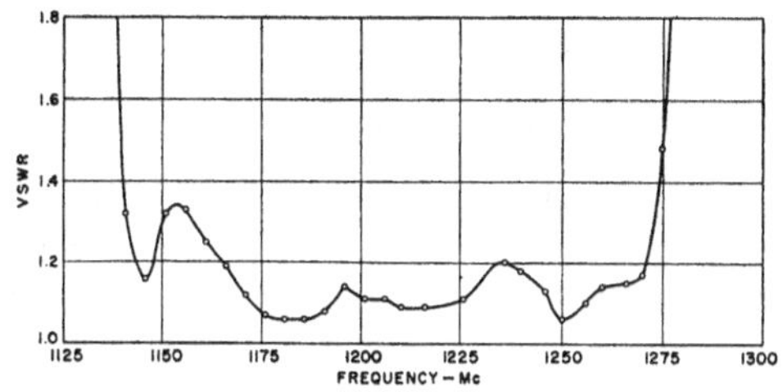

Fig. 12—Measured vswr of parallel-coupled-resonator filter.

Corning XC-271 adhesive, which was applied to the copper and air dried for 45 minutes before pressing the foil onto the polystyrene. The strip circuit then was cut on one surface with a sharp knife and the unwanted foil peeled off.

Data for Experimental Filter

A preliminary model of the parallel-coupled resonator filter was tested first with the resonator ends uncompensated; *i.e.*, with $d_i = 0$. The center frequency was lower than the design value by 4.4 per cent. Then, the open-circuited strip ends all were cut back uniformly a distance $d = 0.220b$, the value of $(C_f'/\epsilon)/(b/2)$ for very thin strips, and the center frequency was higher than the design value by 1.7 per cent. Finally, the filter was reconstructed with the values of d_i given in Table III. The center frequency of this model is 1207 mc, rather than the design value of 1200 mc—an error of only 0.6 per cent. In the three cases the bandwidth and response curves are quite similar. Although the best pass band response was obtained with the last case, it may have been due to more accurate construction.

Fig. 11 shows a comparison between the measured insertion loss of the filter and the exact theoretical curve for the transmission-line network computed for a center frequency of 1207 mc and neglecting dissipation. Above 2 db, the agreement is excellent. The bandwidth error is seen to be only about 1.0 per cent of the bandwidth. In the pass band, however, disagreement is inevitable because of the finite Q of the strip-line resonators. Nevertheless, the measured pass band insertion loss is quite uniform at 0.5 to 0.7 db.

The input vswr of the filter measured with a 50-ohm termination on the output is shown in Fig. 12. Although the vswr exceeds the theoretical ripple level of 1.10, it would be quite acceptable for most applications. The highest vswr peak in the pass band occurs near the low frequency end, where it reaches a value of 1.33, while the vswr elsewhere in the pass band is under 1.20. It is believed that the vswr could be made to conform more closely to theory if tuning adjustments were provided for the resonators. These adjustments also could serve to set the center frequency of the filter exactly on the desired value.

Derivation of Design Formulas

The analysis of the parallel-coupled-resonator filter is based upon the characteristics of the individual section of Fig. 13(a). The image impedance Z_I and image phase shift β of this section are given in a paper by Jones and

69

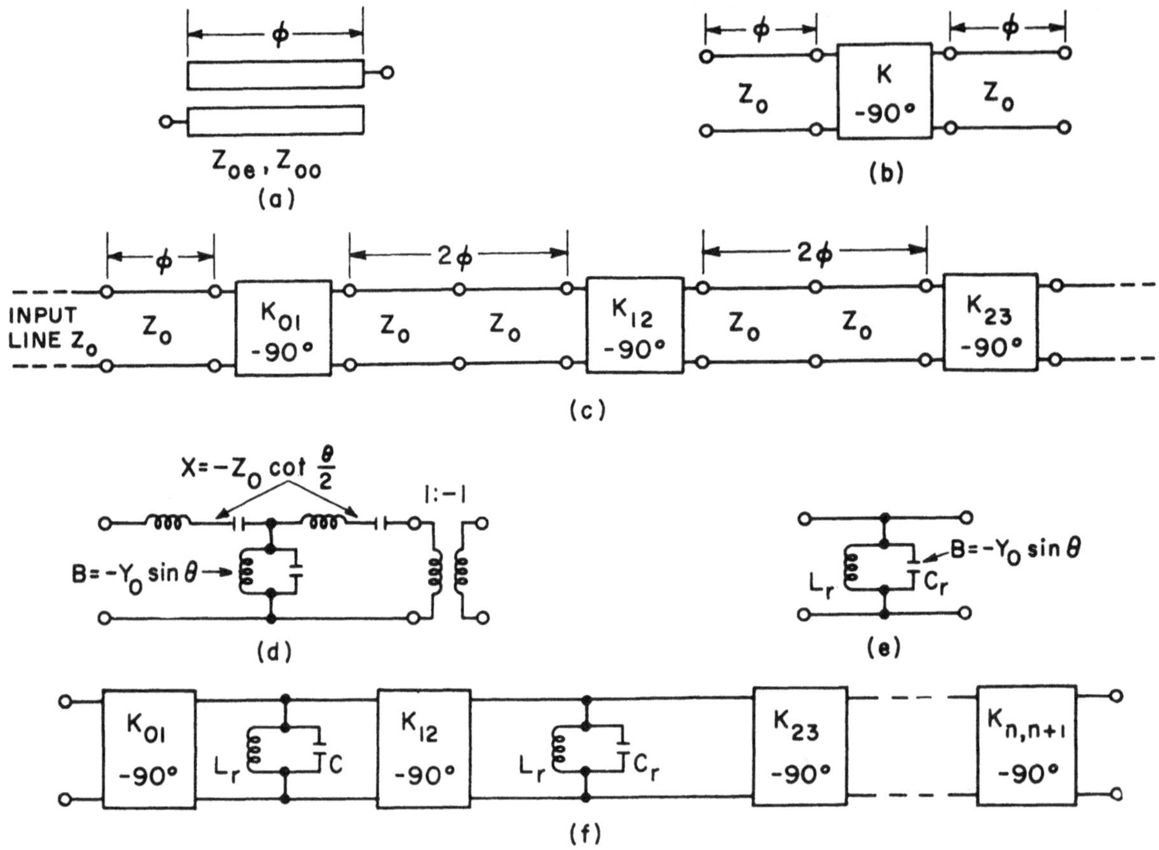

Fig. 13—Development of equivalent circuit of the parallel-coupled-resonator filter.

70

Bolljahn[9] as follows

$$Z_I = \frac{[(Z_{oe} - Z_{oo})^2 - (Z_{oe} + Z_{oo})^2 \cos^2 \phi]^{1/2}}{2 \sin \phi} \tag{3}$$

$$\cos \beta = \left(\frac{Z_{oe} + Z_{oo}}{Z_{oe} - Z_{oo}}\right) \cos \phi, \tag{4}$$

where ϕ is the electrical length of the coupled transmission lines and Z_{oe} and Z_{oo} are the characteristic impedances of each conductor with respect to ground for the even and odd modes, respectively.

It is now shown that the filter section of Fig. 13(b) is approximately equivalent to that of Fig. 13(a). The box represents an ideal impedance inverter having a constant image impedance, K, and a phase shift of $-90°$ at all frequencies. The image impedance, Z_I, and image phase shift, β, of the section will be derived by means of the $ABCD$ matrix parameters.[5,10]

The $ABCD$ matrix of a transmission line of length ϕ and characteristic impedance Z_o is

$$\begin{bmatrix} \cos \phi & jZ_o \sin \phi \\ \dfrac{j \sin \phi}{Z_o} & \cos \phi \end{bmatrix}.$$

[9] E. M. T. Jones and J. T. Bolljahn, "Coupled-strip-transmission-line filters and directional couplers," IRE TRANS., vol. MTT-4, pp. 75–81; April, 1956.

[10] Radio Research Laboratory Staff, "Very High Frequency Techniques," McGraw-Hill Book Co., Inc., New York, N. Y., vol. 2, ch. 26; 1947.

The matrix of the ideal inverter is obtained from this by substituting $\phi = -90°$ and $Z_o = K$, as follows

$$\begin{bmatrix} 0 & -jK \\ \dfrac{-j}{K} & 0 \end{bmatrix}.$$

Therefore, the $ABCD$ matrix of the complete filter section of Fig. 13(b) is

$$\begin{bmatrix} A & B \\ C & D \end{bmatrix} = \begin{bmatrix} \cos \phi & jZ_o \sin \phi \\ \dfrac{j \sin \phi}{Z_o} & \cos \phi \end{bmatrix} \times \begin{bmatrix} 0 & -jK \\ \dfrac{-j}{K} & 0 \end{bmatrix}$$

$$\times \begin{bmatrix} \cos \phi & jZ_o \sin \phi \\ \dfrac{j \sin \phi}{Z_o} & \cos \phi \end{bmatrix}$$

$$= \begin{bmatrix} \left(\dfrac{K}{Z_o} + \dfrac{Z_o}{K}\right) \sin \phi \cos \phi & j\left(\dfrac{Z_o^2}{K} \sin^2 \phi - K \cos^2 \phi\right) \\ j\left(\dfrac{K}{Z_o^2} \sin^2 \phi - \dfrac{\cos^2 \phi}{K}\right) & \left(\dfrac{K}{Z_o} + \dfrac{Z_o}{K}\right) \sin \phi \cos \phi \end{bmatrix}.$$

The image parameters are related to the matrix elements by $Z_I = \sqrt{B/C}$ and $\cos \beta = A$, and, hence,

$$Z_I = Z_o \left\{ \frac{\frac{Z_o}{K}\sin^2\phi - \frac{K}{Z_o}\cos^2\phi}{\frac{K}{Z_o}\sin^2\phi - \frac{Z_o}{K}\cos^2\phi} \right\}^{1/2} = Z_o \left\{ \frac{\frac{Z_o}{K} - \left(\frac{K}{Z_o}+\frac{Z_o}{K}\right)\cos^2\phi}{\left(\frac{K}{Z_o}+\frac{Z_o}{K}\right)\sin^2\phi - \frac{Z_o}{K}} \right\}^{1/2}$$

$$= \frac{1}{2\sin\phi}\left\{ \frac{\frac{4Z_o^3}{K}}{\frac{K}{Z_o}+\frac{Z_o}{K}-\frac{Z_o}{K\sin^2\phi}} - \frac{4Z_o^2\left(\frac{K}{Z_o}+\frac{Z_o}{K}\right)\cos^2\phi}{\frac{K}{Z_o}+\frac{Z_o}{K}-\frac{Z_o}{K\sin^2\phi}} \right\}^{1/2} \tag{5}$$

$$\cos\beta = \left(\frac{K}{Z_o}+\frac{Z_o}{K}\right)\sin\phi\cos\phi. \tag{6}$$

A comparison between (3) and (5), and between (4) and (6), shows that the sections of Fig. 13(a) and 13(b) would be electrically equivalent if the following conditions were met,

$$\left(\frac{K}{Z_o}+\frac{Z_o}{K}\right)\sin\phi = \frac{Z_{oe}+Z_{oo}}{Z_{oe}-Z_{oo}} \tag{7}$$

$$\frac{\frac{4Z_o^3}{K}}{\frac{K}{Z_o}+\frac{Z_o}{K}-\frac{Z_o}{K\sin^2\phi}} = (Z_{oe}-Z_{oo})^2 \tag{8}$$

$$\frac{4Z_o^2\left(\frac{K}{Z_0}+\frac{Z_o}{K}\right)}{\frac{K}{Z_o}+\frac{Z_o}{K}-\frac{Z_o}{K\sin^2\phi}} = (Z_{oe}+Z_{oo})^2. \tag{9}$$

The presence of $\sin\phi$ in the left side of each of the above makes a frequency-independent equality impossible. However, $\sin\phi$ is stationary in the vicinity of $\phi = 90°$, and hence may be replaced by unity with negligible error over a moderate bandwidth. Thus,

$$\frac{K}{Z_o}+\frac{Z_o}{K} = \frac{Z_{oe}+Z_{oo}}{Z_{oe}-Z_{oo}} \tag{10}$$

$$\frac{2Z_o^2}{K} = Z_{oe}-Z_{oo} \tag{11}$$

$$\frac{4Z_o^3}{K}\left(\frac{K}{Z_o}+\frac{Z_o}{K}\right) = (Z_{oe}+Z_{oo})^2. \tag{12}$$

From these equations, formulas for Z_{oe}/Z_o and Z_{oo}/Z_o as functions of Z_o/K may be found. Only two of the three equations are needed for this purpose, and since the three do not form a consistent set, a judicious choice must be made among them. Examination of (3) to (6) indicates that the third relation is the least important of the three. Therefore, (10) and (11) are solved simultaneously to yield

$$\frac{Z_{oe}}{Z_o} = 1 + \frac{Z_o}{K} + \frac{Z_o^2}{K^2} \tag{13}$$

$$\frac{Z_{oo}}{Z_o} = 1 - \frac{Z_o}{K} + \frac{Z_o^2}{K^2}. \tag{14}$$

It may now be noted that, for narrow bandwidth, $Z_o/K \ll 1$ and, therefore, (12) is satisfied approximately by (13) and (14). Thus the effect of the approximations involved in relating the sections of Fig. 13(a) and 13(b) is negligible for narrow bandwidth, but increases in importance as the bandwidth is increased.

When the individual sections of the filter are assembled, it is seen that the original parallel-coupled-resonator filter of Fig. 1(b) or 1(c) is approximately equivalent to the circuit of Fig. 13(c), in which transmission lines approximately $\lambda_o/2$ long are separated by inverters. A lumped-constant equivalent of a line of length θ is shown in Fig. 13(d). This exact equivalent circuit is particularly convenient for θ near 180°. In Fig. 13(e) the equivalent circuit has been simplified by omitting the phase-reversing transformer which has no effect on the insertion-loss response. Also, the series reactances of Fig. 13(f) are small near $\theta = 180°$ and are negligible in comparison with the high characteristic impedances of the inverting elements. Thus, the circuit of Fig. 13(f) is approximately equivalent to that of Fig. 13(c), and hence to the original parallel-coupled-resonator filter.

The analysis will be shortened at this point by comparing the filter circuit of Fig. 13(f) with that shown in Fig. 11(c) of a previous paper.[3] They are seen to be the same, except that the former contains parallel-resonant arms while the latter contains series-resonant arms. Hence, the circuits are duals, and the formulas obtained for the latter may be used in the present analysis if K is replaced by $1/K$, Z_o by $1/Z_o$, and if L and C are interchanged. Thus,

$$\frac{Z_o}{K_{i,i+1}} = \frac{\pi W}{2\omega_1'\sqrt{g_i g_{i+1}}}, \qquad i = 1 \text{ to } n-1, \tag{15}$$

where W is the relative bandwidth defined by (2). For the first and last section of the filter,

$$\frac{Z_o}{K_{01}} = \sqrt{\frac{\pi W}{2\omega_1' g_1}}, \qquad \frac{Z_o}{K_{n,n+1}} = \sqrt{\frac{\pi W r}{2\omega_1' g_n}}. \tag{16}$$

The formulas in Table I follow directly from (13) through (16).

INTERDIGITAL FILTERS III

Interdigital periodic structures had been used as slow-wave periodic structures, for example in microwave vacuum tubes. Bolljahn and Matthaei[8] analyzed such periodic structures and Matthaei (p. 75) Modified the periodicity to make a filter.

The interdigital filter has proved to be very popular. It is compact, has a convenient (almost square) shape, makes good use of the available space, and can be designed for both narrow and wide bandwidths. It has good Q (and therefore low loss) for its size. The two design procedures of Matthaei are approximate, one good for relatively narrow bandwidths of up to 20 or 30 percent (when the two end rods are shorted to ground) and one for wide bandwidths of between about 30 and 70 percent (when the two end rods are open circuited). Dishal (p. 88) has given a simple design procedure for relatively small bandwidths, good up to at least 10 percent. Wenzel (p. 91) has given an exact design theory, and Horton and Wenzel have recently computed tables up to 23 sections, but their work is still unpublished.

Robinson[9] has modified the interdigital filter by capacitively loading the open ends of the rods. This shortens the rods (to below quarter-wave at midband) and reduces the size of the filter. It increases the insertion loss, but it pushed the first spurious pass band from around the third harmonic (in the conventional, unloaded interdigital filter), to higher frequencies (for the capacitively loaded interdigital filter); The spurious passband center frequency is approximately inversely proportional to the rod length. Robinson demonstrated the design of a capacitively loaded interdigital filter in which each rod length was one-eighth wavelength at midband and the passband bandwidth was about 60 percent. It is interesting to compare this filter with the combline filter of Section IV. In brief, the combline filter is easier to construct and tune because the rods are all grounded on the same side, but it is generally narrow band.* The capacitively loaded interdigital filter, on the other hand, is good to large bandwidths. Both filters can be made about the same size (smaller than a regular interdigital filter) and with about the same insertion loss (larger than a regular interdigital filter).

One important paper not reproduced here is by E.G. Cristal,[10] which will be found in Section I, p. 13 in the companion volume on PARALLEL COUPLED LINES AND DIRECTIONAL COUPLERS. Two other papers of general interest on interdigital filters are by Riblet[11] and by Rhodes.[12] The former offers a fresh viewpoint; the latter is a rigorous mathematical treatment.

A recent paper by Sato and Cristal[13] gives equivalent circuits for coupled lines. There is a one-to-one correspondence between coupling terms of the matrix and elements in the equivalent circuit. In an example toward the end of the paper, the interdigital line is treated briefly.

A general remark that applies to all filters, microwave or otherwise, concerns the lining-up procedure. It is similar for all filters, and the reader is particularly referred to Matthaei, Young and Jones[1], Chapter II, and the work of Dishal,[14] Also his Ref. 2, p. 90.

* Wenzel has recently designed a combline filter with 40-percent bandwidth, and has also designed capacitively loaded interdigital filters.

Reprinted from IRE TRANSACTIONS
ON MICROWAVE THEORY AND TECHNIQUES
Volume MTT-10, Number 6, November, 1962

PRINTED IN THE U.S.A.

Interdigital Band-Pass Filters*

GEORGE L. MATTHAEI†, MEMBER, IRE

Summary—**The design of band-pass filters using interdigital arrays of resonator line elements between parallel ground planes is discussed. Two approximate design procedures are described, both of which permit design directly from lumped element, low-pass, prototype filters. Both design procedures will work for either narrow- or wide-band filters, but one procedure gives more practical dimensions for filters having wide bandwidths (such as an octave), while the other gives more practical dimensions for filters having narrow or moderate bandwidth. The resulting filters are very compact, have relatively noncritical manufacturing tolerances, and strong stop bands with the second pass band centered at three times the center frequency of the first pass band. The dimensions and measured performance curves are presented for a 10 per cent bandwidth design and an octave bandwidth design.**

I. Introduction

INTERDIGITAL line structures have in the past been regarded as of interest mainly for use as slow-wave structures. (See, for example, the work of Butcher,[1] Fletcher,[2] and Leblond and Mourier.[3]) However, a recent study[4] has shown that interdigital line structures also have very interesting band-pass filter properties. In that study, various image-impedance and image-propagation phase properties of interdigital line structures were determined, and the results were verified by experimental tests on an interdigital line structure. In this present discussion a different point of view is utilized to obtain approximate design equations for interdigital filters with specified pass-band and cutoff characteristics.

Fig. 1 shows one type of interdigital filter to be discussed. The structure, as shown, consists of TEM-mode strip-line resonators between parallel ground planes. Each resonator element is a quarter-wavelength long at the midband frequency and is short-circuited at one end and open-circuited at the other end. Coupling is achieved by way of the fields fringing between adjacent resonator elements. The design equations to be presented for this case (for which the terminating lines are short-circuited) give filter structural dimensions that will be most practical when the filter is of narrow or moderate bandwidth (say, 30 per cent bandwidth or less). In this structure each line element serves as a resonator, except for the input and output line elements which have an impedance-matching function.

* Received April 23, 1962; revised manuscript received July 30, 1962. This research was sponsored by the U. S. Army Signal Research and Development Laboratory, Ft. Monmouth, N. J., under Contract DA 36-039 SC-87398.

† Stanford Research Institute, Menlo Park, Calif.

[1] P. N. Butcher, "The coupling impedance of tape structures," *Proc. IEE* (*London*), *B*, vol. 104, pp. 177–187; March, 1957.

[2] R. C. Fletcher, "A broad-band interdigital circuit for use in traveling-wave-type amplifiers," PROC. IRE, vol. 40, pp. 951–958; August, 1952.

[3] A. Leblond and G. Mourier, "Etudes lignes a barreaux a structure periodique pour tubes électroniques UHF," *Annales de Radioélectricité*, pt. 1, April, 1954; pt. 2, October, 1955.

[4] J. T. Bolljahn and G. L. Matthaei, "A study of the phase and filter properties of arrays of parallel conductors between ground planes," PROC. IRE, vol. 50, pp. 299–311; March, 1962.

Fig. 2 shows another type of interdigital filter, it differs from the filter in Fig. 1 in that the terminating lines are open-circuited instead of short-circuited. The design equations to be given for this type of interdigital filter will give filter structural dimensions that are most practical when the filter is of moderate to wide bandwidth (say, around 30 per cent bandwidth or greater). In this case all of the line elements serve as resonators, including the open-circuited terminating line element at each end.

Interdigital band-pass filters have a number of very attractive features.

1) They are very compact.
2) The tolerances required in their manufacture are relatively relaxed as a result of the relatively large spacings between resonator elements.
3) The second pass band is centered at three times the center frequency of the first pass band, and there is no possibility of spurious responses in between. (Note that for filters with half-wavelength, parallel-coupled resonators[5,6] even the slightest mistuning will result in narrow spurious pass bands at twice the frequency of the first pass-band center.)

76

4) The rates of cutoff and the strength of the stop bands are enhanced by multiple-order poles of attenuation at dc and at even multiples of the center frequency of the first pass band.
5) These filters can be fabricated in structural forms which are self-supporting so that dielectric material need not be used. Thus dielectric loss can be eliminated.

If all of the coupling effects are accounted for, the mathematics that describe the performance of interdigital filters such as those in Figs. 1 and 2 become quite unwieldy.[4] Since synthesizing a structure to have a prescribed response is a much more difficult problem than analyzing a given structure, and since an exact analysis of this structure is itself very tedious, the prospects of obtaining a usable exact synthesis procedure appear to be dim. Thus the synthesis procedure given here involves a number of simplifying approximations that permit straightforward, easy-to-use design calculations. Although the design formulas are approximate, the results of our trial designs indicate that they are sufficiently accurate for most practical applications.

For the benefit of readers who have filter design requirements but who are not particularly interested in filter theory as such, the design equations for interdigital filters and their practical use will be discussed first.

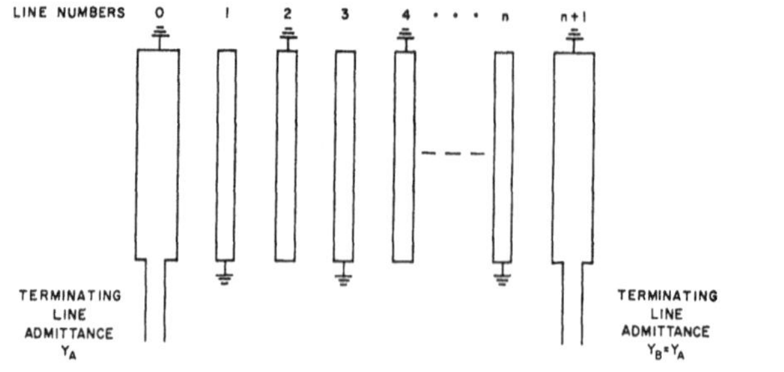

Fig. 1—Interdigital filter with short-circuited lines at the ends.

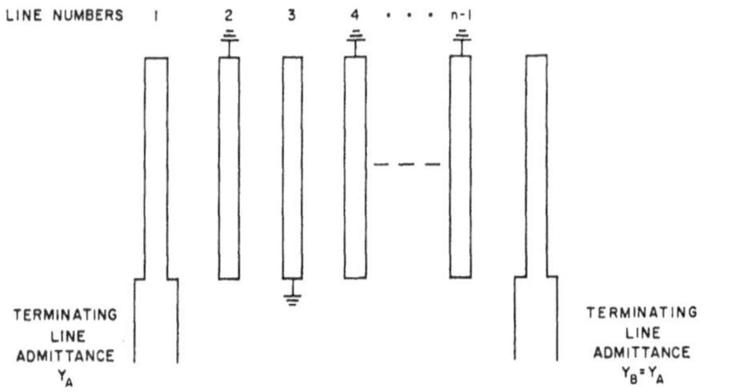

Fig. 2—Interdigital filter with open-circuited lines at the ends.

Readers who are also interested in knowing where the design equations came from will find a discussion of their derivation in Section VI.

II. Low-Pass Prototype Filters and Use of a Low-Pass-to-Band-Pass Transformation

The design procedures to be described are based on the use of low-pass prototype filter element values to give the band-pass filter its desired characteristics. Fig. 3 shows typical low-pass prototypes of the type under discussion, and it also defines the manner in which the element values g_k are to be specified for the purpose of this discussion. The left side of Fig. 4 shows a typical Chebyshev response such as can be achieved by filters of this type. Element values for prototype filters having such responses have been tabulated.[7,8] Notice that n is the number of reactive elements and that, including the resistor terminations, the element values range from g_0 to g_{n+1}. A low-pass prototype with n reactive elements leads to a band-pass filter with n resonators.

The right side of Fig. 4 shows a band-pass filter response which corresponds to the given low-pass prototype response. The band-pass filter response will have the same type of pass-band characteristic as the prototype, but the width of the band-pass filter pass band can

[5] S. B. Cohn, "Parallel-coupled transmission-line-resonator filters," IRE Trans. on Microwave Theory and Techniques, vol. MTT-6, pp. 223–231; April, 1958.

[6] G. L. Matthaei, "Design of wide-band (and narrow-band) band-pass microwave filters on the insertion loss basis," IRE Trans. on Microwave Theory and Techniques, vol. MTT-8, pp. 580–593; November, 1960.

[7] L. Weinberg, "Additional tables for design of optimum ladder networks," *J. Franklin Inst.*, vol. 246, pp. 7–23, 127–138; July and August, 1957.

[8] G. L. Matthaei, *et al.*, "Design Criteria for Microwave Filters and Coupling Structures," Stanford Res. Inst., Menlo Park, Calif., Final Rept., SRI Project 2326, Contract DA 36-039 SC-74862; January, 1961.

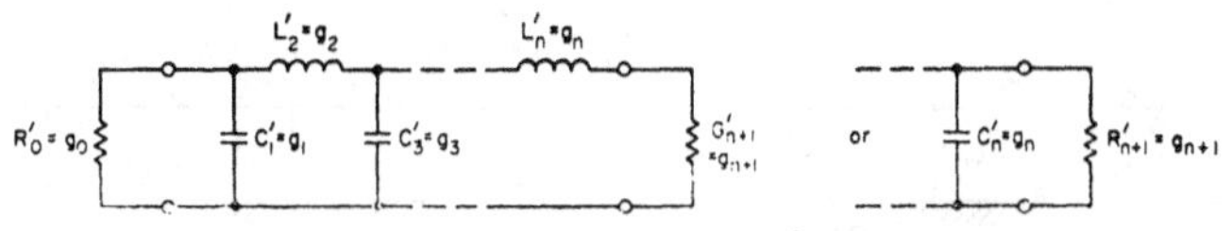

Fig. 3—Definition of prototype filter parameters $g_0, g_1, g_2, \cdots, g_n, g_{n+1}$. An additional prototype parameter ω_1' is defined in Fig. 4.

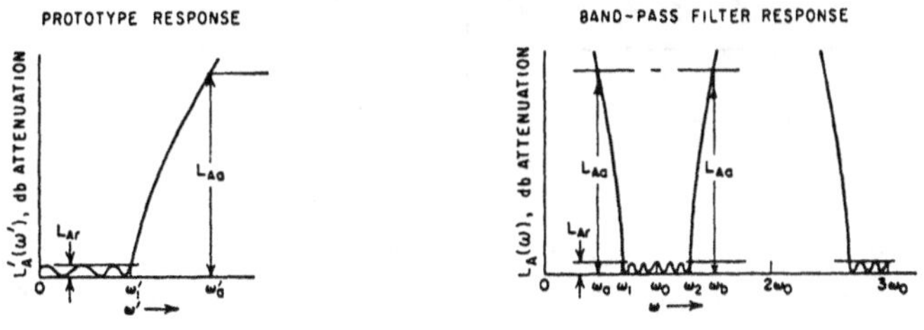

$$L_A(\omega) = L_A'(\omega') \text{ db} \qquad \omega_0 = \frac{\omega_2 + \omega_1}{2}$$

where

$$\omega' = \frac{2\omega_1'}{w} \left| \frac{\omega - \omega_0}{\omega_0} \right| \quad \text{and} \quad w = \frac{\omega_2 - \omega_1}{\omega_0}.$$

Fig. 4—Relation between the low-pass prototype response and the band-pass filter response.

be specified arbitrarily. Note that the band-pass filter response has arithmetic symmetry, and that it will repeat with pass bands centered at ω_0, $3\omega_0$, $5\omega_0$, etc. In Matthaei[6] will be found equations for determining the attenuation characteristics of maximally flat and Chebyshev low-pass prototype filters as a function of the radian frequency variable ω', for specified ω_1' and n. At the bottom of Fig. 4 will be found a low-pass-to-band-pass transformation from which the attenuation characteristic of the low-pass prototype as a function of ω' can be mapped to the band-pass filter attenuation characteristic (centered at ω_0) as a function of the band-pass filter radian frequency variable ω. Since the attenuation is the same for both the low-pass and band-pass filters at frequencies ω' and ω, respectively, which are related as given by the mapping, the band-pass filter attenuation characteristic can be predicted by use of these data. Matthaei[6] contains equations for determining the value of n required in order to achieve a given amount of attenuation, L_{Aa} db, at a given frequency, ω_a.

The low-pass-to-band-pass transformation given in Fig. 4 is an approximate one and is usually considered to be useful only for filters of narrow or moderate bandwidth. However, it appears to work well for interdigital filters even in cases of quite large bandwidth, as is verified by the nearly octave bandwidth design discussed in Section V. The parameter w, which appears in the mapping equation, is the fractional bandwidth of the filter. This parameter will also be used in the filter design equations that are about to be presented. Our experience indicates that, because the design equations are based on various approximations, the actual fractional bandwidth of the filter will be about 7 per cent smaller than the value of w used in the equations. Thus to allow for this shrinkage in bandwidth, it is suggested that a value of w, which is about 7 per cent larger than is actually desired, be used in the design equations in Sections IV and V. However, the actual desired value of w should be used in the mapping in Fig. 4.

III. Parallel-Coupled Lines

As previously mentioned, interdigital filters consist of arrays of parallel lines between ground planes. The design equations for such filters to be given herein will yield the various line capacitances per unit length for the various lines in the filter array. From these values of capacitance per unit length, the actual dimensions of the lines can be determined. Definitions for these various line capacitances will now be given, and also equations for obtaining the line dimensions from the specified capacitances. In order to relate the design methods under discussion to existing design data that are extremely useful when adapted for the design of arrays of parallel lines, we will first consider the case of two unsymmetrical parallel-coupled lines. (The case of two parallel-coupled lines will also be of interest in Section VI, where the derivation of the design equations for interdigital filters will be explained.)

Case of Two Unsymmetrical Parallel-Coupled Lines

Fig. 5 shows an unsymmetrical pair of parallel-coupled lines and various line capacitances. Note that C_a is the capacitance per unit length between Line a and ground, C_{ab} is the capacitance per unit length between Line a and Line b, and C_b is the capacitance per unit length between Line b and ground. When C_a is not equal to C_b, the two lines will have different odd- and even-mode admittances as is indicated by (a) in Table I. In terms of odd- and even-mode capacitances, for Line a[9–11]

$$C_{oo}^a = C_a + 2C_{ab}, \qquad C_{oe}^a = C_a \tag{1}$$

while for Line b

$$C_{oo}^b = C_b + 2C_{ab}, \qquad C_{oe}^b = C_b. \tag{2}$$

For symmetrical parallel-coupled lines the odd-mode impedances are simply the reciprocals of the odd-mode admittances, and analogously for the even-mode impedances and admittances. However, as can be demonstrated from (b) in Table I, this is not the case for unsymmetrical parallel-coupled lines. For unsymmetrical lines, the odd- and even-mode impedances are not simply the reciprocals of the odd- and even-mode ad-

[9] S. B. Cohn, "Shielded coupled-strip transmission line," IRE Trans. on Microwave Theory and Techniques, vol. MTT-3, pp. 29–38; October, 1955.

[10] W. J. Getsinger, "Coupled rectangular bars between parallel plates," IRE Trans. on Microwave Theory and Techniques, vol. MTT-10, pp. 65–72; January, 1962.

[11] E. M. T. Jones and J. T. Bolljahn, "Coupled-strip-transmission-line filters and directional couplers," IRE Trans. on Microwave Theory and Techniques, vol. MTT-4, pp. 75–81; April, 1956.

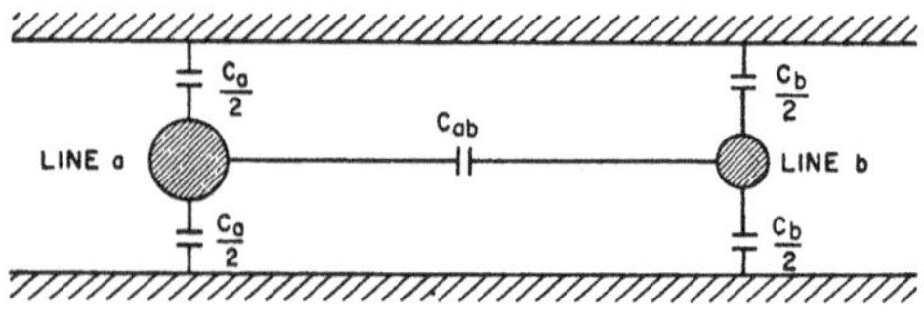

Fig. 5—An unsymmetrical pair of parallel-coupled lines (C_a, C_{ab}, and C_b are line capacitances per input length).

TABLE I

RELATIONS BETWEEN LINE ADMITTANCES, IMPEDANCES, AND CAPACITANCES PER UNIT LENGTH OF UNSYMMETRICAL PARALLEL-COUPLED LINES

v = velocity of light in medium of propagation $= 1.18 \times 10^{10}/\sqrt{\epsilon_r}$ inches/sec.

η_0 = intrinsic impedance of free space = 376.7 ohms

ϵ = dielectric constant $= 0.225\epsilon_r$ $\mu\mu f$/inch

$$Y_{oe}{}^a = vC_a, \qquad Y_{oo}{}^a = v(C_a + 2C_{ab}); \qquad Y_{oe}{}^b = vC_b, \qquad Y_{oo}{}^b = v(C_b + 2C_{ab}) \qquad \text{(a)}$$

$$Z_{oe}{}^a = \frac{C_b + 2C_{ab}}{vF}, \qquad Z_{oo}{}^a = \frac{C_b}{vF}; \qquad Z_{oe}{}^b = \frac{C_a + 2C_{ab}}{vF}, \qquad Z_{oo}{}^b = \frac{C_a}{vF} \qquad \text{(b)}$$

where $F = C_aC_b + C_aC_{ab} + C_bC_{ab}$

$$\frac{C_a}{\epsilon} = \frac{\eta_0 Y_{oe}{}^a}{\sqrt{\epsilon_r}}, \qquad \frac{C_{ab}}{\epsilon} = \frac{\eta_0}{\sqrt{\epsilon_r}}\left(\frac{Y_{oo}{}^a - Y_{oe}{}^a}{2}\right); \qquad \frac{C_b}{\epsilon} = \frac{\eta_0 Y_{oe}{}^b}{\sqrt{\epsilon_r}}, \qquad \frac{C_{ab}}{\epsilon} = \frac{\eta_0}{\sqrt{\epsilon_r}}\left(\frac{Y_{oo}{}^b - Y_{oe}{}^b}{2}\right) \qquad \text{(c)}$$

$$\frac{C_a}{\epsilon} = \frac{\eta_0 2Z_{oo}{}^b}{\sqrt{\epsilon_r}H}, \qquad \frac{C_{ab}}{\epsilon} = \frac{\eta_0}{\sqrt{\epsilon_r}}\left(\frac{Z_{oe}{}^b - Z_{oo}{}^b}{H}\right); \qquad \frac{C_b}{\epsilon} = \frac{\eta_0 2Z_{oo}{}^a}{\sqrt{\epsilon_r}H}, \qquad \frac{C_{ab}}{\epsilon} = \frac{\eta_0}{\sqrt{\epsilon_r}}\left(\frac{Z_{oe}{}^a - Z_{oo}{}^a}{H}\right) \qquad \text{(d)}$$

where $H = Z_{oe}{}^aZ_{oo}{}^b + Z_{oe}{}^bZ_{oo}{}^a$.

mittances. The reason for this lies in the fact that when the odd- and even-mode *admittances* are computed, the basic definition of these admittances assumes that the lines are being driven with *voltages of identical magnitude* with equal or opposite phase, while the *currents* in the lines may be of different magnitudes.[11] When the odd- and even-mode *impedances* are computed, the basic definition of these impedances assumes that the lines are being driven by *currents of identical magnitude* with equal or opposite phases,[11] while magnitudes of the *voltages* on the two lines may be different. These two different sets of boundary conditions can be seen to lead to different voltage-current ratios if the lines are unsymmetrical.

Some unsymmetrical parallel-coupled lines that are quite easy to design are shown in Fig. 6. Note that the capacitance per unit length of each line has been separated into component parts. The capacitance $C_p{}^a$ is the *parallel plate* capacitance per unit length between one side of Line a and the adjacent ground plane, while $C_p{}^b$ is the corresponding *parallel plate* capacitance of Line b. The capacitance C_f' is the fringing capacitance per unit length for any of the outer corners of the striplines. The

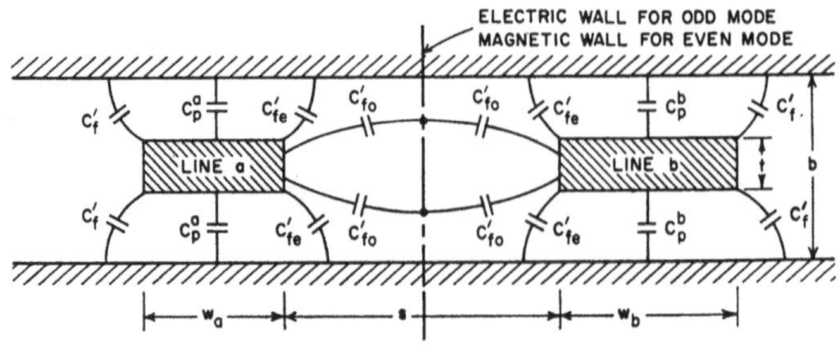

Fig. 6—Cross section of unsymmetrical, rectangular-bar parallel-coupled lines.

fringing capacitances per unit length at the inner corners of each strip are designated C_{fe}' when the strips are excited in the even mode (*i.e.*, with equal voltages of the same phase); they are designated C_{fo}' when the strips are excited in the odd mode (*i.e*, with voltages having equal amplitudes and opposite phases). Both bars have the same height, and both are assumed to be wide enough that the interactions between the fringing fields at the right and left sides of each bar are negligible, or at least small enough to be easily corrected for. On this basis the fringing fields are the same for both bars, and their different capacitances C_a and C_b to ground are due entirely to different parallel plate capacitances $C_p{}^a$ and $C_p{}^b$. For the structure shown

$$C_a = 2(C_p{}^a + C_f' + C_{fe}'), \qquad C_{ab} = (C_{fo}' - C_{fe}'), \qquad C_b = 2(C_p{}^b + C_f' + C_{fe}'). \tag{3}$$

Getsinger[10] has derived equations for the fringing capacitances C_{fe}', C_{fo}', and C_f', and has prepared convenient charts which relate C_{fe}'/ϵ, C_{fo}'/ϵ, and C_f'/ϵ to rectangular-bar strip-line dimensions. Here ϵ is the dielectric constant of the medium of propagation, so that the above ratios are dimensionless and of moderate size. Getsinger gives equations for the design of *symmetrical*, parallel-coupled, rectangular strip-lines, and here we will adapt his equations to fit the unsymmetrical case.

Note that the shape of the lines in Fig. 6 is fixed in terms of the dimensions t, b, s, w_a and w_b. To design a pair of lines such as those in Fig. 6 so as to have specified odd- and even-mode admittances or impedances, first use (c) or (d) in Table I to compute C_a/ϵ, C_{ab}/ϵ, and C_b/ϵ. Select a convenient value for t/b, and noting that

$$\frac{\Delta C}{\epsilon} = \frac{C_{ab}}{\epsilon}, \tag{4}$$

use Getsinger's chart[10] of $\Delta C/\epsilon$ and C_{fe}'/ϵ vs s/b to determine s/b, and also C_{fe}'/ϵ. Using t/b and Getsinger's chart of C_f'/ϵ vs t/b, determine C_f'/ϵ, and then compute

$$\frac{w_a}{b} = \frac{1}{2}\left(1 - \frac{t}{b}\right)\left[\frac{1}{2}\left(\frac{C_a}{\epsilon}\right) - \frac{C_{fe}'}{\epsilon} - \frac{C_f'}{\epsilon}\right] \tag{5}$$

$$\frac{w_b}{b} = \frac{1}{2}\left(1 - \frac{t}{b}\right)\left[\frac{1}{2}\left(\frac{C_b}{\epsilon}\right) - \frac{C_{fe}'}{\epsilon} - \frac{C_f'}{\epsilon}\right]. \tag{6}$$

When the ground plane spacing b is specified, the required bar widths, w_a and w_b, are determined. This procedure also works for the thin-strip case where $t/b=0$. If either w_a/b or w_b/b is less than 0.35 $(1-t/b)$, the width of the bar should be corrected using the approximate formula[10]

$$\frac{w'}{b}=\frac{0.07\left(1-\frac{t}{b}\right)+\frac{w}{b}}{1.20}, \qquad (7)$$

provided that $0.1<(w'/b)/(1-t/b)<0.35$. In (7) w is the uncorrected bar width and w' is the corrected width. The need for this correction arises because of the interaction of the fringing fields at opposite sides of a bar, which will occur when a bar is relatively narrow.

Case of an Array of Parallel-Coupled Lines

Fig. 7 shows an array of parallel-coupled lines such as is used in an interdigital filter. In the structure shown, all of the bars have the same t/b ratio and the other dimensions of the bars are easily obtained by generalizing the procedure just described for designing unsymmetrical parallel-coupled lines. In the structure in Fig. 7 the electrical properties of the structure are characterized in terms of the self-capacitances, C_k, per unit length of each bar with respect to ground, and the mutual capacitances, $C_{k,k+1}$, per unit length between adjacent bars k and $k+1$. This representation is not necessarily always highly accurate; it is conceivable that a significant amount of fringing capacitance could exist between a given line element and, say, the line element beyond the nearest neighbor. However, at least for geometries such as that shown, experience has shown this representation to have satisfactory accuracy.

For the design of the interdigital filter structures discussed herein, equations will be given for the normalized self-and mutual capacitances, C_k/ϵ and $C_{k,k+1}/\epsilon$, per unit length for all the lines in the structure. Then the cross-sectional dimensions of the bars and spacings between them are determined as follows. First choose values for t and b. Then, since

$$\frac{(\Delta C)_{k,k+1}}{\epsilon}=\frac{C_{k,k+1}}{\epsilon}, \qquad (8)$$

Getsinger's charts can be used to determine $s_{k,k+1}/b$. In this manner, the spacings $s_{k,k+1}$ between all the bars are obtained. Also, using Getsinger's charts, the normalized fringing capacitance $(C_{fe}')_{k,k+1}/\epsilon$ associated with the gaps $s_{k,k+1}$ between bars are obtained. Then the normalized width of the kth bar is

$$\frac{w_k}{b}=\frac{1}{2}\left(1-\frac{t}{b}\right)\cdot\left[\frac{1}{2}\left(\frac{C_k}{\epsilon}\right)-\frac{(C_{fe}')_{k-1,k}}{\epsilon}-\frac{(C_{fe}')_{k,k+1}}{\epsilon}\right]. \qquad (9)$$

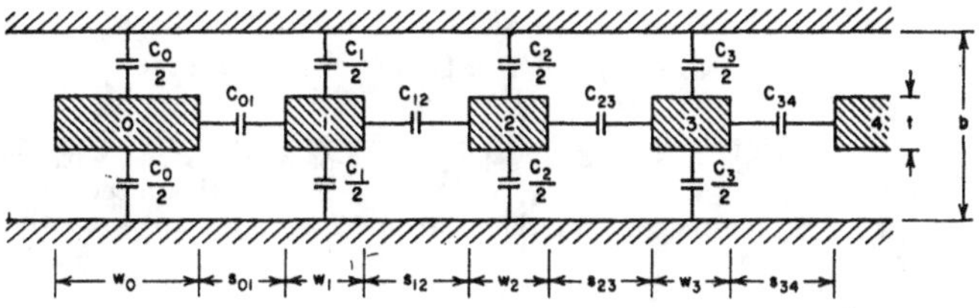

Fig. 7—Cross section of an array of parallel-coupled lines between ground planes.

In the case of the bar at the end of the array (the bar at the far left in Fig. 7), C_{fe}'/ϵ for the edge of the bar that has no neighbor must be replaced by C'_f, which is also obtained from Getsinger's charts. Thus, for example, for Bar 0 in Fig. 7

$$\frac{w_0}{b}=\frac{1}{2}\left(1-\frac{t}{b}\right)\left[\frac{1}{2}\left(\frac{C_0}{\epsilon}\right)-\frac{C_f'}{\epsilon}-\frac{(C_{fe}')_{01}}{\epsilon}\right]. \qquad (10)$$

If $w_k/b<0.35$ $(1-t/b)$ for any of the bars, the width correction given in (7) should be applied to the affected bars. 79

IV. Interdigital Filters of Narrow or Moderate Bandwidth

Table II shows approximate design equations for interdigital filters of the form shown in Fig. 1. This type of design is most practical for filters having narrow or moderate bandwidth. Although no special investigation of this point has been made, it appears probable that one should consider the possibility of using the design equations in Section V when the bandwidth is of the order of 30 per cent or more, instead of those in Table II. Both sets of design equations are valid, however, for either narrow or wide bandwidths. The main drawback in applying the design procedure in this section to filters of wide bandwidth is that the gaps between Lines 0 and 1 and between Lines n and $n+1$ (see Fig. 1) tend to become inconveniently small when the bandwidth is large, and the widths of Bars 1 and n tend to become very small.

To use Table II for the design of an interdigital filter, first use the mapping in Fig. 4 to estimate the number n of reactive elements required in the low-pass prototype in order to give the desired rate of cutoff with the desired pass-band characteristics.[6] When the prototype cutoff frequency ω_1' and element values $g_0, g_1, \cdots, g_{n+1}$ have been obtained from tables[7,8] or by computations,[12] the design computations can begin. It is suggested that the filter fractional bandwidth w be specified to be 6 or 7 per cent larger than is actually desired since, from our experience, there will be some shrinkage in bandwidth due to the approximate nature of the design equations. However, the desired value of w should be used in the mapping in Fig. 4.

After all of the J/Y_A and N parameters in Table II

[12] S. B. Cohn, "Direct-coupled-resonator filters," Proc. IRE, vol. 45, pp. 187–196; February, 1957.

TABLE II

DESIGN EQUATIONS FOR INTERDIGITAL FILTERS OF THE FORM IN FIG. 1

Use mapping in Fig. 4 to select a low-pass prototype with the required value of n. The input and output lines in this filter do not count as resonators, so that there are $n+2$ lines for an n-reactive-element prototype.

Compute

$$\theta_1 = \frac{\pi}{2}\frac{\omega_1}{\omega_0} = \frac{\pi}{2}\left(1 - \frac{w}{2}\right)$$

$$\frac{J_{01}}{Y_A} = \frac{1}{\sqrt{g_0 g_1 \omega_1'}}, \qquad \left.\frac{J_{k,k+1}}{Y_A}\right|_{k=1 \text{ to } n-1} = \frac{1}{\omega_1'\sqrt{g_k g_{k+1}}}$$

$$\frac{J_{n,n+1}}{Y_A} = \frac{1}{\sqrt{g_n g_{n+1}\omega_1'}},$$

$$\left.N_{k,k+1}\right|_{k=1 \text{ to } n-1} = \sqrt{\left(\frac{J_{k,k+1}}{Y_A}\right)^2 + \frac{\tan^2\theta_1}{4}},$$

$$M_1 = Y_A\left[\frac{J_{01}}{Y_A}\sqrt{h} + 1\right], \qquad M_n = Y_A\left[\frac{J_{n,n+1}}{Y_A}\sqrt{h} + 1\right]$$

80 where h is a dimensionless admittance scale factor to be specified arbitrarily so as to give a convenient admittance level in the filter. (See text.)

The normalized self-capacitances C_k/ϵ per unit length for the line elements are

$$\frac{C_0}{\epsilon} = \frac{376.7}{\sqrt{\epsilon_r}}[2Y_A - M_1]$$

$$\frac{C_1}{\epsilon} = \frac{376.7}{\sqrt{\epsilon_r}}\left[Y_A - M_1 + hY_A\left[\frac{\tan\theta_1}{2} + \left(\frac{J_{01}}{Y_A}\right)^2 + N_{12} - \frac{J_{12}}{Y_A}\right]\right]$$

$$\left.\frac{C_k}{\epsilon}\right|_{k=2 \text{ to } n-1} = \frac{376.7}{\sqrt{\epsilon_r}}hY_A\left[N_{k-1,k} + N_{k,k+1} - \frac{J_{k-1,k}}{Y_A} - \frac{J_{k,k+1}}{Y_A}\right]$$

$$\frac{C_n}{\epsilon} = \frac{376.7}{\sqrt{\epsilon_r}}\left[Y_A - M_n + hY_A\left[\frac{\tan\theta_1}{2} + \left(\frac{J_{n,n+1}}{Y_A}\right)^2 + N_{n-1,n} - \frac{J_{n-1,n}}{Y_A}\right]\right]$$

$$\frac{C_{n+1}}{\epsilon} = \frac{376.7}{\sqrt{\epsilon_r}}[2Y_A - M_n]$$

where ϵ is the dielectric constant and ϵ_r is the relative dielectric constant in the medium of propagation.

The normalized mutual capacitances $C_{k,k+1}/\epsilon$ per unit length between adjacent line elements are

$$\frac{C_{01}}{\epsilon} = \frac{376.7}{\sqrt{\epsilon_r}}[M_1 - Y_A],$$

$$\left.\frac{C_{k,k+1}}{\epsilon}\right|_{k=1 \text{ to } n-1} = \frac{376.7hY_A}{\sqrt{\epsilon_r}}\left(\frac{J_{k,k+1}}{Y_A}\right),$$

$$\frac{C_{n,n+1}}{\epsilon} = \frac{376.7}{\sqrt{\epsilon_r}}[M_n - Y_A].$$

have been computed, the admittance scale factor h must be fixed. One of the prime considerations in the choice of h is that the line dimensions must be such that the resonators will have a high unloaded Q. The dimensions that give optimum resonator Q's in structures such as interdigital filters are not known. However, it is known that for air-filled, coaxial-line resonators the optimum Q will result when the line impedance is about 76 ohms, and various approximate studies suggest that the optimum impedance for strip-line resonators is not greatly different. Thus it is suggested that in this case h be chosen to make the quantity

$$\left.\frac{2C_{k-1,k}}{\epsilon} + \frac{C_k}{\epsilon} + \frac{2C_{k,k+1}}{\epsilon}\right|_{\substack{k=n/2 \text{ for } n \text{ even} \\ =(n+1)/2 \text{ for } n \text{ odd}}} = (\text{around } 5.4) \qquad (11)$$

if air dielectric is used. If the quantity in (11) is set equal to 5.4 for the case of $\epsilon_r = 1$, this corresponds to making the line impedance 70 ohms for the resonator lines in the center of the filter, under the conditions that the adjacent lines are being excited with the same amplitude of voltage but with opposite phase (this is a generalized odd-mode admittance condition). Having established a value for h, the remainder of the calculations follow in a straightforward manner. After the normalized capacitances C_k/ϵ and $C_{k,k+1}/\epsilon$ have been computed, the line dimensions are determined as discussed in Section III.

A trial design was worked out using an $n=6$ reactive-element Chebyshev prototype with $L_{A_r}=0.10$ db. The prototype parameters were $g_0=1$, $g_1=1.1681$, $g_2=1.4039$, $g_3=2.0562$, $g_4=1.5170$, $g_5=1.9029$, $g_6=0.8618$, $g_7=1.3554$, and $\omega_1'=1$. The design was worked out for a fractional bandwidth of $w=0.10$ centered at $f_0=1.5$ Gc. Table III summarizes the various parameters used or computed in the calculations. The parameter h was chosen so that (11) would be 5.4. The resulting circuit has symmetry in its dimensions because the Chebyshev prototype is antimetric (*i.e.*, one half of the network is reciprocal to the other half).[7]

TABLE III

TABULATION OF QUANTITIES IN TABLE II AND IN FIG. 7, FOR A 10 PER CENT BANDWIDTH TRIAL DESIGN

k	$J_{k,k+1}/Y_A$	$N_{k,k+1}$	$C_{k,k+1}/\epsilon$	$s_{k,k+1}$ (Inches)	k	C_k/ϵ	w_k (Inches)
0 and 6	0.9253		1.582	0.159*	0 and 7	5.950	0.405†
1 and 5	0.7809	6.401	0.301	0.419	1 and 6	3.390	0.152
2 and 4	0.5886	6.381	0.226	0.512	2 and 5	4.420	0.183
3	0.5662	6.379	0.218	0.520	3 and 4	4.496	0.183

$w=0.10$ $\quad Y_A=0.020$ mho
$\theta_1=1.492$ $\quad M_1=M_6=0.02420$
$h=0.05143$ $\quad b=0.625$ inch
$\epsilon_r=1$ $\quad t=0.187$ inch

* Changed to 0.127 inch after laboratory tests.
† Changed to 0.437 inch after laboratory tests.

Fig. 8 shows a photograph of the completed filter, while Fig. 9 shows those dimensions of the filter not summarized in Table III. The short-circuiting side walls of the structure are spaced exactly a quarter-wavelength apart at the midband frequency $f_0 = 1.5$ Gc ($\lambda_0/4 = 1.968$ in). Because of the capacitance between the open-circuited ends of the resonator elements and the side walls, it was necessary to foreshorten the resonators so as to maintain their resonant frequency at 1.5Gc. No very satisfactory means for accounting for all of the fringing capacitances at the open-circuit ends of the reasonators has been devised, but some rough estimates were made using Getsinger's fringing data[10] and various

Fig. 8—A 10 per cent bandwidth interdigital filter with its cover plate removed.

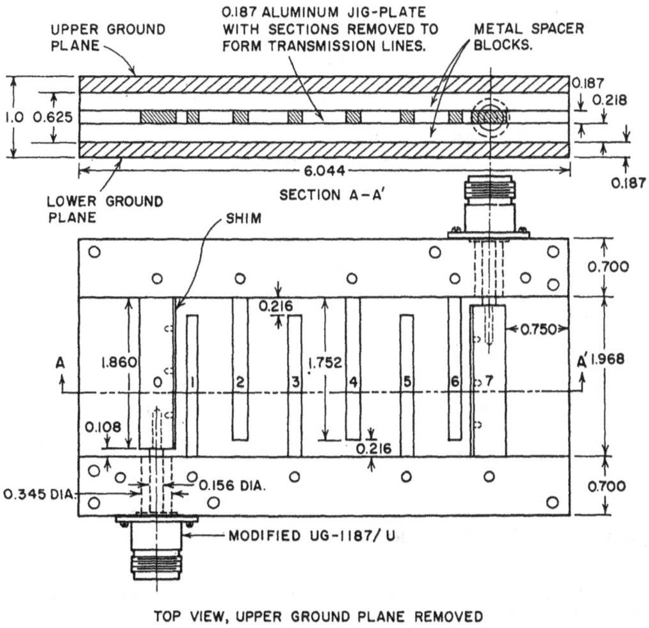

Fig. 9—Drawing of the 10 per cent bandwidth interdigital filter (part of the dimensions are as specified by Table III and Fig. 7).

approximations. The estimated foreshortening for the resonators was 0.216 inch, but laboratory tests showed this to be excessive since the pass-band center was 1.56 instead of 1.50 Gc. Although the filter pass band can always be lowered in frequency by use of tuning screws, if the resonators had been foreshortened by about 0.160 inch instead of 0.216 inch, the pass-band center frequency would probably have been about right.

Although the structure included tuning screws, no effort was made to lower the band center frequency to 1.50 Gc. However, it was found that since Resonators 1 and 6 have different fringing capacitance conditions at their open-circuit ends than do the other resonators, it was necessary to increase the capacitance at their open-circuit ends by inserting the tuning screws. Before this was done, the pass-band response was not symmetrical (this is indicative of mistuning of some of the resonators with respect to the others).

When the filter was first tested the pass-band VSWR reached peaks of 2.2, which is somewhat high since 0.1-db 81
Chebyshev ripples call for VSWR peaks of only 1.36. Such conditions can usually be corrected by altering the couplings between the terminations and the first resonator on each end. Thus, a 0.032-inch-thick brass shim was added to the input and output lines (Lines 0 and 7) to reduce the adjacent gaps from $s_{01} = s_{67} = 0.159$ inch to 0.127 inch. This reduced the input VSWR to 1.30 or less across the band. It appears desirable that in working out the design of trial models of interdigital filters as described herein, some provision should be made for experimental adjustment of the size of coupling gaps at the ends, if the pass-band VSWR is to be closely controlled.

Fig. 10 shows the measured attenuation characteristic of this filter, while Fig. 11 shows the measured VSWR. The measured fractional band width is slightly less than the design value ($w = 0.0935$ instead of 0.100). Using $w = 0.0935$ and $f_0 = \omega_0/(2\pi) = 1.563$ Gc the measured attenuation was compared with that computed by mapping the low-pass prototype attenuation as indicated in Fig. 4. The measured attenuation in the stop bands was found to be somewhat less than that predicted for a Chebyshev filter with $L_{Ar} = 0.1$-db ripple. However, the pass-band VSWR in Fig. 11 is, for the most part, much less than the 1.36 peak value corresponding to a 0.1-db Chebyshev ripple, and some of the VSWR amplitude shown may be due to the connectors and slight mistuning.[13] The attenuation for a $L_{Ar} = 0.01$-db ripple filter (VSWR = 1.1 peak) was also computed, and the results are summarized in Table IV. Note that in all cases the measured stop-band attenua-

[13] The pass-band VSWR is the most sensitive index of the correlation of the actual design as compared to the pass-band characteristic of the prototype. The pass-band attenuation as determined from through transmission measurements includes the additional attenuation due to dissipation loss, which may be markedly greater than 0.1 db depending on the Q's of the resonators.

82

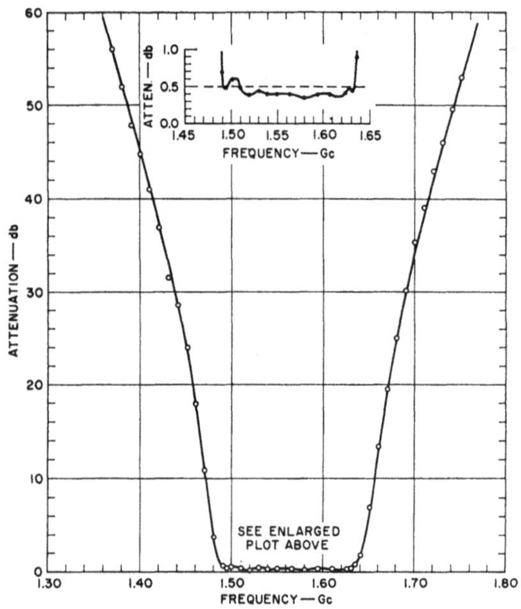

Fig. 10—Measured attenuation characteristic of the filter in Fig. 8.

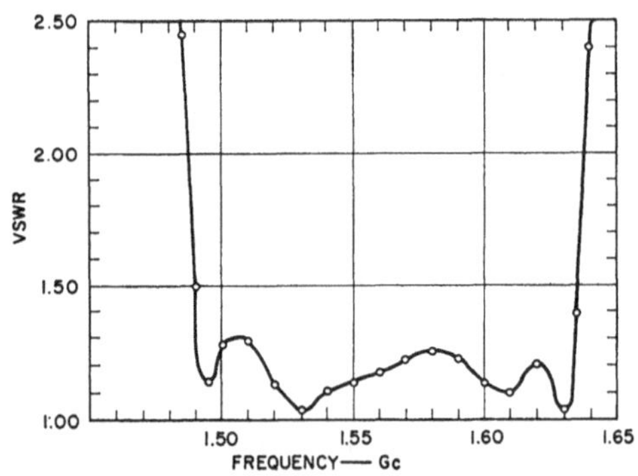

Fig. 11—Measured VSWR of the filter in Fig. 8.

TABLE IV

COMPARISON OF MEASURED ATTENUATION IN FIG. 10 WITH ATTENUATION PREDICTED AS IN FIG. 4

[Computed values are for 0.01- and 0.10-db ripple, $n=6$, $w=0.0935$, and $f_0=\omega_0/(2\pi)=1.563$ Gc]

f, Gc	L_A, db for L_A =0.01	L_A, db measured	L_A, db for $L_{Ar}=0.10$
1.440	25	29	35.5
1.686	25	28	35.5
1.380	49.5	52	59.5
1.746	49.5	52	59.5

tion is less than that computed for a 0.10-db ripple filter, but more than that for a 0.01-db ripple filter. Thus the approximate mapping in Fig. 4 appears to be reasonably consistent with the measured results.

Calculating from the measured pass-band insertion loss, it is estimated that the unloaded Q's of the resonators in this filter are about 1100. Using copper for the structure instead of aluminum would theoretically give a value about 25 per cent higher for the unloaded Q's. It is possible that a different impedance level within the filter might also give higher Q's for a given ground plane spacing.

The lines in the trial interdigital filter (Figs. 8 and 9) were fabricated by machining Lines 0, 2, 4, and 6 in comb form from a single piece of jig plate. Lines 1, 3, 5, and 7 were cut out similarly from a second piece of jig plate. (Actually, both comb structures were machined at once, back to back.) Then, interleaving the two comb structures between ground planes gave the desired interdigital structure.

Fig. 12 shows an alternative form of interdigital filter structure that should be even less expensive to fabricate. In this case the interdigital line structure is photoetched on a copper-clad dielectric card, and the dielectric material removed from the region between the copper-foil lines. In order to provide good support for the lines, the dielectric is not removed at the open-circuit ends of the lines, however. Using this structure the propagation is largely in air, which should permit good performance; also, the dielectric card line structures should be quite inexpensive to mass produce.

Round rods between ground planes also provide an attractive form for fabricating interdigital filters.[4] However, no data are yet available for accurate determination of the rod diameters and spacings from specified line capacitances.[14] Work on this problem is planned at SRI for the near future.

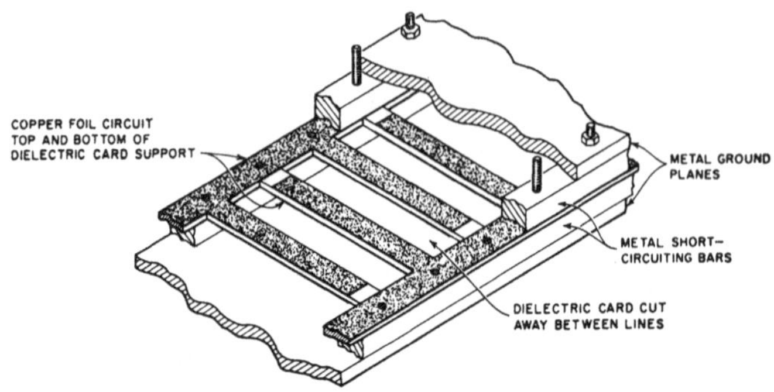

Fig. 12—A proposed low-cost construction for interdigital lines filters.

V. INTERDIGITAL FILTERS OF MODERATE OR WIDE BANDWIDTH

Table V presents approximate design equations for filters of the form in Fig. 2. This type of filter is most practical for designs having moderate or wide bandwidths, although the design procedure given is valid for either narrow- or wide-band filters. The main drawback in applying the procedure in Table V to narrow-band filters is that Lines 1 and n will attain extremely high impedance in such designs.

[14] Bolljahn and Matthaei (4) give an approximate method for design of structures consisting of rods that are all the same size and have the same spacings. A procedure for accurate design where varying diameters and spacings are required has not as yet been obtained.

TABLE V

DESIGN EQUATIONS FOR INTERDIGITAL FILTERS OF THE FORM IN FIG. 2

Use mapping in Fig. 4 to select low-pass prototype with the required value of n. The input and output lines in this filter count as resonators, so that there are n line elements for an n-reactive-element prototype. Compute

$$\theta_1 = \frac{\pi}{2}\frac{\omega_1}{\omega_0} = \frac{\pi}{2}\left(1 - \frac{w}{2}\right)$$

$$\left.\frac{J_{k,k+1}}{Y_A}\right|_{k=2 \text{ to } n-3} = \frac{g_2}{g_0\sqrt{g_k g_{k+1}}}, \qquad \frac{J_{n-2,n-1}}{Y_A} = \frac{1}{g_0}\sqrt{\frac{g_0 g_2}{g_{n-2}g_{n+1}}}$$

$$\left.N_{k,k+1}\right|_{k=2 \text{ to } n-2} = \sqrt{\left(\frac{J_{k,k+1}}{Y_A}\right)^2 + \left(\frac{\omega_1' g_2 \tan\theta_1}{2g_0}\right)^2}$$

$$\frac{Z_1}{Z_A} = \omega_1' g_0 g_1 \tan\theta$$

$$\frac{Y_2}{Y_A} = \frac{\omega_1' g_2}{2g_0}\tan\theta_1 + N_{23} - \frac{J_{23}}{Y_A}$$

$$\left.\frac{Y_k}{Y_A}\right|_{k=3 \text{ to } n-2} = N_{k-1,k} + N_{k,k+1} - \frac{J_{k-1,k}}{Y_A} - \frac{J_{k,k+1}}{Y_A}$$

$$\frac{Y_{n-1}}{Y_A} = \frac{\omega_1'(2g_0 g_{n-1} - g_2 g_{n+1})\tan\theta_1}{2g_0 g_{n+1}} + N_{n-2,n-1} - \frac{J_{n-2,n-1}}{Y_A}$$

$$\frac{Z_n}{Z_A} = \omega_1' g_n g_{n+1} \tan\theta_1.$$

The normalized self-capacitances, C_k/ϵ, per unit length for the line elements are

$$\frac{C_1}{\epsilon} = \frac{367.7}{\sqrt{\epsilon_r}} Y_A \frac{(1-\sqrt{h})}{(Z_1/Z_A)}$$

$$\frac{C_2}{\epsilon} = \frac{376.7}{\sqrt{\epsilon_r}} Y_A h\left(\frac{Y_2}{Y_A}\right) - \sqrt{h}\frac{C_1}{\epsilon}$$

$$\left.\frac{C_k}{\epsilon}\right|_{k=3 \text{ to } n-2} = \frac{376.7}{\sqrt{\epsilon_r}} Y_A h\left(\frac{Y_k}{Y_A}\right)$$

$$\frac{C_{n-1}}{\epsilon} = \frac{376.7}{\sqrt{\epsilon_r}} Y_A h\left(\frac{Y_{n-1}}{Y_A}\right) - \sqrt{h}\frac{C_n}{\epsilon}$$

$$\frac{C_n}{\epsilon} = \frac{376.7}{\sqrt{\epsilon_r}} Y_A \frac{(1-\sqrt{h})}{(Z_n/Z_A)}$$

where ϵ is the dielectric constant, ϵ_r is the relative dielectric constant in the medium of propagation, and h is a dimensionless admittance scale factor whose value should be chosen to give a convenient admittance level in the filter. (See Text.)

The normalized mutual capacitances $C_{k,k+1}/\epsilon$ per unit length between adjacent line elements are

$$\frac{C_{12}}{\epsilon} = \frac{376.7}{\sqrt{\epsilon_r}} Y_A \frac{\sqrt{h}}{(Z_1/Z_A)}$$

$$\left.\frac{C_{k,k+1}}{\epsilon}\right|_{k=2 \text{ to } n-2} = \frac{376.7}{\sqrt{\epsilon_r}} Y_A h\left(\frac{J_{k,k+1}}{Y_A}\right)$$

$$\frac{C_{n-1,n}}{\epsilon} = \frac{376.7}{\sqrt{\epsilon_r}} Y_A \frac{\sqrt{h}}{(Z_n/Z_A)}.$$

Except for the somewhat different design equations, the procedure for the design of filters of the form shown in Fig. 2 is much the same as that described for filters of the form shown in Fig. 1. It is suggested that for use in Table V, w be made about 8 per cent larger than the actual desired fractional bandwidth, in order to allow for some bandwidth shrinkage. Eq. (11) is applicable when selecting a value for the admittance scale factor h.

A trial design was worked out using an $n=8$ reactive-element Chebyshev prototype with $L_{Ar}=0.10$ db. The prototype parameters were $g_0=1$, $g_1=1.1897$, $g_2=1.4346$, $g_3=1.1199$, $g_4=1.6010$, $g_5=2.1699$, $g_6=1.5640$, $g_7=1.9444$ $g_8=0.8778$, $g_9=1.3554$, and $\omega_1'=1$. The design was carried out for a fractional bandwidth of $w=0.70$ centered at 1.50 Gc, and the parameter h was chosen to make (11) equal to 5.86. (This gives a generalized odd-mode impedance of 64.5 ohms for the middle lines.) Table VI summarizes some of the quantities computed in the course of the design of this filter.

Fig. 13 shows the completed filter, while the drawing in Fig. 14 shows additional construction details and additional dimensions not summarized in Table VI. The filter was fabricated in much the same manner as the filter described in Section IV, except that the resonator lines were foreshortened by 0.150 inch. The relatively small cross-sectional dimensions of the resonator elements, unfortunately, made 0.150 inch excessive, so that the measured band center frequency was 1.55 Gc instead of 1.50 Gc. It is probable that foreshortening the line elements about 0.125 inch would have been about right. 83

When this filter was first tested, the VSWR was quite low across the band (about 1.2 or less) except at band center where the VSWR peaked to 1.8. This situation was altered by increasing the s_{12} and s_{78} gaps from 0.087 inch to 0.092 inch, which caused the VSWR peaks across the band to be more nearly even and to be 1.55 or less.

Fig. 15 shows the measured attenuation of this filter, while Fig. 16 shows its measured VSWR. The fractional bandwidth is 0.645 instead of the specified 0.700 value, which indicates a shrinkage of bandwidth of about 8 per cent, as a result of the various approximations involved in the design equations. The attenuation characteristics in Fig. 15 were checked against the attenuation computed using the mapping in Fig. 4 with $w=0.645$, $L_{Ar}=0.10$ db, $n=8$, and $f_0=\omega_0/(2\pi)=1.55$ Gc. The resulting computed values are listed in Table VII along with the corresponding measured values of attenuation. The agreement can be seen to be quite good.

VI. CONCERNING THE DERIVATION OF THE DESIGN EQUATIONS

The design equations presented for interdigital filters were obtained by extending a previously discussed theory[6] for band-pass microwave filter design. Since much of the relevant theory has been explained before, only the necessary extensions of the previous theory will be treated.

Design theory has been presented for filters of the form in Fig. 17 which consist of parallel-coupled resonators which are $\lambda_0/2$ long at the midband frequency.[6] There is a $\lambda_0/4$, short-circuit input and output coupling line at each end of this filter; these are designed to serve as part of an admittance-transforming section. An in-

84

TABLE VI

Tabulation of Some of the Parameters Computed in the Design of the Trial Octave-Bandwidth Interdigital Filter
(The dimensions are as defined in Fig. 7)

k	$J_{k,k+1}/Y_A$	$C_{k,k+1}/\epsilon$	$s_{k,k+1}$ (inches)	k	$\frac{Z_k}{Z_A}$ or $\frac{Y_k}{Y_A}$	C_k/ϵ	w_k (inches)
1 and 7		1.647	0.087*	1 and 8	1.941	2.235	0.126†
2 and 6	0.823	1.115	0.136	2 and 7	1.779	1.463	0.121†
3 and 5	0.779	1.056	0.143	3 and 6	1.235	1.675	0.126†
4	0.770	1.044	0.146	4 and 5	1.258	1.706	0.127†

$w=0.70$ $\quad Y_A=0.020$ mho
$\theta_1=1.021$ $\quad b=0.625$ inch
$h=0.18$ $\quad t=0.063$ inch
$\epsilon_r=1$

* Changed to 0.092 inch after laboratory tests.
† Computed using Eq. (7) for width correction.

Fig. 13—Octave bandwidth interdigital filter with cover plate removed.

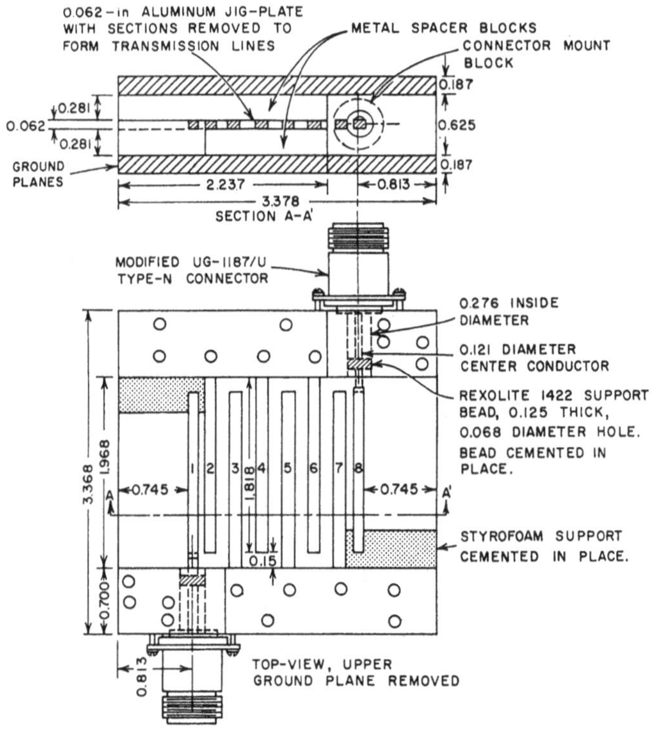

Fig. 14—Drawing of the interdigital filter in Fig. 13 (other dimensions not shown are as defined by Fig. 7 and Table VI).

terdigital filter of the form in Fig. 1 is obtained from the filter in Fig. 17 if each $\lambda_0/2$ line is cut in the middle and folded double to give the structure in Fig. 18. It can be seen from Fig. 19 that this operation has little effect on the currents and voltages on the lines—at least at mid-band. Fig. 19(a) shows the voltages and currents on a short-circuited $\lambda_0/2$ resonator, while Fig. 19(b) shows the voltages and currents after the resonator has been cut and folded. Note that the voltages and currents on the a and b portions of the resonator are the same in either case.

The circuits in Figs. 17 and 18 are clearly not electrically the same. First, if the structure in Fig. 18 has significant fringing capacitances extending beyond nearest-neighbor line elements, the coupling mechanism would become much more complicated than is implied by the simple folding process. Second, the circuit in Fig. 17 can be shown to have only a first-order pole of attenuation at $\omega=0, 2\omega_0, 4\omega_0$, etc.,[6] while the circuit in Fig. 18 has high-order poles of attenuation at these frequencies. For this reason it was believed that folding a filter as in Fig. 18 would have very little effect on its response for frequencies near ω_0 (provided that fringing beyond nearest neighbors is negligible), but that the error might be considerable at frequencies well removed from ω_0 (which would imply that the folding shown in Fig. 18 might considerably disturb the response of a wide-band filter). To check this point, the image cutoff frequency predicted by the approximation in Figs. 17 and 18 was compared with the previous

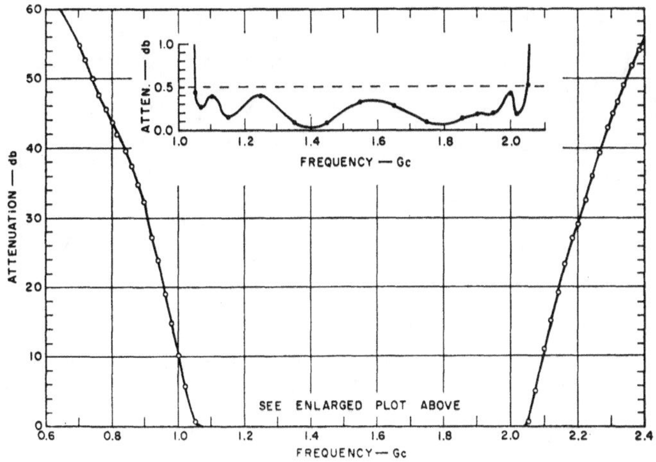

Fig. 15—Measured attenuation characteristic of the filter in Fig. 13.

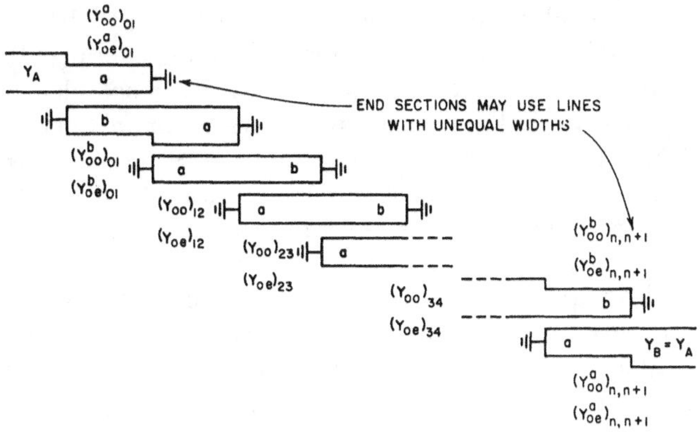

Fig. 17—A parallel-coupled strip-line filter with $\lambda_0/2$ resonators.

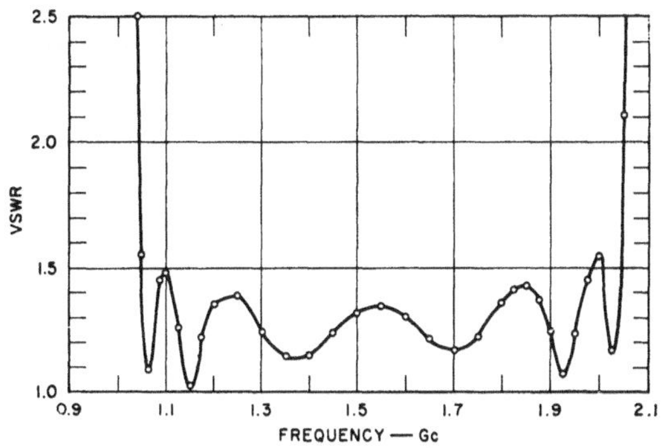

Fig. 16—Measured VSWR of the filter in Fig. 13.

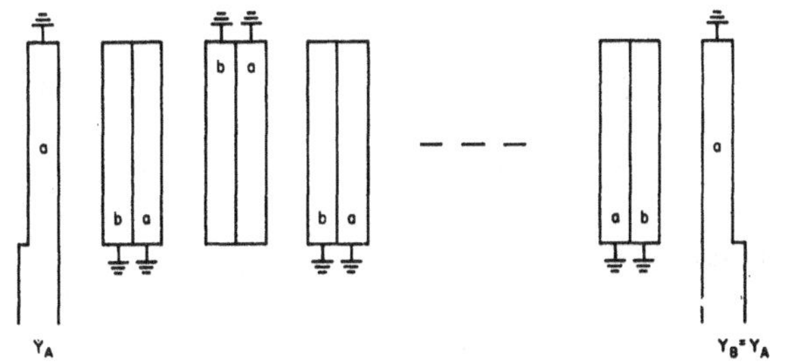

Fig. 18—An interdigital filter formed from the filter in Fig. 17 (each $\lambda_0/2$ resonator has been cut in two in the middle and then folded double).

TABLE VII

COMPARISON OF THE MEASURED ATTENUATION IN FIG. 15 WITH ATTENUATION PREDICTED AS IN FIG. 4

(Computed values are for 0.10-db ripple, $n=8$, $w=0.645$, and $f_0=\omega_0/(2\pi)=1.55$ Gc)

f, Gc	L_A, db for $L_{Ar}=0.10$	L_A, db Measured
0.90	32	31
2.23	32	33
0.70	56	55
2.43	56	57

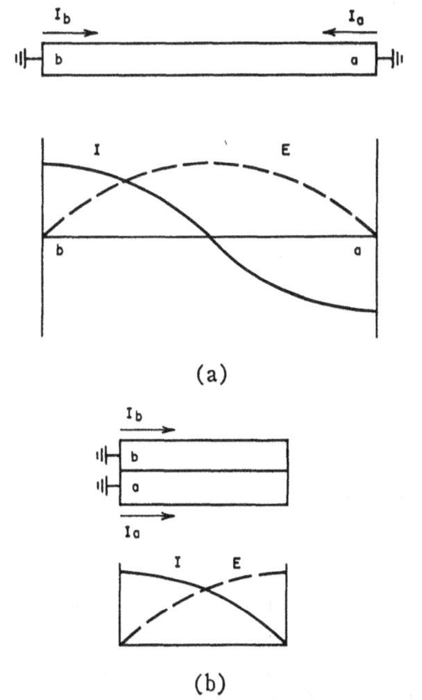

Fig. 19—Effect of folding a $\lambda_0/2$ resonator to make a $\lambda_0/4$ resonator.

interdigital-line exact analysis[4] for the case where there is no fringing beyond nearest neighbors. Surprisingly enough, the image bandwidths were practically the same (to slide rule accuracy) by either theory, even for bandwidths as great as an octave. This unexpected result indicates that the folding process in Fig. 18 should not greatly disturb the response of filters of the form in Fig. 17, even if the bandwidth is quite wide (provided that fringing capacitances beyond nearest-neighbor line elements are negligible). Experimental results show that in typical cases the fringing capacitance beyond nearest neighbor has no serious effect.

The equations in Table II were derived directly from the equations in Table I in Matthaei[6] in dual form (which apply to filters of the form in Fig. 17),[15] but with two generalizations. First, the equations were generalized so that, if desired, they can be used with lumped-element prototypes that are neither symmetrical nor antimetrical (*i.e.*, one half of the network is not the reciprocal of the other half).[7] Second, they were generalized so that the admittance level within the filter can be adjusted as desired. This last feature was made possible by adding an additional degree of freedom in the design of the end sections of the filter. In the procedure discussed in Matthaei[6] all of the parallel-coupled sections of the filter, including the end sections 0, 1 and n, $n+1$, have parallel strips with identical widths. By allowing the widths of the strips in the end sections to be unequal, an additional degree of freedom is obtained, which permits arbitrarily fixing the admittance level in the interior of the filter.

86 Fig. 20 shows a strip-line parallel-coupled section with short circuits, and its exact equivalent in open-wire line.[16] It is interesting to note that the capacitances C_a and C_b per unit length between the strip-lines and ground become the capacitances per unit length of the short-circuited open-wire stubs. The capacitance C_{ab} per unit length between strip-lines becomes the capacitance per unit length of the open-wire connecting line. By use of this equivalence, the filter in Fig. 17 can be converted to an exactly equivalent filter consisting of shunt stubs with connecting lines.[6] If the constraint $Y_{oe}^a+Y_{oo}^a=2Y_A$ is enforced in the design of the end sections, the open-wire equivalent circuit in Fig. 20 reduces exactly to the remarkably simple circuit shown on the right in Fig. 21. The admittance scale factor h in Table II is exactly

$$h=\frac{1}{N^2} \tag{12}$$

where N is the turns ratio of the ideal transformer in Fig. 21. The admitttance scale factor h used in Table II corresponds to s/Y_A in the equations in Table I in Matthaei in dual form.[18] In Matthaei,[6] the scale factor s was so specified as to make the widths of the end strips equal, while in the derivation of the equations herein, $h=s/Y_A$ is arbitrary.

When generalized design equations had been prepared for the filter circuit in Fig. 17, the design equations in

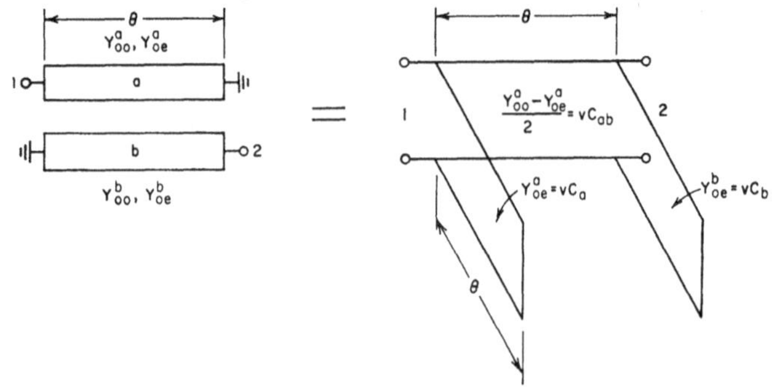

Fig. 20—Short-circuited, parallel-coupled strip-lines and their open-wire-line equivalent circuit (the parallel-coupled lines may be of either equal or unequal widths).

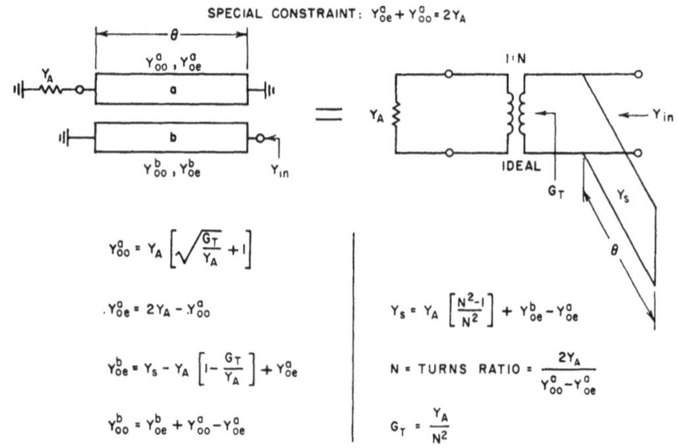

Fig. 21—Special properties of the circuits in Fig. 20 when a special constraint is applied.

Table II were obtained directly by use of the "folding" approximation in Fig. 18. Since the equations for the filter in Fig. 17 were shown to be valid from narrow bandwidths to at least bandwidths of the order of an octave, and since the folding approximation appears to be reasonably good to such bandwidths, the design equations in Table II should, in principle, be good for large as well as for narrow bandwidths. However, the physical dimensions of the wide-band structure are not as desirable as those of the filters shown in Fig. 2.

In order to derive design equations for interdigital filters with open-circuited terminating lines (Fig. 2), design equations were first derived for the type of filter shown in Fig. 22. This filter is nearly the same as the parallel-coupled filter in Fig. 17, except for the manner in which the terminating lines are coupled in. It is readily seen that if the folding process in Fig. 18 is applied to the filter in Fig. 22, an interdigital filter with open-circuited input lines will result. It might seem at first that design equations for interdigital filters with open-circuited terminating lines could have been obtained by folding a parallel-coupled filter that had resonators open-circuited at the ends.[5,6] However, it will be seen that this cannot work, since the voltages at opposite ends of a $\lambda/2$ resonator have opposite polarity.

[15] The equations in Table I in Matthaei[6] are for filters with parallel-coupled resonators with open-circuited ends. Equations for the filter in Fig. 17 were obtained by duality, as is discussed in Matthaei.[6]

[16] The equivalences shown in Figs. 20, 21, and 23 were derived by extension of the work of Jones and Bolljahn,[11] and of Ozaki and Ishii.[17]

[17] Hiroshi Ozaki and Junya Ishii, "Synthesis of a class of strip-line filters," IRE TRANS. ON CIRCUIT THEORY, vol. CT-5, pp. 104–109; June, 1958.

[18] The parameter s in Table I in Matthaei[6] has the dimensions of ohms. In the dual equations, which apply to the circuit in Fig. 17, s has the dimensions of mhos and s/Y_A is dimensionless.

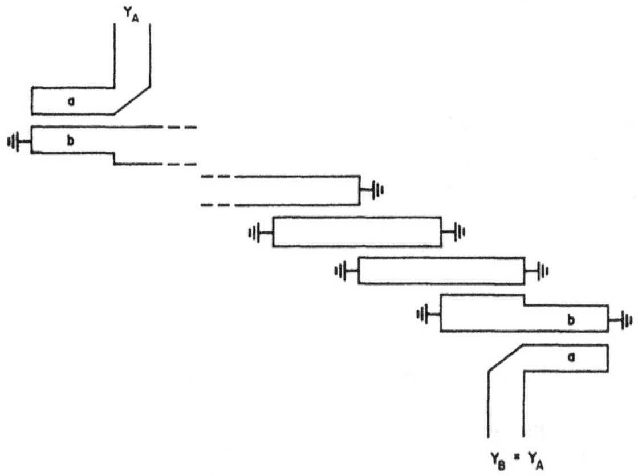

Fig. 22—A parallel-coupled filter with $\lambda_0/2$ short-circuited resonators and open-circuited terminating lines.

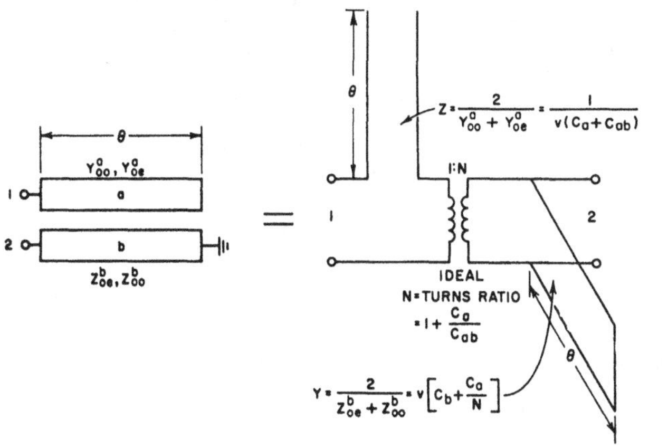

Fig. 23—An additional parallel-coupled strip-line section and its open-wire-line equivalent circuit.

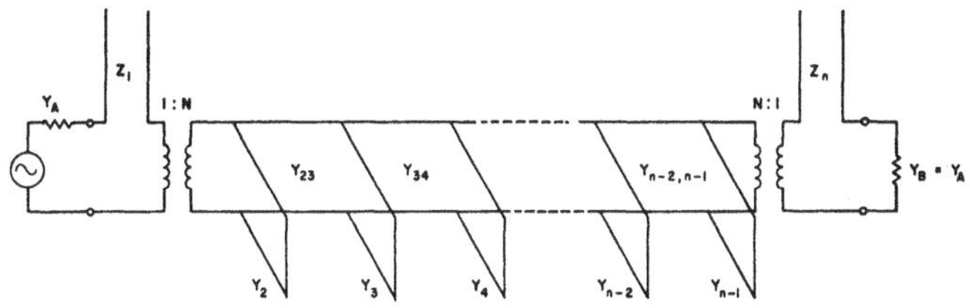

Fig. 24—An open-wire-line equivalent circuit of the filter in Fig. 22 (all stubs and connecting lines are $\lambda_0/4$ long at midband).

Design equations for the filter in Fig. 22 were derived in a manner much like that used for deriving the equations for the filter in Fig. 17, except that the parallel-coupled information summarized in Fig. 23 was used in designing the end sections. Note that this section also has impedance transforming properties [(12) applies], and that the open-wire equivalent circuit in this case has both a series and a shunt stub. In the case of Fig. 21, the constraint equation applied causes one natural mode of vibration to be stifled, and as a result the equivalent circuit has only one stub. Using the end section in Fig. 23, all of the natural modes of vibration of the circuit are fully utilized.

When the strip-line and open-wire line equivalences in Figs. 20 and 23 are used, it will be seen that the strip-line circuit in Fig. 22 is electrically identical to the open-wire line circuit in Fig. 24. The filter circuit in Fig. 24 is very similar to the filter in Fig. 3 in Matthaei,[6] for which design equations were presented in Table III in Matthaei.[6] The filters in Fig. 24 and in Fig. 3 in Matthaei[6] become identical if we set $a = \infty$ for Fig. 3 and if we introduce an ideal transformer at each end of the filter and alter the impedance level within the filter to make the over-all performance the same as before the transformers were introduced. Design equations for the filter in Fig. 24 were obtained in this manner, and also equations for the equivalent filter in Fig. 22.[19] Then the equations for the corresponding interdigital filter were obtained by applying to the filter in Fig. 22 the folding approximation illustrated in Fig. 18. The reader will be interested to note that the $J_{k,k+1}$ in Table V correspond to characteristic admittances $Y_{k,k+1}$ of the connecting lines in Fig. 24, while Z_1, Y_2, Y_3, $\cdots$, Y_{n-1}, Z_n in Table V correspond to the characteristic impedances or admittances of the stubs in Fig. 24, for the limiting case where the transformer turns ratio is $N = 1$. For $N > 1$ the admittances are scaled by the factor h.

VII. Conclusions

The approximate equations developed for the design of interdigital band-pass filters were found to yield reasonably accurate results when used for design of either narrow- or wide-band band-pass filters. By making allowance for about a 7 per cent shrinkage in fractional bandwidth as compared to the value inserted in the equations, and by providing for small adjustment in the gap spacing at each end of the filter, desired performance should be obtainable with high accuracy. The design equations are simple to use, and the trial designs performed very much as expected. The trial designs demonstrated that practical interdigital filters can have the attractive features that were anticipated: they can be made very compact; the relatively large spacings between their resonator elements permit relatively relaxed tolerances; they have strong stop bands with the second pass band at three times the frequency of the first pass band; they make spurious response impossible at or near twice the first pass-band frequency; and they can be fabricated using little or no dielectric material.

Acknowledgment

The assistance of E. J. Scribner, who carried out the design calculations for the two trial designs and made the laboratory tests on the completed filters, is gratefully acknowledged, as is the contribution of Dr. E. M. T. Jones in discussions which helped to crystallize the approximate design procedure described herein.

[19] The design equations in Matthaei[6] were in this present work also modified to include the situation where the low-pass prototype might not be either symmetrical nor antimetrical.

Reprinted from IEEE TRANSACTIONS
ON MICROWAVE THEORY AND TECHNIQUES
Volume MTT-13, Number 5, September, 1965
Pp. 696-698

A Simple Design Procedure for Small Percentage Bandwidth Round-Rod Interdigital Filters

The continued correspondence concerning interdigital filter [1] prompted me to submit this correspondence. The concepts and design procedures to be described have been found to be practically useful for determining the geometry of the coupled-rod structure, and because these procedures are apparently much simpler than those presently available, publication of the subject
88 matter may be helpful to others. It will be apparent that some of this material is tutorial in nature—it is included to clarify the other information being presented.

I. Concerning the Overall Device

However else the device may be described, when used to produce small percentage, i.e., less than 10 per cent, pass bands, the first-order phenomenon involved in a so-called interdigital filter is identical to that occurring in direct coupled-resonator filters with groundings arranged so that the magnetic field coupling is in phase with the electric field coupling; i.e., the so-called "capacity aiding" connection in the $n=2$ IF interstages of old. Thus, the concepts and quantitative procedures described in [2] are directly applicable to the design, adjustment, and alignment of these filters. Specifically the device is adequately designed if, with all other resonances short-circuited, one correctly adjusts the coefficients-of-couplings between adjacent resonances; the singly loaded Q of the input and output resonances; and the resonant frequency of each resonance.

If the foregoing is true, the following helpful questions arise:

1) Why should one bother using different diameter rods (or different width rectangular bars) in the filter?

2) Why should one waste the space to put in rods #0 and #($n+1$) the only purpose of which is to properly couple the resistive generator and resistive load to the input and output resonances, respectively?

The answer to the first question is apparently that different diameter rods and different width bars are used because the presently available designs call for them. Actually there are an infinite number of correct diameter and spacing combinations, and there is no need to use different diameter rods (or different width bars) if the filter is correctly designed and adjusted for uniform diameter rods (or uniform width bars). Section III gives the simple design procedure involved.

The answer to the second question also is that apparently rods #0 and #($n+1$) are used because the presently available designs call for them. Actually in the bandwidth regions considered in this correspondence these rods are not required. In the less than one percent bandwidth region, probe or loop coupling of the generator to resonator #1, and the load to resonator #n is practical; and in the greater than one percent bandwidth region, tapping of the generator onto rod #1, and tapping of the load onto rod #n, is a practical way of producing the required singly loaded Q for the input and output resonators. Section IV gives the simple design procedure involved when tapping is used.

II. The Coefficient-of-Coupling Between Adjacent Uniform Diameter Rods

The definition of coefficient-of-coupling (K) applicable to all small percentage bandwidth coupled-resonator filters is given in Section IV of [2]; also given there is a straightforward experimental procedure for accurately measuring and/or adjusting K. This procedure applied to an actual filter for which $d/h=0.5$ resulted in the three cross-marked points on the graph of Fig. 1. The K measured was that between two adjacent rods which also had equally spaced rods on their other sides; thus, equal far-side coupling was involved.

Within their applicable ranges, either Honey's closed-form equations for Z_{oe} and Z_{oo} [3], or Cristal's graphs for C_m/ϵ and C/ϵ [4] can be used to calculate the coefficient-of-coupling resulting from a given geometry d/h and c/h. It should be noted that the small percentage coupling case herein considered is very forgiving of a multitude of approximation sins, and perhaps a proof of this is the fact that for the less than 10 percent coupling case considered herein, and for normalized rod diameters less than $d/h=0.5$, both procedures give "exactly," i.e., within about 2 per cent, the same numerical answers. If Honey's approximation is used, (1) gives the K between a pair of rods

$$K = \frac{4}{\pi}\left\{\frac{\ln \coth\left(\dfrac{\pi}{2}\dfrac{c}{h}\right)}{\ln \coth\left(\dfrac{\pi}{4}\dfrac{d}{h}\right)}\right\}. \tag{1}$$

If Cristal's graphs are used, (2) is used to obtain the K between a pair of rods

$$K = \frac{4}{\pi}\left\{\frac{1}{\dfrac{C_g/\epsilon}{C_m/\epsilon}+2}\right\}. \tag{2}$$

The graph of Fig. 1, giving the coefficient-of-coupling between two adjacent rods as a function of normalized center to center rod spacing c/h, and normalized rod diameter d/h, results from an application of either (1) or (2). The very excellent agreement on Fig. 1 between the three experimentally determined cross-marked points, and the line for $d/h=0.5$, is gratifying.

As Fig. 1 indicates, in this region of interest, the log of the coefficient-of-coupling is almost a linear function of both the normalized center to center spacing (c/h) and the normalized rod diameter (d/h). A linear equation approximating the straight, almost equally-spaced lines of Fig. 1 is given in (3).

$$\log K \doteq \left\{-1.37\left(\frac{c}{h}\right) + 0.91\left(\frac{d}{h}\right) - 0.048\right\}. \tag{3}$$

As presented, the equations and graph of Fig. 1 apply exactly only to the case of equal (C_m/C) on each side of every equal diameter rod. In a "modern network-theory filter" this is, of course, not the case; however, for small percentage couplings the foregoing simple equations and graph can still be used to design such an unequal C_m filter, using equal diameter rods, because of the following fundamental fact: Fringing phenomenon is such that, when the spacing between rods is changed, the Y_0 of any one line with all other lines short-circuited, e.g. ($Y_{0_{23}}+Y_{e_3}+Y_{0_{34}}$), changes negligibly; e.g., even with spacing changes such that the resultant coefficient of coupling changes from zero up to 10 percent, there is less than 3 percent change in this Y_0, even for d/h as large as 0.5. The value of this essentially constant Y_0 is very closely that for a single rod between parallel ground planes given by (4) (page 592 of [5]).

$$Y_0 \doteq 1/138 \log\left(\frac{4}{\pi}\frac{h}{d}\right) \tag{4}$$

Thus, up to 10 percent couplings, the factor ($C_g/\epsilon\ 2C_m/\epsilon$) in the denominator of (2) is essentially independent of changes in rod spacing; and to a first-order the numerical value for the corresponding numerator of (2), i.e. (C_m/ϵ), is still obtainable from Cristal's Fig. 2 [4], even for unequally spaced rods on each side of a given pair.

It should be noted that a graph similar to Fig. 1 but differing by the factor $4/\pi$, has

Manuscript received May 24, 1965; revised June 2, 1965.

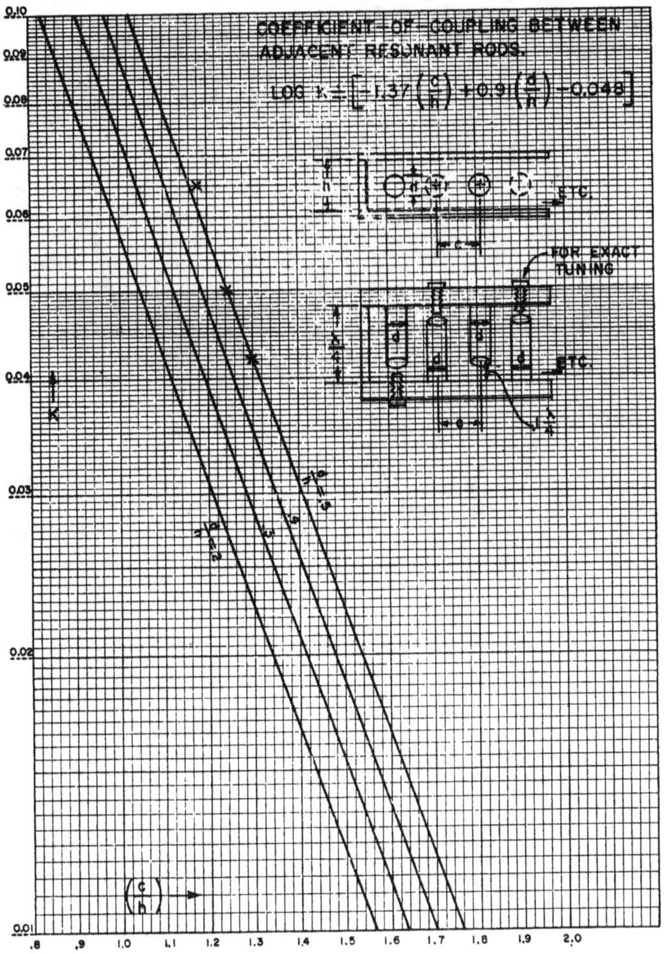

Fig. 1. Coefficient-of-coupling between adjacent resonant rods.

already been presented in [3], but its possible application to an insertion loss type of filter design was not presented.

III. Effect of End Wall Spacing e/h

Because of the method of generator and load coupling to be described in a later section, the end walls can be placed immediately adjacent to the first and last resonators.

With closely spaced end walls the z_0 of the first and last rods with all other rods grounded is given closely by later presented (9).

To obtain the required c_{12}/h, substitution in the basic coefficient of coupling definition gives (5)

$$K_{m_{12}} = K_{12}\sqrt{\frac{Y_{0_1}}{Y_0}} \tag{5}$$

where $K_{m_{12}}$ is the K of Fig. 1, and K_{12} is the value called for by the synthesis procedure, and Y_0 and Y_{01} are given by (4) and (9). It is worth noting that so long as e/h is less than 0.7, the effect of the square root factor is well less than 2 percent for all values of d/h up to 0.5.

IV. Application of Fig. 1 to a Design Example

All design procedures require that a previously accomplished synthesis give the required normalized coefficients-of-coupling, or their equivalents. In Chapter 7 of [5], and in [6], tables of these required, normalized-K's (k) are given for Butterworth, Chebyshev, maximally linear-phase, and Gaussian-amplitude responses; and in Dishal [7] closed-form design equations are given for the normalized-K's required for any n to produce the Butterworth and Chebyshev responses with "lossless" resonators. It is important to realize that in the above three references, the required normalized-K's (k) are normalized to the 3-dB down fractional bandwidth, and to obtain the required actual K from them they are multiplied by $(BW_{3\,\mathrm{dB}}/f_0)$, i.e.,

$$K = k\left(\frac{BW_{3\,\mathrm{dB}}}{f_0}\right). \tag{6}$$

Once one knows the various K's required for a filter, the graph of Fig. 1 [or (1) or, less exactly, (3)] is all that is needed to design the center to center rod spacings in most interdigital filters.

It is apparently standard operating procedure to present a design example for Matthaei's 6 resonator, 10 percent bandwidth, 0.1-dB Chebyshev response shape [8], [4], [1], and we will follow this procedure.

First, let us realize that the 10 percent bandwidth referred to in [8], [4], and [1], is not the 3-dB down-bandwidth, but is the so-called "ripple-bandwidth," which, as Curve 5 on Fig. 9 in Chapter 7 of [5] shows, is 0.914 times the 3-dB down-bandwidth.[1] Thus, the filter involved is to have a 3-dB down percentage bandwidth of 10.95 percent. Next, Fig. 29 in Chapter 7 of [5] shows that for the 0.1-dB ripple response shape, the required 3-dB down normalized-K's are $k_{12}=0.716$, $k_{23}=0.539$, and $k_{34}=0.518$.

The part of the design which is the subject of this section, is then accomplished by the three steps given in the following Table I.

If, the exact K values produced by this design are calculated, we find that the foregoing extremely simple procedure has resulted in errors in K_{23} and K_{34} of less than one percent, errors which are completely tolerable in practice.

Because in Cristal's example only rods 2 and 3 end up with very closely the same diameter, the normalized center-to-center rod spacings (c_{23}/h) and (c_{34}/h) are the ones which should be compared—his values are 1.114 and 1.125, respectively.

V. Correct Loading of the First and Last Resonators by "Tapping"

As indicated in [2] and [7], all coupled-resonator ladder filters of the type considered here require that the first resonator and the last resonator be resistively loaded so that their singly loaded-Q's are set at a value specified by the network synthesis part of the overall design procedure. Along with the required values for the normalized-K's, [2], [5]–[7] give the normalized required values for these singly loaded end Q's. The values given, (q) are normalized to the 3-dB down "fractional mid-frequency" $(f_0/BW_{3\,\mathrm{dB}})$, and to obtain the actual required Q (7) is used

$$Q = q\left(\frac{f_0}{BW_{3\,\mathrm{dB}}}\right). \tag{7}$$

For tap points which are located on the bottom 20 percent of a $\lambda/4$ transmission line, (8) gives the relationship between the tap point and the resulting singly loaded Q (page 576 of [5])

$$Q = \frac{\pi}{4}\left(\frac{R}{Z_0}\right)\left\{1/\sin^2\left(\frac{\pi}{2}\,\frac{l}{L}\right)\right\}. \tag{8}$$

For convenience this equation is graphed on Fig. 2.

It is thus necessary to know the Z_0 of resonator #1 and #n with all other resonators short-circuited. With small percentage couplings the effect of short-circuited rod #2, and short-circuited rod #$(n-1)$ on this Z_0 is, practically speaking, negligible. Thus, with the surface of the end walls a normalized distance e/h from the center of the end rods, (9) can be used to approximate these Z_0's (page 530 of [5]).

$$Z_{0_1} \doteq 138\log\left\{\frac{4}{\pi}\left(\frac{h}{d}\right)\tanh\left(\frac{\pi e}{h}\right)\right\}. \tag{9}$$

[1] Because of the lack of divisions on these figures, three significant figures cannot be read—the exact value was obtained from (15b) of [9] which shows that the relationship between these two bandwidths is

$$\frac{BW_{3\,\mathrm{dB}}}{BW_\beta} = \cosh\left\{\frac{1}{n}\cosh^{-1} 1/[(V_p/V_v)^2 - 1]^{1/2}\right\}.$$

TABLE I

Required K's		From Fig. 1
	Based upon max unloaded-Q, or availability, or etc., pick a normalized rod diameter (d/h): To compare with Cristal we will pick 0.350. Pick a normalized end wall spacing e/h: we will pick 0.6; therefore from (5) $(K_m/K)_{12} = 1.02$	
$K_{12} = 0.716 \times 0.1095 = 0.0784 = K_{56}$		$\left(\frac{c}{h}\right)_{12} = 1.01 = \left(\frac{c}{h}\right)_{56}$
$K_{23} = 0.539 \times 0.1095 - 0.0589 = K_{45}$		$\left(\frac{c}{h}\right)_{23} = 1.11 = \left(\frac{c}{h}\right)_{45}$
$K_{34} = 0.518 \times 0.1095 = 0.0566$		$\left(\frac{c}{h}\right)_{34} = 1.12$

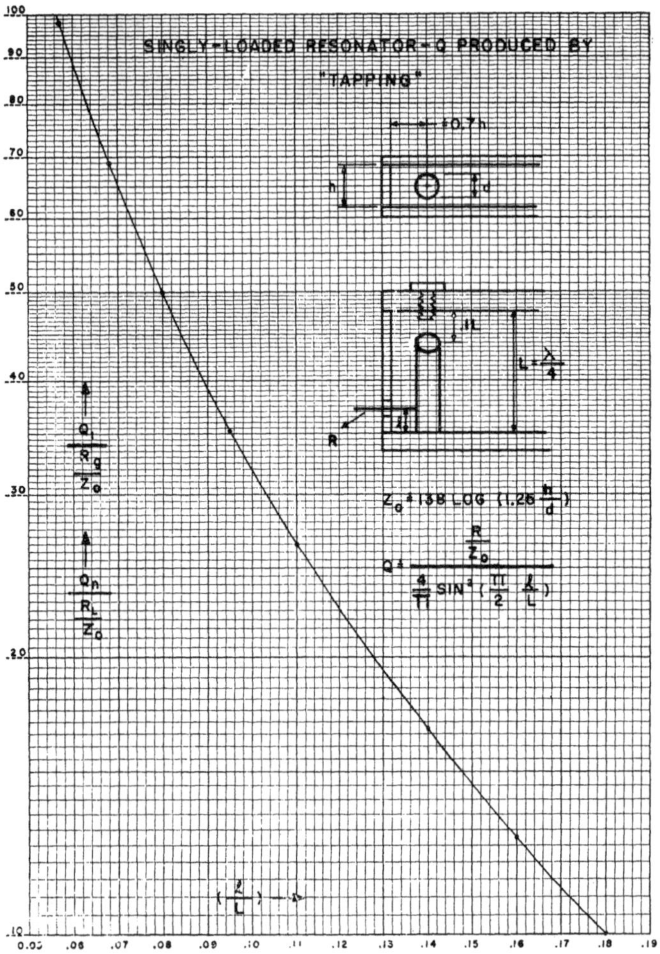

Fig. 2. Singly loaded resonator Q produced by "tapping."

Examination of this equation shows that even with d/h as large as 0.5, the end walls also have negligible practical effect, i.e., less than 2 percent, on Z_0, so long as the e/h used in Table I is greater than 0.75. Within this limit the Z_0 equation simplifies to that given on Fig. 2; if appreciably closer end-wall spacings are used, the full (9) is used. For most practical filter requirements, correction terms for (9) are not required.

Continuing with the $n=6$ design example: the synthesis information on Fig. 29 in Chapter 7 of [5] shows that for the 0.1-dB ripple shape, the 3-dB down singly loaded normalized end-Q's must be $q_1 = q_6 = 1.27$; therefore, $Q_1 = Q_6 = 1.27 \times 9.13 = 11.6$. Substitution of the normalized rod diameter used in our example, i.e., $d/h = 0.350$, in the Z_0 equation on Fig. 2, or in (9), gives $Z_0 = 76$ ohms. If the usual 50-ohm generator is to drive the filter then $Q_1/(R_1/Z_0) = 17.7$; and Fig. 2, or (8), shows that the tap point should be located at $0.136L$. Use of the known load resistance in the step just outlined will then give the required tap point on rod #6.

VI. Correct Tuning of Each Resonator

The final design parameter which must be correctly adjusted is the resonant frequency of each rod (with all other rods short circuited). For good performance each resonance should not differ from the correct frequency by more than 10 percent (approximately) of the 3-dB bandwidth. For small percentage bandwidth filters this is certainly the most important and most critical of the three parameters which must be adjusted. The tuning procedure given in [2] is, of course, directly applicable to interdigital filters. If it is undesirable to mount a loosely coupled alignment probe near the input resonator, it is worth pointing out the perhaps obvious fact that an exactly equivalent alignment procedure is as follows:

1) Feed the f_0 signal to the filter through a slotted line.
2) Short-circuit all rods by means of fine tuning screws.
3) Set the probe of the slotted line very accurately at the location of a zero on the slotted line, and lock the carriage in this position.
4) Tune the first rod for maximum output from the slotted line probe.
5) Tune the second rod for minimum output from the slotted line probe.
6) Third rod for maximum output. Fourth rod for minimum output, etc., etc.

If an interdigital filter having no fine tuning adjustments is desired, then in the engineering model the line lengths must be adjusted so that the preceding phenomenon occurs when the fine tuning screws are just flush with the inner surface of the wall on which they are mounted.

In conclusion I should add (with tongue in cheek?) that one can start, with no geometry calculations whatsoever, and successfully and exactly adjust each K and Q_1 and Q_n by the procedures described in Dishal [2]. In fact, if the pass band performance of the completed filter is not satisfactory, the experimental procedures of [2], or their equivalent, must be used to experimentally check each K and Q_1 and Q_n; this will at least guarantee that the physical filter is truly supplying the element values called for by the synthesis part of the overall design procedure.

M. Dishal
ITT Federal Labs.
Nutley, N. J.

References

[1] P. Vadopalas, with rebuttal by E. G. Cristal, "Coupled rods between ground planes," *IEEE Trans. on Microwave Theory and Techniques (Correspondence)*, vol. MTT-13, pp. 254–255, March 1965.

[2] M. Dishal, "Alignment and adjustment of synchronously tuned multiple-resonant-circuit filters," *Proc. IRE*, vol. 39, pp. 1448–1455, November 1951. Also, *Elec. Commun.*, vol. 29, pp. 154–164, June 1952.

[3] J. T. Bolljahn, and G. L. Matthaei, "A study of the phase and filter properties of arrays of parallel conductors between ground planes," *Proc. IRE*, pp. 299–311, March 1962.

[4] E. G. Cristal, "Coupled circular cylindrical rods between parallel ground planes," *IEEE Trans. on Microwave Theory and Techniques*, vol. MTT-12, pp. 428–439, July 1964.

[5] International Telephone and Telegraph Corp., *Reference Data for Radio Engineers*, 4th Ed. New York: American Book—Stratford Press, Inc.

[6] M. Dishal, "Gaussian-response filter design," *Elec. Commun.*, vol. 36, no. 1, pp. 3–26, 1959.

[7] ——, "Two new equations for the design of filters," *Elec. Commun.*, vol. 30, pp. 324–337, December 1952.

[8] G. L. Matthaei, "Interdigital band-pass filters," *IRE Trans. on Microwave Theory and Techniques*, vol. MTT-10, pp. 479–491, November 1962.

[9] M. Dishal, "Design of dissipative band-pass filters producing desired exact amplitude-frequency characteristics," *Proc. IRE*, vol. 37, pp. 1050–1069, September 1949. Also, *Elec. Commun.*, vol. 27, pp. 56–81, March 1950.

Exact Theory of Interdigital Band-Pass Filters and Related Coupled Structures

R. J. WENZEL, MEMBER, IEEE

Abstract—**An exact theory of interdigital line networks and related coupled structures is presented. The theory of parallel-coupled line arrays is reviewed briefly, and the derivation of exact equivalent circuits from the impedance matrix using modern network synthesis techniques is discussed.**

A simplified theory of equivalent coupled structures is introduced in order to avoid the lengthy analysis required when using the impedance matrix approach. Equivalent networks for the interdigital line are obtained by inspection, using a transformed capacitance matrix associated with the two-dimensional geometry of the conductors and ground planes. The techniques presented are simple to apply and allow a given transmission response to be obtained in a variety of line configurations. A practical design example and experimental results are given to illustrate the simplicity of the approach, along with general criteria for the design of practical filter networks with optimum transmission characteristics.

The paper is directed toward the design of interdigital band-pass filters; however, the techniques presented can be used to analyze and design a much broader class of microwave networks. The relationship of the exact theory to existing approximate theory is discussed.

Glossary

b = ground plane spacing

β = phase constant for TEM propagation along a length of transmission line

$\beta l = \theta$ = electrical length of a transmission line in radians

c' = total static capacitance between conductors per unit length along the conductors

$c = c'/\epsilon$ = ratio of static capacitance between conductors per unit length to the permittivity of the dielectric medium (this ratio is independent of the dielectric medium and depends only on cross-sectional geometry)

C = characteristic admittance of an open-circuited quarter-wavelength line (an S-plane capacitor)

ϵ_r = relative dielectric constant

f = frequency

f_0 = frequency at which a transmission line is a quarter wavelength

I = current

k = an integer used to denote position in a series

$K_{k,k+1}$ = voltage coupling factor between the kth and $(k+1)$st line in an N-line array of parallel conductors

L = characteristic impedance of a short-circuited quarter-wavelength line (an S-plane inductor)

L-C = network elements corresponding to S-plane capacitors C and inductors L

m = number of nonredundant L-C type elements in an optimum filter

n = number of nonredundant unit elements in an optimum filter prototype

N = number of conductors in a parallel array (an integer)

n' = factor used to transform capacitance matrix

η_0 = characteristic impedance of free space (376.7 ohms/square) 91

S-plane = transformed complex frequency plane in which distributed elements are treated as lumped elements

$S = j\tan\ \beta l = j\tan\ \pi f/2f_0$ = transformed frequency variable

s/b = normalized coupled bar line spacing

t/b = normalized coupled bar line thickness

u.e. = unit element, a quarter wavelength of transmission line of characteristic impedance Z_0

V = voltage

w/b = normalized coupled bar line width

z_{11}, z_{22} = open-circuit driving point impedance of a two-port network

z_{12} = open-circuit transfer impedance of a two-port network

Z_0 = characteristic impedance in ohms

Z_{0k} = characteristic impedance of the kth line in the presence of the other open-circuited lines in an N-line parallel array of infinite length conductors.

Manuscript received March 16, 1965; revised May 24, 1965. The work reported in this paper was supported by the U. S. Army Electronics Research and Development Laboratories, Fort Monmouth, N. J., under Contract DA-28-043-AMC-00399(E).

The author is with the Research Laboratories Division, Bendix Corporation, Southfield, Mich.

I. Introduction

APPROXIMATE design techniques for interdigital filters have been presented by Matthaei [1] and have been used to design both narrow-band and wide-band filters. While these techniques are sufficiently accurate for many engineering applications, an exact theory is desirable when optimum network performance or special network configurations are required. The use of exact synthesis procedures allows greater flexibility in design and physical form, and makes possible a more accurate determination of practical circuit limitations.

The work described in this paper is presented in the following manner: In Section II the basic theory of interdigital structures is reviewed briefly. In Section III a general synthesis procedure starting with the impedance matrix of a parallel-coupled line array is described. This procedure can be used to derive an exact cascade equivalent circuit suitable for obtaining optimum network designs. However, because of the tedious nature of this method, equivalent circuits for the interdigital filter are obtained by using a computationally simpler approach, described in Section IV. A simplified theory of equivalent coupled structures based on the two-dimensional geometry of the conductors and ground planes is presented (Section IV). By using the techniques described, exact equivalent circuits of interdigital structures are readily derived. The relatively simple methods can be used to obtain physical realizations of an optimum filter prototype in the form of coaxial structures, interdigital structures, or combinations of the specified network forms.

In Section V a practical design example and experimental results are given which illustrate the simplicity and flexibility of the equivalent network design approach. In Section VI general design criteria for practical network configurations are discussed, along with the relationship of the exact theory to Matthaei's approximate design procedure.

II. Parallel Line Arrays

The basic network to be considered in obtaining an exact equivalent circuit for the interdigital line is the $2N$ port consisting of an array of N-parallel open-circuited conductors between ground planes, as presented in Fig. 1. Bolljahn and Matthaei [2] have shown how to obtain the impedance matrix for an array of parallel conductors all of which have the same characteristic impedance Z_0. The general matrix for an array of conductors with different characteristic impedances Z_{0k} can be obtained by a simple extension of their method. The general network equations in matrix form can be shown to be

$$\begin{bmatrix} V_{1A} \\ V_{2A} \\ \vdots \\ V_{NA} \\ V_{1B} \\ V_{2B} \\ \vdots \\ V_{NB} \end{bmatrix} = \frac{1}{S} \left[\begin{array}{cccc|cccc} Z_{01} & Z_{01}K_{12} & \cdots & Z_{01}K_{1N} & Z_{01}\sqrt{1-S^2} & Z_{01}K_{12}\sqrt{1-S^2} & \cdots & Z_{01}K_{1N}\sqrt{1-S^2} \\ Z_{02}K_{21} & Z_{02} & \cdots & Z_{02}K_{2N} & Z_{02}K_{21}\sqrt{1-S^2} & Z_{02}\sqrt{1-S^2} & \cdots & Z_{02}K_{2N}\sqrt{1-S^2} \\ \vdots & \vdots & \vdots & \vdots & \vdots & \vdots & \vdots & \vdots \\ Z_{0N}K_{N1} & Z_{0N}K_{N2} & \cdots & Z_{0N} & Z_{0N}K_{N1}\sqrt{1-S^2} & Z_{0N}K_{N2}\sqrt{1-S^2} & \cdots & Z_{0N}\sqrt{1-S^2} \\ \hline Z_{01}\sqrt{1-S^2} & Z_{01}K_{12}\sqrt{1-S^2} & \cdots & Z_{0N}K_{1N}\sqrt{1-S^2} & Z_{01} & Z_{01}K_{12} & \cdots & Z_{01}K_{1N} \\ Z_{02}K_{21}\sqrt{1-S^2} & Z_{02}\sqrt{1-S^2} & \cdots & Z_{02}K_{2N}\sqrt{1-S^2} & Z_{02}K_{21} & Z_{02} & \cdots & Z_{02}K_{2N} \\ \vdots & \vdots & \vdots & \vdots & \vdots & \vdots & \vdots & \vdots \\ Z_{0N}K_{N1}\sqrt{1-S^2} & Z_{0N}K_{N2}\sqrt{1-S^2} & \cdots & Z_{0N}\sqrt{1-S^2} & Z_{0N}K_{N1} & Z_{0N}K_{N2} & \cdots & Z_{0N} \end{array}\right] \begin{bmatrix} I_{1A} \\ I_{2A} \\ \vdots \\ I_{NA} \\ I_{1B} \\ I_{2B} \\ \vdots \\ I_{NB} \end{bmatrix} \quad (1)$$

The characteristic impedance Z_0 of a lossless uniform transmission line operating in the TEM mode is related to its shunt capacitance [3] by

$$Z_0 = \frac{\eta_0}{\sqrt{\epsilon_r}(c'/\epsilon)} = \frac{\eta_0}{\sqrt{\epsilon_r}(c)} \cdot \quad (2)$$

The dimensionless ratio $c = c'/\epsilon$ simplifies the symbols required and can be used directly with Getsinger's [3] or Cristal's [4] charts to obtain physical dimensions.

The characteristic impedances Z_{0k} and coupling factors $K_{k,k+1}$ are determined in terms of the dimensionless self and mutual static capacitances per unit length, c, which exist between the open-circuited lines and ground, as shown in Fig. 2(a). The Z_{0k} values are obtained by substituting in (2) the total capacitance c_{0k} from the kth line to ground. The total capacitance can be obtained by series-parallel reduction. The coupling factors between any two lines, k and $k+1$, can

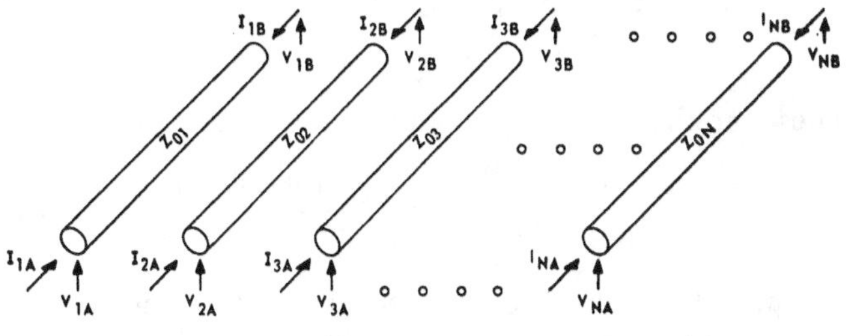

Fig. 1. Basic parallel-coupled line array showing important network variables.

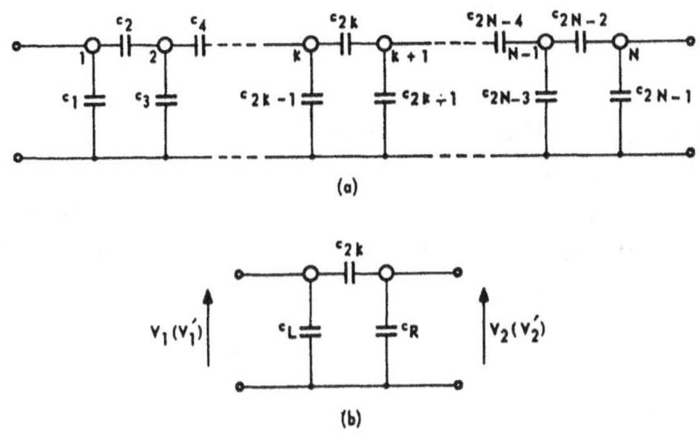

Fig. 2. Two-dimensional capacitance array of the interdigital line.

be computed by replacing the capacitance to the left of line k by an equivalent capacitor c_L, and the capacitance to the right of line $k+1$ by a capacitor c_R, as shown in Fig. 2(b). The values of c_L and c_R are obtained, as before, by series-parallel reduction. The coupling factors are then given by

$$K_{k,k+1} = \frac{V_2}{V_1} = \frac{\frac{1}{c_R}}{\frac{1}{c_R}+\frac{1}{c_{2k}}} = \frac{c_{2k}}{c_{2k}+c_R},$$

$$K_{k+1,k} = \frac{V_1'}{V_2'} = \frac{\frac{1}{c_L}}{\frac{1}{c_L}+\frac{1}{c_{2k}}} = \frac{c_{2k}}{c_{2k}+c_L}, \tag{3}$$

where V_1 is the applied voltage and V_2 is the coupled voltage when determining $K_{k,k+1}$, and V_2' is the applied voltage and V_1' is the coupled voltage when determining $K_{k+1,k}$. Note that in general $K_{k,k+1} \neq K_{k+1,k}$.

The representation of the coupled array by the circuit of Fig. 2(a) assumes that no significant direct coupling exists between nonadjacent lines. Coupling of this type would require the addition of capacitors between nonadjacent lines and would greatly complicate the analysis and synthesis problem. As pointed out by Bolljahn [2] and Matthaei [1], [2], for many practical line configurations, the neglect of coupling of this type is not a serious limitation. Assuming no direct coupling between nonadjacent conductors, the higher-order coupling factors that appear in (1) are obtained by multiplying the lower-order coupling factors. For example,

$$K_{k,k+3} = (K_{k,k+1})(K_{k+1,k+2})(K_{k+2,k+3}). \tag{4}$$

Finally, reciprocity requires that terms located in a manner symmetric about the principal diagonal of the impedance matrix of (1) be equal, i.e., that $z_{ik}=z_{ki}$ for $i \neq k$. Therefore,

$$Z_{01}K_{12} = Z_{02}K_{21}, \qquad Z_{01}K_{13} = Z_{03}K_{31},$$
$$Z_{02}K_{23} = Z_{03}K_{32}, \text{ etc.} \tag{5}$$

Further details concerning the network properties of parallel-coupled line arrays can be found in References [1] and [2].

III. Equivalent Circuit of the Interdigital Line Using Modern Network Synthesis Techniques

The array of conductors in Fig. 1 becomes an interdigital line by alternately shorting and leaving open opposite ends of each conductor. The array can begin with either an open-circuited or a short-circuited line [1]. The input and output ports are on the same side of the array for N odd and on opposite sides for N even. A general network approach that can be used to obtain an exact equivalent circuit for the interdigital filter is as follows: 93

1) The pertinent port conditions, $V=0$ at a short circuit and $I=0$ at an open circuit, are applied to matrix equation (1) to obtain the network equations for the corresponding interdigital structures.
2) The resulting equations are solved simultaneously and reduced to a 2×2 impedance matrix so as to recognize the two-port impedance parameters z_{11}, z_{12}, and z_{22}.
3) Modern network synthesis techniques are used to obtain a network that realizes the specified impedance parameters.

For the general case, the $[z]$ parameters obtained by carrying out steps 1) and 2) cannot be realized in a network form that involves only L's, C's, and ideal transformers (where L and C refer to the impedance and admittance of a short-circuited and open-circuited quarter wavelength of transmission line, respectively [5]). As will be shown, the interdigital filter with open-circuited terminating lines has a third-order pole at dc corresponding to three L-C elements, and the interdigital line with short-circuited terminating lines has a first-order pole at dc corresponding to one shunt L. The remaining N-3 elements for the N-line network with open-circuited terminating lines and N-1 elements for the network with short-circuited terminating lines are unit elements (u.e.).

The analysis and synthesis procedure described in steps 1) through 3) has been carried out for symmetric networks with up to eight lines, and for a general asymmetric network with up to four lines.[1] This process is long and tedious, and becomes intractable for networks with a large number of lines. As network equiv-

[1] A detailed discussion of the synthesis procedure, including examples, is given in Reference [6].

alences were obtained, a definite pattern was noticed in the resulting equivalent relationships; this pattern led to an alternate method of derivation. The main value of the impedance matrix approach is that it provides an alternate derivation method—and thus a check on the technique to be presented in Section IV.

IV. Simplified Theory of Equivalent Coupled Structures

The electrical behavior of an array of parallel conductors can be described either by matrix equation (1) or by the capacitance network of Fig. 2(a). In this section, the capacitance network and its admittance-derived matrix will be used to obtain equivalent circuits in a manner computationally simpler than that of Section III. The techniques to be presented will also provide insight into the physical form of equivalent networks.

A. General Network Relations

Before discussing specific applications, some general relations involving capacitance networks in the form of Fig. 2(a) will be presented.[2] A five-node network will be used to illustrate the method. The extension to more complex networks follows in a straightforward manner. Referring to Fig. 2(a), with $N=5$, the capacitance matrix[3] is given by

$$\begin{bmatrix} c_1+c_2 & -c_2 & 0 & 0 & 0 \\ -c_2 & c_2+c_3+c_4 & -c_4 & 0 & 0 \\ 0 & -c_4 & c_4+c_5+c_6 & -c_6 & 0 \\ 0 & 0 & -c_6 & c_6+c_7+c_8 & -c_8 \\ 0 & 0 & 0 & -c_8 & c_8+c_9 \end{bmatrix}. \tag{6}$$

Multiplying the kth row and column of this matrix by any factor n' corresponds to multiplying the admittance level of the kth node by n'^2, and the transfer admittance between nodes adjacent to the kth node by n'. If this transformation is performed on any interior row and column, the two-port performance of the network is unchanged; if performed on an end row and column, the admittance level of the corresponding port is multiplied by n'^2 and the overall transfer admittance is multiplied by n'. Performing multiple transformations allows an infinite number of equivalent networks and port admittance levels to be obtained. In performing these transformations, care must be taken to insure that the final circuit contains no negative elements. Physical realizability requires that the sum of the elements of any row or column be greater than (or equal to) zero, and gives a limitation on the choice of n'. For example, multiplying the second row and column of matrix (6) by n' leads to the constraints

$$\begin{aligned} n'^2(c_2+c_3+c_4) &\geq n'(c_2+c_4) \\ c_1+c_2 &\geq n'c_2. \end{aligned} \tag{7}$$

These require that n' be confined to the range

$$\frac{c_2+c_4}{c_2+c_3+c_4} \leq n' \leq \frac{c_1+c_2}{c_2}. \tag{8}$$

An important aspect of transformations on the capacitance matrix is that the product of the coupling factors, having indices which are interchanged, remains invariant,[4] i.e.,

$$(K_{k,k+1})(K_{k+1,k}) = K^2(\text{invariant}). \tag{9}$$

Multiplying a row and column by n' corresponds to introducing an admittance level change of n'^2 and results in changing $K_{k,k+1}$ to $n'(K_{k,k+1})$ and $K_{k+1,k}$ to $(1/n')(K_{k+1,k})$. Consequently, once any pair of coupling factors is found, those for all equivalent networks may readily be determined.

B. Forms of Coupling

In carrying out these transformations, the capacitors are treated as though they represent static capacitances in a planar geometry. From a microwave viewpoint, they represent coupling between quarter-wave lines, and if the form of the coupling could be determined, the microwave equivalent circuit could be obtained by inspection. Consider, for example, the single-line network of Fig. 3(a), page 565, and its capacitor representation in Fig. 3(b). The static capacitor of Fig. 3(b) can represent three distinct forms of coupling, depending on the terminal conditions and the terminals between which the electrical performance is observed:

1) An S-plane capacitor[5]—the electrical behavior of the line at either port with the other port open-circuited. The value of the S-plane capacitor is given by $C=1/Z_0=\sqrt{\epsilon_r}c/\eta_0$, where Z_0 is the characteristic impedance of the line as defined by (2).
2) An S-plane inductor—the electrical behavior of the line at either port with the other port short-circuited. The value of the inductor is given by $L=Z_0=\eta_0/\sqrt{\epsilon_r}c$.
3) A unit element—the electrical behavior of the line between the ports. The impedance of the unit element is equal to $Z_0=\eta_0/\sqrt{\epsilon_r}c$.

[2] The basic approach taken is based on the discussion of equivalent networks given in Reference [7], pp. 141–176.

[3] The diagonal terms in the capacitance matrix are given by the sum of the capacitors directly connected to a node, while the off-diagonal terms are the capacitors directly connected between nodes. The matrix is related to the short-circuit admittance matrix, resulting in negative off-diagonal terms [7].

[4] A proof of the invariance of the product of the coupling factors follows in a straightforward manner by applying the transformations to a generalized array and computing the new coupling factors. A discussion of the importance of this property is given in the Appendix.

[5] A detailed discussion of S-plane elements and their properties is given in Reference [5].

If the general array of Fig. 2(a) can be put into a form in which the previously mentioned types of coupling become easily recognizable, an equivalent S-plane network can be found by inspection. Transformations of the capacitance matrix preserve the performance of the network, with the possible exception of a change in impedance level, and offer the possibility of changing the network form into one in which the nature of each coupling is easily recognizable. Furthermore, the capacitance network also represents static capacitors and can be used to obtain dimensions for a physical realization.

To observe the effect that transformations of the capacitance matrix have on the physical form of the network, consider the symmetric capacitor circuit shown in Fig. 4(a) and the interdigital line shown in Fig. 4(b). For simplicity, all the capacitor values are chosen to be unity. The capacitance network indicates that there is coupling from all lines to ground and between adjacent lines as is evident from the physical configuration. The capacitance matrix is

$$\begin{bmatrix} 2 & -1 & 0 \\ -1 & 3 & -1 \\ 0 & -1 & 2 \end{bmatrix}. \tag{10}$$

Multiplying the second row and column of the matrix by n' leaves the two-port properties invariant, but yields an equivalent network with a capacitance matrix given by

$$\begin{bmatrix} 2 & -n' & 0 \\ -n' & 3n'^2 & -n' \\ 0 & -n' & 2 \end{bmatrix}. \tag{11}$$

Physical realizability requires that

$$3n'^2 \geq 2n' \quad \text{and} \quad 2 \geq n'$$

or

$$\frac{3}{2} \leq n' \leq 2. \tag{12}$$

Suppose an n' of 5/3 is chosen. The resulting capacitance matrix is given by

$$\begin{bmatrix} 2 & -\frac{5}{3} & 0 \\ -\frac{5}{3} & \frac{25}{3} & -\frac{5}{3} \\ 0 & -\frac{5}{3} & 2 \end{bmatrix}. \tag{13}$$

The corresponding capacitance network, shown in Fig. 4(c), indicates that the coupling between the ground planes and the outer conductors has decreased, whereas the coupling between conductors, and the coupling between the center conductor and ground, has increased. A physical network that would accomplish this change in coupling is shown in Fig. 4(d).

If a limiting value of $n'=2$ is chosen in the matrix (11), the resulting matrix becomes

$$\begin{bmatrix} 2 & -2 & 0 \\ -2 & 12 & -2 \\ 0 & -2 & 2 \end{bmatrix}. \tag{14}$$

The capacitance network and a physical realization are shown in Figs. 4(e) and 4(f), respectively. The outer lines are no longer coupled to the ground plane but only to the center line. The center line is coupled to both outer lines and to the ground plane. From Fig. 4(f), the form of each coupling is readily determined and an exact equivalent circuit can be written by inspection. The coupling between the outer lines and the center line (terminals ① – ② and ②–③) is that of a series S-plane capacitor. The coupling between the center line and ground (terminals ②–④) is that of an S-plane inductor. The equivalent circuit between the input and output ports is thus that of a three-element ladder, as shown in Fig. 4(g). Since the transformations performed do not change the two-port performance, the circuit of Fig. 4(g) is a general equivalent circuit[6] for networks with any choice of n'. The capacitor values in the planar representation can be linked directly to the L and C values of the S-plane equivalent circuit by (2) and the discussion at the beginning of Section IV-B. The transformation of matrices (10) through (14) can then be worked in reverse order to obtain an interdigital realization of the circuit rather than a coaxial one. The transformed capacitor values can be used directly [3], [4] to obtain line dimensions.

C. Interdigital Line Equivalent Circuits

The equivalent circuits for more complex interdigital lines are obtained in an analogous manner. To facilitate their derivation, it is useful to know a few of the general results concerning capacitance networks of the type under consideration.

A capacitance network containing both self and mutual capacitors can be obtained by transformation from a capacitance network having an odd number of nodes containing all series c's and one shunt c from the center line to the ground plane. Similarly, a capacitance network containing both self and mutual capacitors can be obtained by transforming a capacitance network having an even number of nodes containing all series c's and shunt c's from the two nodes nearest the center.[7] These statements are illustrated in Figs. 5(a) and 5(b) for the odd and even cases, respectively, and can be demonstrated by applying the transformations already outlined. The reason for using a network form containing a small number of shunt c's is that the nature of the coupling between lines is readily found in terms

[6] The circuit of Fig. 4(g) describes the transmission response for all line configurations but does not indicate the impedance level of the interior node.

[7] The transmission properties (for fixed load impedances) of all networks obtained in this manner are identical if the admittance levels of the outer nodes are not changed.

of S-plane elements. A cascade equivalent circuit that describes the transmission characteristics of all equivalent networks then can be obtained by inspection. There is no absolute reason for using the forms of Fig. 5 rather than, for example, those that have all series c's and one shunt c from an arbitrary node. However, the forms shown in Fig. 5 are particularly useful for symmetric networks and can often be used for asymmetric networks without the addition of ideal transformers.

An example derivation of the S-plane equivalent circuit will be given for each of the basic interdigital line configurations. For the first example, consider deriving the equivalent circuit of a five-line network with open-circuited terminating lines. The capacitance network form with all series c's and one shunt c is shown in Fig. 6(a), and a coaxial realization[8] is shown in Fig. 6(b). The lines and port conditions have been added to the capacitance array to aid in reasoning to the equivalent circuit. Referring to Fig. 6, the following three steps are taken to determine the equivalent circuit:

96 1′) Line ① couples only to line ②, and the coupling has the form of a series S-plane capacitor C_1 because terminal 1(b) is open. The distributed capacitor exists between terminals 1(a) and 2(a) and corresponds to a static capacitance c_2, from (2), given by $c_2 = (\eta_0 C_1)/\sqrt{\epsilon_r}$. An identical situation holds between lines ④ and ⑤ and results in $c_8 = (\eta_0 C_2)/\sqrt{\epsilon_r}$.

2′) Line ③ couples to ground, and the coupling has the form of an S-plane inductor of value L because of the shorted terminal 3(a). The inductor exists between terminals 3(b) and the ground plane, and corresponds to a static capacitance c_5, given by $c_5 = \eta_0/\sqrt{\epsilon_r}L$.

3′) The coupling between lines ② and ③ (or ③ and ④) is in the form of a coaxial unit element with a 180° phase reversal. More specifically, consider lines ② and ③:

a′) End 2(a) is an ungrounded terminal at the (a) end of the u.e.
b′) End 3(a) is end 2(a)'s ground terminal.
c′) End 3(b) is the ungrounded terminal at the (b) end of the u.e.
d′) End 2(b) is its respective ground.

A diagram and equivalent circuit of the coupling between lines ② and ③ is shown in Fig. 6(c). Note the presence of the $-1{:}1$ ideal transformer to represent the 180° phase reversal. The static capacitors c_4 and c_6 (corresponding, respectively, to the unit-element characteristic impedance values Z_1 and Z_2) are given by

$$c_4 = \frac{\eta_0}{\sqrt{\epsilon_r}Z_1} \quad \text{and} \quad c_6 = \frac{\eta_0}{\sqrt{\epsilon_r}Z_2}\,.$$

The complete equivalent S-plane circuit is shown in Figure. 6(d). The two $-1{:}1$ ideal transformers can be combined without changing the transmission characteristics and result in the final equivalence of Fig. 6(e).

[8] It is interesting to note the relation of the coaxial interdigital network realization to Cohn's [8] re-entrant section. The interdigital filter is seen to be similar to a multiple re-entrant section with suitable terminal conditions.

The coupling factors and characteristic impedances of the lines can be obtained quite simply from the array of Fig. 6(e) by using the appropriate c's in (2). For example,

$$Z_{01} = \frac{\eta_0}{\sqrt{\epsilon_r}(c_{01})} = \frac{\eta_0}{\sqrt{\epsilon_r}}\left(\frac{1}{c_2} + \frac{1}{c_4} + \frac{1}{c_5}\right) = \frac{1}{C_1} + Z_1 + L$$

and similarly,

$$Z_{06} = Z_2 + L + \frac{1}{C_2}\,.$$

For the coupling factors,

$$K_1^2 = (K_{12})(K_{21}) = \left[\frac{\frac{1}{c_4} + \frac{1}{c_5}}{\frac{1}{c_2} + \frac{1}{c_4} + \frac{1}{c_5}}\right](1) = \frac{Z_1 + L}{\frac{1}{C_1} + Z_1 + L} \tag{15}$$

and similarly,

$$K_2^2 = K_{23}K_{32} = \frac{L}{Z_1 + L}$$

$$K_3^2 = K_{34}K_{43} = \frac{L}{Z_2 + L}$$

$$K_4^2 = K_{45}K_{54} = \frac{Z_2 + L}{\frac{1}{C_2} + Z_2 + L}\,.$$

The results obtained using the capacitance matrix approach are identical to those that were obtained using the method described in Section III. For the second example, consider the six-line network with short-circuited terminating lines. The capacitance network and a physical realization are shown in Figs. 7(a) and 7(b), respectively. The S-plane equivalent circuit with numbered terminal pairs is shown in Fig. 7(c), and the final equivalence is shown in Fig. 7(d). The steps followed in obtaining the equivalent circuit are similar to those of the first example, as are the methods for obtaining the Z_{0k} and coupling coefficients.

The equivalent circuits for interdigital line networks with any number of lines are given in Tables I and II, along with the relationships between the S-plane equivalent circuit and the capacitance network element values. It is interesting to note that the equivalent circuit for the N-2 line network with short-circuited terminating lines can be obtained by letting the input capacitors of the N-line network with open-circuited terminating lines become infinite. An infinite number of equivalent networks can be obtained by transforming the capacitance network, the final result of any set of transformations giving a set of self and mutual capacitors that can be used with Getsinger's [3] or Cristal's [4] charts to obtain physical dimensions. It is obvious that no preferred physical configuration exists. A practical realization may take forms ranging from an all-coaxial structure to an all-interdigital line, or combinations of the two. The physical form that is most practical depends on the network element values required for a given bandwidth and filter type.

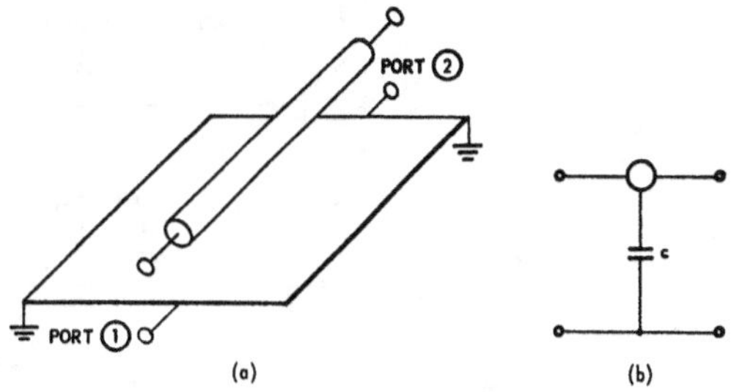

Fig. 3. Determination of S-plane elements for a single line above a ground plane.

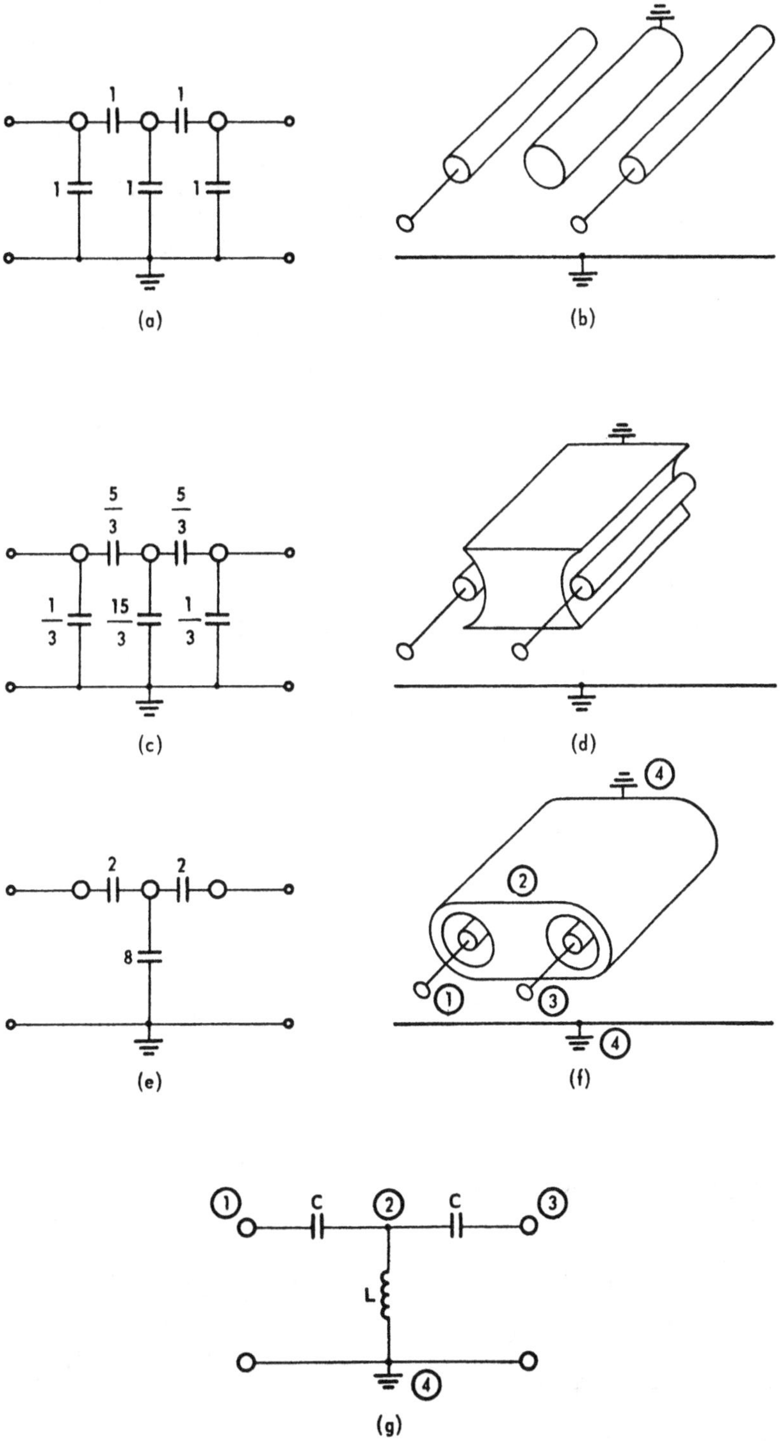

Fig. 4. Effect of transformations of the capacitance matrix on physical form.

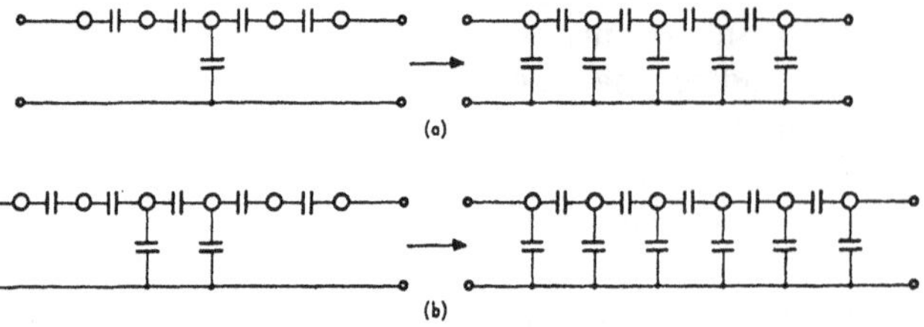

Fig. 5. Capacitance network equivalences.

98

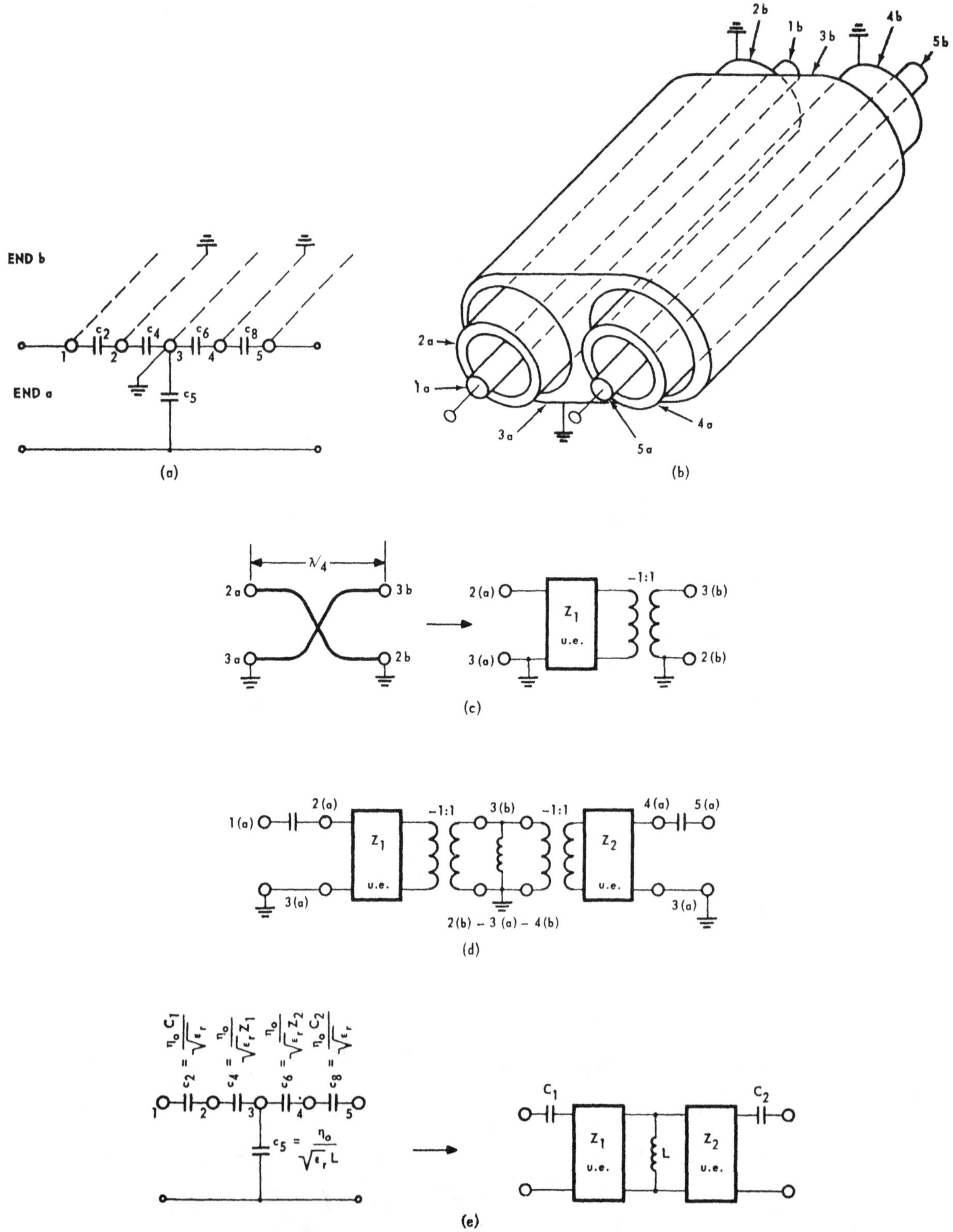

Fig. 6. Derivation of the S-plane equivalent circuit of the five-line interdigital network with open-circuited terminating lines.

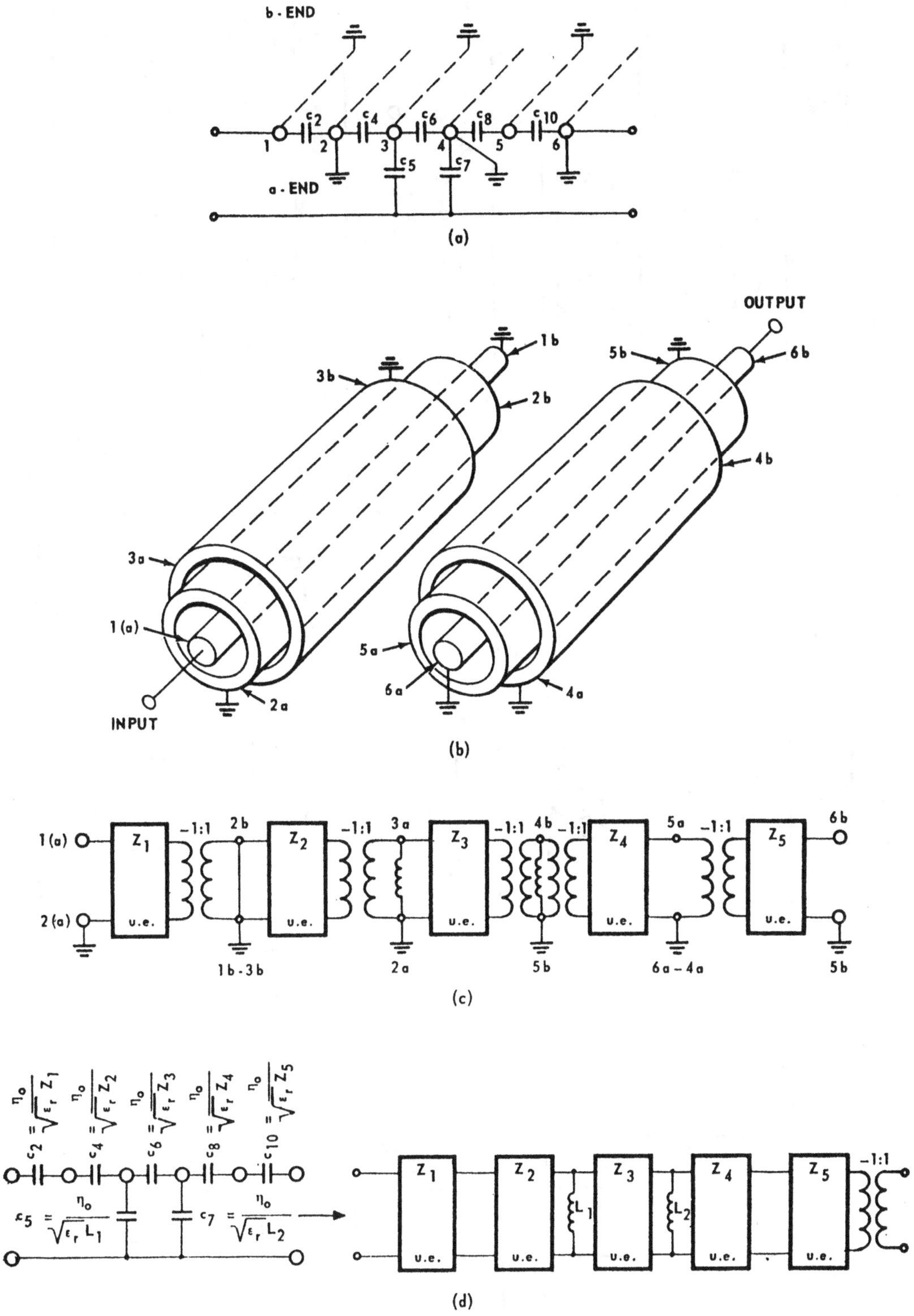

Fig. 7. Derivation of the S-plane equivalent circuit of the six-line interdigital network with short-circuited terminating lines.

100

TABLE I
EQUIVALENT CIRCUITS FOR INTERDIGITAL NETWORKS WITH OPEN-CIRCUITED TERMINATING LINES

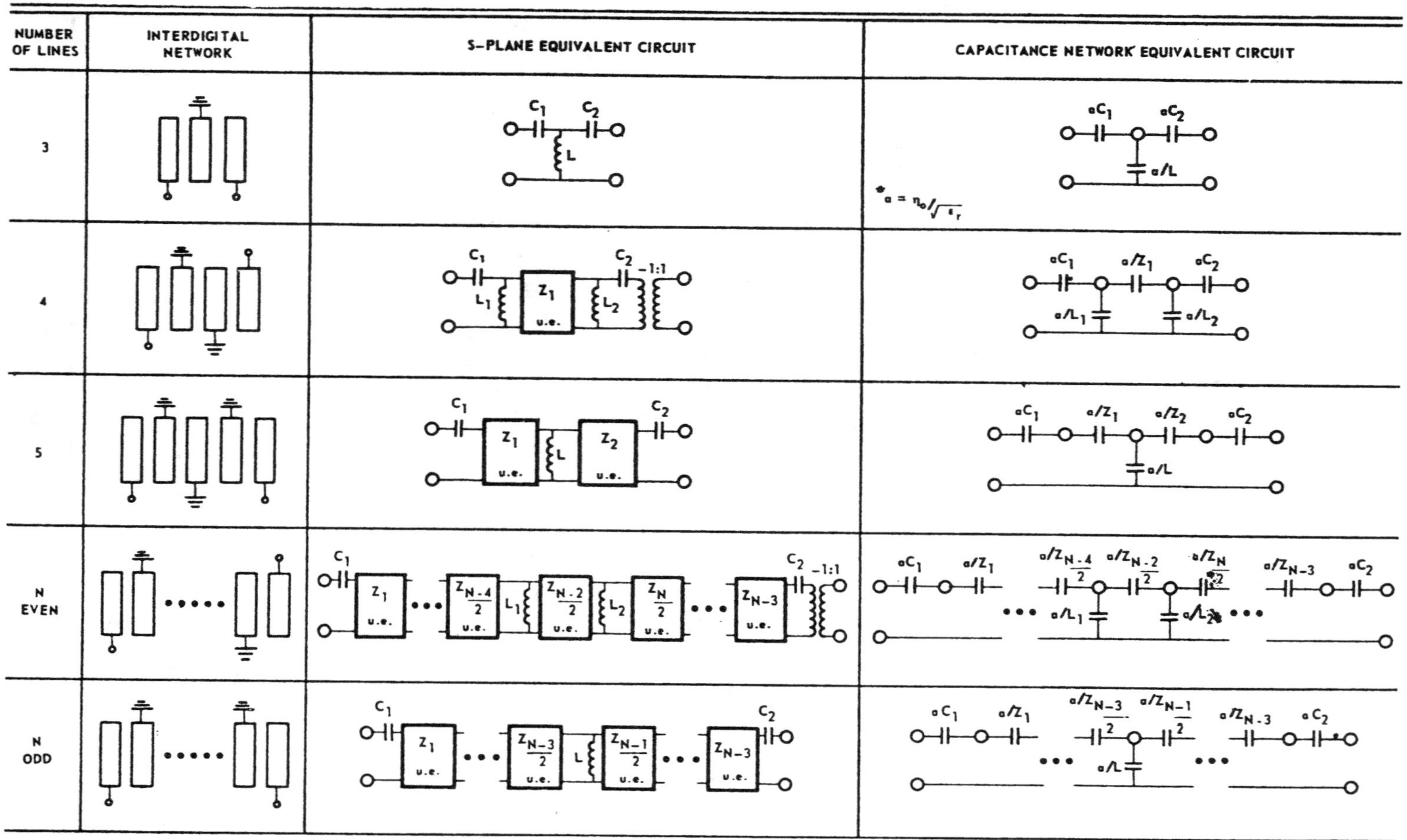

*$a = \eta_0/\sqrt{\epsilon_r} = 7.54/\sqrt{\epsilon_r}$ FOR USE IN OBTAINING 50 OHM DESIGNS USING S-PLANE ELEMENT VALUES NORMALIZED TO 1 OHM

TABLE II
EQUIVALENT CIRCUITS FOR INTERDIGITAL NETWORKS WITH SHORT-CIRCUITED TERMINATING LINES

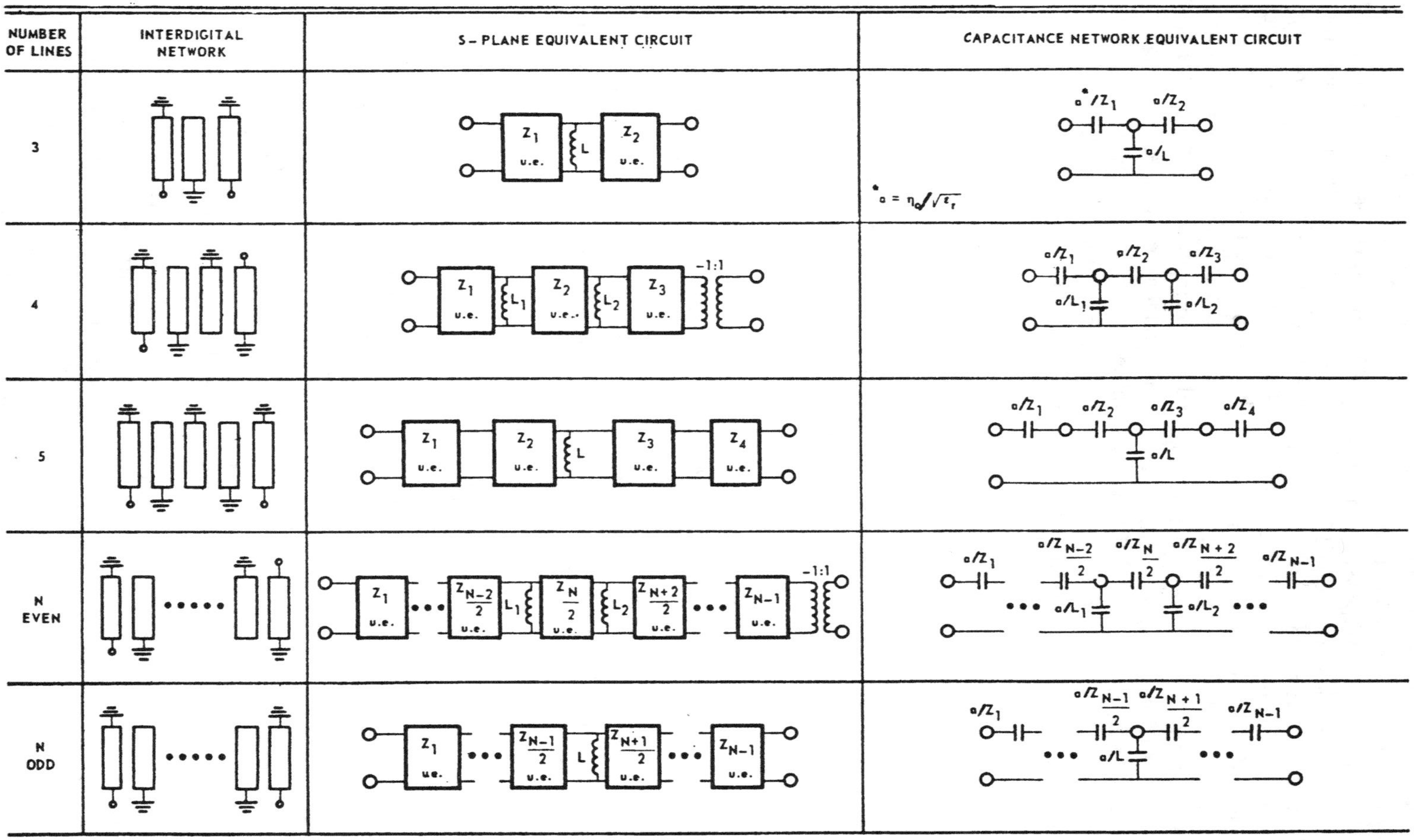

*$a = \eta_0/\sqrt{\epsilon_r} = 7.54/\sqrt{\epsilon_r}$ FOR USE IN OBTAINING 50 OHM DESIGNS USING S-PLANE ELEMENTS NORMALIZED TO 1 OHM

V. Design Example and Experimental Results

The S-plane element values for prototype networks such as those of Tables I and II must be determined in order to obtain a practical filter realizable in an interdigital or related form. The problem of determining element values that give an optimum response (in a Butterworth or Chebyshev sense) has been solved in detail by Horton and Wenzel [9], and the synthesis procedures described can be used to obtain tables of element values for a wide range of bandwidths and filter types.[9] Inspection of Tables I and II shows that the N-line filter with open-circuited terminating lines is an $m=3$, $n=N$-3 prototype, and the N-line filter with short-circuited terminating lines is an $m=1$, $n=N$-1 prototype. The number of nonredundant L-C elements and unit elements is given by m and n, respectively.[10]

To illustrate the simplicity and flexibility of the physical form incorporated in the design procedure, a six-section 3:1 bandwidth 0.05-dB-ripple Chebyshev filter will be described. The element values, shown (with the symmetric prototype) in Fig. 8(a), were obtained by synthesis using the procedure given in Reference [9]. The self and mutual static capacitor values are obtained using the relationships in Table I, and the corresponding capacitance network is shown in Fig. 8(b). The network could be constructed (depending on the practicality of resulting dimensions) in the coaxial form shown in Fig. 7(b) directly from the prototype element values. An interdigital realization is obtained by transforming the capacitance matrix corresponding to the array of Fig. 8(b). This matrix, obtained by inspection, is

$$\frac{7.54}{\sqrt{\epsilon_r}}\begin{bmatrix} 1.00 & -1.00 & 0 & 0 & 0 & 0 \\ -1.00 & 2.80 & -1.80 & 0 & 0 & 0 \\ 0 & -1.80 & 9.32 & -3.60 & 0 & 0 \\ 0 & 0 & -3.60 & 9.32 & -1.80 & 0 \\ 0 & 0 & 0 & -1.80 & 2.80 & -1.00 \\ 0 & 0 & 0 & 0 & -1.00 & 1.00 \end{bmatrix}. \tag{16}$$

$$\frac{7.54}{\sqrt{\epsilon_r}}\begin{bmatrix} 1 & -n'_1 & 0 & 0 & 0 & 0 \\ -n'_1 & 2.80n'^2_1 & -1.80n'_1n'_2 & 0 & 0 & 0 \\ 0 & -1.80n'_1n'_2 & 9.32n'^2_2 & -3.60n'^2_2 & 0 & 0 \\ 0 & 0 & -3.60n'^2_2 & 9.32n'^2_2 & -1.80n'_1n'_2 & 0 \\ 0 & 0 & 0 & -1.80n'_1n'_2 & -2.80n'^2_1 & -n_1 \\ 0 & 0 & 0 & 0 & -n'_1 & 1 \end{bmatrix}\begin{matrix} \\ \leftarrow \times n'_1 \\ \leftarrow \times n'_2 \\ \leftarrow \times n'_2 \\ \leftarrow \times n'_1 \\ \\ \end{matrix} \tag{17}$$

$$\begin{matrix} \uparrow & \uparrow & \uparrow & \uparrow \\ \times n'_1 & \times n'_2 & \times n'_2 & \times n'_1 \end{matrix}$$

The steps required to design an optimum filter in an interdigital or related form can be summarized as follows:

1) From tables[9] or by synthesis [9], obtain the S-plane element values (L, C, and Z).
2) Using the relationships in Tables I and II, obtain static capacitors for use directly with Getsinger's [3] or Cristal's [4] charts.
3) Equivalent networks that realize the same transmission response as the prototype network are obtained by applying the transformation presented in Section IV.

An all-interdigital realization can be obtained by multiplying the interior rows and columns by constants, as shown in matrix (17). The values of these constants are restricted by realizability conditions as discussed in Section IV, but are otherwise arbitrary. In general, the factors (n_i') that multiply the interior nodes need not be chosen in a symmetric manner; this is necessary only if a symmetric physical structure is desired. Different load admittance levels can be accommodated by changing the admittance levels of either or both outer nodes.

For the 3:1 bandwidth filter under consideration, the convenient value of the series capacitors in the prototype circuit allows them to be incorporated coaxially in the terminating lines of a four-line interdigital realization of the interior prototype elements. By incorporating the capacitors coaxially in the end elements, as shown in Fig. 8(c), relaxed mechanical tolerances are obtained as compared with those required for an all-interdigital realization, and the filter is more compact. With the capacitance matrix approach, the designer has

[9] Tables of element values for symmetric filters with equal terminating impedances are presently being computed and will be included in the Final Report of USAERDL Contract DA-28-043-AMC-00399(E). Element values for single terminated networks required in multiplexer [10] design can be obtained by letting the functions described in Reference [9] represent transfer impedances or admittances.

[10] A detailed discussion of optimum filter theorv and the parameters m and n is given in Reference [9].

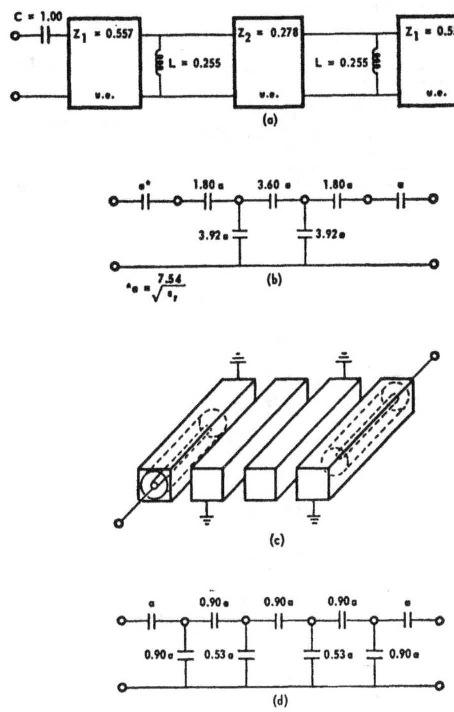

Fig. 8. Six-section 3:1 bandwidth 0.05-dB-ripple filter design. (a) Synthesized prototype. (b) Prototype capacitance matrix. (c) Coaxial-interdigital configuration. (d) Transformed capacitance matrix.

the freedom of using an essentially all-coaxial configuration, a combination coaxial-interdigital structure, or an all-interdigital structure as dicated by the practicality of the resulting physical dimensions.

Incorporating the series capacitors in the end elements of the interdigital interior section is equivalent to choosing $n'_1=1$. The interdigital sections can be designed by referring to the 4×4 interior submatrix of matrix (17) and choosing a suitable value of n'_2. One possibility is to choose n'_2 so that the line spacings for a rectangular bar realization are equal. This requires [3] the interior mutual capacitors to be equal, or from matrix (17)

$$3.60 n_2'^2 = 1.80 n'_2$$
$$n'_2 = \tfrac{1}{2}. \tag{18}$$

The final capacitance matrix is given in (19), and the capacitance network is shown in Fig. 8(d). The dotted submatrices of matrices (17) and (19) show the interior 4×4 matrix pertinent to the interdigital sections:

For a rectangular bar realization, the normalized line spacings s/b are determined [1] directly from Getsinger's [3] chart and the normalized line widths w/b by the equations

$$\frac{w_1}{b} = \frac{1}{2}\left(1 - \frac{t}{b}\right)\left[\frac{1}{2}(c_1) - c_{f'} - (c_{fe'})_{12}\right]$$
$$\frac{w_k}{b} = \frac{1}{2}\left(1 - \frac{t}{b}\right)\left[\frac{1}{2}(c_k) - (c_{fe'})_{k-1,k} - (c_{fe'})_{k,k+1}\right]$$
$$\frac{w_N}{b} = \frac{1}{2}\left(1 - \frac{t}{b}\right)\left[\frac{1}{2}(c_N) - c_{f'} - (c_{fe'})_{N-1,N}\right] \tag{20}$$

where c_k is the capacitance from the kth line to ground, and $c_{k,k+1}$ is the capacitance between the kth and $(k+1)$st lines. (Note the use of small c's consistent with the Glossary definitions.) A thorough discussion of the method of obtaining dimensions from a set of self and mutual capacitors, along with a detailed definition of the symbols in (20), is given in References [1], [3], and [11]. The dimensions of the coaxial end sections are readily determined from convenient graphs [12] and correspond on a normalized 1-ohm basis to

$$Z_0 = \frac{\eta_0}{\sqrt{\epsilon_r}\, C/2} = \frac{7.54}{\sqrt{\epsilon_r}\left(\dfrac{7.54}{\sqrt{\epsilon_r}}\right)} = 1$$

A summary of the data used in designing the 3:1 bandwidth filter, along with calculated dimensions, is given in Table III.

An experimental model was constructed based on the data in Table III. The interdigital line sections were arbitrarily shortened by 0.200 inch, and tuning screws were added to obtain the proper resonator length. The desired length of the coaxial end elements was determined by temporarily shortening the interior lines with shorting blocks. This moved the singularity produced by the interior sections at twice the center frequency to a higher frequency, and allowed the series capacitors to be adjusted to proper length. After performing these operations the filter was tested. The bandwidth agreed with the theoretical to within 1 percent, but a reflection peak of 1.8 VSWR appeared near both upper and lower band edges. The addition of a 0.020-inch shim under the gap between the center lines and the adjustment of the tuning screws resulted in the final response, shown in Fig. 9. A detailed drawing of the filter is given in Fig. 10, and a photograph appears in Fig. 11. As can

$$\frac{7.54}{\sqrt{\epsilon_r}}\begin{bmatrix} 1 & -1 & 0 & 0 & 0 & 0 \\ -1 & 2.80 & -.90 & 0 & 0 & 0 \\ 0 & -.90 & 2.33 & -.90 & 0 & 0 \\ 0 & 0 & -.90 & 2.33 & -.90 & 0 \\ 0 & 0 & 0 & -.90 & 2.80 & -1 \\ 0 & 0 & 0 & 0 & -1 & 1 \end{bmatrix} \tag{19}$$

TABLE III

DESIGN DATA FOR A 3:1 BANDWIDTH INTERDIGITAL FILTER

k*	c_k	w_k (INCHES)	$c_{k,k+1}$	$s_{k,k+1}$ (INCHES)	NOTES
1	0	-	$\frac{7.54}{\sqrt{\epsilon_r}}$	-	c_{12} WAS REALIZED AS A TEFLON LOADED SERIES COAXIAL STUB, $\sqrt{\epsilon_r} = 1.44$, SEE FIG. 10
2	6.79	0.248	6.79	0.058	
3	4.02	0.195	6.79	0.058	THE DIMENSIONS ARE SYMMETRIC ABOUT THE CENTER OF THE FILTER

b = 0.625 INCHES
t = 0.375 INCHES

*THE k's REFER TO THE LINE NUMBERS IN FIG. 10. c_k IS THE CAPACITANCE FROM THE kth LINE TO GROUND AND $c_{k,k+1}$ IS THE CAPACITANCE BETWEEN THE kth AND $k+1^{st}$ LINES

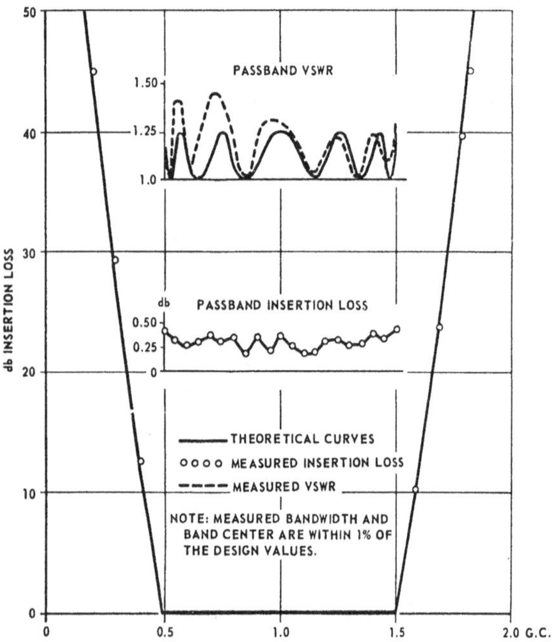

Fig. 9. Theoretical and measured performance of a six-section 3:1 bandwidth interdigital filter.

be seen from Fig. 9, the experimental response agrees very closely with the theoretical. The bandwidth and center frequency are within 1 per cent of the theoretical, and the maximum VSWR was 1.45:1 as compared with a theoretical maximum of 1.25:1. The close agreement between measured and theoretical VSWR curves and experimental bandwidth is a good check on the accuracy of the design theory.

VI. CRITERIA FOR PRACTICAL NETWORK DESIGN

The techniques presented in Section V can be used to design a wide variety of filter types and to obtain many different physical realizations. No one physical form is best suited to all bandwidths and all filter types. It may be necessary to investigate a number of equivalent forms to obtain practical dimensions. Fortunately, by using the techniques presented, this can be accomplished in a relatively simple manner. The following general criteria[11] are given to aid in obtaining a practical design.

[11] The criteria presented are directed toward symmetric filters, but similar results hold true for asymmetric networks.

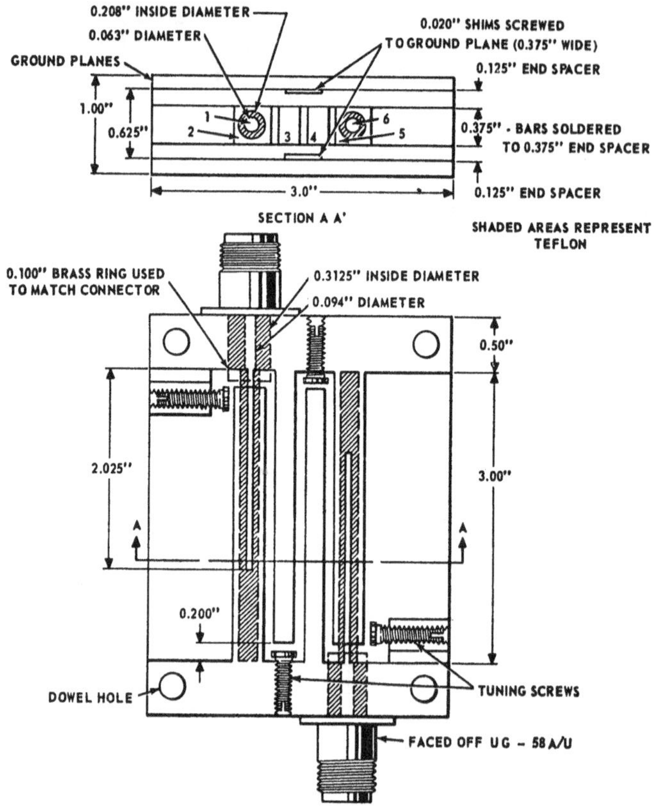

Fig. 10. Drawing of an experimental 3:1 bandwidth interdigital filter. (Other dimensions are given in Table III.)

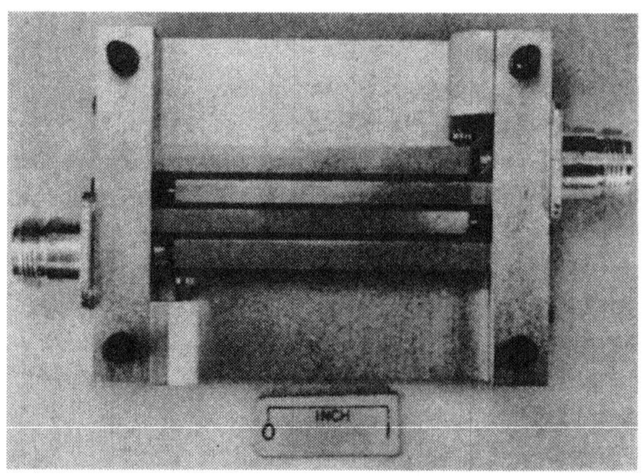

Fig. 11. A 3:1 bandwidth interdigital filter.

1) *Filters of moderate to wide bandwidths:* As pointed out in Reference [9], the filter properties of *L-C* type elements are superior to unit elements, the improvement becoming greater as filter bandwidth increases. The interdigital filter form with open-circuited terminating lines contains a third-order *L-C* type pole at dc as compared with a single *L-C* pole for the interdigital filter with short-circuited terminating lines. This would indicate a preference for the open-circuited terminating line configuration. However, this form has a serious disadvantage in that the Z_{0k} are often large for both wide and moderate band-

104

width filters. This requires the use of conductors with a small cross-sectional area, and for the case of wide-band filters it requires extremely small line spacings.

In addition to the poor mechanical properties of this physical configuration, the assumption that no significant coupling exists between nonadjacent conductors is not always accurate. The interdigital filter with short-circuited terminating lines often has lower values of Z_{0k} than the comparable interdigital form with open-circuited terminating lines. Series S-plane capacitors sometimes can be incorporated into the end lines, resulting in a prototype identical to the interdigital filter with open-circuited terminating lines in a mechanically superior physical form. For the design of wide-band filters, both network types should be investigated. In many instances, a combination coaxial-interdigital realization such as that of the design example is extremely practical.

2) *Filters of narrow bandwidths:* Both basic interdigital filter forms provide similar performance for narrow bandwidths in that the filtering properties of L-C elements are, for all practical purposes, identical to those of unit elements [9]. In both filter types, however, use of the full number of elements available leads to impractical dimensions. The problem is solved by choosing the end elements arbitrarily to obtain a convenient impedance level throughout the filter. Referring to Tables I and II, this requires arbitrarily choosing C for the configuration with open-circuited terminating lines and Z_1 for the network with short-circuited terminating lines.[12] Thus, for narrow bandwidths, the N-line filter realizes N-2 sections. The values required for C and Z_1, respectively, lead to responses that are very broad so that they do not affect significantly the narrow response produced by the N-2 interior sections.[13] By removing the effects of C and Z_1 to obtain practical dimensions, the interior sections of both basic network configurations become identical. The proper prototype element values for the interior sections of an N-line filter with an optimum attenuation characteristic are obtained as outlined in Reference [9] with $m=1$ and $n=N$-3.

For filters of narrow bandwidth and/or a large number of sections, the element values for the nonredundant prototype become extremely small, especially in the center of the filter. Many transformations of the capacitance matrix are then required to obtain a practical network realization. This difficulty can be overcome by the initial synthesis of a redundant network that has convenient values, using a procedure similar to that given in Reference [5], pp. 109–110. Variations in network form can be obtained by transforming the corresponding redundant capacitance matrix. Tables of element values are presently being computed for both redundant and nonredundant prototypes. (See footnote 9.)

A subject of importance in the practical design aspects of interdigital filters is the relationship of the theory and design methods presented here to Matthaei's approximate design procedure [1], [11]. The advantages and disadvantages of the two methods will be discussed with respect to accuracy, simplicity, flexibility, and theoretical considerations.

1) *Accuracy:* The theory presented in this paper is exact[14] and should be capable of providing a high degree of accuracy. Negligible bandwidth shrinkage was obtained in the experimental model, as compared with a nominal shrinkage of about 7 percent produced by the approximate method. As in the design of any practical microwave component, slight adjustments are always required, but it is anticipated that the exact design approach will provide better results, especially for wide-bandwidth designs. A further advantage of the exact approach is that it allows the use of optimum filter theory [9] and thus makes possible the realization of an optimum transmission characteristic. 105

2) *Simplicity:* The approximate design method makes use of readily available tables, while the exact method requires the use of special transfer functions and a knowledge of modern network synthesis techniques. The use of tables of element values presently being computed makes the exact design procedure very simple, but until these are readily available, the exact design procedure is limited to those people who have access to a digital computer. A further disadvantage of the exact approach is that each new network type, i.e., doubly terminated designs, singly terminated designs, linear phase designs, etc., requires a new set of computer data while the approximate design methods can use existing tables.

3) *Flexibility:* Great flexibility of network configuration is achieved by the use of the capacitance matrix. Fortunately the advantages of this approach apply equally well to both the exact and the approximate design techniques. Once a set of self and mutual capacitors is obtained, by either the exact or the approximate method, the capacitance matrix can be written by inspection and the equivalent networks determined.

4) *Theoretical Considerations:* It is expected that the application of exact synthesis methods and the

[12] Again, these statements apply directly to symmetric filters, with similar results holding for the asymmetric case.

[13] For the filter with short-circuited terminating lines, if Z_1 is chosen equal to the terminating impedance, the transmission power response realized is exactly that of the N-2 interior sections.

[14] It is assumed that no direct coupling exists between nonadjacent lines. See Section II.

capacitance matrix approach to coupled line networks will be useful in the design of many microwave networks other than the interdigital filter. The methods presented provide detailed insight into possible physical realizations and enable the designer to obtain networks with optimum performance.

VII. Conclusions

The synthesis method discussed based on the impedance matrix provides a general procedure for obtaining exact equivalent circuits. Given an N-port impedance matrix, an equivalent circuit can be derived which exactly describes the network behavior between any two ports. While this procedure is of a very general nature, it is often tedious to apply and becomes practically intractable for complicated networks.

For the case of an array of parallel-coupled lines, the use of the capacitance matrix associated with the two-dimensional network geometry allows an exact equivalent circuit to be obtained in a simplified manner. Transformations of the capacitance matrix which preserve the two-port transmission properties are applied to obtain a network in which an equivalent circuit can be obtained by inspection.[15]

106

The use of the capacitance matrix method allows the realization of an infinite number of equivalent circuits in a wide variety of physical forms. The examples presented demonstrate the simplicity and flexibility of this design approach. Use of the capacitance matrix in conjunction with tables of prototype element values will make possible the simplified design of filter networks in a large number of different physical forms. The techniques presented might also be applied to more complicated coupled line array configurations, including those in which significant coupling exists between nonadjacent conductors.

The specific examples given are directed toward the design of interdigital band-pass filters; however, the general techniques presented can be used to analyze and design a much broader class of microwave networks. Exact equivalent circuits for other configurations such as the meander line and hairpin line can be obtained by applying different boundary conditions to the basic array of parallel conductors. Once the exact S-plane equivalent circuit is determined, the techniques of modern network synthesis can be used to design matching networks, phase shifters, phase equalizers, and many other microwave devices.

The measured performance of a 3:1 bandwidth interdigital filter showed very close agreement with the theoretical performance. Relatively wide line spacings were obtained by using a combination coaxial-interdigital realization.

[15] The transformation of the capacitance matrix presented preserves the two-port performance for interdigital line networks but does not necessarily do so for arrays with other than open-circuited or short-circuited interior port boundary conditions.

Appendix

The invariance of the products of coupling factors insures that transformations of the capacitance matrix preserve the two-port transmission properties of the corresponding interdigital line. The validity of this statement may be demonstrated as follows.

The two-port impedance parameters of a general N-line interdigital array as determined by the method of Section III are functions of the coupling factors and end line impedances as given in (21):

$$\begin{aligned} z_{11} &= Z_{01} f_{11}(K_i^2, S) \qquad i = 1, 2, \cdots, N-1 \\ z_{22} &= Z_{0N} f_{22}(K_i^2, S) \qquad K_i^2 = (K_{k,k+1})(K_{k+1,k}) \\ z_{12} &= Z_{01} \prod_{k=1}^{N-1} K_{k,k+1} f_{12}(K_i^2, S) \\ z_{21} &= Z_{0N} \prod_{k=1}^{N-1} K_{k+1,k} f_{12}(K_i^2, S). \end{aligned} \tag{21}$$

From (21) it follows that maintaining the impedance levels of the outer lines and performing transformations that leave the products of the coupling factors (K_i^2) invariant does not change the open-circuit driving point impedances z_{11} and z_{22}.

For the transfer impedances z_{12} and z_{21}, reciprocity requires that

$$Z_{01} \prod_{k=1}^{N-1} K_{k,k+1} = Z_{0N} \prod_{k=1}^{N-1} K_{k+1,k}. \tag{22}$$

Maintaining Z_{01} and Z_{0N} gives

$$\frac{\prod_{k=1}^{N-1} K_{k,k+1}}{\prod_{k=1}^{N-1} K_{k+1,k}} = \frac{Z_{0N}}{Z_{01}} = \text{constant}. \tag{23}$$

But

$$\left(\prod_{k=1}^{N-1} K_{k,k+1}\right)\left(\prod_{k=1}^{N-1} K_{k+1,k}\right) = \prod_{i=1}^{N-1} K_i^2 = \text{constant}. \tag{24}$$

Equations (23) and (24) require that both products in (22) be constant, resulting in the invariance of the transfer impedances.

Acknowledgment

A definite pattern was found in the relationship between the parameters of the interdigital filter and its cascade equivalent circuit as obtained by the synthesis technique of Section III. This pattern, described by M. C. Horton, encouraged the author to investigate alternate methods of obtaining the equivalent circuit. The author is further indebted to Mr. Horton for a number of suggestions which added to the clarity of this paper.

Appreciation is also expressed to T. E. Gordon for his work in testing the experimental filter, and to N. O. Tiffany for programming the digital computer.

References

[1] G. L. Matthaei, "Interdigital band-pass filters," *IEEE Trans. on Microwave Theory and Techniques*, vol. MTT-10, pp. 479–491, November 1962.

[2] J. T. Bolljahn and G. L. Matthaei, "A study of the phase and filter properties of arrays of parallel conductors between ground planes," *Proc. IRE*, vol. 50, pp. 299–311, March 1962.

[3] W. J. Getsinger, "Coupled rectangular bars between parallel plates," *IRE Trans. on Microwave Theory and Techniques*, vol. MTT-10, pp. 65–72, January 1962.

[4] E. G. Cristal, "Coupled circular cylindrical rods between parallel ground planes," *IEEE Trans. on Microwave Theory and Techniques*, vol. MTT-12, pp. 428–439, July 1964.

[5] R. J. Wenzel, "Exact design of TEM microwave networks using quarter-wave lines," *IEEE Trans. on Microwave Theory and Techniques*, vol. MTT-12, pp. 94–111, January 1964.

[6] "Exact design techniques for TEM microwave filters," U. S. Army Electronics Research and Development Labs., Ft. Monmouth, N. J., First Quarterly Rept., Contract DA-28-043-AMC-00399(E), February 1965.

[7] E. A. Guillemin, *Synthesis of Passive Networks*. New York: Wiley, 1957.

[8] S. B. Cohn, "The re-entrant cross section and wide-band 3-dB hybrid couplers," *IEEE Trans. on Microwave Theory and Techniques*, vol. MTT-11, pp. 254–258, July 1963.

[9] M. C. Horton and R. J. Wenzel, "General theory and design of optimum quarter-wave TEM filters," *IEEE Trans. on Microwave Theory and Techniques*, vol. MTT-13, pp. 316–327, May, 1965.

[10] R. J. Wenzel, "Application of exact synthesis methods to multichannel filter design," *IEEE Trans. on Microwave Theory and Techniques*, vol. MTT-13, pp. 5–15, January 1965.

[11] G. L. Matthaei, L. Young, and E. M. T. Jones, *Microwave Filters, Impedance Matching Networks, and Coupled Structures*. New York: McGraw-Hill, 1964.

[12] *The Microwave Engineers' Handbook and Buyers' Guide*. Brookline, Mass.: Horizon House—Microwave, Inc., 1963.

Reprinted from IEEE TRANSACTIONS
ON MICROWAVE THEORY AND TECHNIQUES
Volume MTT-13, Number 5, September, 1965
Pp. 559-575

COMB-LINE BAND-PASS FILTERS OF NARROW OR MODERATE BANDWIDTH

GEORGE L. MATTHAEI
STANFORD RESEARCH INSTITUTE
Menlo Park • California

111

General

Figure 1 shows a comb-line band-pass filter in strip-line form. The resonators in this type of filter consist of TEM-mode transmission-line elements that are short-circuited at one end and have a lumped capacitance, C_k^s, between the other end of each resonator line element and ground. In Figure 1, lines 1 to n and their associated capacitances, C_1^s to C_n^s, constitute resonators, while lines 0 and n+1 are not resonators but simply part of impedance-transforming sections at the ends of the filter. Coupling between resonators is achieved in this type of filter by way of fringing fields between resonator lines. With the capacitors C_k^s present, the resonator lines will be less than $\lambda_0/4$ long at resonance (where λ_0 is the wavelength in the medium of propagation at midband), and the coupling between resonators is predominantly magnetic. Interestingly enough, if the capacitors C_k^s were not present, the resonator lines would be a full $\lambda_0/4$ long at resonance, *and the structure would have no pass-band.*[1] Without some kind of reactive loading at the ends of the resonator line elements, the magnetic and electric coupling effects would cancel each other, and the comb-line structure would become an all-stop structure.*

For reasons described above, it is usually desirable to make the capacitances C_k^s in this type of filter sufficiently large so that the resonator lines will be $\lambda_0/8$ or less long at resonance. Besides permitting efficient coupling between resonators (with sizeable spacings between adjacent resonator lines), the resulting filter will be relatively small. In this type of filter the second pass-band occurs when the resonator line elements are somewhat over one-half wavelength long, so that if the resonator lines are $\lambda_0/8$ long at the pri-

* *However, if every other unloaded $\lambda_0/4$ resonator were turned end for end so that the structure had open- and short-circuited ends alternating, the band-stop structure would become a band-pass structure. The resulting configuration is that of the interdigital filters previously discussed (References 1 and 2).*

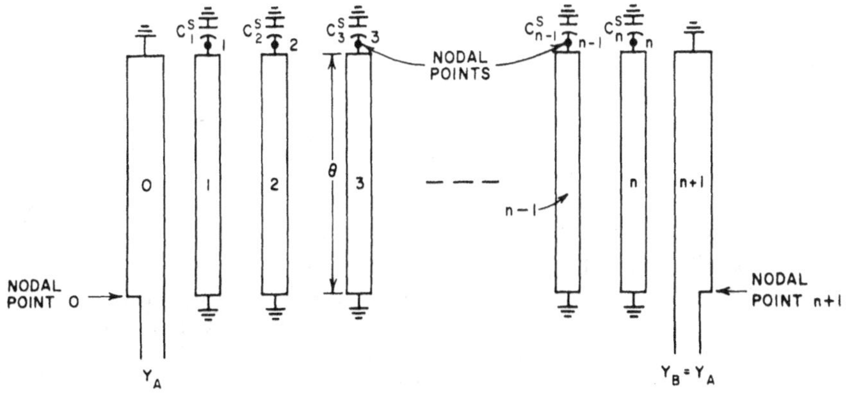

Figure 1 — A comb-line band-pass filter (The nodal points are defined for use in the design equation derivations discussed in Section E).

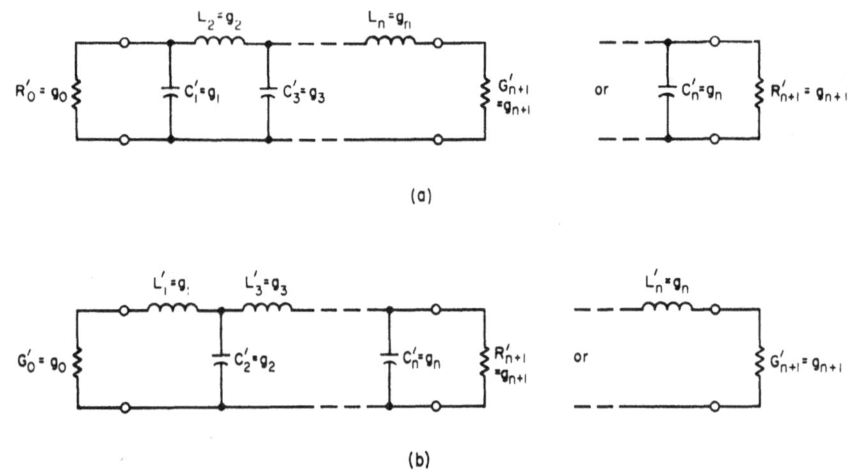

Figure 2 — Definition of prototype filter parameters g_0, g_1, g_2, g_{n+1} (A prototype circuit is shown at (a) and its dual is shown at (b). Either form will give the same response. An additional prototype parameter, ω'_1, is defined in Figure 3).

mary pass-band, the second pass-band will be centered slightly over four times the frequency of the center of the first pass-band. If the resonator line elements are made to be less than $\lambda_0/8$ long at the primary pass-band, the second pass-band will be even further removed from the primary pass-band.

Another property of this type of filter is that, in theory, the attenuation through the filter will be infinite at the frequency for which the resonator lines are one-quarter wavelength long. Because of this property, the attenuation above the primary pass band will be very high and, depending on what electrical length the resonator lines have at the pass-band center, the rate of cutoff on the upper side of the pass-band can be made to be unusually steep. (The closer to $\lambda_0/4$ long the resonators are at the pass-band center, the steeper the rate of cutoff will be above the pass-band.)

From the above discussion we see that comb-line filters have the following attractive features:

(1) They are compact.
(2) They have strong stop-bands and the stop-band above the primary pass-band can be made to be very broad.
(3) If desired, they can be designed to have an unusually steep rate of cutoff on the high side of the pass-band.
(4) Adequate coupling can be maintained between resonator elements with sizeable spacings between resonator lines. (This feature means that the proper couplings can be maintained in manufactured filters without unreasonable tolerance requirements.)
(5) Filters of this type can usually be fabricated without the use of dielectric support materials so that, if desired, dielectric losses can be eliminated.

The interdigital filters[2] previously discussed have some of these advantages, but comb-line filters are somewhat smaller and are capable of a broader stop-band above the primary pass-band. On the other hand, for the same cross-sectional dimensions of the resonator lines, the unloaded Q's of the interdigital-line resonators should be somewhat higher than the Q's of the comb-line resonators.

The design procedure to be described involves several simplifying approximations. Because of these approximations the accuracy should be best for narrow-band designs. Although the question of accuracy for the design procedure given has not been fully explored, it is expected that reasonably good results should be possible for bandwidths as large as perhaps 15 per cent. The design procedure will be discussed first, and the derivation of the equations will be discussed at the end of this paper.

Low-Pass Prototype Filters and Use of a Low-Pass to Band-Pass Transformation

The design procedures to be described are based on the use of low-pass prototype filter element values to give the band-pass filter its desired characteristics. Figure 2 shows typical low-pass prototypes of the type under discussion, and they also define the manner in which the element values g_k are to be specified for the purpose of this discussion. The left side of Figure 3 shows a typical Chebyshev response such as can be achieved by filters of this type. Element values for prototype filters having such responses have been tabulated.[3,4] Notice that n is the number of reactive elements and that, including the resistor terminations, the element values range from g_0 to g_{n+1}. A low-pass prototype with n reactive elements leads to a band-pass filter with n resonators.

The right side of Figure 3 shows a band-pass filter response that corresponds to the given low-pass prototype response. The band-pass filter response will have the same type of pass-band characteristic as the prototype, but the width of the band-pass filter pass-band can be specified arbitrarily. On the left side of Figure 3, equations are given for determining the attenuation characteristic of the low-pass prototype as a function of the radian frequency variable ω', for specified ω_1' and n. On the right side of Figure 3, a low-pass to band-pass transformation is given from which the attenuation characteristic of the low-pass prototype as a function of ω' can be mapped to the band-pass filter attenuation characteristic (centered at ω_0) as a function of the band-pass filter radian frequency variable ω. Since the attenuation is the same for both the low-pass and band-pass filters at frequencies ω' and ω, respectively, which are related as given by the mapping, the band-pass filter attenuation characteristic can be predicted by use of these data. At the bottom of Figure 3 are equations for determining the value of n required in order to achieve a given amount of attenuation, L_{Aa} db, at a given frequency, ω_a.

The mapping given in Figure 3 is in greatly simplified form, and is most accurate for filters of quite narrow bandwidth. A more accurate mapping would have to account for such things as the fact that the attenuation of this type of

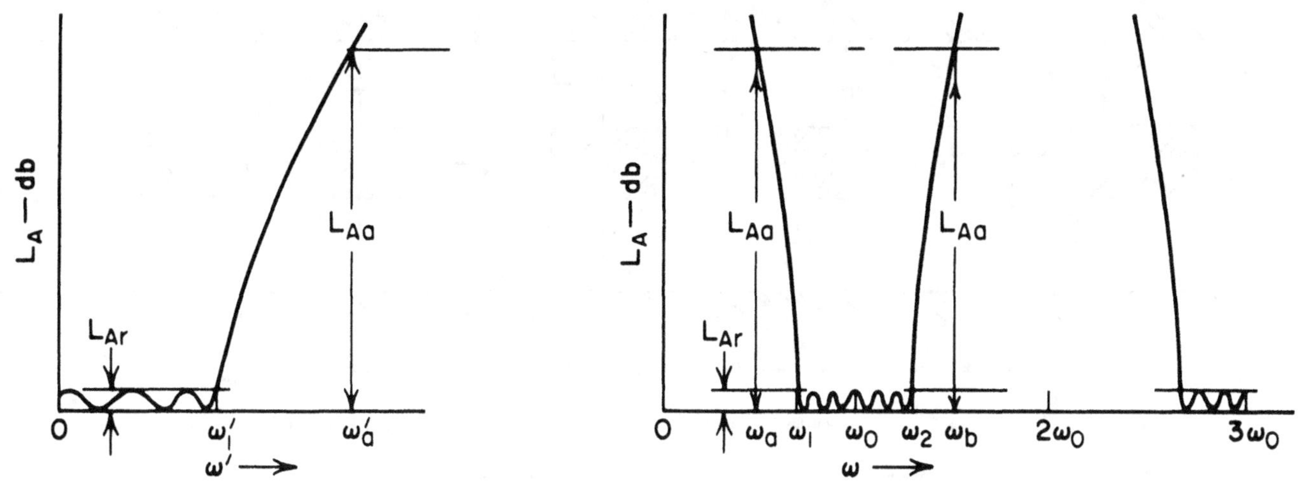

$$L'_A(\omega') = 10 \log_{10} \left\{ 1 + \epsilon \cos^2 \left[n \cos^{-1} \left(\frac{\omega'}{\omega'_1} \right) \right] \right\} \text{ db}$$

for

$$\omega' \leqq \omega'_1$$

$$L'_A(\omega') = 10 \log_{10} \left\{ 1 + \epsilon \cosh^2 \left[n \cosh^{-1} \left(\frac{\omega'}{\omega'_1} \right) \right] \right\} \text{ db}$$

for

$$\omega' \geqq \omega'_1$$

where

$$\epsilon = \left\{ \left[\text{antilog}_{10} \left(\frac{L_{Ar}}{10} \right) \right] - 1 \right\}$$

$$L_A(\omega) = L'_A(\omega') \text{ db}$$

where

$$\omega' = \frac{2\omega'_1}{w} \left| \frac{\omega - \omega_0}{\omega_0} \right|$$

$$\omega_0 = \frac{\omega_2 + \omega_1}{2}$$

and

$$w = \frac{\omega_2 - \omega_1}{\omega_0}$$

To determine n required for given values of ω_1/ω_0, L_{Ar}, ω_a/ω_0 and L_{Aa}, find smallest integer n value which satisfies

$$n \geqq \frac{\cosh^{-1} \sqrt{\dfrac{\text{antilog}_{10}\left(\dfrac{L_{Aa}}{10}\right) - 1}{\epsilon}}}{\cosh^{-1} \left(\dfrac{\omega'_a}{\omega'_1} \right)}$$

where

$$\omega'_a = \frac{2\omega'_1}{w} \left| \frac{\omega_a - \omega_0}{\omega_0} \right|$$

Figure 3 — Equations and parameters for Chebyshev response.

filter will go to infinity (theoretically) when the resonator lines are $\lambda_0/4$ long. Although the given mapping is oversimplified in some respects, it is believed to represent a useful practical compromise. For the 10-per cent-bandwidth trial filter to be described later, this mapping predicted the attenuation on the low side of the pass band quite accurately, but it underestimated the amount of attenuation on the high side of the pass-band.

Because the design procedure is approximate, the fractional bandwidth of the completed filter may not come out exactly as specified. Judging from the trial filter constructed in our laboratory, there is a tendency for the actual bandwidth to be somewhat larger than the specified bandwidth. (In the trial design, the measured fractional bandwidth was $w=0.116$ instead of the specified value of $w=0.100$). It may be desirable to make some allowance for this fact when planning the design of a filter.

Definitions of the Resonator Line Parameters

As previously mentioned, comb-line filters contain arrays of parallel lines between ground planes. The design equa-

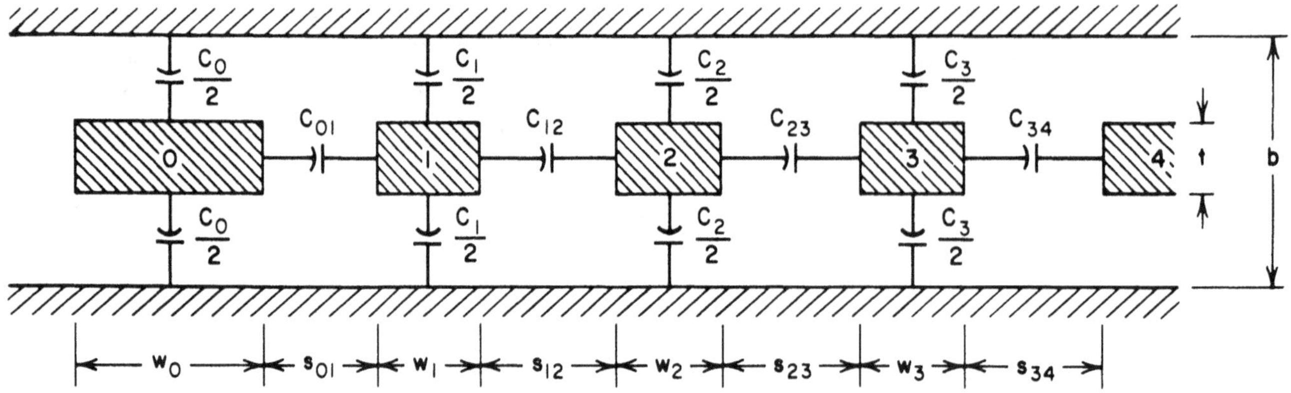

Figure 4 — Definitions of the line capacitances and some of the dimensions involved in comb-line filter design.

tions given herein will yield the various line capacitances per unit length for the various lines in the filter array. From these values of capacitance per unit length, the actual dimensions of the lines can be determined. Definitions for these various line capacitances will now be given, and also equations for obtaining the line dimensions from the specified capacitances.

Figure 4 shows the cross-section of an array of parallel-coupled lines such as is used in a comb-line filter. In the structure shown, all of the bars have the same t/b ratio and the other dimensions of the bars are easily obtained by generalizing Getsinger's procedure[5] for designing rectangular-bar parallel-coupled lines. In the structure in Figure 4 the electrical properties of the structure are characterized in terms of the capacitances, C_k, per unit length of each bar with respect to ground, and the mutual capacitances, $C_{k,k+1}$, per unit length between adjacent bars k and k+1. This representation is not necessarily always highly accurate; it is conceivable that a significant amount of fringing capacitance could exist between a given line element and, say, the line element beyond the nearest neighbor. However, at least for geometries such as that shown, experience has shown this representation to have satisfactory accuracy.

For design of the comb-line-filter structures to be discussed, equations will be given for the normalized self and mutual capacitances, C_k/ϵ and $C_{k,k+1}/\epsilon$, per unit length for all the lines in the structure (where ϵ is the dielectric constant of the medium of propagation). Then the cross-sectional dimensions of the bars and spacings between them are determined as follows. First, choose values for t and b. Then, since

$$\frac{(\Delta C)_{k,k+1}}{\epsilon} = \frac{C_{k,k+1}}{\epsilon} \tag{1}$$

Getsinger's charts[5] can be used to determine $s_{k,k+1}/b$. In this manner, the spacings $s_{k,k+1}$ between all the bars are obtained. Also, using Getsinger's charts, the normalized fringing capacitance $(C'_{fe})_{k,k+1}/\epsilon$ associated with each gap, $s_{k,k+1}$, between bars is obtained. The normalized width of the kth bar is then

$$\frac{w_k}{b} = \frac{1}{2}\left(1-\frac{t}{b}\right)\left[\frac{1}{2}\left(\frac{C_k}{\epsilon}\right) - \frac{(C'_{fe})_{k-1,k}}{\epsilon} - \frac{(C'_{fe})_{k,k+1}}{\epsilon}\right]. \tag{2}$$

In the case of the bar at the end of the array (the bar at the far left in Figure 4), C'_{fe}/ϵ for the edge of the bar that has no neighbor must be replaced by C'_f/ϵ, which is also obtained from Getsinger's charts.[5] Thus, for example, for Bar 0 in Figure 4,

$$\frac{w_0}{b} = \frac{1}{2}\left(1-\frac{t}{b}\right)\left[\frac{1}{2}\left(\frac{C_0}{\epsilon}\right) - \frac{C'_f}{\epsilon} - \frac{(C'_{fe})_{01}}{\epsilon}\right]. \tag{3}$$

If $w_k/b<0.35(1-t/b)$ for any of the bars, the width of the bar should be corrected using the approximate formula[5]

$$\frac{w'_k}{b} = \frac{0.07\left(1-\frac{t}{b}\right)+\frac{w_k}{b}}{1.20}, \tag{4}$$

provided that $0.1<(w'_k/b)/(1-t/b)<0.35$. In Equation (4) w_k is the uncorrected bar width and w'_k is the corrected width. The need for this correction arises because of the interaction of the fringing fields at opposite sides of a bar, which will occur when a bar is relatively narrow.

Comb-Line Filter Design

Using the design formulas in Table I, the distributed line capacitances can be computed in normalized form to give C_k/ϵ and $C_{k,k+1}/\epsilon$ values, where ϵ is the absolute dielectric constant of the medium of propagation. Then, using Equations (1) to (4) and Getsinger's charts,[5] the corresponding rectangular-bar line dimensions w_k and $s_{k,k+1}$ in Figure 4 can be determined for specified t and b.

To carry out the design of a comb-line filter by use of Table I, the low-pass prototype filter parameters g_0, g_1, . . . , g_{n+1} and ω'_1 are selected as previously discussed. Next the terminating line admittance, Y_A, the midband electrical length, θ_0, of the resonator lines, the fractional bandwidth, w, and the normalized line admittances, Y_{ak}/Y_A, must all be specified. As indicated above, it is usually desirable to make $\theta_0=\pi/4$ radians or less. The choice of the resonator-line admittances, Y_{ak}, fixes the admittance level within the filter, and this is important in that it influences the unloaded Q's that the resonators will have. At the time of this writing, the line characteristic admittances to give optimum unloaded Q's for structures of this type have not been determined. However, choosing the Y_{ak} in Equation (1) of Table I to correspond to about 0.0143 mho (i.e., about 70 ohms), appears to be a reasonable choice. (The admittance

TABLE I

Design Equations for Comb-Line Filters

Choose the normalized characteristic admittances Y_{ak}/Y_A so as to give good resonator unloaded Q's. (See text.) Then compute:

$$\left.\frac{b_k}{Y_A}\right|_{k=1\text{ to }n} = \frac{Y_{ak}}{Y_A}\left(\frac{\cot\theta_0 + \theta_0 \csc^2\theta_0}{2}\right) \qquad (1)$$

where θ_0 is the electrical length of the resonator elements at the midband frequency ω_0.

Compute:

$$\frac{G_{T1}}{Y_A} = \frac{w\dfrac{b_1}{Y_A}}{g_0 g_1 \omega'_1} \qquad (2)$$

$$\left.\frac{J_{k,k+1}}{Y_A}\right|_{k=1\text{ to }n-1} = \frac{w}{\omega'_1}\sqrt{\frac{(b_k/Y_A)\,(b_{k+1}/Y_A)}{g_k g_{k+1}}} \qquad (3)$$

$$\frac{G_{Tn}}{Y_A} = \frac{\left(w\dfrac{b_n}{Y_A}\right)}{g_n g_{n+1}\omega'_1} \qquad (4)$$

where w is the fractional bandwidth defined below.

The normalized capacitances per unit length between each line and ground are

$$\frac{C_0}{\epsilon} = \frac{376.7\,Y_A}{\sqrt{\epsilon_r}}\left(1-\sqrt{\frac{G_{T1}}{Y_A}}\right)$$

$$\frac{C_1}{\epsilon} = \frac{376.7\,Y_A}{\sqrt{\epsilon_r}}\left(\frac{Y_{a1}}{Y_A} - 1 + \frac{G_{T1}}{Y_A} - \frac{J_{12}}{Y_A}\tan\theta_0\right) + \frac{C_0}{\epsilon}$$

$$\left.\frac{C_k}{\epsilon}\right|_{k=2\text{ to }n-1} = \frac{376.7\,Y_A}{\sqrt{\epsilon_r}}\left(\frac{Y_{ak}}{Y_A} - \frac{J_{k-1,k}}{Y_A}\tan\theta_0 - \frac{J_{k,k+1}}{Y_A}\tan\theta_0\right) \qquad (5)$$

$$\frac{C_n}{\epsilon} = \frac{376.7\,Y_A}{\sqrt{\epsilon_r}}\left(\frac{Y_{an}}{Y_A} - 1 + \frac{G_{Tn}}{Y_A} - \frac{J_{n-1,n}}{Y_A}\tan\theta_0\right) + \frac{C_{n+1}}{\epsilon}$$

$$\frac{C_{n+1}}{\epsilon} = \frac{376.7\,Y_A}{\sqrt{\epsilon_r}}\left(1-\sqrt{\frac{G_{Tn}}{Y_A}}\right)$$

where ϵ is the absolute dielectric constant of the medium of propagation, and ϵ_r is the relative dielectric constant.

The normalized mutual capacitances per unit length between adjacent lines are:

$$\frac{C_{01}}{\epsilon} = \frac{376.7\,Y_A}{\sqrt{\epsilon_r}} - \frac{C_0}{\epsilon}$$

$$\left.\frac{C_{k,k+1}}{\epsilon}\right|_{k=1\text{ to }n-1} = \frac{376.7\,Y_A}{\sqrt{\epsilon_r}}\left(\frac{J_{k,k+1}}{Y_A}\tan\theta_0\right) \qquad (6)$$

$$\frac{C_{n,n+1}}{\epsilon} = \frac{376.7\,Y_A}{\sqrt{\epsilon_r}} - \frac{C_{n+1}}{\epsilon}$$

The lumped capacitances C_k^s are:

$$\left.C_k^s\right|_{k=1\text{ to }n} = Y_A\left(\frac{Y_{ak}}{Y_A}\right)\frac{\cot\theta_0}{\omega_0} \qquad (7)$$

Y_{ak} in Table I is interpreted physically as the admittance of line k with the adjacent lines $k-1$ and $k+1$ grounded.) The remainder of the calculations proceed in a straightforward manner as presented in the table. As mentioned above, having the C_k/ϵ and $C_{k,k+1}/\epsilon$, the required line dimensions are obtained from Getsinger's charts and Equations (1) to (4).

Table II summarizes various parameters used and computed in the design of a trial four-resonator, comb-line filter designed for a fractional bandwidth of w=0.10, and 0.1-db Chebyshev ripple; and Figure 5 shows the completed filter. Because of a misprint in the table of prototype filter-element values that were used for the design of this filter, the g_1 element value is low by about 10 per cent. However, a computed response for this filter revealed that this error should not have any sizable effect on the shape of the response. In this design $\theta_0=\pi/4$ radians so that the resonator lines are $\lambda_0/8$ long at the midband frequency, which was to be 1.5 Gc. Note that $Y_{ak}/Y_A=0.677$ which with $Y_A=0.020$ mho makes $Y_{ak}=0.0135$ mho, or $1/Y_{ak}=74$ ohms. The electrically important dimensions of this filter are summarized in Table II along with Figures 4 and 6.

The filter was tuned using the alternating short-circuit, and open-circuit procedure described by Dishal.[6] To adjust the capacitive loading of each resonator line, its adjacent sliding block (shown in Figure 6) was first adjusted to give

Figure 5 — A four-resonator comb-line filter with its cover plate removed.

TABLE II

Various Parameters That Were Specified or Computed in the Design of the Trial, Four-Resonator, Comb-Line Filter

k	$\frac{J_{k,k+1}}{Y_A}$	$\frac{C_{k,k+1}}{\epsilon}$	$s_{k,k+1}$ (inches)	k	$\frac{C_k}{\epsilon}$	w_k (inches)
0 and 4		2.130	0.116	0 and 5	5.404	0.362
1 and 3	0.0730	0.550	0.337	1 and 4	3.022	0.152
2	0.0572	0.431	0.381	2 and 3	4.119	0.190

$g_0=1$ $\quad$ $g_3=1.7703$ $\quad$ $w=0.10$ $\quad$ $\theta_0=\pi/4$ radian

$g_1=1.0880$* $\quad$ $g_4=0.8180$ $\quad$ $Y_A=0.020$ mho $\quad$ $b=0.625$ inch

$g_2=1.3061$ $\quad$ $g_5=1.3554$ $\quad$ $b_k/Y_A=0.870$ $\quad$ $t=0.188$ inch

$\omega_1=1$ $\quad$ $Y_{ak}/Y_A=0.677$ $\Big\}$ $k=1$ to 4

* This value should have been $g_1=1.1088$ for a true 0.1-db ripple prototype.

slightly less than the required resonator capacitance, and then the tuning screws on the resonator were used to bring the resonator to the exact desired frequency. In this case the bandwidth was sufficiently large so that the alternating short-circuit and open-circuit procedure did not give entirely satisfactory results, as evidenced by some lack of symmetry in the pass-band response. However, it was found that this could be easily corrected by readjustment of the
116
tuning screws on the end resonators* while using a sweep generator and recording reflectometer set up. After the tuning was completed, the measured input VSWR was as shown in Figure 7 and the measured attenuation as shown in Figure 8. As indicated in Figure 8, the attenuation above the pass band is very high for frequencies up to about 6.3 Gc.

* *Since the end resonators have adjacent couplings that are quite different from those of the interior resonators, the end resonators may cause difficulties when using the alternating short-circuit, open-circuit tuning procedure.*

The VSWR characteristic in Figure 7 corresponds to roughly 0.2 db Chebyshev attenuation ripple rather than 0.1 db. The discrepancy is believed to be due to the fact that coupling effects beyond nearest-neighbor lines have been neglected in the design procedure in Table I. If a smaller ripple were necessary, it could be achieved by a small adjustment of spacing s_{01} between the input line and Resonator 1, and spacing s_{45} between Resonator 4 and the output line. A similar phenomenon occurred in the 10-per cent bandwidth interdigital-line filter example discussed previously.[2] In that case the size of the ripples was easily reduced by decreasing the sizes of end-gap spacings s_{01} and $s_{n,n+1}$. In the case of Figure 7, the ripples were not considered to be sufficiently oversized to warrant taking the time to make further adjustments.

From the VSWR characteristic in Figure 7, the measured fractional bandwidth at the equal-VSWR-ripple level is found to be $w=0.116$ instead of the specified $w=0.100$. This somewhat oversize bandwidth may also be due to coupling effects beyond nearest-neighbor line elements, which were neglected in the derivation of the design equa-

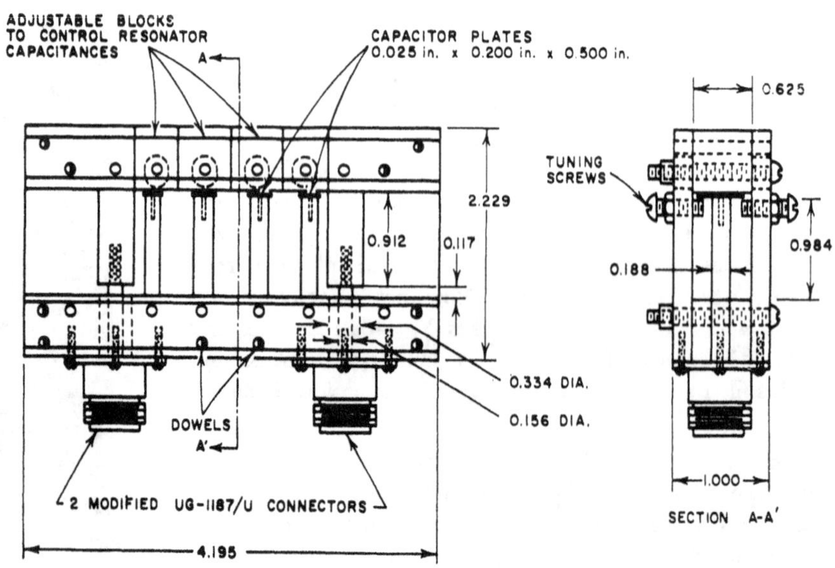

Figure 6 — Drawing of the trial four-resonator, comb-line filter (Additional dimensions of electrical importance are given in Table II).

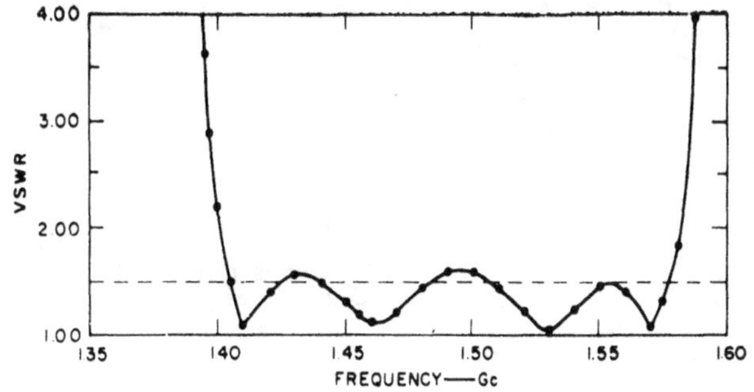

Figure 7 — Measured VSWR of the filter in Figure 5.

tions in Table I. Table III compares attenuation values computed by use of the mapping in Figure 3 to the actual measured values. Conditions A are for the original specifications; Conditions B are for the w=0.116 fractional bandwidth and approximately 0.2-db attenuation ripple indicated by the VSWR characteristic in Figure 7. Note that in both Case A and Case B, the attenuation predicted by the mapping for f=1.25 Gc (f below f_0) has come out reasonably close to correct, while the attenuation predicted by the mapping for f=1.70 Gc (f above f_0) is somewhat low, for reasons previously discussed. For the actual measured conditions (i.e., Conditions B), the mapping was very accurate on the low side of the band, but was quite pessimistic on the high side.

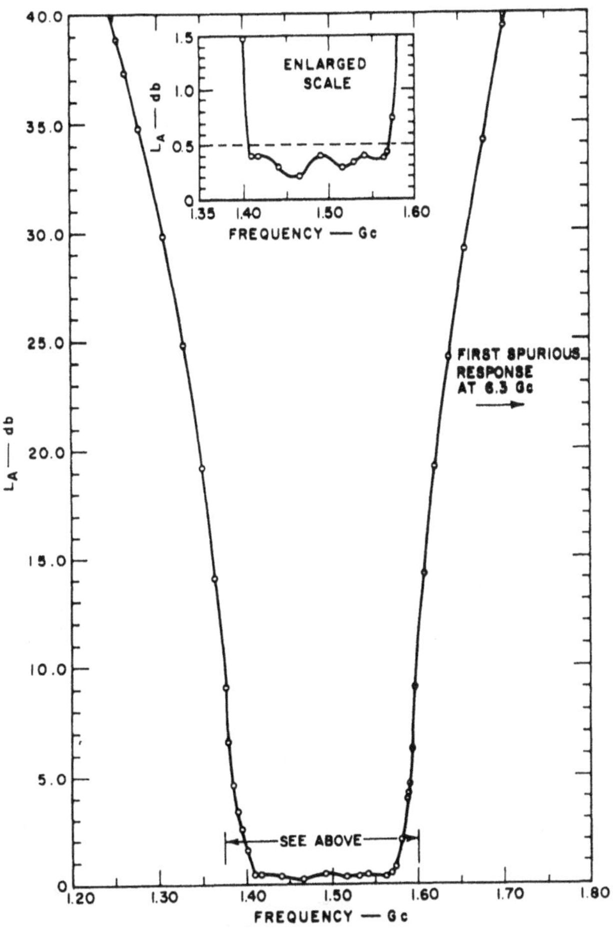

Figure 8 — Measured attenuation of the filter in Figure 5.

TABLE III

Comparison of Attenuation Values Obtained by Mapping and by Measurement

MAPPING CONDITIONS	f (Gc)	L_A BY MAPPING (db)	L_A MEASURED (db)
A			
Original Specifications, w=0.10, 0.10-db	1.25	41.5	39
Chebyshev Ripple, f_0=1.491 Gc	1.70	36.5	39
B			
Measured Specifications, w=0.116, 0.20-db	1.25	39.5	39
Chebyshev Ripple, f_0=1.491 Gc	1.70	34	39

Derivation of the Design Equations

The point of view used to obtain the design equations in Table I is that due to Cohn.[7] Cohn's design point of view will now be briefly outlined, by using a somewhat different and more general terminology than that in Cohn's paper.[7]

Figure 9 shows a band-pass filter such as can be designed directly from a low-pass prototype filter using the equations indicated in the figure. Note that the prototype parameters $g_0, g_1, \ldots, g_{n+1}$ are as defined in Figure 2, and the frequencies ω'_1, ω_0, ω_1, and ω_2 are as defined in Figure 3. This filter has both series and shunt lumped-element resonators, which are characterized by their resonant frequency ω_0 and a slope parameter. For shunt resonators, the "susceptance slope parameter" b_j applies, and it is simply the susceptance of the shunt-resonator capacitance at the resonant frequency. For the series resonators, the "reactance slope parameter" x_j applies, and it is simply the reactance of the series resonator inductance at the resonant frequency.*

Any filter of the form in Figure 9, which uses a ladder configuration of shunt and series resonators, can be converted to the form in Figure 10, which uses only series resonators.[7] The boxes marked $K_{j,j+1}$ are what Cohn has referred to as "impedance inverters." An ideal impedance inverter is defined as a device which operates just like a quarter-wavelength line of characteristic impedance $K_{j,j+1}$ ohms at all frequencies, where, $K_{j,j+1}$ is, in the hypothetical ideal case, independent of frequency. If the impedance inverters in a circuit such as that in Figure 10 were ideal, its response would be identical to that of an equivalent filter of the form in Figure 9. Cohn used the term *impedance inverter* because if an impedance Z_b is attached to one end of an inverter, the impedance seen at the other end is

$$Z_a = \frac{K^2_{j,\,j+1}}{Z_b} \,. \tag{5}$$

* *These slope parameters also have more general definitions which apply for all forms of resonators. The general definition of b_j is given in Equation (1) of Figure 11. The parameter of x_j has an analogous definition on a dual basis.*

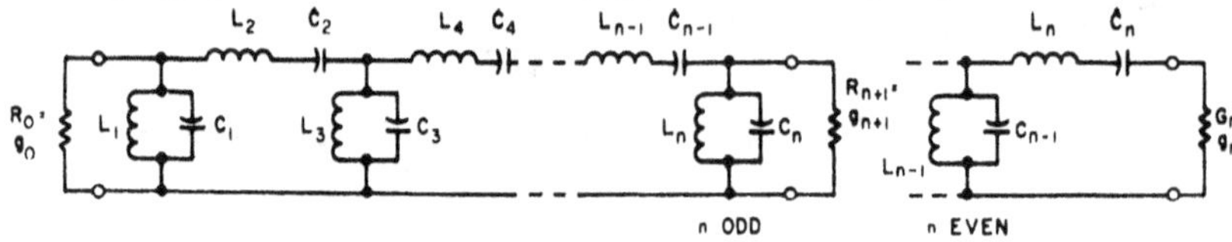

FOR SHUNT RESONATORS:

$$b_j = \omega_0 C_j = \frac{1}{\omega_0 L_j} = \frac{\omega_1' g_j}{w} \qquad (1)$$

= SUSCEPTANCE SLOPE PARAMETER

FOR SERIES RESONATORS:

$$x_k = \omega_0 L_k = \frac{1}{\omega_0 C_k} = \frac{\omega_1' g_k}{w} \qquad (2)$$

= REACTANCE SLOPE PARAMETER

$$w = \frac{\omega_2 - \omega_1}{\omega_0}$$

$$\omega_0 = \sqrt{\omega_1 \omega_2}$$

Figure 9 — Band pass filters and their relation to low-pass prototypes.

A quarter-wavelength line is, of course, a narrow-band impedance inverter and has frequently been used as such for filter design.[8] However, as Cohn points out[7] numerous other types of circuits have inverting properties, and many of the other circuits have much more broadband inverter operation than does a quarter-wavelength line.

The equations whose derivation is about to be described are based on the dual representation to that in Figure 10; i.e., in the representation to be used the resonators are all shunt resonators. A generalized circuit of this form is shown in Figure 11(a). The boxes marked $B_k(\omega)$ represent resonators, which may be of any form (i.e., they may be formed from lumped elements, transmission elements, or combinations thereof) as long as they are resonant at ω_0 and exhibit a susceptance characteristic like that of a shunt resonator in the vicinity of ω_0. The required form of susceptance characteristic is shown in Figure 11(b). If the shunt resonators are of the shunt L and C form, the susceptance slope parameter for the jth resonator is simply $b_j = \omega_0 C$. If the resonators are of some other form, the slope parameters can be computed by the more general equation given in Equation (1) of Figure 11.

The boxes marked $J_{j,j+1}$ in Figure 11(a) are herein defined as ***admittance inverters.*** An admittance inverter is the same as an impedance inverter, except it is described in terms of its characteristic ***admittance*** (or image admittance), $J_{j,j+1}$ mhos, instead of its characteristic impedance, $K_{j,j+1}$. The introduction of the J-inverter definition is convenient so that the case of shunt resonators will be mathematically the same as that for series resonators, except that admittances will be used in place of impedances.*

* *Actually, there is another less superficial difference between K- and J-inverters in that some types of inverters are, in practice, useful only with series resonators (where by the definition above they are K-inverters), while some other types are useful only with shunt resonators (where by the definition above they are J-inverters). Thus, most inverters can be specifically identified as K-inverters or J-inverters.*

In the design equations in Figure 11, the terminating conductances G_A and G_B, along with the resonator slope parameters $b_1, b_2, \ldots, b_n$, can be specified arbitrarily. Then if the inverter parameters $J_{j,j+1}$ are specified as indicated in Equations (2) to (4) of Figure 11, the transmission response will be identical to that of an equivalent

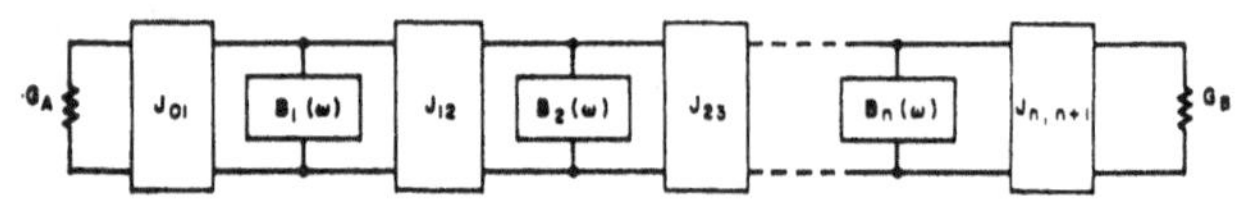

(a) A GENERALIZED, BAND-PASS FILTER CIRCUIT USING ADMITTANCE INVERTERS

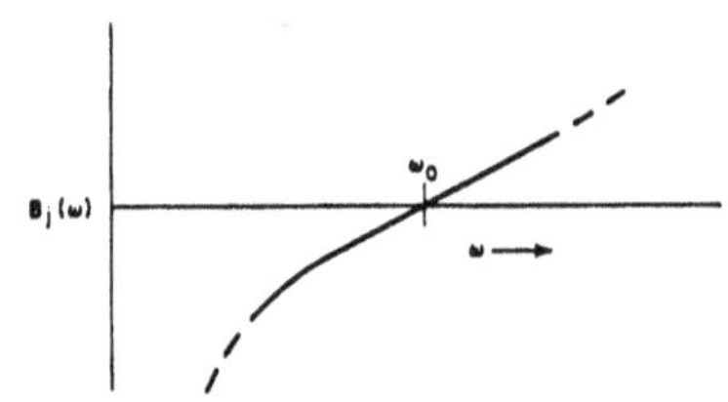

(b) SUSCEPTANCE OF k th RESONATOR

$$b_k = \frac{\omega_0}{2} \frac{d B_k(\omega)}{d\omega}\Big|_{\omega=\omega_0} \quad \text{mhos} \qquad (1)$$

= SUSCEPTANCE SLOPE PARAMETER

$$J_{01} = \sqrt{\frac{G_A b_1 w}{g_0 g_1 \omega_1'}} \quad \text{mhos} \qquad (2)$$

$$J_{k,k+1}\Big|_{k=1 \text{ to } n-1} = \frac{w}{\omega_1'}\sqrt{\frac{b_k b_{k+1}}{g_k g_{k+1}}} \quad \text{mhos} \qquad (3)$$

$$J_{n,n+1} = \sqrt{\frac{G_B b_n w}{\omega_1' g_n g_{n+1}}} \quad \text{mhos} \qquad (4)$$

$$w = \text{FRACTIONAL BANDWIDTH} \quad \text{OR} \quad w = \frac{\omega_2 - \omega_1}{\omega_0} \qquad (5)$$

Figure 11 — Generalized equations for design of band-pass filters from low-pass prototypes (Case of filters having resonators with only shunt-type resonances. The J-inverters represent the couplings).

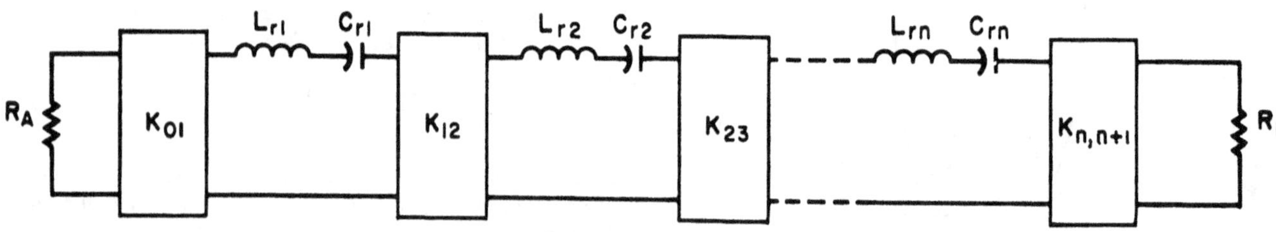

Figure 10 — The band-pass filter in Figure 9 converted to use only series resonators and impedance inverters.

filter in the form in Figure 9 (assuming that the inverters are ideal).

The J-inverters used in the derivation of the comb-line filter equations in Table I are of the form shown in Figure 12. This type of inverter operates as an inverter at all frequencies, but it falls short of being an ideal inverter since its inverter parameter J varies with frequency. Note that this inverter involves shunt lines having negative characteristic admittances. As will be seen, these cause no difficulty since they will be paralleled with other shunt lines having a positive characteristic admittance, and the total characteristic admittance of the shunt lines will be positive.

Figures 13, 14, and 15 show equivalences between parallel-coupled line circuits and open-wire line circuits. These equivalences were used in the derivation of the comb-line filter equations in Table I, and were obtained by extension of the work of Jones and Bolljahn[9] and of Ozaki and Ishii.[10] In Figures 13 to 15, Y^a_{oo} and Y^a_{oe} are the odd- and even-mode admittances for strip a,[9,10] Y^b_{oo} and Y^b_{oe} are the odd- and even-mode admittances for strip b, C_a and C_b are the capacitances per unit length between strip a and ground and between strip b and ground, C_{ab} is the capacitance per unit length between strips, and v is the velocity of propagation. Note that the circuit in Figure 15 is of the same form as that in Figure 14, but that a very special termination value Y_A has been specified.

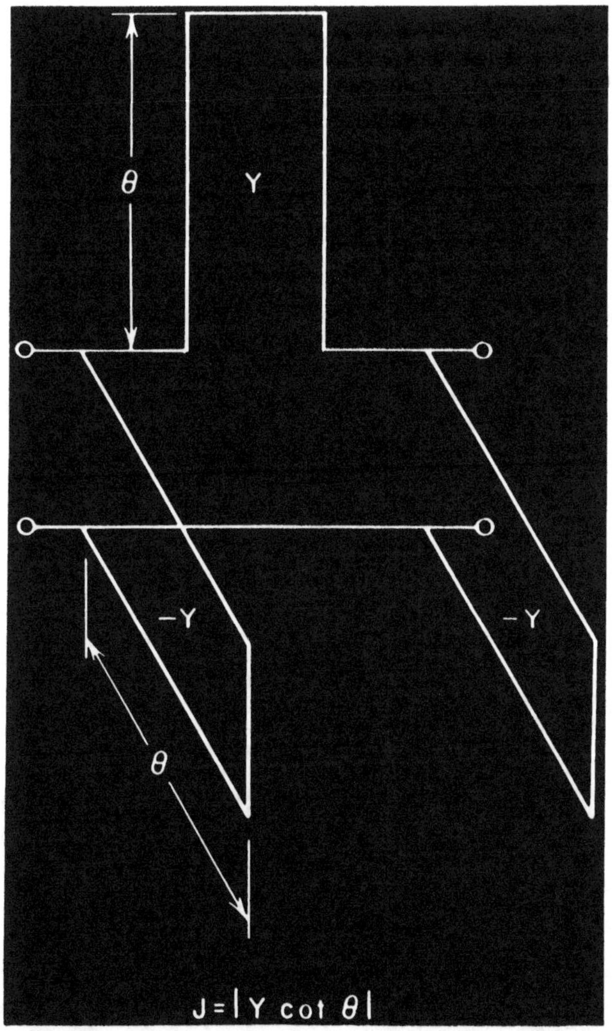

Figure 12 — An admittance inverter formed from stubs of electrical length θ.

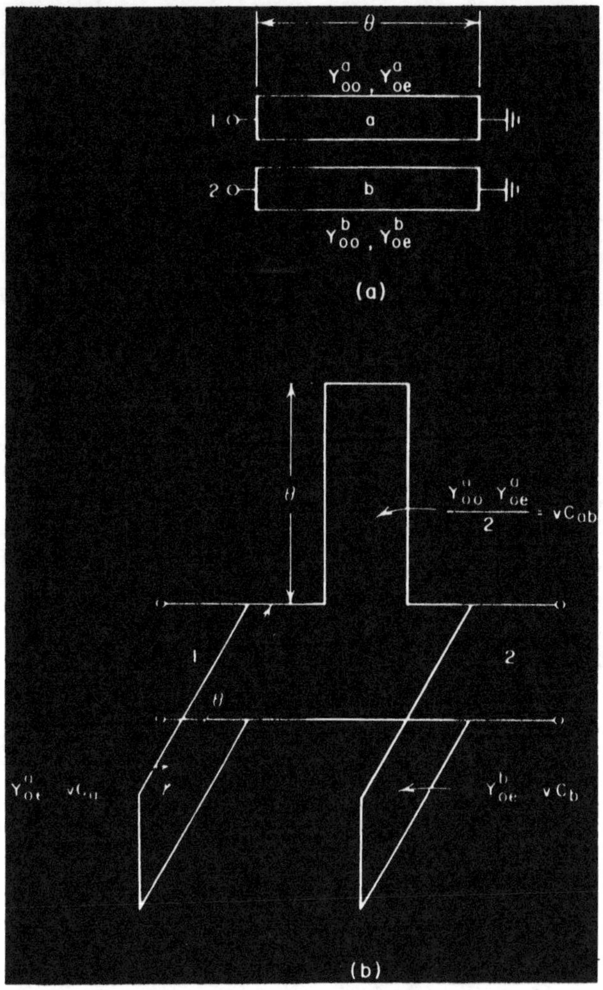

Figure 13 — A parallel-coupled line circuit and an open-wire line circuit that is electrically equivalent.

The comb-line filter design equations in Table I were derived by use of the approximate, open-wire-line representation in Figure 16. Note that in Figure 16 nodal points are indicated that correspond to nodal points indicated in Figure 1. The circuitry between Nodal Points 0 and 1 in Figure 16 was specified with aid of the equivalences in Figure 14 to correspond to the parallel-coupled strip lines between Nodal Points 0 and 1 in Figure 1. The circuitry between Nodal Points 1 and 2, 2 and 3, . . . , n−1 and n, was specified with the aid of the equivalences in Figure 13(a), (b) to correspond to the

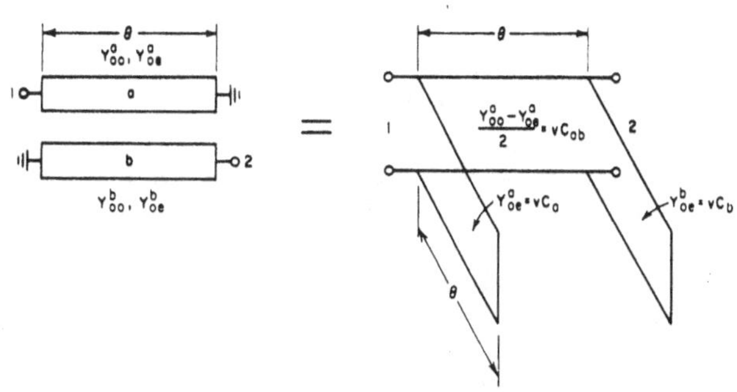

Figure 14 — Short-circuited, parallel-coupled strip lines and their open-wire-line equivalent circuit (The parallel-coupled lines may be of either equal or unequal widths).

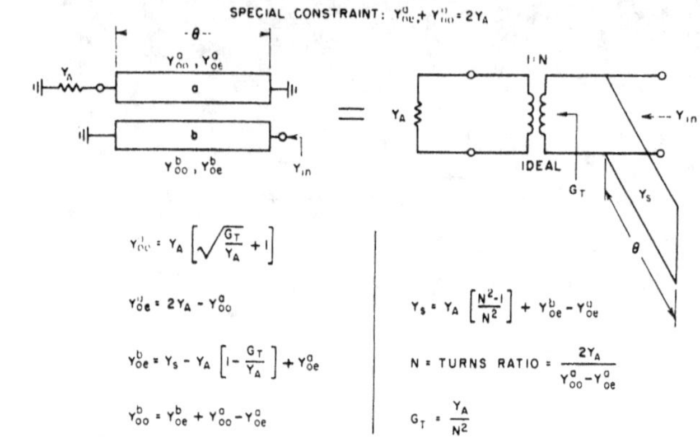

Figure 15 — Special properties of the circuits in Figure 14 when a special constraint is applied.

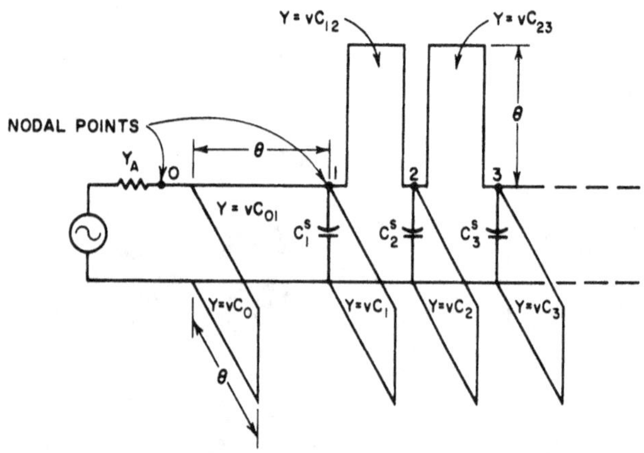

Figure 16 — An approximate equivalent circuit of the comb-line filter in Figure 1 (The capacitances C_k and $C_{k,\,k+1}$ are the self and mutual capacitances per unit length of the lines in Figure 4, while the v is the velocity of propagation).

pairs of parallel-coupled sections formed by the strip lines connected to Nodal Points 1 and 2, 2 and 3, etc. Note that the line admittances in Figure 16 are defined in terms of the line capacitances per unit length, C_k and $C_{k,k+1}$ as defined in Figure 4, times the velocity of propagation (which gives the dimensions of admittances). This representation of a comb-line filter is approximate, and neglects the effects of fringing capacitances beyond nearest neighbors.[1]

In order to modify the circuit in Figure 16 to a form such that the data in Figure 11 can be easily applied, the series stubs between Nodal Points 1 and 2, 2 and 3, etc., in Figure 16 were incorporated into J-inverters of the form in Figure 12, which gave the result shown in Figure 17. Since each of the inverters $J_{k,k+1}$ consists of a pi configuration of a series stub of characteristic admittance Y, and two shunt stubs of characteristic admittance $-Y$, it was necessary to increase the characteristic admittances of the actual shunt stubs on each side in order to compensate for the negative admittances ascribed to the inverters. This is why Shunt Stubs 2 to $n-1$ in Figure 17 now have the admittances $Y_{ak}=v(C_k+C_{k-1,k}+C_{k,k+1})$ instead of just vC_k. The portion of the circuit in Figure 16 between Nodal Points 0 and 1 has been converted to the form shown in Figure 17 by use of the simplifying constraint that brings about the properties summarized in Figure 15.

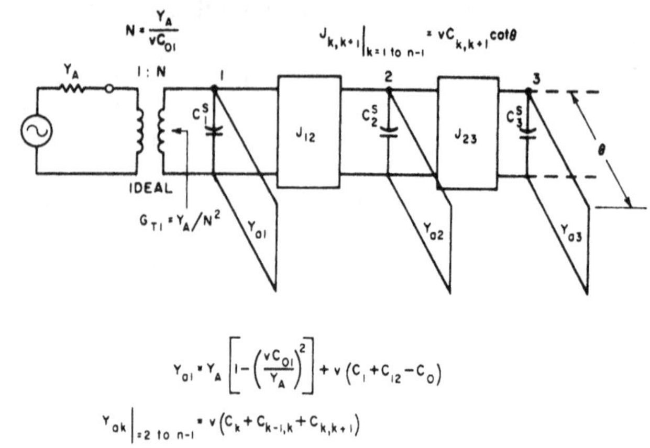

Figure 17 — The circuit in Figure 16 reorganized to include admittance inverters (The constraint $vC_0=Y_A-vC_{01}$ has also been applied).

When applying the general relations in Figure 11 to the circuit in Figure 17 to derive design equations for comb-line filters, the admittance-inverter parameters $J_{k,k+1}$ are evaluated at midband. The resonator functions $B_k(\omega)$ are

$$B_k(\omega)=\omega C_k{}^s-Y_{ak}\cot\left(\theta_0\frac{\omega}{\omega_0}\right) \qquad (6)$$

where θ_0 is the midband electrical length of the resonator lines and the Y_{ak} are as indicated in Figure 17. The resonator slope parameters are evaluated as indicated by Equation (1) of Figure 11, the terminating admittance G_{T1} in Figure 17 is specified so that $G_{T1}=J^2{}_{01}/Y_A$, where the required value for J_{01} is as indicated in Figure 11. The derivation of the design equations for the right end section of the filter is entirely analogous to that for the left end section.

Acknowledgment

The design calculations, shop drawings, and laboratory tests for the trial comb-line filter design discussed in this paper were made by Mr. York Sato.

References

1. Bolljahn, J. T. and G. L. Matthaei, "A Study of the Phase and Filter Properties of Arrays of Parallel Conductors Between Ground Planes," *Proc. IRE 50*, pp. 299-311 (March 1962).
2. Matthaei, G. L., "Interdigital Filters," *IRE Trans. PGMTT*-10, pp. 479-491 (November 1962).
3. Weinberg, L., "Additional Tables for Design of Optimum Ladder Networks," *Journal of Franklin Institute 246*, pp. 7-23 and 127-138 (July and August 1957).
4. Matthaei, G. L., et al, "Design Criteria for Microwave Filters and Coupling Structures," pp. 201-205, Final Report, SRI Project 2326, Contract DA 36-039 SC-74862, Stanford Research Institute, Menlo Park, California (January 1961).
5. Getsinger, W. J., "Coupled Rectangular Bars Between Parallel Plates," *IRE Trans. PGMTT-10*, pp. 65-72 (January 1962).
6. Dishal, M., "Alignment and Adjustment of Synchronously Tuned Multiple-Resonator-Circuit Filters," *Proc. IRE 39*, pp. 1448-1455 (November 1951).
7. Cohn, S. B., "Direct-Coupled-Resonator Filters," *Proc. IRE 45*, pp. 187-196 (February 1957).
8. Mumford, W. W., "Quarter-Wave Coupled Waveguide Filter," *Bell Syst. Tech. Jour. 4*, pp. 684-713 (October 1948).
9. Jones, E. M. T. and J. T. Bolljahn, "Coupled-Strip-Transmission-Line Filters and Directional Couplers," *IRE Trans. PGMTT-4*, pp. 75-81 (April 1956).
10. Ozaki, H. and J. Ishii, "Synthesis of a Class of Strip-Line Filters," *IRE Trans. PGCT-5*, pp. 104-109 (June 1958).

121

Reprinted from IEEE TRANSACTIONS
ON MICROWAVE THEORY AND TECHNIQUES
Volume MTT-14, Number 7, July, 1966
Pp. 351-353

Design of Comb-Line Band-Pass Filters

Comb-line band-pass filters are filters using direct-coupled quarter-wave TEM resonators with adjacent resonators having the *same* open- and short-circuit reference planes. This is in contrast to interdigital band-pass filters where adjacent resonators *alternate* the open- and short-circuit reference planes. Comb-line band-pass filters can be realized using either strip (rectangular center conductors) or *slab* (round center conductors) transmission line resonators. When partitions are employed between adjacent resonators, the comb-line filter structure evolves into a coaxial filter structure. Consequently, comb-line and coaxial nomenclature will be used interchangeably herein. In this correspondence, the published design procedure for comb-line band-pass filters will be related to existing narrow-band filter theory. Certain aspects of the evolution of comb-line structures into coaxial structures will also be discussed.

Manuscript received April 4, 1966.

A procedure has been presented applicable to the design of comb-line band-pass filters of narrow or moderate bandwidth [1]. In this filter structure, use of coupled quarter-wave lines of uniform cross section has been shown to result in an all-stop network. Adjacent resonators are decoupled due to the cancellation of equal magnetic and electric fields in phase opposition. To achieve band-pass filter behavior, resonator lines are foreshortened to electrical lengths less than ninety degrees, and resonance is obtained using lumped-capacitances at the resonator open-ends. Appreciable coupling between adjacent resonators is realized because the magnetic and electric couplings are no longer equal. In this case, the net coupling will be magnetic.

For narrow-band filters (i.e., bandwidth less than ten percent), it is convenient to describe the interstage couplings by coefficients of coupling [2] which can be readily measured by simple experimental procedures [3]. Assuming all interstage couplings to be equal, Matthaei's design equations [1] can be rewritten as follows, when $Y_{AK}=Y_A$:

$$\frac{C_{ij}}{\epsilon}=\frac{377\,Y_A}{\sqrt{\epsilon_r}}\left(\frac{J_{ij}}{Y_A}\right)\tan\theta_0 \tag{1}$$

$$\frac{J_{ij}}{Y_A}=\frac{w}{\omega_1{}^1}\left(\frac{b}{Y_A}\right)\frac{1}{\sqrt{g_ig_j}} \tag{2}$$

$$\frac{b}{Y_A}=\left(\frac{\cot\theta_0+\theta_0\csc^2\theta_0}{2}\right). \tag{3}$$

Letting $w=1/Q_T$, $\omega_1{}^1=1.0$, and

$$K_{ij}=\frac{1}{Q_T}\sqrt{\frac{1}{g_ig_j}},$$

and combining (1), (2), and (3), it can be shown that

$$\frac{C_{ij}}{\epsilon}=\frac{377\,Y_A}{\sqrt{\epsilon_r}}K_{ij}f(\theta_0) \tag{4}$$

where

$$f(\theta_0) = \frac{\tan\theta_0(\cot\theta_0 + \theta_0 \csc^2\theta_0)}{2} = \frac{1}{2}\left[1 + \frac{2\theta_0}{\sin 2\theta_0}\right] \quad (5)$$

$$Q_T = f_0/\Delta f_{3dB} = \frac{\text{filter center frequency}}{\text{filter 3-dB bandwidth}}$$

= filter total Q, and

K_{ij} = absolute coefficient of coupling between ith and jth resonators.

If $C = C_i = C_j$, from Matthaei's [1] equations,

$$\frac{C}{\epsilon} = \frac{377\,Y_A}{\sqrt{\epsilon_r}}\left[1 - 2\left(\frac{J_{ij}}{Y_A}\right)\tan\theta_0\right]. \quad (6)$$

Combining (2), (3), (4), and (6), it can be shown that

$$\frac{C}{\epsilon} = \frac{377\,Y_A}{\sqrt{\epsilon_r}}\left[1 - K_{ij} f(\theta_0)\right] \quad (7)$$

where $f(\theta_0)$ is defined per (5). Dividing (4) by (7) and rearranging terms,

$$K_{ij} = \frac{1}{f(\theta_0)}\left[\frac{1}{\dfrac{C/\epsilon}{C_{ij}/\epsilon} + 2}\right]. \quad (8)$$

Using slab-line construction [4] (i.e., round center conductors between ground planes),

$$\frac{C}{\epsilon} = \frac{C_g}{\epsilon} \quad \text{and} \quad \frac{C_{ij}}{\epsilon} = \frac{C_m}{\epsilon}.$$

Then

$$K_{ij} = \frac{1}{f(\theta_0)}\left[\frac{1}{\dfrac{C_g/\epsilon}{C_m/\epsilon} + 2}\right]. \quad (9)$$

$1/f\ (\theta_0)$ is tabulated below for common values of θ_0:

Degrees	$1/f(\theta_0)$
0	1.000
15	0.978
30	0.906
45	0.778
60	0.586
75	0.320
90	0.000

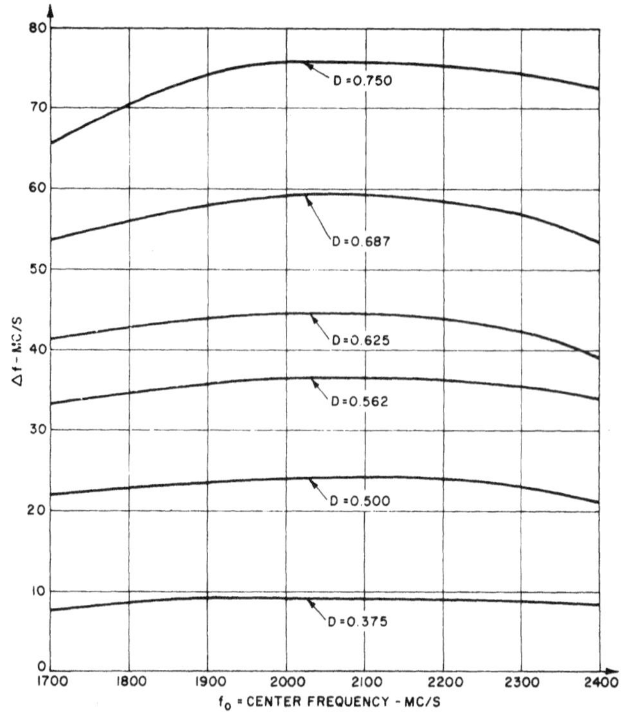

Fig. 1. Aperture coupling data. Coupling bandwidth vs. aperture diameter. Circular apertures 0.650" from plane of short coaxial resonators: 1.375"×0.750"/0.375 diameter.

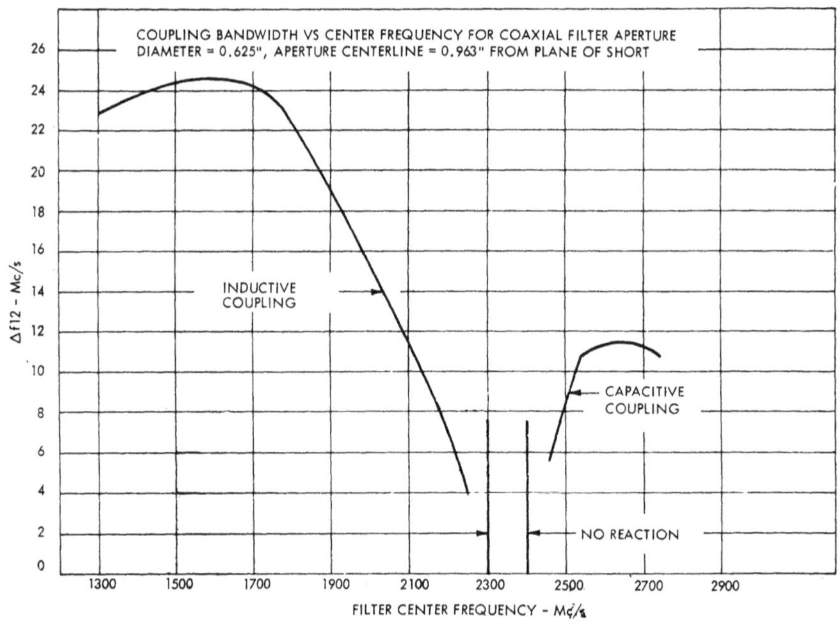

Fig. 2. Frequency sensitivity of coupling aperture.

Use of the recommended comb-line structure with foreshortened resonators results in a filter package that is smaller in one dimension; however, this foreshortening of the resonator line lengths degrades the resonator unloaded Q's and increases pass band insertion losses due to dissipation. Matthaei [1] has mentioned that foreshortened comb-line resonators have lower unloaded Q's than quarter-wave interdigital resonators of the same cross section. Experimental data for comb-line band-pass filters for $30° < \theta_0 < 90°$ has shown the following empirical relationship to be applicable for comb-line reson tors of constant cross section:

$$Q_{UL}(\text{effective}) \cong \sin^2(\theta_0) Q_{UL}, \quad (10)$$

where Q_{UL} (effective) is the unloaded Q of the foreshortened resonator, and Q_{UL} is the unloaded Q of the quarter-wave resonator. When $\theta_0 = 45$ degrees, the effective unloaded Q is reduced by a factor of 2. To use comb-line filters with full quarter-wave resonators, different coupling techniques can be employed.

Interstage coupling in comb-line filters can be realized using partitions between adjacent resonators with coupling apertures in the partitions [5], [6]. Circular coupling apertures are usually located for minimum frequency sensitivity [7], while aperture diameters are determined experimentally to

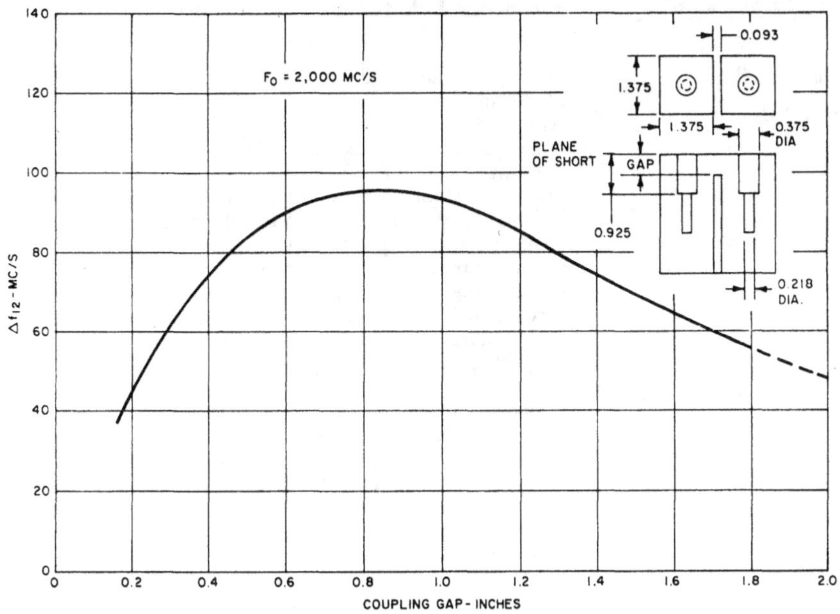

Fig. 3. Coaxial filter coupling data (square cross section outer conductors).

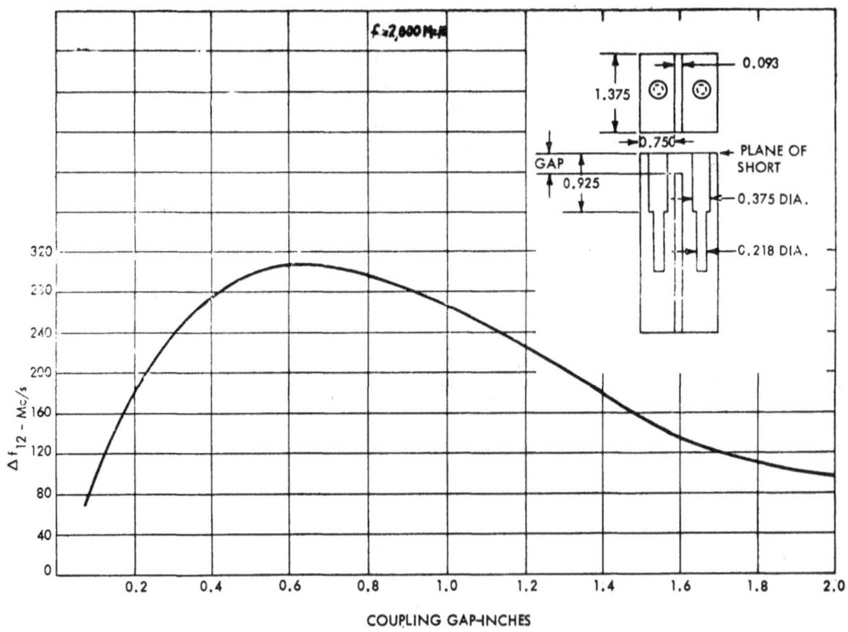

Fig. 4. Coaxial filter coupling data (rectangular cross section outer conductors).

obtain the desired coefficients of coupling. Typical experimental curves of coupling bandwidth vs. center frequency for various aperture diameters are shown in Fig. 1 for a comb-line filter structure. For this aperture location, the net coupling will be magnetic (i.e., inductive). It should be noted that absolute coefficient of coupling $K_{12} \cong \Delta f_{12}/f_0$, where Δf_{12} is the coupling bandwidth, vs. center frequency is shown in Fig. 2 for a somewhat different aperture location. When the aperture is a certain electrical distance from the plane of the short circuit, a cancellation of magnetic and electric fields will occur, and adjacent resonators are effectively decoupled. This is the region of no reaction of Fig. 2. For comb-line filters using partitions between adjacent resonators, some degradation in resonator unloaded Q's will occur due to resistive losses in the partition walls.

Wall losses can be substantially avoided using partial partitions that shield (decouple) adjacent resonators only in the regions of strong electric fields. Figures 3 and 4 show coupling bandwidth Δf_{12} as a function of coupling gap for comb-line filter structures of two different resonator cross sections. With resonators using square outer conductor cross section (see Fig. 3), a maximum coupling bandwidth of 4.75 percent has been obtained. With resonators using rectangular outer conductor cross section, (see Fig. 4), a maximum coupling bandwidth of 14.7 percent has been obtained. In both cases, as the coupling gap increases, Δf_{12} increases until the incremental increase in electric coupling overtakes the incremental increase in magnetic coupling. With no partitions, the net coupling is still magnetic, and complete cancellation of magnetic and electric couplings does not occur. This is due to the use of compound center conductors making resonator cross sections nonuniform, and magnetic and electric couplings unequal.

Under some circumstances, use of partitions can have other advantages. A family of comb-line band-pass filters with different response shapes and/or bandwidths can be developed with overall filter size depending only upon center frequency and number of resonators. Direct-coupled comb-line filters will require variations in resonator foreshortening and/or adjacent resonator separation to implement changes in filter response shapes and/or bandwidths. The use of partitions can sometimes help in suppressing some of the filter spurious responses.

The comments of Dishal [8] with respect to narrow-band interdigital filters are also applicable to narrow-band comb-line filters. Use of auxiliary rods for filter input/output couplings are unnecessary; and coupling mechanisms such as loops, probes, or direct taps should be employed in obtaining the required singly loaded Q's for the input/output resonators [5].

For filters of moderate percentage bandwidth (i.e., ten percent to twenty percent), use of Matthaei's design equations is more appropriate than utilizing the narrow-band concepts of coefficients of coupling and singly loaded Q. In this case, reductions in unloaded Q's, due to the foreshortening of resonator lengths, will be less significant.

R. M. Kurzrok
Communications Systems Labs.
RCA
New York, N. Y.

References

[1] G. L. Matthaei, "Comb-line band-pass filters of narrow or moderate bandwidth," *Microwave J.*, pp. 82–91, August 1963.
[2] M. Dishal, "Design of dissipative band-pass filters producing desired exact amplitude-frequency characteristics," *Proc. IRE*, vol. 37, pp. 1050–1069, September 1949.
[3] ——, "Alignment and adjustment of synchronously tuned multiple-resonator-circuit filters," *Proc. IRE*, vol. 39, pp. 1448–1455, November 1951.
[4] E. G. Cristal, "Coupled circular cylindrical rods between parallel ground planes," *IEEE Trans. on Microwave Theory and Techniques*, vol. MTT-12, pp. 428–439, July 1964.
[5] Radio Research Lab. Staff, "Very high frequency techniques," Vol. II. New York: McGraw-Hill, 1947, chap. 28, pp. 769–787.
[6] R. M. Kurzrok, "Direct coupled coaxial and waveguide band-pass filters," *IRE Trans. on Microwave Theory and Techniques (Correspondence)*, vol. MTT-10, pp. 218–219, May 1962.
[7] ——, "Design of interstage coupling apertures for narrow-band tunable coaxial band-pass filters," *IRE Trans. on Microwave Theory and Techniques (Correspondence)*, vol. MTT-10, pp. 143–144, March 1962.
[8] M. Dishal, "A simple design procedure for small percentage bandwidth round-rod interdigital filters," *IEEE Trans. on Microwave Theory and Techniques (Correspondence)*, vol. MTT-13, pp. 696–698, September 1965.

Reprinted from the IEEE Trans. on Microwave Theory and Techniques, Vol. MTT-16, pp. 888-889 (October 1968).

124

Frequency Behavior of Post-Coupled TEM Comb-Line Resonators

In prior published work [1], [2], the use of a transverse coupling post between adjacent TEM comb-line resonators was presented. Such a post acts as an electric shield reducing the electric coupling which is in phase opposition with the magnetic coupling. In this correspondence, additional experimental data will be presented on post-coupled comb-line resonators. It will be shown that a properly positioned coupling post can result in some useful couplings that are quite insensitive to filter center frequency. Similar coupling behavior has been previously obtained with optimally located coupling apertures [3] in partitions between adjacent resonators.

A comb-line filter structure using slab-transmission line resonators has been employed for interstage coupling measurements in the 1.6-GHz to 2.4-GHz frequency range (see Fig. 1). Coupling bandwidths ΔF_{12} were measured using the method of Dishal [4] for various center frequencies F_0 and center-to-center resonator spacings c. The absolute coefficient of coupling K_{12} is approximately equal to the coupling bandwidth divided by the center frequency.

Measured data of coupling bandwidths versus center frequency are plotted in Fig. 2 for the direct-coupled case (i.e., without a coupling post). Coupling bandwidths are quite frequency sensitive. (*Note;* A bandpass filter displaying constant *absolute* bandwidth over an appreciable tuning range should have coupling bandwidths that are independent of

Manuscript received March 1, 1968; revised May 6, 1968.

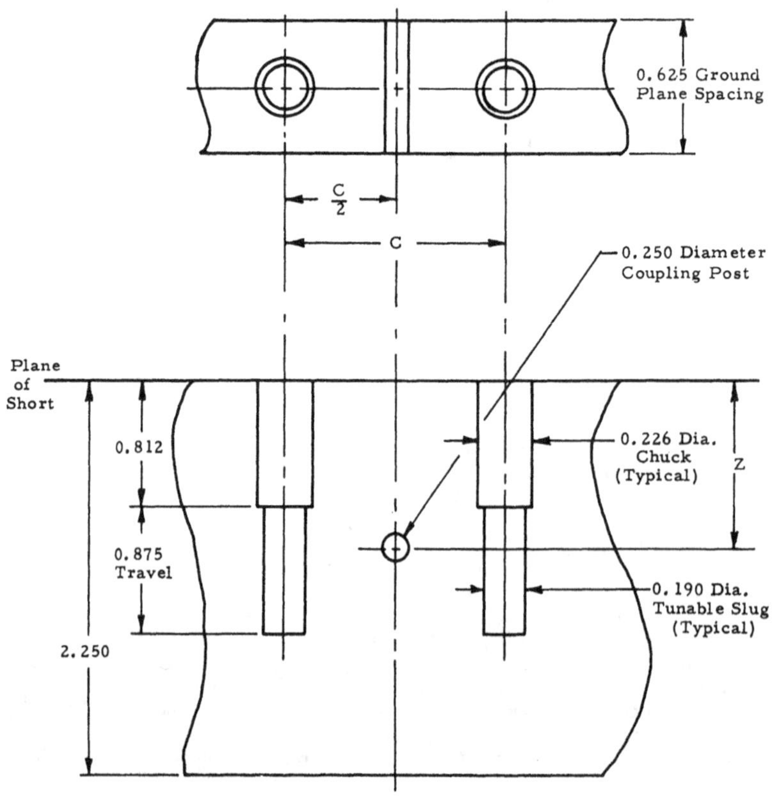

Fig. 1. Mechanical details of post-coupled comb-line filter structure. *Note:* All dimensions in inches.

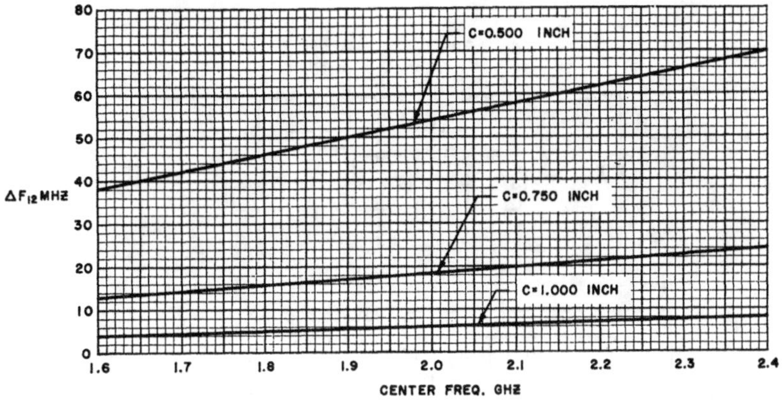

Fig. 2. Direct coupled resonators. Measured data of coupling bandwidth versus center frequency.

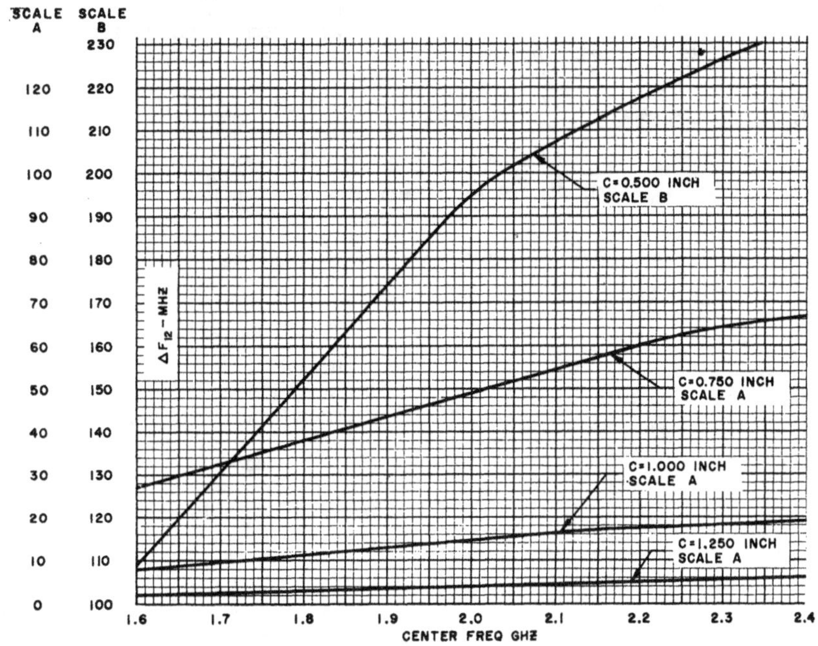

Fig. 3. Post-coupled resonators. Measured data of coupling bandwidth versus center frequency for $Z=1.000$ inch.

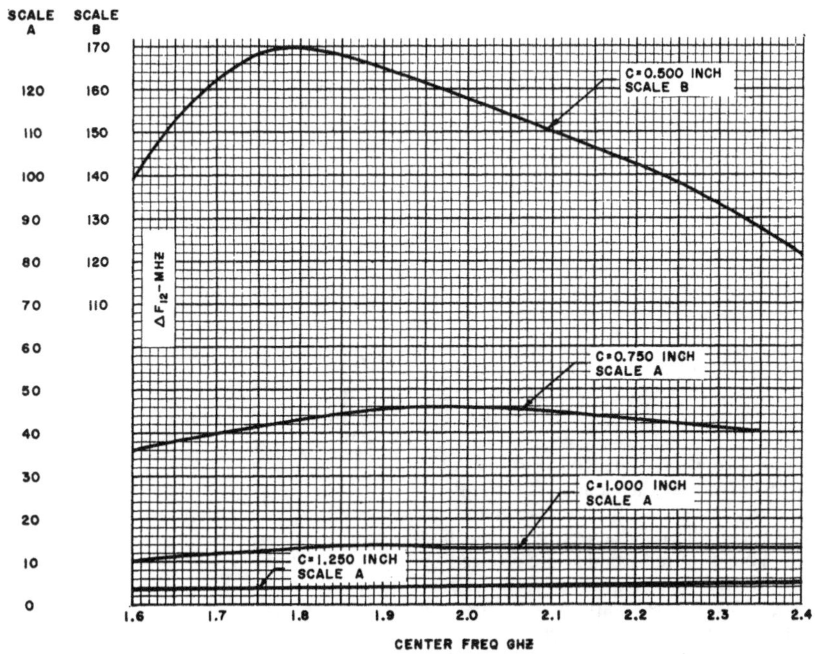

Fig. 4. Post-coupled resonators. Measured data of coupling bandwidth versus center frequency for $Z=1.312$ inch.

frequency. This is *not* the same as obtaining constant *fractional* bandwidth.) The addition of a transverse coupling post was considered as a means of reducing the frequency sensitivity of the interstage couplings. Two different post locations were arbitrarily chosen and coupling bandwidths were measured for different center frequencies.

Upon adding a coupling post located $Z=1.000$ inch from the plane of the short, the data of Fig. 3 were obtained. Tighter couplings are achieved using the coupling post. Coupling bandwidths seem less frequency sensitive for the higher center frequencies. Upon moving the coupling post to a new position ($Z=1.312$ inch), the data of Fig. 4 were obtained. For resonator spacings between 0.75 inch and 1.25 inches, coupling bandwidths are relatively independent of frequency. This is a new and useful interstage coupling technique. It is applicable to low-cost bandpass filters of small percentage bandwidths that are designed for an appreciable tuning range.

R. M. Kurzrok
Radio Engrg. Labs.
Long Island City, N. Y. 11101

References

[1] R. M. Kurzrok, "Additional considerations in comb-line bandpass filter interstage couplings," *IEEE Trans. Microwave Theory and Techniques* (Correspondence), vol. MTT-15, pp. 262–263, April 1967.

[2] ——, "Coupling posts in comb-line boost filter performance," *Electronic Design*, vol. 15, p. 20, May 10, 1967.

[3] ——, "Design of interstage coupling apertures for narrow-band tunable coaxial bandpass filters," *IEEE Trans. Microwave Theory and Techniques* (Correspondence), vol. MTT-10, pp. 143–144, March 1962.

[4] M. Dishal, "Alignment and adjustment of synchronously tuned multiple-resonator-circuit filters," *Proc. IRE*, vol. 39, pp. 1448-1455, November 1951.

Reprinted from the IEEE Trans. on Microwave Theory and Techniques, Vol. MTT-18, pp. 278-280 (May 1970).

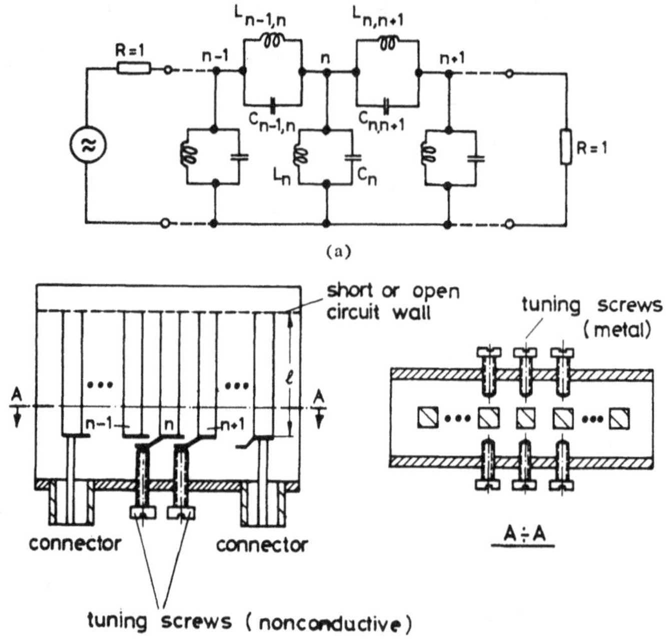

Fig. 1. (a) Lumped-element prototype circuit, for the design of the microwave filters (b).

Microwave Filters of Coupled Lines and Lumped Capacitances

***Abstract*—Two frequency transformations are given which enable microwave filters to be realized with coupled lines and lumped capacitances from a characteristic prototype lumped network to a good approximation.**

The procedure includes filters originally proposed by Matthaei and Cristal, and enables realizations for large bandwidths. A number of examples demonstrate the efficiency of the procedures.

I. Introduction

Filters consisting of lines with uniform length l [1]–[4] can be derived according to the transformation $p=\tanh \gamma l$ from lumped-element networks as stated by Richards [5]. These filters have a periodic attenuation versus frequency characteristic. For some applications such a periodic characteristic is undesirable. Nonperiodic characteristic will be obtained when lumped capacitances are incorporated in the filter. In this correspondence a whole class of filters consisting of coupled lines and lumped capacitances are described, which can approximately be derived by simple frequency transformations from lumped-filter networks, and which are of similar structure as the filters in [4].

II. The Filter Configuration

Many filter networks made of lumped elements can be derived as special cases of the general prototype circuit according to Fig. 1(a). If all capacitances in the shunt branches are zero, a high-pass filter is obtained. If in the series branches, either the capacitances or the inductances are omitted, then the inductively and capacitively coupled resonant-filter will be obtained, respectively. According to Richards' transformation, capacitances will become open-circuit lines and inductances will become short-circuit lines. If only the capacitances or only the inductances are replaced by transmission lines, hybrid circuits are obtained, for which new frequency transformations must be found. The lumped elements in the microwave filters are always capacitances. In the case of a microwave filter with short-circuited conductors, these capacitances are calculated from the capacitances of the prototype circuit. In the case of a microwave filter with open-circuited conductors, these capacitances are calculated from the inductances of the prototype circuit.

Manuscript received October 6, 1969; revised December 1, 1969.

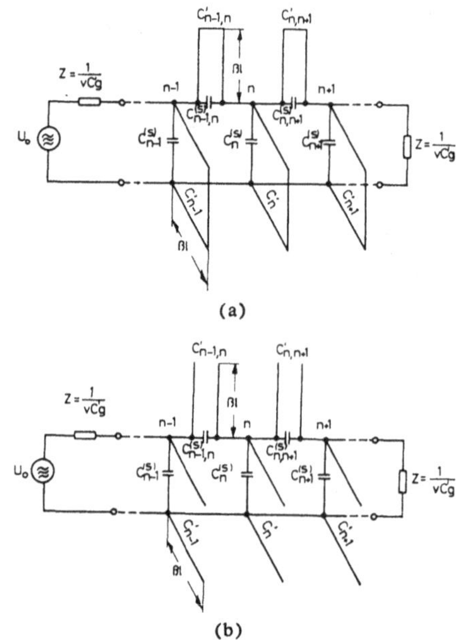

Fig. 2. Equivalent circuits of the microwave filters (a) With short-circuited conductors. (b) With open-circuited conductors.

The network in Fig. 1(a) may thus be constructed as a microwave filter of the configuration shown in Fig. 1(b). If the inductances of Fig. 1(a) are replaced by coupled conductors, these conductors will be short-circuited by a metal wall at their ends. The shunt inductances are realized by the distributed capacitances of the conductors to the ground plates. The series inductances are realized by the distributed capacitances between the conductors. The lumped capacitances may be adjusted by tuning screws. If realizing the capacitances by distributed parameters, these must be open-circuited lines. Fig. 2 shows the equivalent networks of the two microwave filters.

III. The Frequency Transformations

In order to derive the frequency transformations it is necessary to set up the difference equation for the voltages in the points $n-1$, n, and $n+1$ for both networks. In the case of short-circuited conductors, the equation for one network may be converted into the equation for the other by letting

$$\omega^2 = \beta l \tan \beta l \tag{1}$$

where βl is the electrical length of the lines.

The distributed capacitances of the microwave filter are related to the inductances of the lumped-element network, and the lumped capacitances $C^{(s)}$ divided by l of the microwave filter are related to the capacitances of the lumped-element network.

$$\frac{C'_{n-1,n}}{\epsilon}=\frac{1}{L_{n-1,n}}; \quad \frac{C'_{n,n+1}}{\epsilon}=\frac{1}{L_{n,n+1}}; \quad \frac{C_n'}{\epsilon}=\frac{1}{L_n}$$
$$\frac{C'^{(s)}_{n-1,n}}{\epsilon}=C_{n-1,n}; \quad \frac{C'^{(s)}_{n,n+1}}{\epsilon}=C_{n,n+1}; \quad \frac{C_n'^{(s)}}{\epsilon}=C_n. \tag{2}$$

In the case of open-circuited conductors, the following frequency transformation must be applied:

$$\omega^2 = -\frac{\tan \beta l}{\beta l}. \tag{3}$$

Here the distributed capacitances and the lumped capacitances divided by l are related to the elements of the lumped network as follows:

$$\frac{C'_{n-1,n}}{\epsilon}=C_{n-1,n}; \quad \frac{C'_{n,n+1}}{\epsilon}=C_{n,n+1};$$
$$\frac{C_n'}{\epsilon}=C_n$$
$$\frac{C'^{(s)}_{n-1,n}}{\epsilon}=\frac{1}{L_{n-1,n}}; \quad \frac{C'^{(s)}_{n,n+1}}{\epsilon}=\frac{1}{L_{n,n+1}};$$
$$\frac{C_n'^{(s)}}{\epsilon}=\frac{1}{L_n}. \tag{4}$$

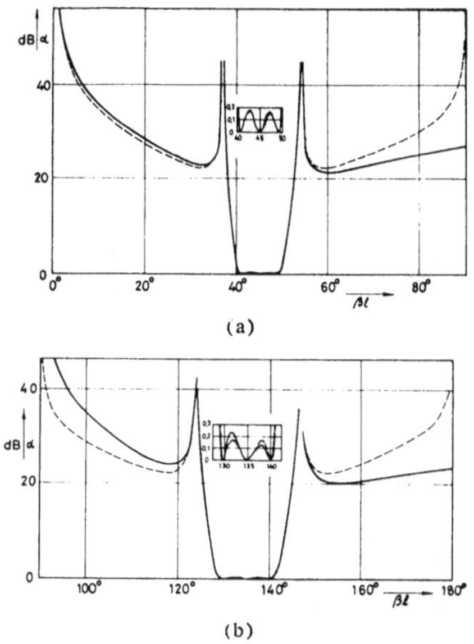

Fig. 3. Attenuation characteristic of a microwave filter. (a) With short-circuited conductors. (b) With open-circuited conductors. Broken lines indicate expected characteristic from the frequency transformations. Solid lines indicate actual theoretical characteristic.

Equation (1) transforms the passband of the lumped-element filter to the range $0° < \beta l < 90°$; (3) transforms the passband to the range $90° < \beta l < 180°$ for the microwave filter.

The characteristics of the input and output sections of the networks are transformed precisely at one singular frequency only. This frequency should be chosen in the middle of the passband. The distributed capacitance of the input and output lines in the normalized filter must be

$$\frac{C_g'}{\epsilon} = \sqrt{\frac{\beta l_0}{\tan \beta l_0}} \tag{5}$$

for the filter with short-circuited conductors, and

$$\frac{C_g'}{\epsilon} = \sqrt{-\beta l_0 \tan \beta l_0} \tag{6}$$

for the filter with open-circuited conductors. βl_0 is the angle at the center frequency.

Finally, the normalized filter must be converted to the specified impedance level Z_g. To this end, all distributed and lumped capacitances must be multiplied by the factor

$$K = \frac{120\pi\Omega}{\sqrt{\epsilon_r} Z_g C_g'/\epsilon}. \tag{7}$$

Z_g is the characteristic impedance of the input line. The synthesis is given in detail in [7], [8].

IV. Results

The accuracy of this filter design has been demonstrated by calculating the attenuation characteristics for typical examples. The attenuation characteristics which will result from the frequency transformations (1) and (3) are compared to the theoretical attenuation characteristic of the original microwave filters. Fig. 3 shows the results for a microwave filter with three short-circuited conductors (a) and three open-circuited conductors (b). The corresponding lumped-element network has the configuration of Fig. 1(a) and has been synthesized according to the catalog in [6].

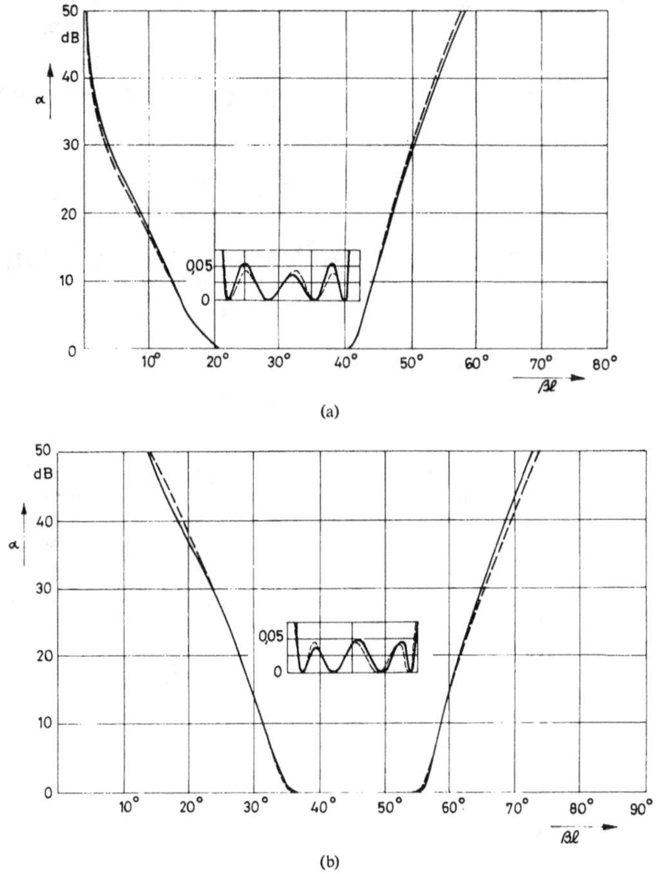

Fig. 4. Attenuation characteristic of comb-line filters. (a) Comb-line filter according to Matthaei [9]. (b) Capacitive coupled comb-line filter similar to Cristal's filter [10]. Broken lines indicate expected characteristic from frequency transformation (1). Solid lines indicate actual theoretical characteristic.

Fig. 4 shows results for comb-line filters [9], [10] with four short-circuited conductors. The corresponding lumped-element prototype circuits are inductively coupled resonant filters. The expected and actual attenuation characteristics are in close agreement over a relatively wide band. Further results are given in [7], [8].

Reinhold Pregla
Institut für Hochfrequenztechnik
Technische Universität Braunschweig
Braunschweig, Germany

References

[1] R. J. Wenzel, "Exact design of TEM microwave networks using quarter-wave lines," *IEEE Trans. Microwave Theory and Techniques*, vol. MTT-12, pp. 94–111, January 1964.

[2] M. C. Horton and R. J. Wenzel, "General theory and design of optimum quarter-wave TEM filter," *IEEE Trans. Microwave Theory and Techniques*, vol. MTT-13, pp. 316–327, May 1965.

[3] G. L. Matthaei, L. Young, and E. M. T. Jones, *Microwave Filters, Impedance Matching Networks, and Coupling Structures*. New York: McGraw-Hill, 1964.

[4] M. C. Horton and R. J. Wenzel, "The digital elliptic filter—A compact sharp-cutoff design for wide bandstop or bandpass requirements," *IEEE Trans. Microwave Theory and Techniques*, vol. MTT-15, pp. 307–314, May 1967.

[5] P. I. Richards, "Resistor-transmission-line circuits," *Proc. IRE*, vol. 36, pp. 217–220, February 1948.

[6] A. I. Zverev, *Handbook of Filter Synthesis*. New York: Wiley, 1967.

[7] R. Pregla, "Ein Syntheseverfahren für Kammfilter für kleine und grosse Bandbreiten," *Arch. Elek. Übertragung*, vol. 23, no. 1, pp. 13–19, 1969.

[8] ——, "Mikrowellenfilter aus verkoppelten Leitungen und konzentrierten Elementen," *Arch. Elek. Übertragung*, vol. 23, no. 12, pp. 593–600, 1969.

[9] G. L. Matthaei, "Comb-line band-pass filters of narrow or moderate bandwidth," *Microwave J.*, vol. 6, pp. 82–191, 1963.

[10] E. G. Cristal, "Capacity coupling shortens comb-line filters," *Microwaves*, vol. 6, pp. 44–50, December 1967.

Sonderdruck
aus
ARCHIV DER ELEKTRISCHEN ÜBERTRAGUNG

Ein Syntheseverfahren für Kammfilter für kleine und große Bandbreiten

von REINHOLD PREGLA *

Mitteilung aus dem Institut für Hochfrequenztechnik der Technischen Universität Braunschweig

(A.E.Ü. **23** [1969], Heft 1, 13–19; eingegangen am 27. Juni 1968)

DK 621.372.54

Eine Frequenztransformation wird angegeben, die die Realisierung von Kammfiltern aus Resonanzfiltern mit konzentrierten Elementen in guter Näherung ermöglicht. Das Verfahren ist für das von MATTHAEI und für das von CRISTAL angegebene Filter geeignet. Es liefert auch bei großen Bandbreiten noch gute Ergebnisse. Einige Beispiele zeigen die Güte des Syntheseverfahrens.

A Synthesis Method for Comb Filters of Narrow and Large Bandwidths

A frequency transformation is devised which allows comb filters to be realized to a good approximation using resonance type filters with lumped elements. The method is suitable for the filter devised by MATTHAEI and for that devised by CRISTAL. Good results are obtained even in the case of large bandwidths. Some examples prove the quality of the synthesis method.

1. Einleitung

Filter mit Leitungen konstanter Länge haben die Eigenschaft, daß ihr Verhalten mit der Frequenz entsprechend der von RICHARDS [1] angegebenen Transformation periodisch wiederkehrt. In manchen Anwendungsfällen ist diese Eigenschaft nicht erwünscht oder störend. Die strenge Periodizität wird verhindert durch gemischte Filter aus Leitungen und konzentrierten Elementen. Ein Filter dieser Art ist das von MATTHAEI [2] angegebene Kamm-(Comb-line)-Filter (Bild 1). Eine Variante dazu ist von CRISTAL untersucht worden [3]. Es handelt sich um ein kapazitiv angekoppeltes Kammfilter (Bild 2). Ein weiteres Filter ergibt sich, wenn man in Bild 2 die Kapazitäten C_1^s und C_n^s wegläßt. Auf diese Variante wird in dieser Arbeit noch speziell eingegangen werden.

Bild 3 zeigt den Querschnitt der Anordnungen, wenn rechteckige Leiter verwendet werden. Für die Dimensionierung wird in den angegebenen Arbeiten von einem normierten Polynom-Tiefpaß mit konzentrierten Elementen ausgegangen. Nach der Transformation in einen frequenzsymmetrischen Bandpaß wird die Schaltung in eine Schaltung mit Invertoren umgerechnet. Die Invertoren werden dann näherungsweise realisiert. Dieses Verfahren ist brauchbar für Filter mit einer kleinen bis mittleren relativen Bandbreite. Mit wachsender Bandbreite weicht die tatsächliche Filterkurve von der durch die Transformation erhaltenen immer mehr ab. Die Ursache dafür liegt in zwei Punkten, die miteinander zusammenhängen:

a) Das Kammfilter hat keine frequenzsymmetrische Bandpaßkurve. Die Dämpfungspole bei $f = 0$ und f_0 ($l = \lambda/4$) sind von unterschiedlicher Ordnung.

b) Die Realisierung des Invertors ist nur in der Bandmitte genau.

* Dr. R. PREGLA, im Institut für Hochfrequenztechnik der Technischen Universität, 33 Braunschweig, Postfach 7050.

In dieser Arbeit wird ein Syntheseverfahren beschrieben, das auch für relativ große Bandbreiten noch gute Ergebnisse liefert.

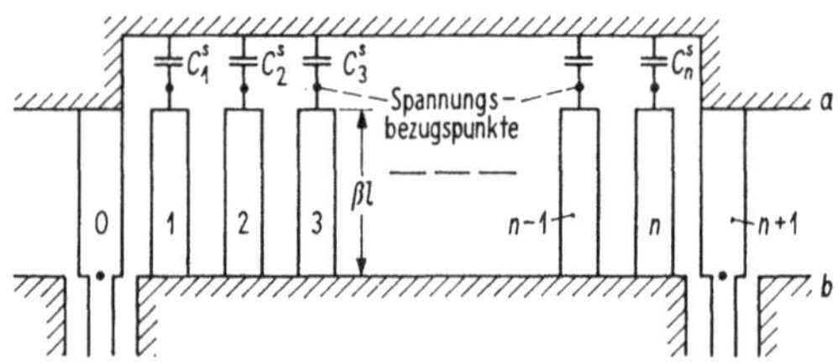

Bild 1. Kammfilter mit Transformationsleitungen.

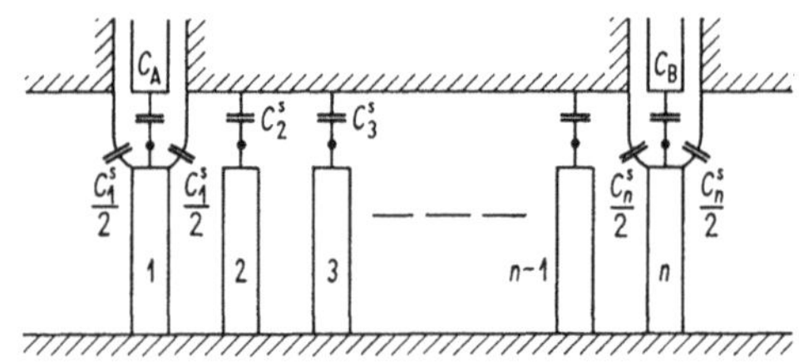

Bild 2. Kammfilter mit kapazitiver Ankopplung.

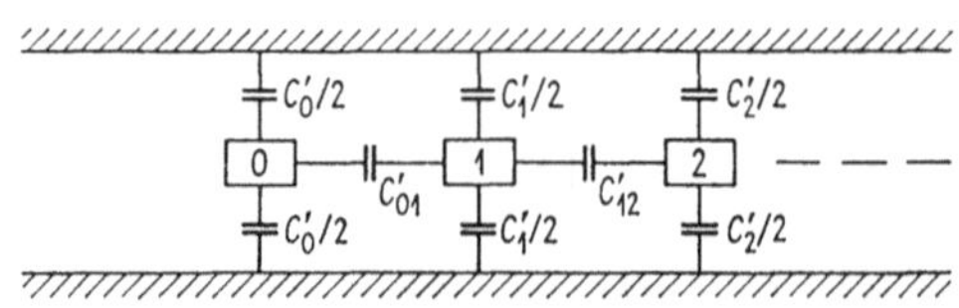

Bild 3. Querschnitt durch die Kammfilteranordnung.

2. Ersatzschaltbilder für die Kammfilter

In der Arbeit von MATTHAEI [2] wird für das Kammfilter das Ersatzschaltbild entsprechend Bild 4 angegeben. Es ist exakt, wenn nur Teilkapazitäten zwischen unmittelbar benachbarten Leitern vorhanden sind (Bild 3). Im mittleren Bereich ist dieses Ersatzschaltbild ohne weiteres einleuchtend; denn zwischen zwei benachbarten Bezugspunkten

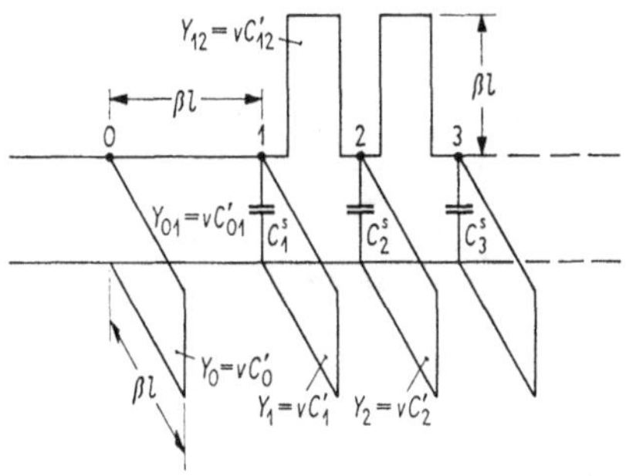

Bild 4. Ersatzschaltbild für das Kammfilter nach Bild 1.

wirkt eine kurzgeschlossene Leitung durch die kurzgeschlossenen Stäbe und den zwischen ihnen bestehenden Kapazitätsbelag. Außerdem bildet jeder Stab für sich mit dem Gehäuse eine parallelgeschaltete Kurzschlußleitung. Die Streukapazitäten C^s wirken parallel an den Bezugspunkten. Das Ersatzglied am Anfang kann man der Tabelle 1 in [4] entnehmen. Das Glied am Anfang bzw. der Leiter Null dient nicht der Filterwirkung, sondern zur Impedanztransformation. Um dies zu zeigen, wird der Anfangsteil der Schaltung in Bild 4 mit einer der von KURODA angegebenen Identitäten, die in Bild 5 wiedergegeben ist (siehe [4]), umgewandelt in die Schaltung nach Bild 6.

Für die Parameter in Bild 5 gelten die Beziehungen

$$Z'_{01} = \frac{Z_{01}}{N}, \quad Z'_0 = \frac{Z_{01}}{N(N-1)}, \qquad N = 1 + \frac{Z_{01}}{Z_0} = 1 + \frac{C'_0}{C'_{01}}. \tag{1}$$

Die Wellenleitwerte in Bild 6 ergeben sich somit zu

$$Y'_{01} = N\,Y_{01}, \quad Y'_1 = \frac{N-1}{N}\,Y_{01} + Y_1. \tag{2}$$

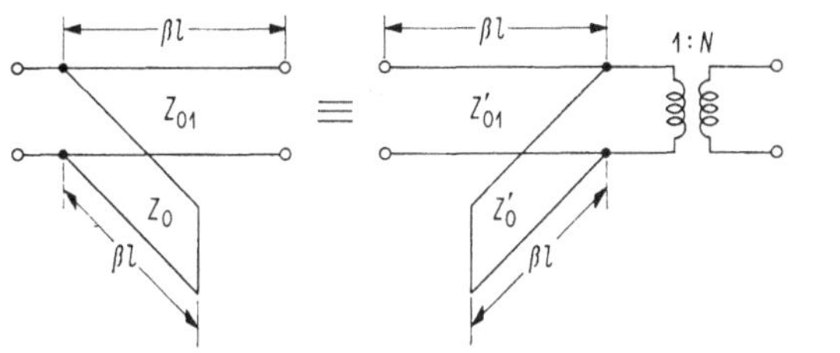

Bild 5. Äquivalente Umwandlung des Eingangsgliedes (Identität nach KURODA).

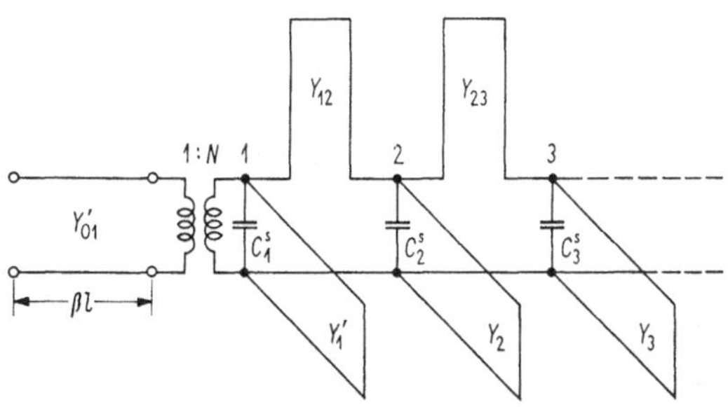

Bild 6. Ersatzschaltbild für das Kammfilter nach Bild 1 nach Umwandlung des Eingangsgliedes.

Hat $1/Y'_{01}$ den gleichen Wert wie der Wellenwiderstand der Anschlußleitung, dann wird die Filtereigenschaft nur durch die Schaltung rechts vom Transformator bestimmt. Für das Kammfilter nach Bild 2, das keine Leiter für die Impedanztransformation hat, gilt die Ersatzschaltung nach Bild 7.

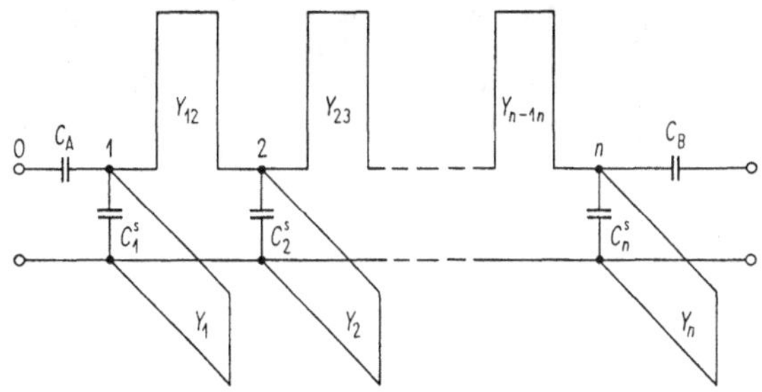

Bild 7. Ersatzschaltbild für das Kammfilter mit kapazitiver Ankopplung.

3. Vergleich des Kammfilters mit einem geeigneten Filter aus konzentrierten Elementen

Der Aufbau der Ersatzschaltungen in den Bildern 6 und 7 für die Kammfilter legt den Vergleich nahe mit den Schaltungen in Bild 8; denn das Kammfilter hat seinen ersten und nur wichtigen Durchlaßbereich bei $\beta l < \pi/2$. In diesem Bereich haben die kurzgeschlossenen Leitungen einen induktiven Eingangswiderstand. 129

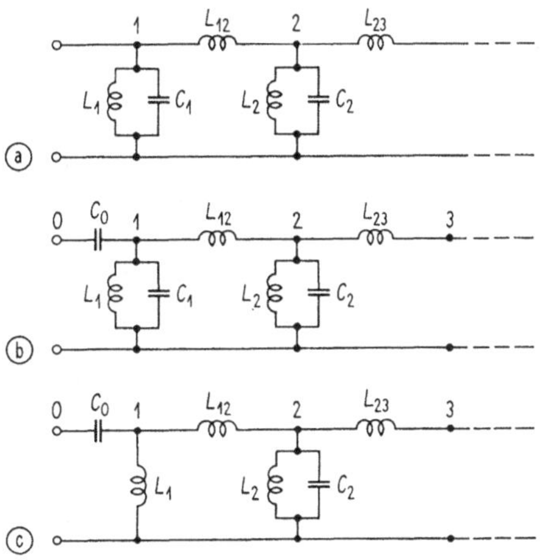

Bild 8. Bandfilter aus konzentrierten Elementen; Schaltung a zum Vergleich mit dem Kammfilter nach Bild 1, Schaltung b zum Vergleich mit dem Kammfilter nach Bild 2, Schaltung c zum Vergleich mit dem Kammfilter nach Bild 2 mit $C_1^s = C_n^s = 0$.

Für einen Kettenleiter mit den Leitwerten $G_{i-1,i}$; $G_{i,i+1}$ in den Längszweigen zwischen den Knotenpunkten $i-1$ und i bzw. i und $i+1$ und dem Querleitwert G_i am Punkt i kann man für die Spannungen in den Punkten $i-1$, i und $i+1$ die Differenzengleichung

$$0 = -G_{i-1,i}\,U_{i-1} + (G_i + G_{i-1,i} + G_{i,i+1})\,U_i - G_{i,i+1}\,U_{i+1} \tag{3}$$

aufstellen. Mit den Elementen aus den Bildern 6

und 7 gilt für $1 < i < n$

$$0 = -\frac{C'_{i-1,i}}{\varepsilon} U_{i-1} + \left(\frac{C'_i}{\varepsilon} + \frac{C'_{i-1,i}}{\varepsilon} + \frac{C'_{i,i+1}}{\varepsilon} - \beta l \tan\beta l \frac{C_i^{s\prime}}{\varepsilon}\right) U_i - \frac{C'_{i,i+1}}{\varepsilon} U_{i+1} . \tag{4}$$

Die Kapazität C_i^s wurde hier auf die Stablänge l bezogen. Es ist $C_i^{s\prime} l = C_i^s$. Alle Kapazitätsbeläge wurden mit der Dielektrizitätskonstante ε dividiert, um dimensionslose Werte für diese Konstanten zu bekommen.

Mit den Elementen aus Bild 8 ergibt sich aus Gl. (3)

$$0 = -\frac{1}{L_{i-1,i}} U_{i-1} + \left(\frac{1}{L_i} + \frac{1}{L_{i-1,i}} + \frac{1}{L_{i,i+1}} - \omega^2 C_i\right) U_i - \frac{1}{L_{i,i+1}} U_{i+1} ; \quad 1 < i < n . \tag{5}$$

Die Gl. (5) kann mit der Frequenztransformation

$$\omega^2 = \beta l \tan\beta l \tag{6}$$

in die Gl. (4) überführt werden. Die Werte für die Kapazitätsbeläge können direkt aus den L- und C-Werten der Gl. (5) abgelesen werden:

$$\frac{C'_{i-1,i}}{\varepsilon} = \frac{1}{L_{i-1,i}}, \quad \frac{C'_i}{\varepsilon} = \frac{1}{L_i}, \quad \frac{C'_{i,i+1}}{\varepsilon} = \frac{1}{L_{i,i+1}}, \quad \frac{C_i^{s\prime}}{\varepsilon} = C_i . \tag{7}$$

Die Werte L, C der konzentrierten Schaltung und die Frequenz sind normiert und daher dimensionslos.

Untersucht werden muß nun, ob die in der Gl. (6) angegebene Transformation auch für den Anfang bzw. das Ende der Schaltungen gilt. Da die Untersuchungen auf das Betriebsübertragungsmaß hinzielen, sollen die Filterschaltungen von einem Generator mit der Leerlaufspannung U_0 und dem Innenwiderstand Z_g bzw. R gespeist werden. Für die Schaltung nach Bild 6 werden die Generatordaten auf die rechte Seite des Transformators umgerechnet. Die neuen Werte seien U'_0 und Z'_g. Für die Kammfilter wird der Innenwiderstand durch den Kapazitätsbelag der Verbindungsleitung angegeben

$$Z_g = \frac{\sqrt{\varepsilon_r}}{c\, C'_g}, \quad Z'_g = \frac{\sqrt{\varepsilon_r}}{c\, C''_g} . \tag{8}$$

Durch Anwendung der Gl. (3) auf den Punkt $i = 1$ erhält man für die Schaltung nach Bild 6:

$$\mathrm{j} \tan\beta l (U'_0 - U_1) \frac{C''_g}{\varepsilon} = \left(\frac{C''_1}{\varepsilon} + \frac{C'_{12}}{\varepsilon} - \beta l \tan\beta l \frac{C_1^{s\prime}}{\varepsilon}\right) U_1 - \frac{C'_{12}}{\varepsilon} U_2 , \tag{9}$$

nach Bild 8a:

$$\mathrm{j}\, \omega (U_0 - U_1) \frac{1}{R} = \left(\frac{1}{L_1} + \frac{1}{L_{12}} - C_1 \omega^2\right) U_1 - \frac{1}{L_{12}} U_2 , \tag{10}$$

nach Bild 7:

$$-\frac{C'_A}{\varepsilon} \beta l \tan\beta l \frac{U_0 - U_1}{1 + \mathrm{j} \dfrac{C'_A/\varepsilon}{C'_g/\varepsilon} \beta l} = \left(\frac{C'_1}{\varepsilon} + \frac{C'_{12}}{\varepsilon} - \beta l \tan\beta l \frac{C_1^{s\prime}}{\varepsilon}\right) U_1 - \frac{C'_{12}}{\varepsilon} U_2 \tag{11}$$

und nach Bild 8b:

$$-C_0 \omega^2 \frac{U_0 - U_1}{1 + \mathrm{j}\, C_0 R \omega} = \left(\frac{1}{L_1} + \frac{1}{L_{12}} - \omega^2 C_1\right) U_1 - \frac{1}{L_{12}} U_2 . \tag{12}$$

Der Fall nach Bild 7 ($C_1^s = 0$) und Bild 8c ist in den Gl. (11) und (12) enthalten. C'_A in Gl. (11) ergibt sich aus C_A durch Division mit l. Man erkennt, daß eine Überführung der zueinander gehörenden Gleichungen mit der Transformation nach Gl. (6) nicht in allen Teilen gelingt, weil auf der linken Seite $\tan\beta l$, βl und ω teilweise allein auftreten. Für $\beta l \ll \pi/2$ geht die Transformation der Gl. (6) über in $\omega \approx \beta l$, denn es ist $\tan\beta l \approx \beta l$. Dann ist wieder eine Überführung der Gleichungen ineinander möglich. Wichtig ist aber, daß die Gleichungen sich möglichst im Durchlaßbereich der Filter ineinander überführen lassen. Um dies zu erreichen, wird in den Gleichungen an der entsprechenden Stelle jeweils mit dem fehlenden Faktor erweitert und der Ko-Faktor in die Abschlußgröße hineingezogen. Für die Überführung der Gl. (10) in die Gl. (9) gilt.

$$\frac{C''_g}{\varepsilon} = \frac{\beta l}{\omega R} = \frac{1}{R} \sqrt{\frac{\beta l}{\tan\beta l}} \tag{13}$$

und für die Überführung der Gl. (12) in die Gl. (11)

$$\frac{C'_g}{\varepsilon} = \frac{\omega}{\tan\beta l\, R} = \frac{1}{R} \sqrt{\frac{\beta l}{\tan\beta l}} . \tag{14}$$

Auf der rechten Seite dieser beiden Gleichungen steht ein frequenzabhängiger Ausdruck, die linke Seite dagegen kann nur einen konstanten Wert annehmen. Für den βl-Wert, der auf der rechten Seite eingesetzt wird, lassen sich die konzentrierten Schaltungen exakt in die Kammfilter überführen. βl wird daher gleich $(\beta l)_D$ gewählt und $(\beta l)_D$ soll in der Mitte des Durchlaßbereiches der Filter liegen, damit die Unexaktheit an beiden Bandgrenzen etwa gleiche Störungen liefert.

Mit $R = 1$ gilt dann

$$\frac{C''_g}{\varepsilon} = \frac{C'_g}{\varepsilon} = \sqrt{\frac{(\beta l)_D}{\tan(\beta l)_D}} . \tag{15}$$

Für den Ausgang der Schaltungen gilt das Entsprechende.

Mit den gefundenen Beziehungen sollen Kammfilterrealisierungen aus den konzentrierten Schaltungen versucht werden. Hierzu müssen allerdings erst für die konzentrierten Schaltungen die Elemente gefunden werden.

4. Tschebyscheff-Approximation für den Übertragungsfaktor des Resonanzfilters aus konzentrierten Elementen

Das induktiv gekoppelte Resonanzfilter nach Bild 8a mit n-Resonanzkreisen hat in seiner Dämpfungskurve einen einfachen Dämpfungspol bei $\omega = 0$ und einen $(2n-1)$-fachen Dämpfungspol bei $\omega = \infty$. Das Betragsquadrat des Betriebsübertragungsfaktors, das das Verhältnis der im Abschlußwiderstand verbrauchten zu der vom Generator verfügbaren Leistung angibt, hat die Form

$$|H'|^2 = \frac{\omega^2(1-\delta^2)}{\omega^2 + \delta^2 P_{2n}(\omega^2)}. \tag{16}$$

δ^2 ist eine Konstante, die die Größe der Schwankung im Durchlaßbereich bestimmt. P_{2n} ist ein Polynom $2n$-ten Grades in ω^2. Es muß so bestimmt werden, daß es im Bereich $\omega_1 \leqq \omega \leqq \omega_2$ zwischen den Grenzkurven $G_1 = \omega^2$ und $G_2 = -\omega^2$ schwankt. Für $|H'|^2$ ergibt sich dann der Verlauf nach Bild 9. Das Resonanzkreisfilter nach Bild 8b mit $(n-1)$ Resonanzkreisen und das Filter nach Bild 8c mit n Querzweigen haben einen dreifachen Dämpfungspol bei $\omega = 0$ und einen $(2n-3)$-fachen bei $\omega = \infty$. Das Betragsquadrat des Betriebsübertragungsfaktors hat die Form

$$|H'|^2 = \frac{\omega^6(1-\delta^2)}{\omega^6 + \delta^2 P_{2n}(\omega^2)}. \tag{17}$$

Das Polynom P muß im Bereich $\omega_1 \leqq \omega \leqq \omega_2$ zwischen den Grenzkurven $G_{1,2} = \pm\omega^6$ schwanken, um ein Verhalten nach Bild 9 zu erhalten. Die

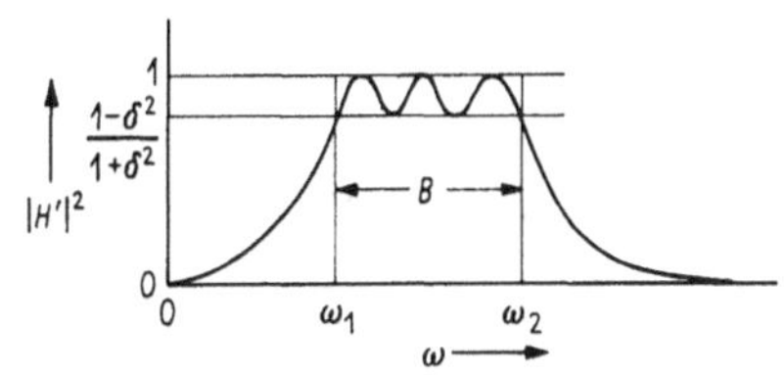

Bild 9. Prinzipieller Verlauf des Betragsquadrates des Betriebsübertragungsfaktors für die Bandfilter nach Bild 8.

eigentliche Aufgabe dieses Kapitels besteht also in der Bestimmung des Polynoms P_{2n}. Etwas allgemeiner formuliert lautet sie: Es wird ein Polynom $P_{2n}^{(m)}$ gesucht, das dem Betragsquadrat des Betriebsübertragungsfaktors

$$|H'|^2 = \frac{\omega^{2m}(1-\delta^2)}{\omega^{2m} + \delta^2 P_{2n}^{(m)}(\omega^2)} \tag{18}$$

ein Verhalten nach Bild 9 erteilt. Das Polynom $P_{2n}^{(m)}$ muß dann im Bereich $\omega_1 \leqq \omega \leqq \omega_2$ zwischen den Grenzkurven $G_{1,2} = \pm\omega^{2m}$ schwanken. Es wird hier nur für die Schaltungen nach Bild 8a und c bestimmt, weil es für die Schaltung nach Bild 8b nicht die nachstehenden Bedingungen für die Lage der Nullstellen erfüllt. Für die Schaltungen nach Bild 8a und c ist $P_{2n}^{(m)}$ ein gerades Polynom, dessen Nullstellen reell sind und ausnahmslos in dem erwähnten Bereich und in dem dazu entsprechenden negativen Bereich liegen. Setzt man $\omega^2 = x$, dann liegen alle Nullstellen im positiven Bereich zwischen den Grenzen x_1 und x_2. Die angeführten Bedingungen lassen sich nicht nur von einem Polynom $P_{2n}(x)$, d.h. mit geradem höchsten Exponenten, sondern auch von einem Polynom P_{2n-1} mit ungeradem höchsten Exponenten erfüllen. Es wird daher zu-

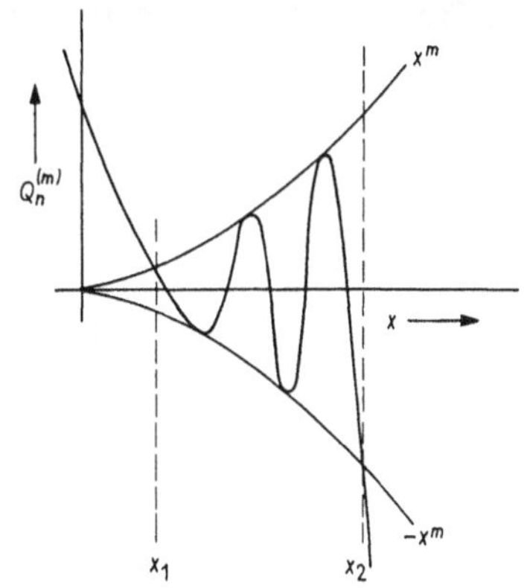

Bild 10. Verlauf des Polynoms $Q_n^{(m)}$ zur Erzielung des $|H'|^2$ nach Bild 9.

nächst ein Polynom $Q_n^{(m)}(x)$, das ein Verhalten nach Bild 10 zeigt, gesucht. Das Polynom $Q_n^{(m)}(x)$ genügt der Differentialgleichung

$$(Q_n^{(m)2} - x^{2m})(x + a_n)^2 = \tag{19}$$
$$= K^2[x^2 - (x_1+x_2)x + x_1x_2]\left(\frac{\mathrm{d}Q_n^{(m)}}{\mathrm{d}x}x - m\,Q_n^{(m)}\right)^2,$$

denn $(Q_n^{(m)} - x^m)(Q_n^{(m)} + x^m)$ hat einfache Nullstellen bei x_1 und x_2 und doppelte Nullstellen an den Berührungsstellen des $Q_n^{(m)}$ mit den Grenzkurven. Bildet man $Q_n^{(m)}/x^m$, so werden aus den Berührungsstellen Extremstellen. Sie sind gegeben durch

$$0 = \frac{\mathrm{d}(Q_n^{(m)}/x^m)}{\mathrm{d}x} = \frac{Q_n^{(m)\prime}x - m\,Q_n^{(m)}}{x^{m+1}}, \tag{20}$$

d.h. durch die Nullstellen des Zählers der rechten Seite. Der Zähler ist ein Polynom n-ten Grades, $Q_n^{(m)}$ hat aber nur $(n-1)$ Berührungsstellen mit den Grenzkurven. Auf der linken Seite der Gl. (19) mußte daher noch ein Faktor hinzugefügt werden, der den überflüssigen Nullstellenfaktor in $Q_n^{(m)\prime}x - mQ_n^{(m)}$ heraushebt. Die Konstanten K und a_n erhält man durch Koeffizientenvergleich für die höchsten und niedrigsten Potenzen. Es ist

$$K^2 = \frac{1}{(n-m)^2}, \quad a_n = \frac{m}{n-m}. \tag{21}$$

Setzt man für $Q_n^{(m)\prime}x - mQ_n^{(m)}$ in Gl. (19)

$$x^{m+1}\frac{\mathrm{d}(Q_n^{(m)}/x^m)}{\mathrm{d}x},$$

so erhält man eine Differentialgleichung für $Q_n^{(m)}/x^m$:

$$-\left[1 - \left(\frac{Q_n^{(m)}}{x^m}\right)^2\right][(n-m)x + m]^2 = \tag{22}$$
$$= x^2[x^2 - x(x_1+x_2) + x_1x_2]\left[\frac{\mathrm{d}(Q_n^{(m)}/x^m)}{\mathrm{d}x}\right]^2.$$

x_1 und x_2 sind frei wählbar; es wird

$$x_1 x_2 = 1\,, \qquad x_1 + x_2 = \omega_a^2 \tag{23}$$

gesetzt.

ω_a^2 ist mit der Bandbreite B (Bild 9) durch die Gleichung

$$\omega_a^2 = B^2 + 2 \tag{24}$$

verknüpft und kann daher niemals kleiner als zwei werden. Die Trennung der Veränderlichen in Gl. (22) ergibt nun

$$\frac{\mathrm{d}\,(Q_n^{(m)}/x^m)}{\sqrt{1-(Q_n^{(m)}/x^m)^2}} = \pm\,\mathrm{j}\,\frac{(n-m)\,x+m}{x\sqrt{1-\omega_a^2 x + x^2}}\,, \tag{25}$$

und die Integration beider Seiten führt auf

Elemente der konzentrierten Schaltung gewonnen werden.

Die ersten beiden Polynome im Fall $m = 3$ sind

$$Q_3^{(3)} = \frac{1}{\sqrt{\omega_a^4 - 4^3}} \times$$
$$\times\,[32 - 48\,\omega_a^2 x + (18\,\omega_a^4 + 24)\,x^2 - \omega_a^2(\omega_a^4 + 12)\,x^3]\,,$$
$$Q_4^{(3)} = \frac{2x - \omega_a^2}{\sqrt{\omega_a^4 - 4}}\,Q_3^{(3)} - \tag{34}$$
$$- \frac{(4x^2 - 4\,\omega_a^2 x + 4)\,[16 - 16\,\omega_a^2 x + (3\,\omega_a^4 + 4)\,x^2]}{(\omega_a^4 - 4)^2}\,.$$

$$Q_n^{(m)} = x^m \cos\left[(n-m)\arccos\frac{2x-\omega_a^2}{\sqrt{\omega_a^4-4}} - m\arccos\frac{\frac{2}{x}-\omega_a^2}{\sqrt{\omega_a^4-4}}\right]. \tag{26}$$

Diese Gleichung kann man auch mit den Tschebyscheff-Funktionen 1. und 2. Art [5] umschreiben in

$$Q_n^{(m)} = x^m\,T_m\left(\frac{\frac{2}{x}-\omega_a^2}{\sqrt{\omega_a^4-4}}\right) T_{n-m}\left(\frac{2x-\omega_a^2}{\sqrt{\omega_a^4-4}}\right) + x^m\,U_m\left(\frac{\frac{2}{x}-\omega_a^2}{\sqrt{\omega_a^4-4}}\right) U_{n-m}\left(\frac{2x-\omega_a^2}{\sqrt{\omega_a^4-4}}\right). \tag{27}$$

Setzt man $U_k(x) = \sqrt{1-x^2}\,V_k(x)$, so ist

$$V_k(x) = \frac{1}{k}\,T_k'(x)$$

und aus Gl. (27) wird

$$Q_n^{(m)} = x^m\,T_m\left(\frac{\frac{2}{x}-\omega_a^2}{\sqrt{\omega_a^4-4}}\right) T_{n-m}\left(\frac{2x-\omega_a^2}{\sqrt{\omega_a^4-4}}\right) - \frac{4x^2-4\,\omega_a^2 x+4}{\omega_a^4-4}\,x^{m-1}\,V_m\left(\frac{\frac{2}{x}-\omega_a^2}{\sqrt{\omega_a^4-4}}\right) V_{n-m}\left(\frac{2x-\omega_a^2}{\sqrt{\omega_a^4-4}}\right). \tag{28}$$

Für die Tschebyscheff-Polynome T_k und V_k gelten die Rekursionsformeln

$$\begin{aligned} T_{n+1}(x) - 2x\,T_n(x) + T_{n-1}(x) &= 0\,,\\ V_{n+1}(x) - 2x\,V_n(x) + V_{n-1}(x) &= 0\,. \end{aligned} \tag{29}$$

Für die Polynome $Q_n^{(m)}$ erhält man dann die Rekursionsformel

$$Q_{n+1}^{(m)} = 2\,\frac{2x-\omega_a^2}{\sqrt{\omega_a^4-4}}\,Q_n^{(m)} - Q_{n-1}^{(m)}\,. \tag{30}$$

Für $m = 1$ lauten die ersten beiden Polynome

$$\begin{aligned} Q_1^{(1)} &= \frac{2}{\sqrt{\omega_a^4-4}} - \frac{\omega_a^2}{\sqrt{\omega_a^4-4}}\,x\,,\\ Q_2^{(1)} &= -\frac{2}{\omega_a^2-2} + \frac{\omega_a^2+2}{\omega_a^2-2}\,x - \frac{2}{\omega_a^2-2}\,x^2\,. \end{aligned} \tag{31}$$

Alle weiteren Polynome erhält man mit der Rekursionsformel in Gl. (30).

Da der Koeffizient der höchsten Potenz von x negativ ist, muß man anstelle der Gl. (18) bzw. (16)

$$|H'|^2 = \frac{\omega^2(1-\delta^2)}{\omega^2 - \delta^2\,Q_{2n}^{(1)}(\omega^2)} \tag{32}$$

schreiben. Erweitert man diese Beziehung auf die gesamte p-Ebene, so wird

$$H'(p)\,H'(-p) = \frac{-p^2(1-\delta^2)}{-p^2 - \delta^2\,Q_{2n}^{(1)}(-p^2)}\,. \tag{33}$$

Aus dieser Beziehung können nach den bekannten Verfahren der Betriebsdämpfungssynthese [6] die

Alle weiteren erhält man mit der Gl. (30). Die der Gl. (33) entsprechende Beziehung lautet

$$H'(p)\,H'(-p) = \frac{-p^6(1-\delta^2)}{-p^6 - \delta^2\,Q_{2n}^{(3)}(-p^2)}\,. \tag{35}$$

n gibt hier die Zahl der Querzweige an.

5. Beziehungen und Bemerkungen zur Synthese der konzentrierten Schaltungen nach Bild 8a und c

Zur Ermittlung der Schaltelemente bestimmt man zunächst den Reflexionsfaktor, dann den Eingangswiderstand und daraus die Vierpolparameter. Für verlustlose Vierpole besteht der Zusammenhang

$$S_{11}(p)\,S_{11}(-p) + H'(p)\,H'(-p) = 1\,. \tag{36}$$

Setzt man

$$S_{11} = \frac{m_r + n_r}{m_o + n_o} \tag{37}$$

mit n_o, n_r ungerade Polynomanteile,

m_r, m_o gerade Polynomanteile,

so ergibt sich aus den Gl. (33) und (36)

$$\frac{m_r^2 - n_r^2}{m_o^2 - n_o^2} = \frac{-\delta^2[Q_{2n}^{(1)}(-p^2) + p^2]}{\delta^2\,Q_{2n}^{(1)}(-p^2) - p^2}\,. \tag{38}$$

Der Zähler der rechten Seite läßt sich als Quadrat eines geraden Polynoms schreiben. Mit Gl. (28) erhält man

$$n_r = 0$$

und

$$m_{rn}^{(1)} = \delta\left[\sqrt{\frac{\omega_a^2+2}{2}}\,T_n\left(\frac{2p^2+\omega_a^2}{\sqrt{\omega_a^2-4}}\right) - \sqrt{\frac{\omega_a^2-2}{2}}\,T_{n-1}\left(\frac{2p^2+\omega_a^2}{\sqrt{\omega_a^4-4}}\right)\right]. \tag{39}$$

Die beiden ersten Polynome lauten

$$m_{r1}^{(1)} = \delta\left(\sqrt{\frac{2}{\omega_a^2-2}}\,p^2 + \sqrt{\frac{2}{\omega_a^2-2}}\right),$$
$$m_{r2}^{(1)} = \delta[4p^4+(2+3\omega_a^2)p^2 + (\omega_a^2+2)]\frac{\sqrt{2}}{(\omega_a^2-2)\sqrt{\omega_a^2+2}}. \tag{40}$$

Alle weiteren erhält man mit der Rekursionsformel

$$m_{rn} = \frac{4p^2+2\omega_a^2}{\sqrt{\omega_a^2-4}}\,m_{r\,n-1} - m_{r\,n-2}. \tag{41}$$

Der Nenner in Gl. (38) muß aufgeteilt werden in das Produkt $(m_o+n_o)(m_o-n_o)$, dabei muß m_o+n_o ein Hurwitz-Polynom sein. Hier wurde keine explizite Formel gefunden. Es ist aber

$$(m_o^2-n_o^2) = m_r^2 - p^2(1-\delta^2) = \left(m_r + p\sqrt{1-\delta^2}\right)\left(m_r - p\sqrt{1-\delta^2}\right).$$

Da die Nullstellen von $m_r - p\sqrt{1-\delta^2}$ die an der imaginären Achse gespiegelten von $m_r + p\sqrt{1-\delta^2}$ sind, erhält man die Nullstellen von m_o+n_o, indem man die Nullstellen von $m_r + p\sqrt{1-\delta^2}$ bestimmt und diejenigen der rechten p-Halbebene an der imaginären Achse in die linke Halbebene spiegelt. Hat man m_r, m_o und n_o bestimmt, dann ergeben sich die Vierpolparameter ($R = 1$) zu

$$z_{11} = z_{22} = \frac{n_o}{m_o+m_r}, \quad z_{12} = \frac{p\sqrt{1-\delta^2}}{m_o+m_r}. \tag{42}$$

Der Vierpol ist symmetrisch ($z_{11} = z_{22}$, da $n_r = 0$), man kann ihn aus zwei gleichen Hälften zusammensetzen. Für eine Hälfte gilt nach dem Halbierungstheorem von Bartlett

$$z_{11\mathrm{h}} = z_{11} + z_{12}, \quad 1/y_{11\mathrm{h}} = z_{11} - z_{12}. \tag{43}$$

$z_{11\mathrm{h}}$ und $y_{11\mathrm{h}}$ sind der Eingangswiderstand bzw. der Eingangsleitwert des Vierpols bei Leerlauf bzw. Kurzschluß in der Mitte des Vierpols. Aus $z_{11\mathrm{h}}$ und $y_{11\mathrm{h}}$ können die Schaltelemente gewonnen werden.

Die Realisierung des Filters nach Bild 8c geht entsprechend vor sich. Es wird hier von der Gl. (35) ausgegangen. Man erhält.

$$n_r^{(3)} = 0,$$

$$m_{rn}^{(3)} = \delta\frac{\sqrt{2(\omega_a^2+2)}}{\omega_a^4-4}\left\{[4+(3\omega_a^2+2)p^2 + (\omega_a^2+2)p^4]\,T_{n-2}(y) + (y^2-1)\sqrt{\frac{\omega_a^4-4}{2}}\,[4+p^2(\omega_a^2+2)]\,V_{n-2}(y)\right\} \tag{44}$$

mit $$y = \frac{2p^2+\omega_a^2}{\sqrt{\omega_a^4-4}}.$$

Die ersten beiden Polynome lauten

$$m_{r2}^{(3)} = \frac{\sqrt{2}\,\delta}{(\omega_a^2-2)\sqrt{\omega_a^2+2}} \times [4+(3\omega_a^2+2)p^2+(\omega_a^2+2)p^4],$$
$$m_{r3}^{(3)} = \frac{\sqrt{2}\,\delta}{(\omega_a^2-2)\sqrt{\omega_a^2-2}} \times [4+3(\omega_a^2+2)(p^2+p^4)+4p^6]. \tag{45}$$

Die weiteren erhält man mit der Rekursionsformel in Gl. (41). Anstelle Gl. (42) setzt man

$$z_{11} = \frac{n_o}{m-m_r} = z_{22}, \quad z_{12} = \frac{p^3\sqrt{1-\delta^2}}{m-m_r}. \tag{46}$$

Nach dem Halbierungstheorem von Bartlett genügt die Synthese des halben Filters.

6. Synthese der Kammfilter

Die Werte für die Kammfilter erhält man aus den Werten der konzentrierten Schaltung mit den in den Gl. (7) und (15) angegebenen Beziehungen. Außerdem ist $C'_\mathrm{A}/\varepsilon$ gleich C_0 zu setzen. Alle Kapazitätsbeläge müssen zur Erzielung des erforderlichen Impedanzniveaus mit einem konstanten Faktor multipliziert werden. Dieser Faktor ergibt sich aus dem erforderlichen

$$\frac{C'_\mathrm{g}}{\varepsilon} = \frac{120\,\pi\,\Omega}{\sqrt{\varepsilon_\mathrm{r}}\,Z_0};$$

Z_0 ist der Wellenwiderstand der Anschlußleitung.

Da N in den Gl. (1) und (2) in einem gewissen Bereich frei wählbar ist, läßt sich das Impedanzniveau der Filterelemente nach Bild 1 in Richtung guter Realisierungsmöglichkeit der Kapazitätsbeläge steuern.

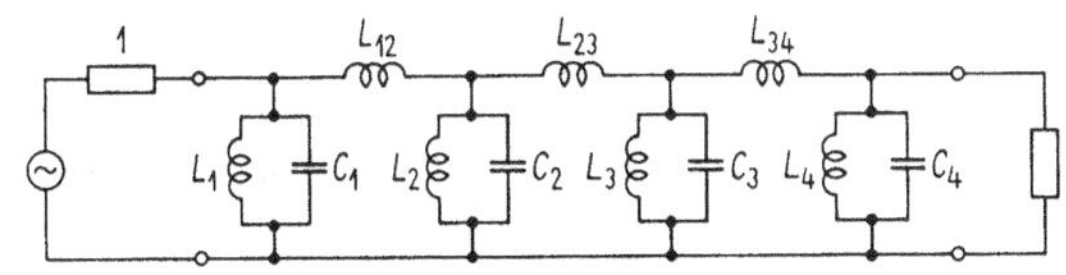

Bild 11. Synthesebeispiel für das Filter nach Bild 8a: Bandbreite 66,6%;
$C_4 = C_1 = 1{,}315177$, $C_2 = C_3 = 1{,}254266$,
$L_1 = L_{12} = 1{,}123228 = L_{34} = L_4$,
$L_2 = L_3 = 5{,}181756$, $L_{23} = 1{,}566207$.

Für die Demonstration der Güte des hier angegebenen Syntheseverfahrens für Kammfilter nach Bild 1, wurde ein Filter durchgerechnet mit

$$\omega_a^2 = 2{,}5; \quad \delta^2 = 0{,}005; \quad n = 4.$$

Mit diesen Werten erhält man die Schaltung nach Bild 11. Die Bandbreite dieser Schaltung beträgt 66,6%. Ersetzt man nun die Induktivitäten durch Leitungen mit den Kapazitätsbelägen $C'/\varepsilon = 1/L$ und ändert den Generatorinnen- und den Belastungswiderstand gemäß der Gl. (15), so hat das sich ergebende Kammfilter ein Betriebsverhalten nach Bild 12a. Der Verlauf, der sich durch die Transformation $\omega^2 = \beta l \tan \beta l$ ergibt, ist gestrichelt eingezeichnet. Die Abweichung ist bei dieser großen Bandbreite erträglich. Der Durchlaßbereich liegt außerdem bei hohen βl-Werten. Verschiebt man den Durchlaßbereich zu niedrigeren βl-Werten, indem man zunächst in der konzentrierten Schaltung eine lineare Frequenztransformation vornimmt, so wird die Übereinstimmung besser. Bei der Transformation werden alle Elemente mit dem Faktor a und die Frequenz mit $1/a$ multipliziert. Die Kurven b in Bild 12 zeigen das Ergebnis für $a = 2$. Die Bandbreite der Kammfilter ist kleiner als die Bandbreite der konzentrierten Schaltung.

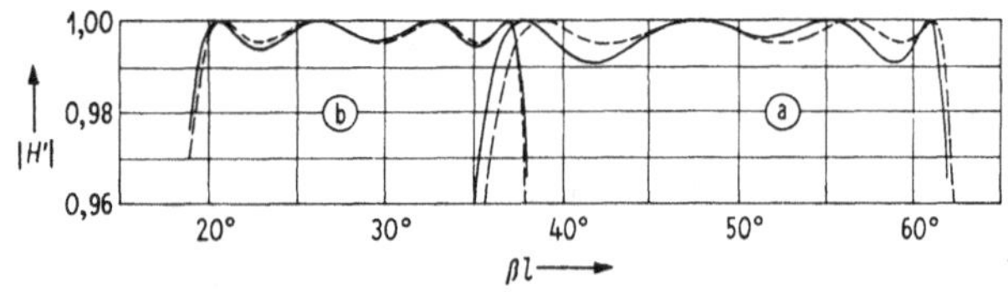

Bild 12. Verhalten des aus Bild 11 erhaltenen Kammfilters mit vier Resonatorstäben bei verschiedenen Lagen des Durchlaßbereiches auf der βl-Achse;
– – – erwarteter Verlauf des $|H'|$ gemäß Transformation nach Gl. (6),
——— wirklicher Verlauf des $|H'|$.

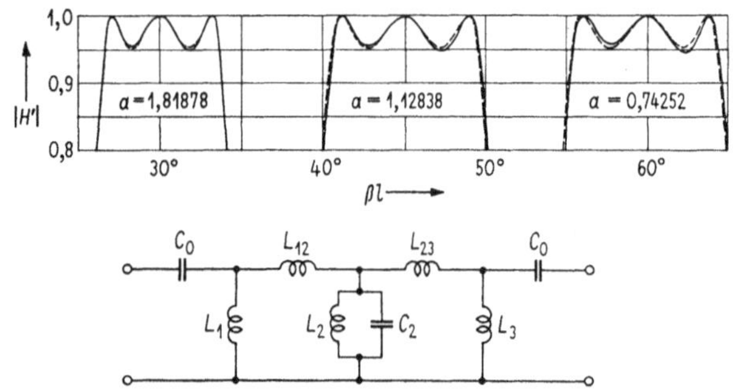

Bild 13. Verhalten eines Kammfilters nach Bild 2 mit drei Stäben und $C_1^s = C_3^s = 0$ und Ausgangsschaltung mit konzentrierten Elementen;
$C_0 = C_2 = 0{,}1692855\,a$, $L_1 = L_3 = 7{,}371067\,a$, $L_{12} = L_{23} = 29{,}744244\,a$, $L_2 = 8{,}665556\,a$;
– – – erwarteter Verlauf des $|H'|$ gemäß Transformation nach Gl. (6),
——— wirklicher Verlauf des $|H'|$.

Zur Demonstration der Güte des Syntheseverfahrens für Kammfilter nach Bild 2 ohne Streukapazitäten am Eingangs- und Ausgangsleiter wurden die Beispiele der Bilder 13 und 14 berechnet. Die Bandbreiten wurden hier nicht so groß gewählt, die Übereinstimmung des erwarteten mit dem tatsächlichen

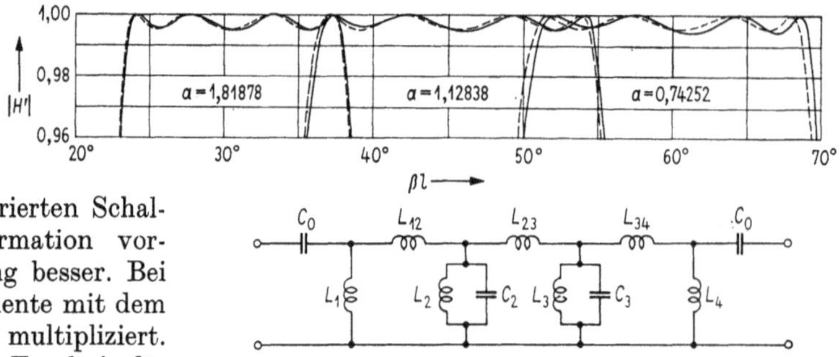

Bild 14. Verhalten eines Kammfilters nach Bild 2 mit vier Stäben und $C_1^s = C_4^s = 0$ und Ausgangsschaltung mit konzentrierten Elementen;
$C_0 = 0{,}570726\,a$, $C_2 = C_3 = 0{,}589659\,a$,
$L_1 = L_2 = L_3 = L_4 = 3{,}470967\,a$,
$L_{12} = L_{34} = 3{,}617102\,a$, $L_{23} = 4{,}381637\,a$;
– – – erwarteter Verlauf des $|H'|$ gemäß Transformation nach Gl. (6),
——— wirklicher Verlauf des $|H'|$.

Verlauf des Übertragungsfaktorbetrages ist daher besser als in Bild 12. Die Bandbreiten der mittleren Kurven betragen etwa 20% bzw. 40%. Der Korrekturfaktor wurde mit $(\beta l)_D = 30°$, $45°$ bzw. $60°$ berechnet.

Für die Dimensionierung in der Praxis wird man den Weg umgekehrt gehen, wie hier im ersten Beispiel beschrieben: Man wählt den Durchlaßbereich auf der βl-Skala, transformiert die Bandgrenzen in den ω-Bereich, bestimmt a und ω_a^2 und kann dann die Werte für die Elemente finden.

7. Schlußbemerkung

Das hier dargelegte Verfahren liefert sicherlich deshalb gute theoretische Ergebnisse, weil in der eingeführten Frequenztransformation sowohl die Eigenschaften der als konzentriert angenommenen Streukapazitäten als auch die der Leitungen zum Tragen kommen. Weitere Untersuchungen müssen nun noch zeigen, inwieweit große Bandbreiten experimentell realisierbar sind.

Schrifttum

[1] RICHARDS, P. I., Resistor-transmission-line-circuits. Proc. Inst. Radio Engrs. **36** [1948], 217–220.

[2] MATTHAEI, G. L., Comb-line band-pass-filters of narrow or moderate bandwidth. Microwave J. **6** [1963], Nr. 8, 82–91.

[3] CRISTAL, E. G., Capacity coupling shortens comb-line filters. Microwaves [1967], Dez., 44–50.

[4] WENZEL, R. J., Exact design of TEM microwave networks using quarter-wave-lines. Transact. Inst. Elect. Electron. Engrs. MTT-**12** [1964], 94–111.

[5] RYSHIK, I. M. U. und GRADSTEIN, I. S. Tafeln ΣΠ∫. Deutscher Verlag der Wissenschaften, Berlin 1957.

[6] z.B. BALABANIAN, N., Network synthesis. Prentice-Hall, Englewood Cliffs, N. J., 1958, oder andere Lehrbücher über Netzwerksynthese.

Sonderdruck
aus
ARCHIV DER ELEKTRISCHEN ÜBERTRAGUNG

Mikrowellenfilter aus verkoppelten Leitungen und konzentrierten Kapazitäten *

von Reinhold Pregla **

Mitteilung aus dem Institut für Hochfrequenztechnik der Technischen Universität Braunschweig

(A.E.Ü. **23** [1969], Heft 12, 593 – 600; eingegangen am 13. August 1969)

DK 621.372.852.1

Zwei Frequenztransformationen werden angegeben, welche die Realisierung von Mikrowellenfiltern aus verkoppelten kurzgeschlossenen oder leerlaufenden Leitungsstäben und konzentrierten Kapazitäten aus leicht zu bestimmenden konzentrierten Filterschaltungen ermöglichen. Das Verfahren liefert auch bei relativ großen Bandbreiten noch gute Ergebnisse. Einige Beispiele zeigen die Güte des Syntheseverfahrens.

Microwave Filters Consisting of Coupled Lines and Lumped Capacitances

Two frequency transformations are devised which allow microwave filters consisting of coupled short-circuited or open-circuited lines and lumped capacitances to be realized from lumped-element filter circuit configurations which can easily be determined. The method yields good results even in the case of relatively wide bands. A number of examples show the quality of the synthesis method.

1. Einleitung

Durch die von Richards [1] angegebene Transformation $p = \tanh \gamma l$ ist es möglich, Filter mit den konzentrierten Elementen R, L und C und der komplexen Frequenzvariablen p in Schaltungen mit Leitungen einheitlicher Länge l und Widerständen mit der Frequenzvariablen γl zu überführen. Diese neuen Schaltungen haben gemäß der angegebenen Frequenztransformation periodisches Verhalten und lassen sich in verschiedenen Formen realisieren [2], [3], [4]. Interessant sind vor allem diejenigen mit verkoppelten Stäben, weil auf diese Weise kleine Baugrößen erreicht werden. Filterrealisierungen mit sehr steilen Flanken sind von Horton und Wenzel [5], [6] angegeben worden. Ihre Dimensionierung ist außerdem sehr einfach, weil sie mit Hilfe eines Kataloges normierter Tiefpässe erfolgen kann. Nachteilig in der Realisierung ist, daß Leiter ineinandergeschachtelt werden müssen. Handelt es sich bei

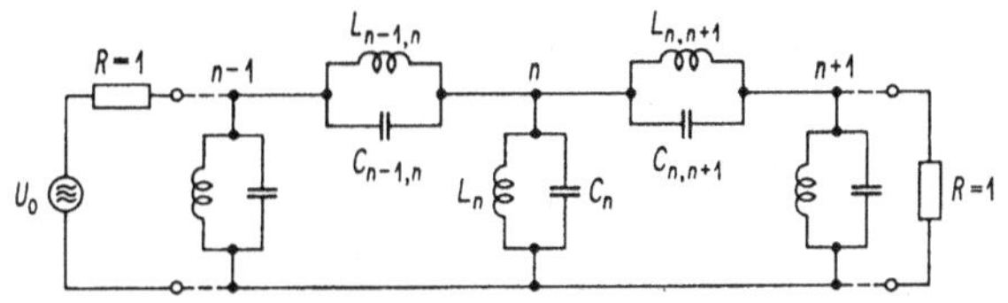

Bild 1. Allgemeine konzentrierte Ausgangsschaltung.

den in den Stäben liegenden Leitungen um solche, die am Ende leerlaufen, dann ergibt sich die Frage, ob es nicht möglich ist, diese durch konzentrierte Kapazitäten zu ersetzen. Das Ergebnis sind Filter aus Leitungen und konzentrierten Elementen. Ein weiterer Gesichtspunkt, der für derartige Filter spricht, ist der, daß in manchen Anwendungen das periodische Verhalten reiner Leitungsfilter (mit Leitungen einheitlicher Länge) störend wirkt. In dieser Arbeit wird daher eine ganze Klasse solcher gemischter Filter, die eine Struktur ähnlich der in [5] und [6] haben und die sich durch einfache Frequenztransformationen aus konzentrierten Filterschaltungen ableiten lassen, untersucht.

2. Die konzentrierte Ausgangsschaltung und die von ihr abgeleiteten Leitungsschaltungen mit konzentrierten Kapazitäten

Eine ganze Reihe von Filterschaltungen mit konzentrierten Elementen läßt sich einordnen in das Schema nach Bild 1, das eine Kettenfilterschaltung darstellt. Dabei können beispielsweise alle Kapazitäten in den Querzweigen Null sein (Hochpaß) oder alle Induktivitäten unendlich werden (Tiefpaß). Läßt man alle Kapazitäten in den Längszweigen gegen Null gehen, so ergibt sich das induktiv gekoppelte Resonanzkreisfilter. Bei unendlichen Induktivitäten in den Längszweigen ergibt sich das kapazitiv gekoppelte Resonanzkreisfilter. Natürlich sind auch gemischte Anordnungen möglich.

Durch die Transformation nach Richards [1] werden aus Kapazitäten leerlaufende und aus Induktivitäten kurzgeschlossene Leitungen. Die Kettenschaltung nach Bild 2a geht über in die Schaltung nach Bild 2b. Die Realisierung dieser Schaltung ist mit der Anordnung in Bild 2c möglich. Es handelt sich um am Ende kurzgeschlossene Stäbe. Zwischen den unmittelbar benachbarten Stäben soll ein bedeutender Kapazitätsbelag vorhanden sein, nicht aber zum jeweils übernächsten und auch nicht zu allen weiteren. Die kurzgeschlossenen Leitungen zwischen den Punkten $n-1$, n und n, $n+1$ werden durch die zwischen den Stäben $n-1$ und n bzw. n und $n+1$ bestehenden Leitungen mit dem Kapazitätsbelag $C'_{n-1,n}$ bzw. $C'_{n,n+1}$ realisiert, während

* Auszug aus der von der Fakultät für Maschinenbau und Elektrotechnik der Technischen Universität Braunschweig genehmigten Habilitationsschrift.

** Dr. R. Pregla, im Institut für Hochfrequenztechnik der Technischen Universität, 33 Braunschweig, Postfach 7050.

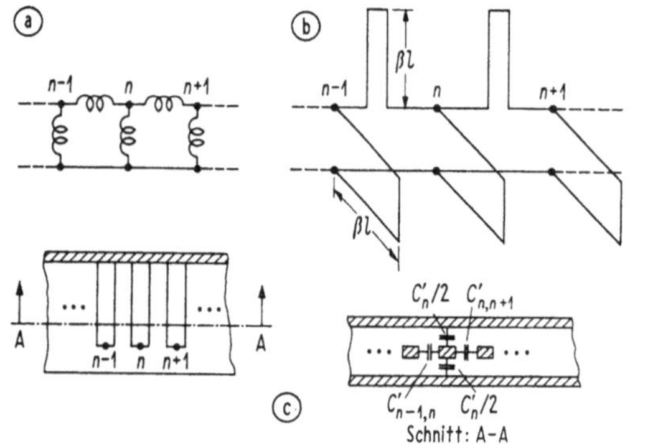

Bild 2. (a) induktiver Kettenleiter,
(b) mit der Richards-Transformation erzeugte Leitungsschaltung,
(c) Realisierung als Mikrowellenschaltung durch verkoppelte Stäbe, die am Ende kurzgeschlossen sind.

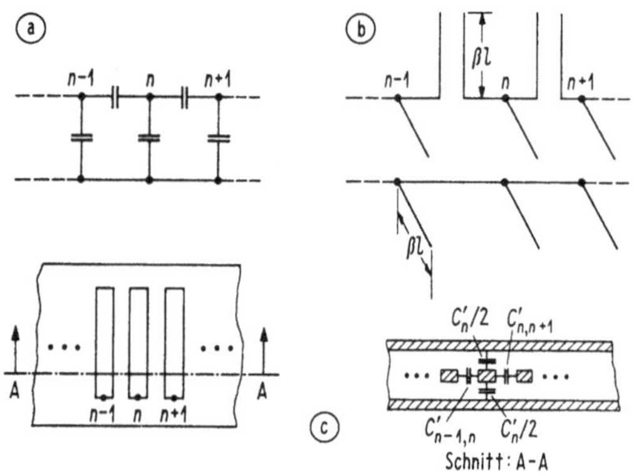

Bild 3. (a) kapazitiver Kettenleiter,
(b) mit der Richards-Transformation erzeugte Leitungsschaltung,
(c) Realisierung als Mikrowellenschaltung durch verkoppelte Stäbe, die am Ende leerlaufen.

die Kurzschlußleitung am Punkt n und der gemeinsamen Masse durch die Leitung zwischen dem Stab n und den Grund- und Deckplatten mit dem Kapazitätsbelag C_n' gebildet wird. Bild 3 zeigt die Realisierung eines kapazitiven Kettenleiters. Die im Bild 1 im Vergleich zu den Bildern 2 und 3 zusätzlich auftretenden Elemente sollen in dieser Arbeit in den Anordnungen nach Bild 2c und Bild 3c durch konzentrierte Kapazitäten realisiert werden. Bild 4

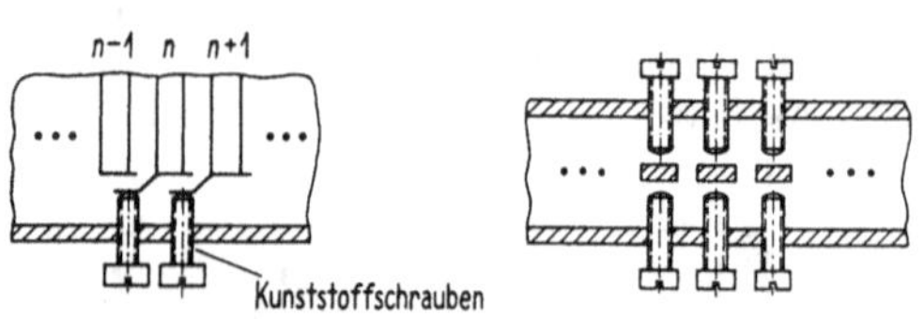

Bild 4. Realisierung der konzentrierten Kapazitäten im Mikrowellenfilter.

zeigt, wie diese Kapazitäten abstimmbar angebracht werden können. Die Kapazitäten zur Grund- und Deckplatte der Anordnung werden beispielsweise durch eindrehbare Metallschrauben und die Kapazitäten zwischen den Stabenden durch Metallblättchen, deren Abstand mit Kunststoffschrauben reguliert werden kann, eingestellt. Die Ersatzschaltungen, die sich auf diese Weise ergeben haben, sind in Bild 5a und b dargestellt. Sie sind mit der Schal-

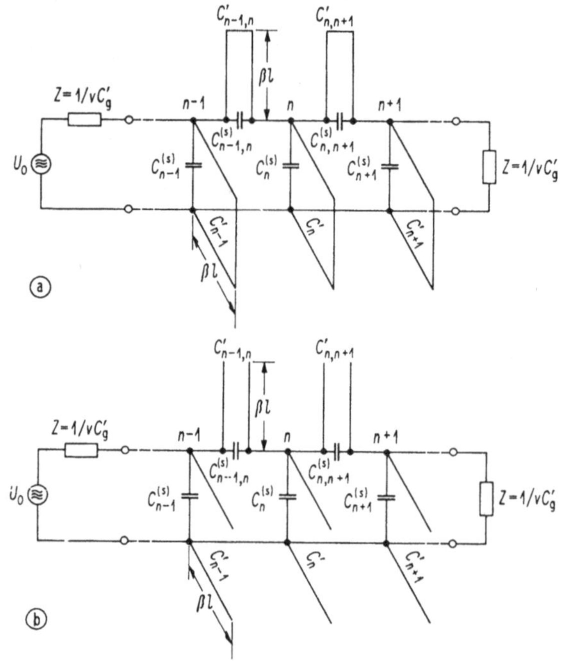

Bild 5. Ersatzschaltungen der Mikrowellenfilter;
(a) mit kurzgeschlossenen Stäben,
(b) mit leerlaufenden Stäben.

tung in Bild 1 zu vergleichen. Es lassen sich nun einfache Frequenztransformationen ableiten, die eine gute Übertragung des Dämpfungsverhaltens der Schaltung nach Bild 1 für relativ große Bereiche auf die Schaltungen nach Bild 5 ermöglichen und damit eine Dimensionierung der Mikrowellenfilter aus den Werten der konzentrierten Schaltung erlauben.

3. Ableitung der Frequenztransformationen

Zur Ableitung der Frequenztransformation betrachtet man die Spannungsübertragung vom Knotenpunkt $n-1$ auf den Knotenpunkt $n+1$. Diese Spannungsübertragung wird bei Kettenleitern, bei denen kein Knotenpunkt überbrückt wird, beschrieben durch eine Differenzengleichung 2. Ordnung. Man gewinnt sie, indem man im Punkt n die Kirchhoffsche Knotenregel erfüllt und dabei die Ströme durch die Spannungen in den Knoten $n-1, n, n+1$ ersetzt. Für einen Kettenleiter mit den Leitwerten $G_{n-1,\,n}$ und $G_{n,\,n+1}$ zwischen den Punkten $n-1$, n bzw. n, $n+1$ und dem Querleitwert G_n im Punkt n erhält man

$$G_{n-1,\,n}\,U_{n-1} - (G_{n-1,\,n} + G_{n,\,n+1} + G_n)\,U_n + G_{n,\,n+1}\,U_{n+1} = 0\,. \qquad (1)$$

Ersetzt man in dieser Gleichung die Leitwerte G durch die entsprechenden Leitwerte der Schaltung in Bild 1, so wird daraus nach Multiplikation mit $\mathrm{j}\,\omega$

136

$$\left(\frac{1}{L_{n-1,n}} - \omega^2 C_{n-1,n}\right) U_{n-1} - \left[\frac{1}{L_{n-1,n}} + \frac{1}{L_{n,n+1}} + \right.$$
$$\left. + \frac{1}{L_n} - \omega^2 (C_{n-1,n} + C_{n,n+1} + C_n)\right] U_n +$$
$$+ \left(\frac{1}{L_{n,n+1}} - \omega^2 C_{n,n+1}\right) U_{n+1} = 0 . \tag{2}$$

Ganz entsprechend lassen sich Differenzengleichungen für die Schaltungen im Bild 5 aufstellen. Man erhält für die Schaltung nach Bild 5a nach Multiplikation mit $\mathrm{j} \tan \beta l$

$$\left(\frac{C'_{n-1,n}}{\varepsilon} - \beta l \tan \beta l \frac{C'^{(s)}_{n-1,n}}{\varepsilon}\right) U_{n-1} -$$
$$- \left[\frac{C'_{n-1,n}}{\varepsilon} + \frac{C'_{n,n+1}}{\varepsilon} + \frac{C'_n}{\varepsilon} - \right.$$
$$\left. - \beta l \tan \beta l \left(\frac{C'^{(s)}_{n-1,n}}{\varepsilon} + \frac{C'^{(s)}_{n,n+1}}{\varepsilon} + \frac{C'^{(s)}_n}{\varepsilon}\right)\right] U_n + \tag{3}$$
$$+ \left(\frac{C'_{n,n+1}}{\varepsilon} - \beta l \tan \beta l \frac{C'^{(s)}_{n,n+1}}{\varepsilon}\right) U_{n+1} = 0$$

und für die Schaltung nach Bild 5b nach Division mit $\mathrm{j} \beta l$

$$\left(\frac{C'^{(s)}_{n-1,n}}{\varepsilon} + \frac{\tan \beta l}{\beta l} \frac{C'_{n-1,n}}{\varepsilon}\right) U_{n-1} -$$
$$- \left[\frac{C'^{(s)}_{n-1,n}}{\varepsilon} + \frac{C'^{(s)}_{n,n+1}}{\varepsilon} + \frac{C'^{(s)}_n}{\varepsilon} + \right.$$
$$\left. + \frac{\tan \beta l}{\beta l} \left(\frac{C'_{n-1,n}}{\varepsilon} + \frac{C'_{n,n+1}}{\varepsilon} + \frac{C'_n}{\varepsilon}\right)\right] U_n + \tag{4}$$
$$+ \left(\frac{C'^{(s)}_{n,n+1}}{\varepsilon} + \frac{\tan \beta l}{\beta l} \frac{C'_{n,n+1}}{\varepsilon}\right) U_{n+1} = 0 .$$

Es wird angenommen, daß alle Größen in Gl. (2) in der üblichen Weise normiert und daher dimensionslos sind. Die auf die Dielektrizitätskonstante ε bezogenen Kapazitätsbeläge in den Gl. (3) und (4) sind ebenfalls dimensionslose Größen. Ferner soll $C'^{(s)} = C^{(s)}/l$ bedeuten.

Durch Vergleich der Beziehungen (2) und (3) findet man die Frequenztransformation

$$\omega^2 = \beta l \tan \beta l , \tag{5}$$

die für einen Spezialfall bereits in [11] gefunden wurde. Für die Elemente liest man die Zusammenhänge

$$\frac{C'^{(s)}_{n-1,n}}{\varepsilon} = C_{n-1,n} , \quad \frac{C'^{(s)}_{n,n+1}}{\varepsilon} = C_{n,n+1} ,$$
$$\frac{C'^{(s)}_n}{\varepsilon} = C_n ; \quad \frac{C'_{n-1,n}}{\varepsilon} = \frac{1}{L_{n-1,n}} , \tag{6}$$
$$\frac{C'_{n,n+1}}{\varepsilon} = \frac{1}{L_{n,n+1}} , \quad \frac{C'_n}{\varepsilon} = \frac{1}{L_n}$$

ab. Mit den Beziehungen in den Gl. (5) und (6) ist eine exakte Überführung der Gl. (2) in die Gl. (3) möglich.

Vergleicht man die Differenzengleichungen (2) und (4) miteinander, so findet man die Frequenztransformation

$$\omega^2 = - \frac{\tan \beta l}{\beta l} \tag{7}$$

und für die Transformation der Elemente gilt

$$\frac{C'^{(s)}_{n-1,n}}{\varepsilon} = \frac{1}{L_{n-1,n}} , \quad \frac{C'^{(s)}_{n,n+1}}{\varepsilon} = \frac{1}{L_{n,n+1}} ,$$
$$\frac{C'^{(s)}_n}{\varepsilon} = \frac{1}{L_n} ; \quad \frac{C'_{n-1,n}}{\varepsilon} = C_{n-1,n} , \tag{8}$$
$$\frac{C'_{n,n+1}}{\varepsilon} = C_{n,n+1} , \quad \frac{C'_n}{\varepsilon} = C_n .$$

Mit den Beziehungen in den Gl. (7) und (8) ist eine exakte Überführung der Gl. (2) in die Gl. (4) möglich.

Mit den bisherigen Ergebnissen ist klargelegt, daß in dem Geltungsbereich der Differenzengleichungen das Verhalten der Schaltungen mit Leitungen und konzentrierten Kapazitäten nach Bild 5 aus dem Verhalten der Schaltung nach Bild 1 berechnet werden kann. Für den Punkt 1 der jeweiligen Schaltung muß eine neue Differenzengleichung aufgestellt werden, ebenso im Endpunkt der Schaltung, die im Prinzip jedoch der im Punkt 1 entspricht. Da Eingang und Ausgang der Schaltungen von der übrigen Struktur abweichen, werden die Spannungsbeziehungen auch anders ausfallen. Es ergibt sich für den Eingang der Schaltung nach Bild 1

$$(U_0 - U_1) \frac{\mathrm{j}\omega}{R} -$$
$$- \left[\frac{1}{L_{1,2}} + \frac{1}{L_1} - \omega^2 (C_{1,2} + C_1)\right] U_1 + \tag{9}$$
$$+ \left(\frac{1}{L_{1,2}} - \omega^2 C_{1,2}\right) U_2 = 0 ,$$

für den Eingang der Schaltung nach Bild 5a

$$(U_0 - U_1) \frac{\mathrm{j} \tan \beta l}{Z v \varepsilon} -$$
$$- \left[\frac{C'_{1,2}}{\varepsilon} + \frac{C'_1}{\varepsilon} - \beta l \tan \beta l \left(\frac{C'^{(s)}_{1,2}}{\varepsilon} + \frac{C'^{(s)}_1}{\varepsilon}\right)\right] U_1 +$$
$$+ \left(\frac{C'_{1,2}}{\varepsilon} - \beta l \tan \beta l \frac{C'^{(s)}_{1,2}}{\varepsilon}\right) U_2 = 0 \tag{10}$$

und für den Eingang der Schaltung nach Bild 5b

$$(U_0 - U_1) \frac{1}{\mathrm{j} \beta l Z v \varepsilon} -$$
$$- \left[\frac{C'^{(s)}_{1,2}}{\varepsilon} + \frac{C'^{(s)}_1}{\varepsilon} + \frac{\tan \beta l}{\beta l} \left(\frac{C'_{1,2}}{\varepsilon} + \frac{C'_1}{\varepsilon}\right)\right] U_1 +$$
$$+ \left(\frac{C'^{(s)}_{1,2}}{\varepsilon} + \frac{\tan \beta l}{\beta l} \frac{C'_{1,2}}{\varepsilon}\right) U_2 = 0 . \tag{11}$$

Bis auf jeweils das erste Glied der Gleichungen läßt sich die Transformation hier auch nach den Beziehungen der Gl. (5) bis (8) vornehmen. Faßt man Z als Wellenwiderstand der Anschlußleitungen auf, so kann man Z durch den entsprechenden Kapazitätsbelag ausdrücken, und es wird

$$\frac{1}{Z v \varepsilon} = \frac{C'_g}{\varepsilon} . \tag{12}$$

137

Dabei werden der Einfachheit halber gleiche Dielektrizitätskonstanten angesetzt.

Soll nun die Gl. (9) in die Gl. (10) bzw. Gl. (11) übergeführt werden, so muß mit $R = 1$ im ersten Fall

$$\frac{C_g'}{\varepsilon} = \frac{\beta l}{\omega} = \sqrt{\frac{\beta l}{\tan \beta l}} \tag{13}$$

und im zweiten Fall

$$\frac{C_g'}{\varepsilon} = \frac{\tan \beta l}{\omega} = \sqrt{-\beta l \tan \beta l} \tag{14}$$

gewählt werden.

Während die linken Seiten feste Werte sein müssen, sind die Wurzeln der rechten Seiten frequenzabhängig. Man muß sich daher auf einen Wert $\beta l = \beta l_0$ festlegen, für den eine exakte Transformation möglich sein soll. Es hat sich als sinnvoll erwiesen, βl_0 jeweils in die Mitte des Durchlaßbereiches zu legen.

Wie noch zu zeigen sein wird, kommt für das Filter nach Bild 5a als Durchlaßbereich sinnvollerweise ein Teil des Gebietes zwischen $0 \leqq \beta l \leqq \pi/2$ in Frage. βl und $\tan \beta l$ sind positiv. Das Filter nach Bild 5b hat den interessierenden Durchlaßbereich im Bereich $\pi/2 \leqq \beta l \leqq \pi$. Der $\tan \beta l$ ist in diesem Bereich negativ, so daß das negative Vorzeichen des Radikanten in Gl. (14) kompensiert wird.

4. Die Eigenschaften der Frequenztransformationen

Die Frequenztransformation nach Gl. (5) bildet die gesamte ω-Achse ($-\infty \leqq \omega \leqq \infty$) in den Bereich $-\pi/2 \leqq \beta l \leqq \pi/2$ ab. Der Bereich um den Nullpunkt wird nahezu äquivalent abgebildet; denn für kleine βl ist der $\tan \beta l \approx \beta l$ und somit $\omega \approx \beta l$. Da der Tangens eine periodische Funktion ist, wird die ω-Achse auch noch auf andere βl-Bereiche abgebildet. Es sind dies Bereiche, in denen der Tangens positiv ist auf der positiven βl-Achse und in denen er negativ ist auf der negativen βl-Achse. Der Durchlaßbereich des konzentrierten Ausgangsfilters, der irgendwo auf der ω-Achse liegt, ergibt Durchlaßbereiche auf der βl-Achse in den Abschnitten

$$\begin{gathered} 0 < \beta l < \pi/2 \\ \pi < \beta l < \frac{3}{2}\pi \\ \vdots \\ n\pi < \beta l < \left(n + \frac{1}{2}\right)\pi, \quad n = 0, 1, 2, \ldots \end{gathered} \tag{15}$$

und spiegelbildlich dazu auf der negativen βl-Achse.

Der Abstand der einzelnen Durchlaßbereiche voneinander ist nicht konstant, da βl kontinuierlich ansteigt, während $\tan \beta l$ periodisch ist. In die Bereiche $\left(n - \frac{1}{2}\right)\pi \leqq \beta l \leqq n\pi$ wird die reelle Achse der $p (= \sigma + j\omega)$-Ebene transformiert. In diesen Bereichen zeigen die Leitungsschaltungen daher kein ausgesprochenes Filterverhalten, sondern mehr oder weniger gute Dämpfungseigenschaften. Die Eingangswiderstände der Leitungen sind in diesem Bereich kapazitiv. Es liegt daher insgesamt eine kapazitive Spannungsteilung vor.

Mit der Frequenztransformation nach Gl. (7) wird die reelle Frequenzachse der p-Ebene in die Bereiche

$$\begin{gathered} \frac{\pi}{2} < \beta l < \pi \\ \vdots \\ \left(n - \frac{1}{2}\right)\pi < \beta l < n\pi, \quad n = 1, 2, \ldots \end{gathered} \tag{16}$$

und die dazu spiegelbildlich auf der negativen βl-Achse liegenden Bereiche transformiert. In die dazwischen liegenden Bereiche wird wieder die reelle Achse der p-Ebene transformiert.

Der erste Durchlaßbereich der hier behandelten zweiten Filterform liegt bei βl-Werten zwischen $\pi/2$ und π. Bei gleicher Stablänge liegen die Durchlaßbereiche bei höheren Frequenzen. Es müßte also möglich sein, Filter dieser Form für höhere Frequenzen zu realisieren, als bisher üblich. Besonders hingewiesen werden soll auf die Eigenschaft der beiden Frequenztransformationen, daß sie die Dämpfungspole bei endlichen ω-Werten frequenzgetreu transformieren. Die Dämpfungspole bei endlichen ω-Werten werden durch die Pole der Impedanzen der Parallelresonanzkreise in den Längszweigen verursacht. Für diesen Bereich der Schaltung läßt sich aber die exakte Transformation durchführen. Zu dem gleichen Ergebnis gelangt man, wenn man die Impedanzen der Kreise in den Längszweigen für die konzentrierte Schaltung und die Schaltungen nach Bild 5 bestimmt. Die Polstellen dieser Impedanzen transformieren sich gemäß den Gl. (5) und (7).

Diese Eigenschaft und die exakte Transformation in dem Punkt βl_0 lassen gute Ergebnisse im Durchlaßbereich und in der Umgebung der Dämpfungspole erwarten.

5. Die konzentrierte Ausgangsschaltung

Die Elemente für Tiefpaßschaltungen in normierter Form findet man tabelliert [7], [8]. Diese Schaltung läßt sich direkt in die Mikrowellenfilterform mit leerlaufenden Stäben übertragen. Die Elemente der normierten Hochpaßschaltungen erhält man durch Transformation aus den Elementen der normierten Tiefpaßschaltung.

Ein Hochpaß ist die Ausgangsschaltung für einen Mikrowellenhochpaß mit kurzgeschlossenen Stäben.

Symmetrische Bandpässe, d. h. Bandpässe, deren Dämpfungskurve eine geometrische Symmetrie zur Bandmittenfrequenz aufweist, erhält man mit der Transformation

$$p_T = \frac{1}{B_r}\left(p_B + \frac{1}{p_B}\right).$$

Der Index B steht für Bandpaß, der Index T für Tiefpaß, B_r ist die relative Bandbreite. Die Bandmitte liegt bei $\omega_B = 1$.

An einem einfachen Beispiel sollen diese Transformation und die anschließende Umwandlung demonstriert werden. Aus dem Tiefpaß in Bild 6a er-

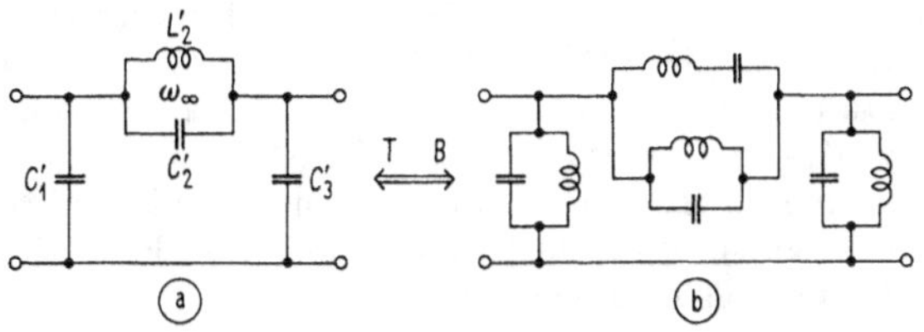

Bild 6. Transformation des Tiefpasses (a) in einen Bandpaß (b).

gibt sich der Bandpaß in Bild 6b. Dieser muß nun so umgeformt werden, daß er in das Schema von Bild 1 paßt. Für die beiden Querzweige ist dies bereits der Fall. Der Längszweig läßt sich umrechnen in die Hintereinanderschaltung zweier Parallelkreise. Die normierten Werte in Bild 7 ergeben sich

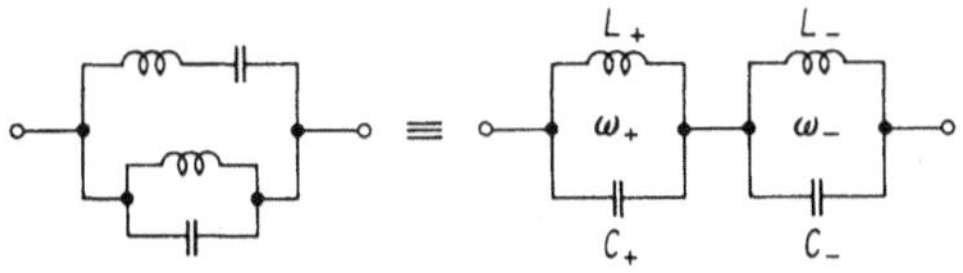

Bild 7. Äquivalente Zweipole.

aus den normierten Werten L_2' und C_2' in Bild 6a nach den Gleichungen [8]

$$C_+ = \frac{1}{L_-} = \frac{C_2'}{B_r}(1+\omega_-^2),$$
$$C_- = \frac{1}{L_+} = \frac{C_2'}{B_r}(1+\omega_+^2) \tag{17}$$

mit

$$\omega_\pm = \sqrt{1+\left(\frac{\omega_\infty B_r}{2}\right)^2} \pm \frac{\omega_\infty B_r}{2} \quad \text{und} \quad \omega_\infty^2 = \frac{1}{L_2' C_2'}.$$

Durch einen weiteren Umformungsschritt muß nun noch zwischen den beiden Resonanzkreisen im Längszweig ein Resonanzkreis im Querzweig erzeugt werden. Hierzu kann die von NORTON angegebene Transformation erster Art benutzt werden. In Bild 8

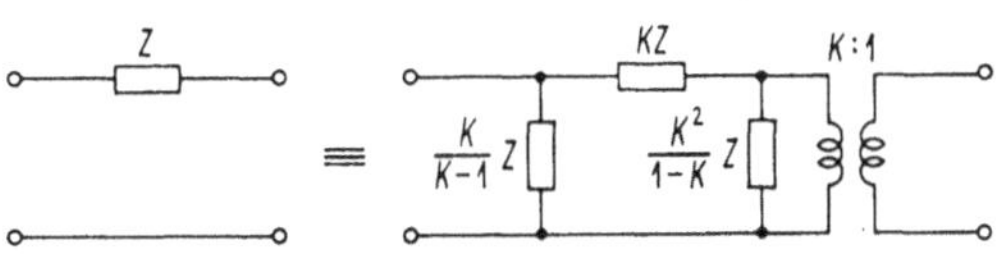

Bild 8. NORTON-Transformation 1. Art.

ist diese mit den Umrechnungsgleichungen wiedergegeben. Für Z wird jeweils einer der beiden Resonanzkreise des Längszweiges eingesetzt. Die Umrechnung erfolgt so, daß beide Transformatoren unmittelbar hintereinander liegen. Wählt man für beide K gleich groß, dann heben sich ihre Wirkungen auf, d.h. sie können weggelassen werden. Das Endergebnis ist in Bild 9 dargestellt. Für die Elemente gilt für den Fall, daß der Tiefpaß in Bild 6a symmetrisch ist ($C_3' = C_1'$),

$$\begin{aligned}
C_1 &= \frac{1}{L_5} = \frac{1}{B_r}\left[C_1' + \frac{K-1}{K}C_2'(1+\omega_-^2)\right],\\
C_2 &= \frac{1}{L_4} = \frac{1}{KB_r}C_2'(1+\omega_-^2),\\
C_3 &= \frac{1}{L_3} = \frac{1-K}{K^2 B_r}C_2'(2+\omega_-^2+\omega_+^2),\\
C_4 &= \frac{1}{L_2} = \frac{1}{KB_r}C_2'(1+\omega_+^2),\\
C_5 &= \frac{1}{L_1} = \frac{1}{B_r}\left[C_1' + \frac{K-1}{K}C_2'(1+\omega_+^2)\right].
\end{aligned} \tag{18}$$

Der Wert K ist innerhalb der Grenzen

$$\frac{1+\omega_+^2}{1+\omega_+^2+C_1'/C_2'} \leq K \leq 1 \tag{19}$$

frei wählbar. Das Verfahren, das hier zur Synthese eines Bandpasses mit drei Querzweigen vorgeführt wurde, gilt natürlich genauso für Filter höherer Ordnung.

Für die von MATTHAEI [9] und CRISTAL [10] angegebenen Kammfilter dient als konzentrierte Ausgangsschaltung das induktiv gekoppelte Resonanzkreisfilter. Über die Dimensionierung der Kammfilter aus diesen Ausgangsschaltungen wurde bereits in [11] berichtet. 139

Für Kammfilter mit leerlaufenden Stäben benötigt man als Ausgangsschaltung das kapazitiv gekoppelte Resonanzkreisfilter. Man erhält es aus dem induktiv gekoppelten Resonanzkreisfilter mit der Tiefpaß-Hochpaß-Transformation.

6. Numerische Ergebnisse

Da die im Kapitel 3 abgeleiteten Frequenztransformationen nicht exakt für die ganze Schaltung gelten, sollen in diesem Kapitel die Dämpfungsverläufe, die sich gemäß den Transformationen ergeben, verglichen werden mit den wirklichen theoretischen Dämpfungsverläufen der Mikrowellenfilter, um so eine Aussage machen zu können über die Güte der Transformationen und des damit verknüpften Dimensionierungsverfahrens.

6.1. Mikrowellenhochpässe

Als Ausgangsfilter für das Beispiel dient ein Tiefpaß 5. Ordnung mit einer Sperrdämpfung von 42,38 dB und einem maximalen Reflexionsfaktor von 20% im Durchlaßbereich (siehe [8], S. 218, $\Theta = 45$). Dieser Tiefpaß hat zwei Parallelresonanzkreise im Längszweig und drei kapazitive Querzweige. Infolge der Parallelresonanzkreise hat er zwei Dämpfungspole bei endlichen ω-Werten. Der zu diesem Tiefpaß gehörende Hochpaß ergibt einen Mikrowellenhochpaß mit drei kurzgeschlossenen Stäben. In Bild 10 sind die Dämpfungsverläufe dargestellt. Das Mikrowellenfilter hat ein besseres Sperr-

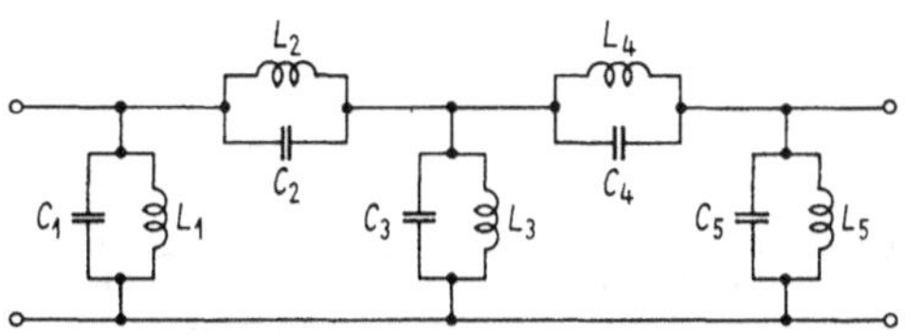

Bild 9. Aus Bild 6b umgewandelter Bandpaß in die Form der Schaltung nach Bild 1.

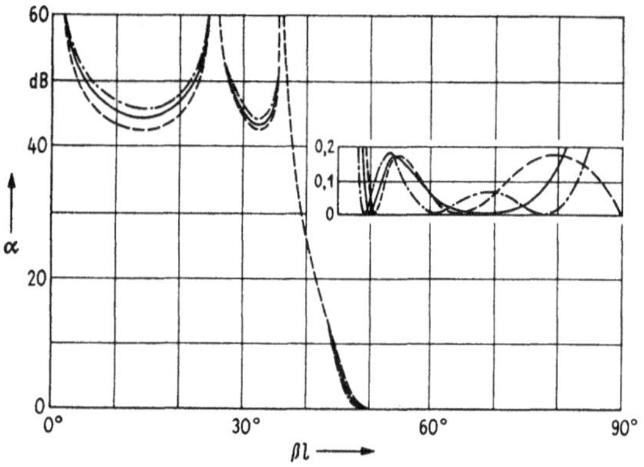

Bild 10. Dämpfungsverlauf eines Mikrowellenhochpasses mit drei Kurzschlußstäben;
— — — erwarteter Verlauf gemäß Frequenztransformation nach Gl. (5),
——— wirklicher Verlauf bei $\beta l_0 = 60°$,
—·—·— wirklicher Verlauf bei $\beta l_0 = 50°$.

verhalten als erwartet. Der Durchlaßbereich kann bis etwa $\beta l = 80°$, 85° oder gar bis 90°, je nach den Anforderungen, angenommen werden. Im Durchlaßbereich ist der Verlauf von βl_0 abhängig, jedoch sind im Grunde die Verläufe für beide βl_0-Werte zufriedenstellend. Mit wachsendem βl_0 wird die Durchlaßbereichsgrenze etwas nach links verschoben. Oberhalb von $\beta l = 90°$ bis $\beta l = 180°$ verhält sich das Mikrowellenfilter wie ein kapazitiver Spannungsteiler. Die Dämpfung steigt ganz allmählich von kleinen Werten bei $\beta l = 90°$ auf ∞ bei $\beta l = 180°$ an. Im Bild 11 sind die Dämpfungsverläufe für den Mikrowellenhochpaß mit drei leerlaufenden Stäben gezeichnet, der sich aus dem konzentrierten Tiefpaß ergibt. Hier ist das Dämpfungsverhalten im Bereich $\beta l = 0°$ bis 90° nicht besonders günstig, da im Grunde wieder ein kapazitiver Spannungsteiler vorliegt.

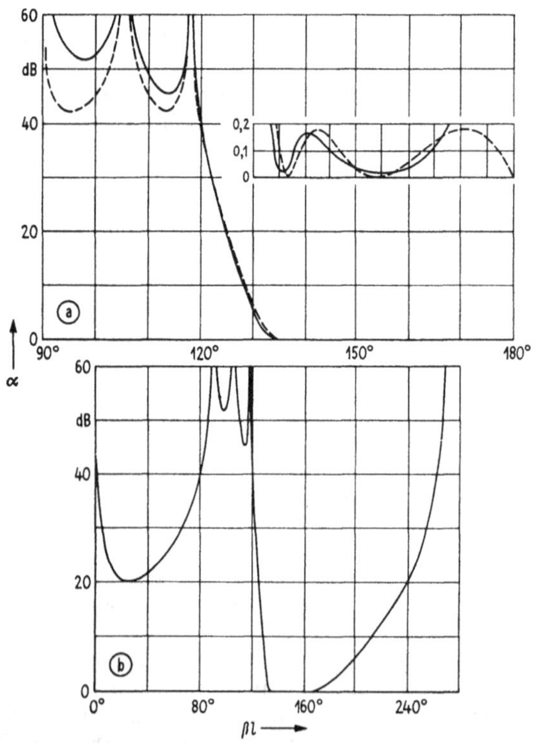

Bild 11. Dämpfungsverlauf eines Mikrowellenhochpasses mit drei Leerlaufstäben;
— — — erwarteter Verlauf gemäß Frequenztransformation nach Gl. (7),
——— wirklicher Verlauf bei $\beta l_0 = 150°$.

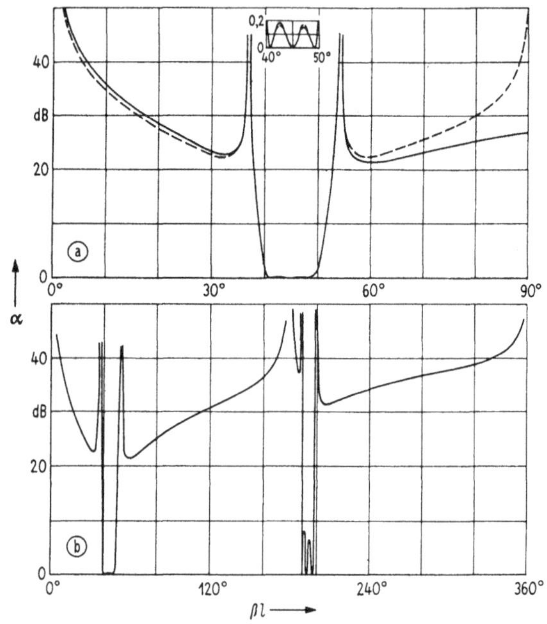

Bild 12. Dämpfungsverlauf eines symmetrischen Mikrowellenbandfilters mit drei Kurzschlußstäben;
— — — erwarteter Verlauf gemäß Frequenztransformation nach Gl. (5),
——— wirklicher Verlauf bei $\beta l_0 = 45°$.

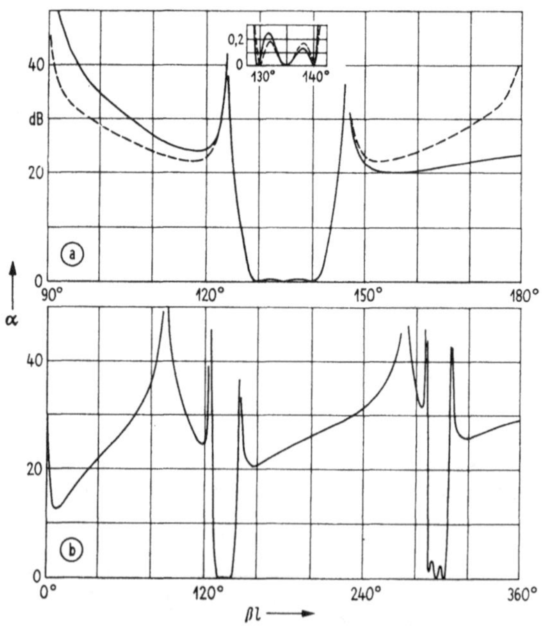

Bild 13. Dämpfungsverlauf eines symmetrischen Mikrowellenbandfilters mit drei Leerlaufstäben;
— — — erwarteter Verlauf gemäß Frequenztransformation nach Gl. (7),
——— wirklicher Verlauf bei $\beta l_0 = 135°$.

6.2. Symmetrische Mikrowellenbandpässe

Als Ausgangsfilter für das Beispiel dient ein Tiefpaß dritter Ordnung mit einer minimalen Sperrdämpfung von 22,29 dB und einem maximalen Reflexionsfaktor von 20% im Durchlaßbereich (siehe [8], S. 176, $\Theta = 35$). Mit der im Kapitel 5 beschriebenen Prozedur wurde er mit $B_r = 0{,}25$ und $K = 0{,}5$ umgerechnet in den symmetrischen Bandpaß gewünschter Struktur. Außerdem wurde die Mitte des Durchlaßbereiches verschoben nach ω_0, gegeben durch (20), (21)

$$\omega_0 = \sqrt{\beta l_0 \tan \beta l_0} \quad \text{oder} \quad \omega_0 = \sqrt{\frac{-\tan \beta l_0}{\beta l_0}},$$

wobei βl_0 die Bandmitte des Mikrowellenfilters mit kurzgeschlossenen bzw. leerlaufenden Stäben sein soll. βl_0 wurde zu $45° \cdot \pi/180°$ bzw. $135° \cdot \pi/180°$ gewählt. In Bild 12a sind die Dämpfungsverläufe für das Filter mit kurzgeschlossenen Stäben und in Bild 13a für das Filter mit leerlaufenden Stäben gezeichnet. Links des Durchlaßbereiches ist das Verhalten der Mikrowellenfilter stets besser als erwartet, rechts davon dagegen etwas schlechter. Beim Mikrowellenfilter mit leerlaufenden Stäben ist die Abweichung auf beiden Seiten stärker ausgeprägt. Auch im Durchlaßbereich weicht der wirkliche Verlauf der Dämpfung beim Filter mit leerlaufenden Stäben stärker vom erwarteten Verlauf ab. In den Bildern 12b und 13b sind die Dämpfungsverläufe über einem größeren βl-Bereich dargestellt. Man erkennt den jeweils nächsten Durchlaßbereich bei $\beta l_{02} = 195° \cdot \pi/180°$ bzw. $\beta l_{02} \approx 295° \cdot \pi/180°$. In den βl-Bereichen mit schlechtem Sperrverhalten liegt wieder kapazitive Spannungsteilung, d.h. kein ausgesprochenes Filterverhalten vor.

Will man bei gleich steilen Flanken eine höhere Dämpfung im Sperrbereich erreichen, so ist das nur möglich durch Erhöhung der Resonatorzahl. Am Prinzip der Realisierung und im grundsätzlichen Verhalten ändert sich dabei nichts.

6.3. Kammfilter

Für Kammfilter mit Kurzschlußstäben wurden bereits in [11] Beispiele angegeben. Es hat sich gezeigt, daß das angegebene Dimensionierungsverfahren auch bei relativen Bandbreiten von 50% noch gute Ergebnisse liefert. Bedingt durch den einfachen Dämpfungspol bei $\omega = 0$ ist die linke Filterflanke bei Kammfiltern mit Ankopplung durch einen Impedanz-transformierenden Kurzschlußstab nicht sehr steil. Bessere Filterflanken auf der linken Seite des Durchlaßbereiches erhält man bei kapazitiver Ankopplung des Anschlußleiters. Für Kammfilter mit Leerlaufstäben zeigen die Bilder 14 und 15 Beispiele für den Dämpfungsverlauf. Die konzentrierten Ausgangsschaltungen wurden durch Tiefpaß-Hochpaß-Transformation aus den konzentrierten Schaltungen in [11] bestimmt. Die Übereinstimmung zwischen tatsächlichem und erwartetem Dämpfungsverlauf im Durchlaßbereich ist nicht so gut wie bei den ersteren. Trotzdem ist das Gesamtverhalten der Filter zufriedenstellend. Für die Filter

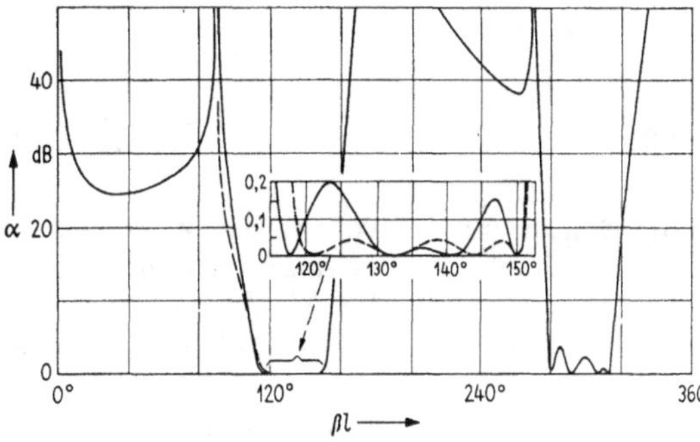

Bild 14. Dämpfungsverlauf eines Kammfilters mit vier Leerlaufstäben, bestimmt aus der konzentrierten Schaltung nach Bild 11 in [11];
– – – erwarteter Verlauf gemäß Frequenztransformation nach Gl. (7),
——— wirklicher Verlauf bei $\beta l_0 = 135°$.

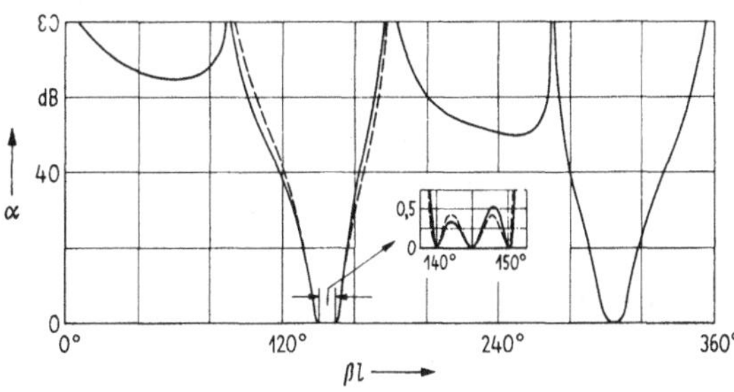

Bild 15. Dämpfungsverlauf eines kapazitiv angekoppelten Kammfilters mit drei Leerlaufstäben, bestimmt aus der konzentrierten Schaltung nach Bild 13 in [11];
– – – erwarteter Verlauf gemäß Frequenztransformation nach Gl. (7),
——— wirklicher Verlauf bei $\beta l_0 = 145°$.

mit kurzgeschlossenen Stäben ist zu bemerken, daß die Durchlaßbereichgrenzen kaum oder nur wenig verschoben wurden. Für die Kurven der Filter mit leerlaufenden Stäben ist die Verschiebung größer, bezogen auf die Bandmitte beträgt die Verschiebung aber beispielsweise in Bild 14 auch nur 1,85%.

7. Transformation des Impedanzniveaus und Bemerkungen zur Konstruktion der Mikrowellenfilter

In der Schaltung nach Bild 5a kann durch Einführung eines Transformators mit dem Übersetzungsverhältnis N das Impedanzniveau der Schaltung verändert werden, weil hier die Realisierung dieses Transformators mit Hilfe eines zusätzlichen Stabes gelingt [9], [11]. Die Realisierung wird so vorgenommen, wie es in Bild 16 angegeben ist. In dieser Schaltung übernimmt der Stab 0 die Impedanztransformation. Er trägt zum Dämpfungsverhalten der Schaltung nichts bei. Die Kapazitätsbeläge der Stäbe 0 und 1 erhält man aus

$$\begin{aligned} C_0' &= (N-1)\, C_g'/N, \\ C_{01}' &= C_g'/N, \\ C_1'' &= C_1' - (N-1)\, C_g'/N^2. \end{aligned} \tag{22}$$

C_1'' ist der neue Kapazitätsbelag des Stabes 1. Durch

die Einführung des Transformators erhöht sich der Kapazitätsbelag der Anschlußleitung gegenüber Gl. (13) um den Faktor N^2. Für positive Werte der Kapazitätsbeläge nach Gl. (22) muß N innerhalb der Grenzen

$$1 \leqq N \leqq \frac{C_1'}{\varepsilon}\sqrt{\frac{\tan\beta l_0}{\beta l_0}} + 1 \tag{23}$$

liegen.

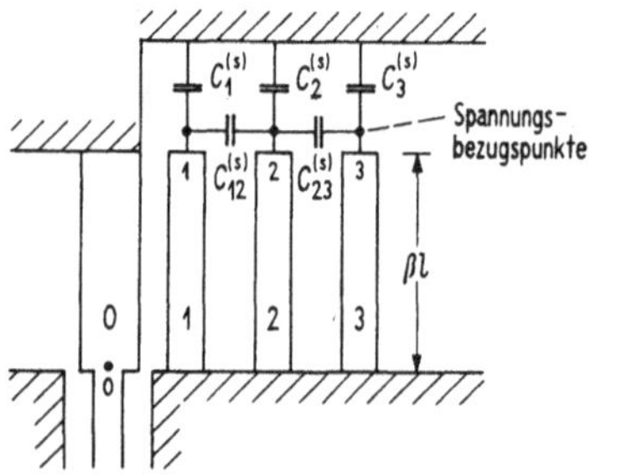

Bild 16. Filter mit Zusatzstab 0 zur Impedanztransformation.

Wählt man die Übersetzungsverhältnisse der Transformatoren am Eingang und am Ausgang der Schaltung unterschiedlich, dann erhält man aus

einer vorgegebenen Schaltung mit gleichen Abschlußwiderständen eine Leitungsschaltung mit ungleichen Abschlußwiderständen.

Die im Kapitel 3 erhaltenen Kapazitätsbeläge müssen noch auf das erforderliche Impedanzniveau umgerechnet werden, das dem vorgegebenen Wellenwiderstand Z_g der Anschlußleitungen entspricht. Der zu Z_g gehörende Kapazitätsbelag ist

$$\frac{C'_{\text{g erf}}}{\varepsilon} = \frac{120\,\pi\,\Omega}{\sqrt{\varepsilon_r}\,Z_g}. \tag{24}$$

ε_r ist die relative Dielektrizitätskonstante des Filterdielektrikums. Das bisher vorhandene Impedanzniveau wird durch die Gl. (13) mit Faktor N^2 und (14) angegeben. Alle Kapazitätsbeläge und Streukapazitäten $C^{(s)}$ müssen daher im Fall des Filters mit Kurzschlußstäben mit dem Faktor

$$K_i = \frac{120\,\pi\,\Omega}{\sqrt{\varepsilon_r}\,Z_g N^2}\sqrt{\frac{\tan\beta l_0}{\beta l_0}} \tag{25}$$

und im Fall des Filters mit Leerlaufstäben mit dem Faktor

$$K_i = \frac{120\,\pi\,\Omega}{\sqrt{\varepsilon_r}\,Z_g\sqrt{-\beta l_0 \tan\beta l_0}} \tag{26}$$

multipliziert werden. Für verlustarme Filter verwendet man kein festes Dielektrikum, sondern Luft ($\varepsilon_r \approx 1$). Für die leerlaufenden Stäbe muß allerdings mit dünnen Scheiben aus festem Isolierstoff, die die Stäbe halten, für mechanische Stabilität gesorgt werden [6]. Die Abmessungen der Stäbe können bei rechteckigem Querschnitt mit den Diagrammen in [12] und bei Kreisquerschnitt mit dem Verfahren in [13] bestimmt werden. In Bild 16 ist eine Möglichkeit der Verbindung des Filters mit einer Anschlußleitung angegeben worden. Bild 17 zeigt schematisch eine Möglichkeit für die Ausführung eines kapazitiv angekoppelten Kammfilters. In diesem Bild sind außerdem die Flächen der Stabenden durch aufgelötete Metallplättchen vergrößert, da-

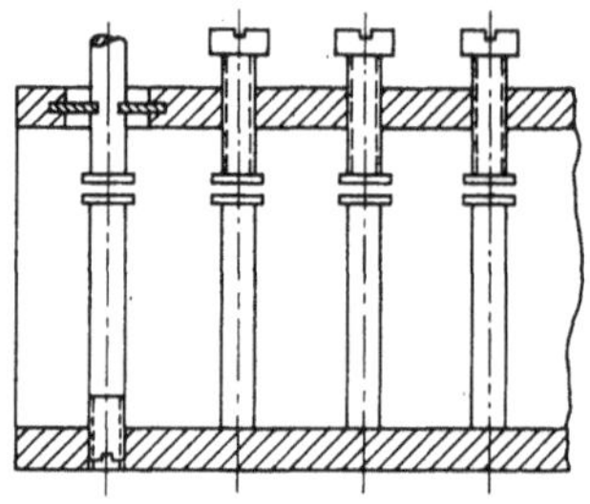

Bild 17. Schema der Konstruktion eines kapazitiv angekoppelten Kammfilters.

mit die Streukapazitäten $C^{(s)}$ gut realisierbar werden. Weitere Einzelheiten über die Konstruktion kann man dem Buch [4] entnehmen, in dem Konstruktionszeichnungen für verschiedene ausgeführte Leitungsfilter angegeben sind.

8. Schlußbemerkung

Das hier dargelegte Verfahren liefert sicherlich deshalb gute theoretische Ergebnisse, weil in den eingeführten Frequenztransformationen sowohl die Eigenschaften der als konzentriert angenommenen Streukapazitäten als auch die der Leitungen zum Tragen kommen. Praktische Untersuchungen müssen zeigen, inwieweit sich die Ergebnisse in den angegebenen Formen realisieren lassen. Sicherlich wird für manche Bandbreiten und in bestimmten βl-Bereichen die Realisierung nur schwer oder gar nicht möglich sein.

Schrifttum

[1] Richards, P. I., Resistor-transmission-line-circuits. Proc. Inst. Radio Engrs. **36** [1948], 217–220.

[2] Wenzel, R. J., Exact design of TEM microwave networks using quarter-wave-lines. Transact. Inst. Elect. Electron. Engrs. MTT-**12** [1964], 94–111.

[3] Horton, M. C. und Wenzel, R. J., General theory and design of optimum quarter-wave TEM filter. Transact. Inst. Elect. Electron. Engrs. MTT-**13** [1965], 316–327.

[4] Matthaei, G. L., Young, L. und Jones, E. M. T., Microwave filters, impedance matching networks and coupling structures. McGraw-Hill Book Co., New York 1964.

[5] Horton, M. C. und Wenzel, R. J., Realization on microwave filters with equal ripple response in both pass and stop bands. Polytechnic Institute of Brooklyn, Symposium on Generalized Networks, New York 1966.

[6] Horton, M. C. und Wenzel, R. J., The digital elliptic filter – a compact sharp-cut off design for wide bandstop or band-pass requirements. Transact. Inst. Elect. Electron. Engrs. MTT-**15** [1967], 307–314.

[7] Saal, R., Der Entwurf von Filtern mit Hilfe des Kataloges normierter Tiefpässe. Telefunken AG, Backnang 1963.

[8] Zverev, A. I., Handbook of filter synthesis. J. Wiley & Sons, New York 1967.

[9] Matthaei, G. L., Comb-line band-pass-filters of narrow or moderate bandwidth. Microwave J. **6** [1963], 82–91.

[10] Cristal, E. G., Capacity coupling shortens comb-line filters. Microwaves [1967], Dez., 44–50.

[11] Pregla, R., Ein Syntheseverfahren für Kammfilter für kleine und große Bandbreiten. A.E.Ü. **23** [1969], 13–19.

[12] Getsinger, W. J., Coupled rectangular bars between parallel plates. Transact. Inst. Radio Engrs. MTT-**10** [1962], 65–72.

[13] Cristal, E. G., Coupled circular cylindrical rods between parallel ground planes. Transact. Inst. Elect. Electron. Engrs. MTT-**12** [1964], 428–439.

BANDSTOP FILTERS V

Bandstop filters are usually designed using stubs; see Matthaei, Young and Jones,[1] Chapter 12. Schiffman and Matthaei (p. 145) show that such stubs are equivalent to parallel coupled lines, with or without end-point contact between the two lines.

See also a later contribution by Schiffman on the same subject.[16]

Exact Design of Band-Stop Microwave Filters*

B. M. SCHIFFMAN†, MEMBER, IEEE, AND G. L. MATTHAEI†, MEMBER, IEEE

Summary—**An exact method for the design of band-stop filters which adapts synthesis techniques due to Ozaki and Ishii is discussed. This method places no theoretical limit on the width of the stop-band, although, for practical reasons, different (but equivalent) circuit configurations are used for stop-bands of different widths. These configurations include a form having open-circuited shunt stubs separated by lengths of line; a second form using resonators which are separate from the main line but parallel to it, so that coupling takes place by way of fringing fields; and a third form in which the resonators are attached directly to the main line, but are folded parallel to it so that coupling is both by direct connection and by fringing fields. Easy to use formulas are given for the exact design of band-stop filters from low-pass prototype filters and equations are given for converting from one form of filter structure to any of the other equivalent forms. The experimental results of trial designs are presented.**

I. General

VARIOUS TECHNIQUES for the exact synthesis of microwave filters have existed for some time.[1–5] In many cases, these procedures have involved the synthesis of special transfer functions and then the determination of the filter elements from the transfer function. Although these procedures are mathematically very elegant, they have found little actual application in microwave engineering because of the very tedious nature of the design process. However, one procedure, described by Ozaki and Ishii,[4] utilizes Kuroda's transformation[4] applied in a simple way to lumped-element, low-pass filters in order to design a certain type of transmission-line filter (which happens to be useful as a practical band-stop filter). In this case, the use of a low-pass prototype along with Kuroda's transformation makes it possible to avoid most of the complexity inherent in many of the exact procedures.

Though the procedure of Ozaki and Ishii mentioned above[4] was described in mathematical form quite some time ago, no investigation of its practical nature appears to have been made. In this study, the practical usefulness of this procedure as applied to band-stop filter design has been verified.[6–8] In addition, tables of design equations have been prepared to make the results as easy as possible to use. With the aid of these tables, a designer can obtain the filter designs he requires with a minimum of computational effort and without having to go into the theory of how the equations were derived. However, the theory is discussed in Section VI.

In Section II, there is a table of transmission-line filter formulas for filters with from one to five stubs, which are well-suited to band-stop filter design. The formulas have been derived by means of these new techniques. Each filter consists of either a tandem array of quarter-wavelength-long open-circuited transmission line shunt stubs (referred to stop-band center frequency ω_0) connected by quarter-wavelength-long transmission lines, or its dual in which open-circuited shunt stubs are replaced by short-circuited series stubs. The shunt stub arrangement, called a basic type band-stop filter in this paper, is suitable for both balanced and unbalanced transmission-line filters operating in the TEM mode while the series stub arrangement is preferred for waveguide filters.

The characteristic impedance of each transmission-line section of the filter can be obtained by direct substitution of the element values of a low-pass ladder prototype filter into the formulas. The resulting band-stop filter response will have the same properties as that of the prototype (*i.e.*, Chebyshev ripples, etc.). An example of a three-resonator filter design is worked out and theoretical and measured results are given.

There is no theoretical limitation on the number of $\lambda_0/4$ sections (so long as the number is odd) that can be used to separate each stub (or resonator) from its immediate neighbors and, in the case of waveguide filters, it has been found to be desirable to separate the stubs by $3\lambda_0/4$ in order to avoid undesirable interaction effects because of the fringing fields in the vicinity of the resonator stubs. In Section III, design equations are

* Received May 27, 1963; revised manuscript received July 31, 1963. This work was supported by the U. S. Army Electronics Research and Development Laboratory, Fort Monmouth, N. J., as part of Contract DA 36-039 SC-87398.

† Stanford Research Institute, Menlo Park, Calif.

[1] P. I. Richards, "Resistor-transmission-line circuits," Proc. IRE, vol. 36, pp. 217–220; February, 1948.

[2] H. Ozaki and J. Ishii, "Synthesis of transmission-line networks and the design of UHF filters," IRE Trans. on Circuit Theory, vol. CT-2, pp. 325–336; December, 1955.

[3] E. M. T. Jones, "Synthesis of wide-band microwave filters to have prescribed insertion loss," IRE 1956 National Convention Record, pt. 5, pp. 119–128.

[4] H. Ozaki and J. Ishii, "Synthesis of a class of strip-line filters," IRE Trans. on Circuit Theory, vol. CT-5, pp. 104–109; June, 1958.

[5] A. I. Grayzel, "A synthesis procedure for transmission line networks," IRE Trans. on Circuit Theory, vol. CT-5, pp. 172–181; September, 1958.

[6] B. M. Schiffman, P. S. Carter, Jr., and G. L. Matthaei, "Microwave Filters and Coupling Structures," Stanford Research Institute Menlo Park, Calif., Quarterly Progress Rept. 7, Section II, SRI Project 3527, Contract DA 36-039 SC-87398; October, 1962.

[7] G. L. Matthaei, Leo Young, and E. M. T. Jones, "Design of Microwave Filters, Impedance Matching Networks, and Coupling Structures," Stanford Research Institute, Menlo Park, Calif., SRI Project 3527, Contract Da 36-039 SC-87398; January, 1963. See Sec. 12.09.

[8] G. L. Matthaei, B. M. Schiffman, E. G. Cristal and L. A. Robinson, "Microwave Filters and Coupling Structures," Stanford Research Institute, Menlo Park, Calif., Final Rept., Sec. IV, SRI Project 3527, Contract DA 36-039 SC-87398; February, 1963.

given for filters having $3\lambda_0/4$ spacings between resonators, for the cases of two and three resonators.

Although these basic stub forms of band-stop filters are conceptually simple and easy to design, they are not suitable for filters requiring very narrow stop-bands. The reason for this limitation is the difficulty of constructing stub lines with the high value of characteristic impedance required in such a filter. To overcome this difficulty and also to allow greater freedom to the designer in general, other configurations of the band-stop filter are described here, together with design equations for converting the basic configuration to one of the newer types.

Two of these newer forms, both of which employ parallel-coupled lines, are completely consonant with the exact design procedure. A filter section of the first type, which is described in Section IV, has a stub resonator that is open-circuited on one end and short-circuited on the other[8,9] and is separate from the main line. This type is particularly useful for filters with very narrow stop-bands, because the stop-band width is essentially a function of the degree of coupling between the stub and main line.

The second type of filter employing coupled lines is suitable for wide bandwidths. This type of filter, described in Section V, is similar to the first except that the stub is open-circuited on one end and directly connected to the main line at the other end.[8] Because of its form, it is called a spur-line type of filter here. In both types, there is electromagnetic coupling between the stub and the main line along the full length of the stub.

In each of the two described cases employing coupled lines, it is required that a basic band-stop filter be designed as a first step. Then, from this basic circuit (or its dual), one of the three alternative forms can be designed with the aid of the given equations.

In Section VI, the derivation of the formulas is fully described. The designer can thereby extend the table of formulas to greater values of n, which is the number of reactive elements in the prototype filter, and also derive formulas for transmission-line filters having connecting lines any desired odd number of quarter wavelengths long.

II. Design Formulas and Their Use

The relationship of the low-pass prototype response to that of the basic transmission-line filter is illustrated in Fig. 1 and Fig. 2. One notes that the transmission-line filter response has, in addition to a low pass band, an infinite number of higher pass bands, each centered on an even multiple of the design frequency ω_0. The peaks of attenuation at ω_0, $3\omega_0$, etc., each have a multiple-order pole of attenuation, the order of the pole being equal to the number n of resonators used in the band-stop filter structure (which is equal to the number of reactive elements in the low-pass prototype).

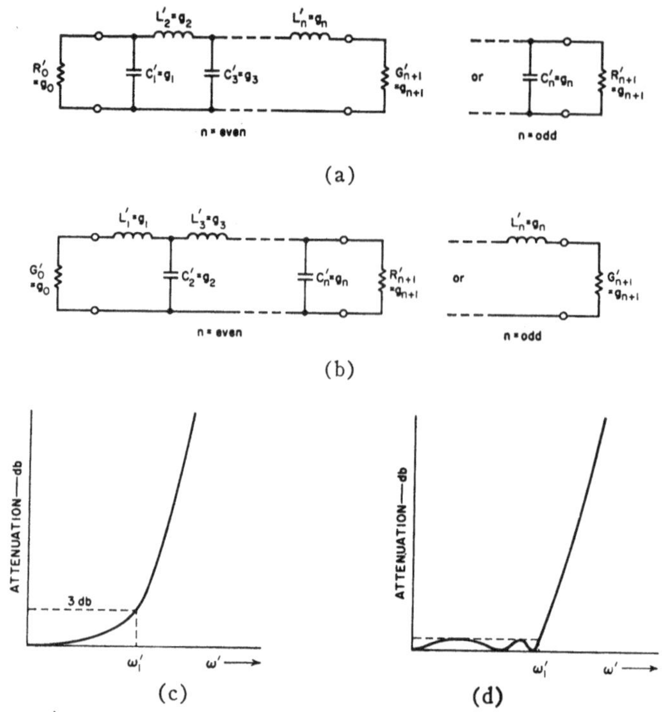

Fig. 1—Low-pass prototype filter. (a), (b) Four basic circuit types defining the parameters $g_0, g_1, \cdots, g_{n+1}$. (c), (d) Maximally flat and equiripple characteristics defining the band-edge ω_1'.

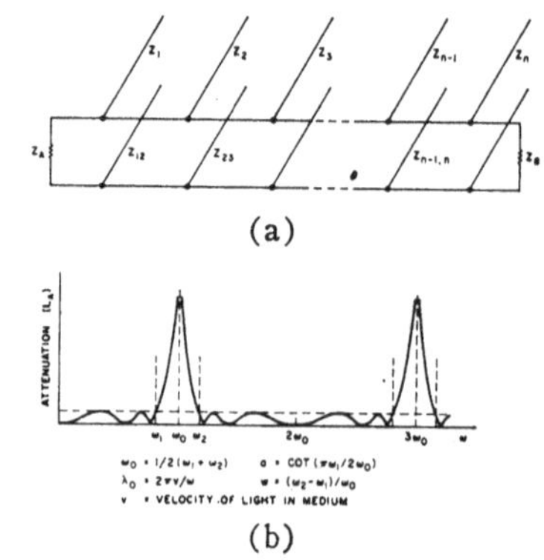

Fig. 2—Band-stop filter. (a) Transmission-line filter derived from n-element low-pass prototype. (b) Equiripple characteristic defining center frequency ω_0, parameter a and stop-band fractional bandwidth. All stubs and connecting lines are $\lambda_0/4$ long.

Table I relates specific frequencies of the lumped-element low-pass prototype filter in the ω'-plane to corresponding frequencies of the transmission-line filter in the ω-plane. The mapping function which maps the ω'-plane into the ω-plane is also given in Table I. Two of the three critical frequencies of the low-pass prototype $\omega'=0$ and $\omega'=\infty$ are fixed and the third ω_1' is the cutoff frequency of the prototype filter.

The formulas for transforming the prototype networks of Fig. 1 to the transmission-line network of Fig. 2 (or to the dual of that network) are given in Table II.

[9] E. M. T. Jones and J. T. Bolljahn, "Coupled-strip-transmission-line filters and directional couplers," IRE Trans. on Microwave Theory and Techniques, vol. 4, pp. 75–81; April, 1956.

TABLE I

DEFINITION OF PARAMETERS AND CORRESPONDENCE OF CRITICAL FREQUENCIES IN ω'-PLANE AND ω-PLANE

Transformation

$$\omega' = \Lambda \tan\left(\frac{\pi}{2}\frac{\omega}{\omega_0}\right)$$

Definition of Parameters

ω' = Prototype frequency variable

ω = Transmission line filter frequency variable

$\Lambda = a\omega_1'$

$$a = \cot\left(\frac{\pi}{2}\frac{\omega_1}{\omega_0}\right)$$

Frequency Correspondences

ω'		ω	
0		0	
ω_1'	$(0 < \omega_1' < \infty)$	$m\omega_0 \pm \omega_1$	(m even)
∞		$n\omega_0$	(n odd)

To use Table II, one must determine the following quantities:

1) The left-hand terminating impedance Z_A for the shunt stub type filter or Y_A of the series stub type filter (dual case),
2) The element values of the prototype circuit g_j ($j=0$ to $n+1$) and the cutoff frequency ω_1' of the prototype,
3) The stop-band center frequency ω_0,
4) The bandwidth parameter a.

Numbers 1) and 3) are arbitrary and 2) may be obtained from tables (see Weinberg,[10,11] or Matthaei, Young and Jones[12]) or from any suitable prototype ladder circuit consisting of alternate series inductors and shunt capacitors. The element values g_j ($j=1$ to n) are the inductance of the series inductors in henries and the capacitance of the shunt capacitors in farads. The first and last elements g_0 and g_{n+1} are the prototype filter terminations given as resistance in ohms or conductance in mhos according to Fig. 1. Number 4) is calculated from the formula

$$a = \cot\left(\frac{\pi}{2}\frac{\omega_1}{\omega_0}\right) \tag{1}$$

[10] L. Weinberg, "Network Design by Use of Modern Synthesis Techniques and Tables," Hughes Aircraft Company, Research Laboratories, Culver City, Calif., Tech. Memo 427; April, 1956. Also *Proc. of the Natl. Electronics Conf.*, vol. 12; 1956.

[11] L. Weinberg, "Additional tables for design of optimum ladder networks," *J. Franklin Inst.*, vol. 264, pt. I, p. 7, July, 1957; and pt. II, p. 127, August, 1957.

[12] G. L. Matthaei, L. Young, and E. M. T. Jones, *op. cit.*, ch. IV.

TABLE II

EXACT EQUATIONS FOR BAND-STOP FILTERS WITH $\lambda_0/4$ SPACING BETWEEN STUBS

The filter structure is as shown in Fig. 2. For the dual case having short-circuited series stubs, replace all impedances in the equations below by corresponding admittances.

n = number of stubs
Z_A, Z_B = terminating impedances
Z_j ($j=1$ to n) = impedances of open-circuited shunt stubs
$Z_{j-1,j}$ ($j=2$ to n) = connecting line impedances
g_j = values of the elements of the low-pass prototype network as defined in Fig. 1.
$\Lambda = \omega_1' a$ where ω_1' = low-pass prototype cutoff frequency and a = bandwidth parameter defined in (1).
(In all cases the left terminating impedance Z_A is arbitrary.)

Case of n=1

$$Z_1 = \frac{Z_A}{\Lambda g_0 g_1}, \qquad Z_B = \frac{Z_A g_2}{g_0}:$$

Case of n=2

$$Z_1 = Z_A\left(1 + \frac{1}{\Lambda g_0 g_1}\right), \qquad Z_{12} = Z_A(1 + \Lambda g_0 g_1),$$

$$Z_2 = \frac{Z_A g_0}{\Lambda g_2}, \qquad Z_B = Z_A g_0 g_3.$$

Case of n=3

Z_1, Z_{12} and Z_2—same formulas as Case $n=2$.

$$Z_3 = \frac{Z_A g_0}{g_4}\left(1 + \frac{1}{\Lambda g_3 g_4}\right), \qquad Z_{23} = \frac{Z_A g_0}{g_4}(1 + \Lambda g_3 g_4),$$

$$Z_B = \frac{Z_A g_0}{g_4}.$$

Case of n=4

$$Z_1 = Z_A\left(2 + \frac{1}{\Lambda g_0 g_1}\right), \qquad Z_{12} = Z_A\left(\frac{1 + 2\Lambda g_0 g_1}{1 + \Lambda g_0 g_1}\right),$$

$$Z_2 = Z_A\left(\frac{1}{1 + \Lambda g_0 g_1} + \frac{g_0}{\Lambda g_2(1 + \Lambda g_0 g_1)^2}\right), \qquad Z_{23} = \frac{Z_A}{g_0}\left(\Lambda g_2 + \frac{g_0}{1 + \Lambda g_0 g_1}\right),$$

$$Z_3 = \frac{Z_A}{\Lambda g_0 g_3}, \qquad Z_{34} = \frac{Z_A}{g_0 g_5}(1 + \Lambda g_4 g_5),$$

$$Z_4 = \frac{Z_A}{g_0 g_5}\left(1 + \frac{1}{\Lambda g_4 g_5}\right), \qquad Z_B = \frac{Z_A}{g_0 g_5}.$$

Case of n=5

Z_1, Z_{12}, Z_2, Z_{23}, Z_3—same formulas as Case $n=4$.

$$Z_4 = \frac{A}{g_0}\left(\frac{1}{1 + \Lambda g_5 g_6} + \frac{g_6}{\Lambda g_4(1 + \Lambda g_5 g_6)^2}\right), \qquad Z_{34} = \frac{Z_A}{g_0}\left(\Lambda g_4 + \frac{g_6}{1 + \Lambda g_5 g_6}\right),$$

$$Z_5 = \frac{Z_A g_6}{g_0}\left(2 + \frac{1}{\Lambda g_5 g_6}\right), \qquad Z_{45} = \frac{Z_A g_6}{g_0}\left(\frac{1 + 2\Lambda g_5 g_6}{1 + \Lambda g_5 g_6}\right),$$

$$Z_B = \frac{Z_A g_6}{g_0}.$$

where ω_1 is the lower cutoff frequency of the transmission-line filter which corresponds with ω_1' of the low-pass prototype filter.

The following example will illustrate the use of the design formulas and the practical realization of a strip-line filter employing open-circuited stubs. A band-stop filter having 50-Ω input impedance, 0.1-db pass-band ripple and 60 per cent stop-band bandwidth is desired and a three-stub design is chosen. The center of the stop band is to be 1.6 Gc. The prototype element values (obtained from Table 4.05-2(a) in Matthaei, Young and Jones[12] or Weinberg[10,11] are $g_0 = g_4 = 1$, $g_1 = g_3 = 1.0315$

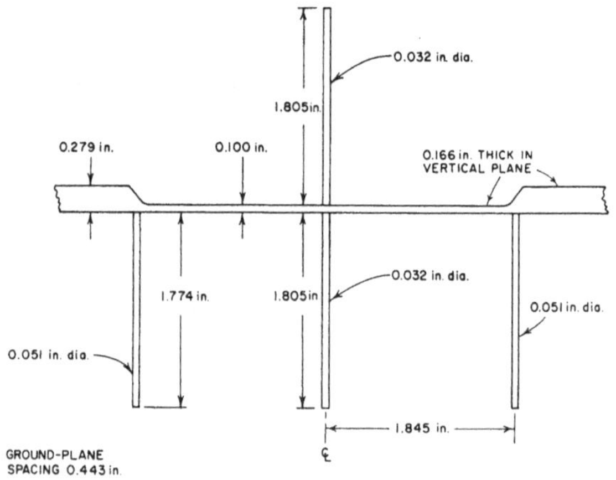

Fig. 3—A strip-line wide-stop-band filter.

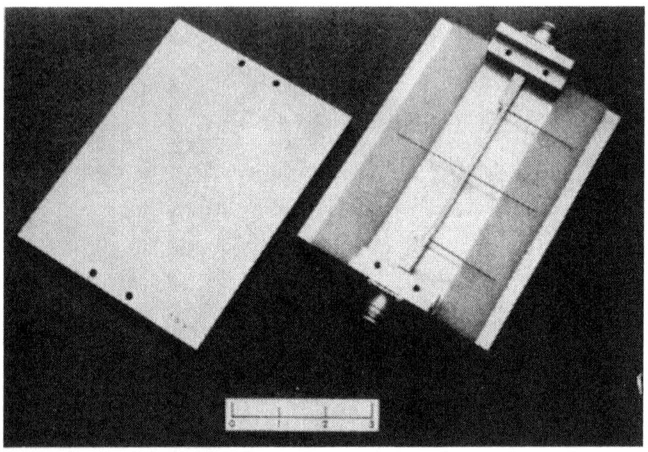

Fig. 4—Photograph of the strip-line wide-stop-band filter.

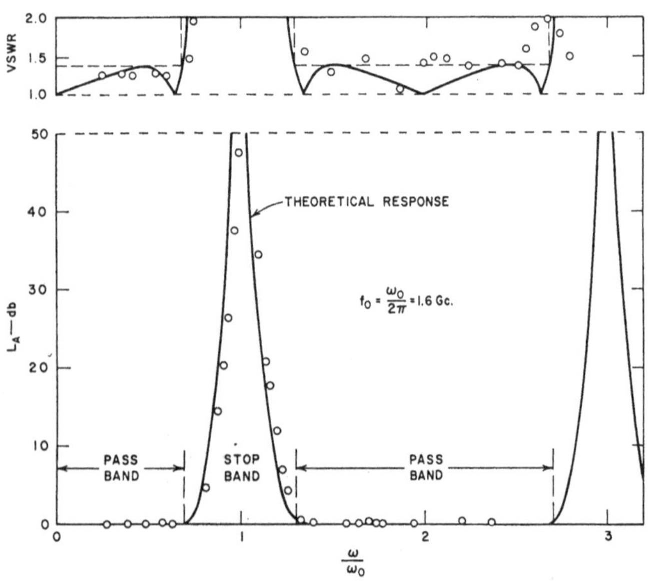

Fig. 5—Computed and measured performance of the wide-stop-band filter in Fig. 4.

and $g_2 = 1.1474$. The quantitity $\Lambda = \omega_1' a$ is next computed. For the prototype chosen, $\omega_1' = 1$ radian per second. From the fractional bandwidth $w = 0.6$, as defined in Fig. 2, we obtain $\omega_1/\omega_0 = 0.7$, and the bandwidth parameter $a = \cot (\pi\omega_1/2\omega_0)$ is found to be 0.50953, thus $\Lambda = 0.50953$. Now using $Z_A = 50\ \Omega$ as the source impedance, we calculate each stub and connecting-line impedance. Since n is odd and the prototype design is symmetrical, the design formulas here yield a symmetrical transmission line design so that the right-hand terminating impedance $Z_B = 50\ \Omega$. The stub impedances are calculated from the design formulas of Table II for $n = 3$ to be $Z_1 = Z_3 = 145.1\ \Omega$ and $Z_2 = 85.5\ \Omega$. The two connecting-line impedances are $Z_{12} = Z_{23} = 76.3\ \Omega$. Each stub and connecting line is then made one-quarter wavelength long at ω_0.

A filter based on the foregoing design was fabricated in strip transmission line. With air as the dielectric, $\lambda_0/4$ is found to be 1.845 inches. The center stub was replaced by two parallel stubs of twice the impedance, *i.e.*, with $Z = 2Z_2 = 171.0\ \Omega$, in order to minimize the junction size and thereby reduce junction effect. The essential dimensions are given in Fig. 3. The stub lengths were reduced by an amount slightly more than one diameter to allow for fringing capacitance at the ends of the stubs.

A photograph of the filter is shown in Fig. 4. The rectangular center-conductor dimensions were obtained from Fig. 6 in Bates[13] and the cylindrical stubs were designed with the aid of the following approximate formula for the impedance of a round conductor between parallel planes:

$$Z_0 = \frac{138}{\sqrt{\epsilon}} \log_{10} \frac{4b}{\pi d} \tag{2}$$

where d is the diameter of center conductor and b is the plate-to-plate spacing.

The computed and measured performance of this filter is given in Fig. 5. The measured results agree well with the theoretical predictions over a rather broad band; however, some improvement could have been obtained by using tuning screws to make the stubs resonate precisely at the desired stop-band center frequency.

III. Formulas for Filters with $3\lambda_0/4$ Spacing Between Resonators

In the case of waveguide band-stop filters, the dual form of filter to that in Fig. 2 a) (*i.e.*, the form using short-circuited series stubs and connecting lines) is usually preferable. If the bandwidth of the filter is very narrow, the impedances of the series stubs will be very low, so that it is desirable to replace the stubs with half-wavelength cavity resonators and coupling irises[6]

[13] R. H. Bates, "The characteristic impedance of shielded slab lines," IRE Trans. on Microwave Theory and Techniques, pp. 28–33; January, 1956.

TABLE III

EXACT EQUATIONS FOR BAND-STOP FILTERS WITH $3\lambda_0/4$ SPACING BETWEEN STUBS OR RESONATOR IRISES

The filter structure is of the form in Fig. 6 (a). For the dual case in Fig. 6. (b) replace all admittances in the equations below with corresponding impedances.

n = number of stubs
Y_A, Y_B = terminating admittances
$Y_j(j=1$ to $n)$ = admittances of short-circuited series stubs
$(Y_{j-1,j})_k$ = admittance of kth ($k=1$, 2 or 3) connecting line from the left between stubs $j-1$ and j
g_j = values of the elements of the low-pass prototype network as defined in Fig. 1
$\Lambda = \omega_1'a$ where ω_1' = low-pass cutoff frequency and a = bandwidth parameter defined in (1).
(In all cases the left terminating impedance is arbitrary)

Case of $n=2$

$$Y_1 = Y_A\left(1+\frac{1}{\Lambda g_0 g_1}\right), \qquad (Y_{12})_1 = Y_A(1+\Lambda g_0 g_1),$$

$$Y_2 = Y_A\frac{g_0}{\Lambda g_2}(1+2\Lambda g_2 g_3), \qquad (Y_{12})_2 = Y_A g_0 g_3\left(\frac{1}{1+\Lambda g_2 g_3}\right),$$

$$Y_B = Y_A g_0 g_3, \qquad (Y_{12})_3 = Y_A g_0 g_3\left(\frac{1+2\Lambda g_2 g_3}{1+\Lambda g_2 g_3}\right).$$

Case of $n=3$

$$Y_1 = Y_A\left(3+\frac{1}{\Lambda g_0 g_1}\right), \qquad (Y_{12})_1 = Y_A\left(\frac{1+3\Lambda g_0 g_1}{1+2\Lambda g_0 g_1}\right),$$

$$(Y_{12})_2 = Y_A\left(\frac{1+\Lambda g_0 g_1}{1+2\Lambda g_0 g_1}\right),$$

$$Y_2 = Y_A\frac{g_0}{\Lambda g_2}, \qquad (Y_{12})_3 = Y_A(1+\Lambda g_0 g_1),$$

$$(Y_{23})_1 = Y_A\frac{g_0}{g_4}(1+\Lambda g_3 g_4),$$

$$Y_3 = Y_A\frac{g_0}{g_4}\left(3+\frac{1}{\Lambda g_3 g_4}\right), \qquad (Y_{23})_2 = Y_A\frac{g_0}{g_4}\left(\frac{1+\Lambda g_3 g_4}{1+2\Lambda g_3 g_4}\right),$$

$$Y_B = Y_A\frac{g_0}{g_4}, \qquad (Y_{23})_3 = Y_A\frac{g_0}{g_4}\left(\frac{1+3\Lambda g_3 g_4}{1+2\Lambda g_3 g_4}\right).$$

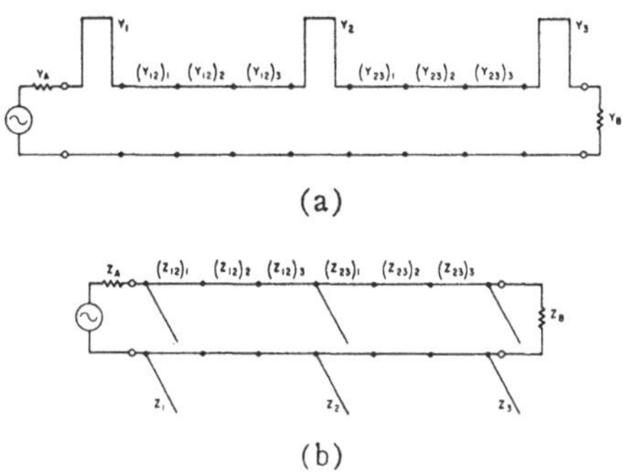

Fig. 6—Band-stop filters with $3\lambda_0/4$ spacing between $\lambda_0/4$ stubs.

with the coupling-iris sizes adjusted to give the required resonator slope parameters.[14 6] This introduces an approximation into the design, but the results are generally more accurate than is achieved by use of the approximate procedure described in Young, Matthaei and Jones.[14]

[14] L. Young, G. L. Matthaei, and E. M. T. Jones, "Microwave band-stop filters with narrow stop bands," IRE TRANS. ON MICROWAVE THEORY AND TECHNIQUES, vol. 10, pp. 416–427; November, 1962.

Previous experience with waveguide band-stop filters has shown that it is desirable to separate the resonators by $3\lambda_0/4$ instead of $\lambda_0/4$ in order to avoid interaction effects due to the fringing fields in the vicinity of the coupling irises. For this reason, equations for the exact design of stub filters having $3\lambda_0/4$ spacings between stubs were prepared in this study. It is expected that these equations will be used as the basis for the design of waveguide band-stop filters by replacing the stubs called for by the equations with iris-coupled resonators as mentioned above, and by replacing λ_0 by λ_{g0}.

Table III presents equations for the design of filters with series-connected stubs with $3\lambda_0/4$ spacings between stubs, for the cases of $n=2$ and $n=3$ stubs. The configuration under consideration is shown in Fig. 6 a). However, by duality, the same equations can be made to apply also to the structure in Fig. 6 b) by simply redefining all of the admittances in Table III as impedances. The resonator susceptance slope parameters[14] for the series stubs in Fig. 6 a) are given by

$$b_j = \frac{\pi}{4}Y_j. \tag{3}$$

Having the admittances of the connecting line sections and also the slope parameters required for the various resonators, the designer can carry out the remainder of the waveguide filter design using the procedure that was discussed previously.[14]

IV. PARALLEL-COUPLED-RESONATOR FILTER

A single section of the parallel-coupled-resonator type filter suitable for narrow stop-bands is shown in Fig. 7. Also shown is the corresponding portion of the basic stub form of filter and design equations that enable the basic filter to be converted, section by section, into the parallel-coupled type. The coupled lines are completely specified by their even- and odd-mode impedances Z_{oe}^a, Z_{oe}^b, Z_{oo}^a, Z_{oo}^b (or their odd- and even-mode admittances).[4 9] The superscripts a and b pertain to lines a and b in Fig. 7.

Before the formulas can be applied, it is necessary to partition the basic filter, shown in Fig. 2 a), so that each section has one length of connecting line and one stub, according to Fig. 7. Since there are more stubs (one more) than connecting lines, one quarter-wavelength section of connecting line must be inserted between the filter and one of the terminations. The characteristic impedance of the inserted section must, of course, equal the terminating impedance; although the filter amplitude response will be unaltered by this change, some time delay will be added to the signal.

The conversion formulas yield values for the distributed capacitances of the coupled lines as illustrated in Fig. 8. These distributed capacitances C_a, C_b and C_{ab}, pertaining to lines a and b shown in the form of a general π-network, are directly related to the even- and odd-mode admittances (or impedances) of the coupled

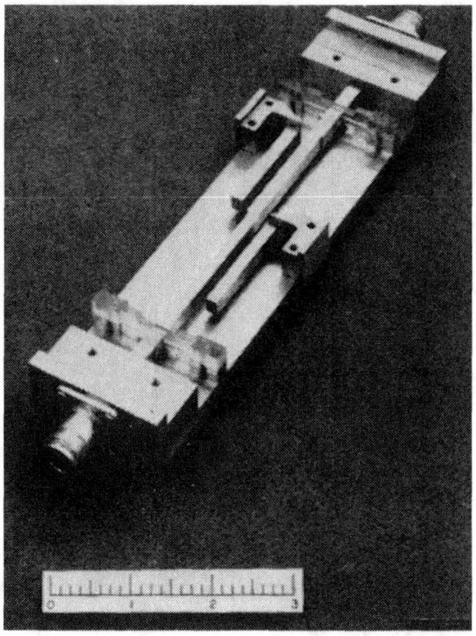

Fig. 10—A parallel-coupled-line, narrow-band filter with cover plate removed.

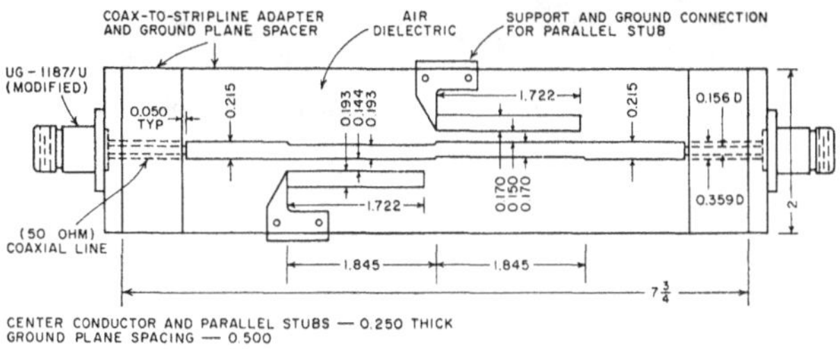

Fig. 11—Sketch of a parallel-coupled-line, narrow-band filter with cover plate removed, showing dimensions in inches.

ance. A quarter-wavelength section of 50-Ω impedance line was inserted between Stub 2 and the 50-Ω termination, thus yielding two complete L-sections ready to convert to two sections of parallel-coupled lines. The symmetrical type of coupled-line construction was chosen for the filter. The values of distributed capacitance (normalized to the permittivity of the medium) were then calculated with the aid of the formulas of Fig. 7 and found to be 1) for Section 1 (corresponding to Stub 1 and its connecting line), $C_a/\epsilon = C_b/\epsilon = 5.80$ and $C_{ab}/\epsilon = 1.73$ and 2) for Section 2 (corresponding to Stub 2 and the added 50-Ω section), $C_a/\epsilon = C_b/\epsilon = 6.12$ and $C_{ab}/\epsilon = 1.82$. With the aid of Getsinger's graphs[15] the strip-line dimensions shown in Fig. 11 were obtained.

Each parallel-coupled resonator and its grounding wedge were milled from a solid piece of aluminum. Each resonator length is slightly less than $\lambda_0/4$, as determined experimentally, to allow for fringing capacity. The measured and computed responses of the filter are shown in Fig. 12. The stop-band center frequency was found, after the final adjustments of the lengths of the resonators, to be $f_0 = 1.608$ Gc and the curves of Fig. 12 are plotted with frequency normalized to f_0. The measured attenuation loss points are seen to be in excellent

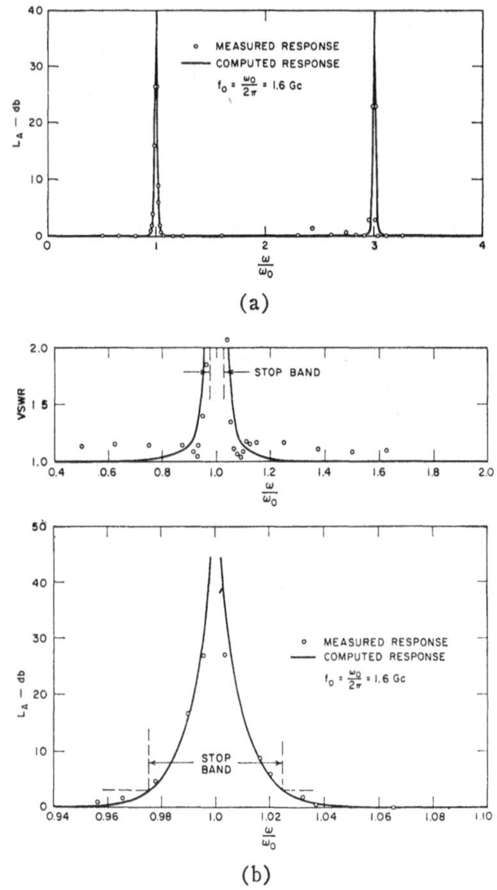

Fig. 12—Measured and computed response of filter in Fig. 11.

agreement with the computed values over a very wide frequency range, even including the second stop-band at $3f_0$.

V. Spur-Line Type of Filter

A second type of band-stop filter section employing coupled lines is shown together with its design formulas in Fig. 13. Because of the direct connection (and the resulting strong coupling) of the stub to the line, this type of filter is similar to the basic type. In the limit, as the coupled lines are separated until there is very little distributed coupling, the structure becomes almost identical to the basic shunt-stub filter section. The spur-line type of construction is thus suitable for filters with wide stop-bands. In general, the energy stored in the resonant structure is determined mainly by the electromagnetic fields of the odd mode of propagation. This energy can be reduced and the bandwidth narrowed by reducing Z_{oo}—that is, by bringing the coupled lines close together—but this process has a practical limit. Furthermore, copper losses would tend to increase and thereby reduce the maximum attainable attenuation in the stop-band.

A two-resonator spur-line-type band-stop filter was designed with the aid of the formulas of Fig. 13 for 60 per cent stop-band bandwidth and 1.6-Gc stop-band center frequency. Here again, as for the narrow-stop-band filter described in Section IV, symmetrical coupled

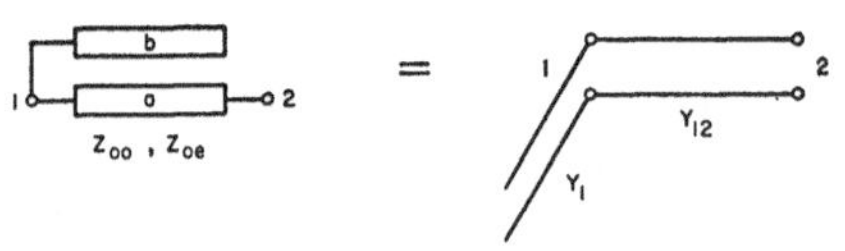

$$\frac{C_a}{\epsilon} = \frac{C_b}{\epsilon} = \frac{\eta_0(Y_1 + Y_{12})}{2\epsilon_r} \qquad \frac{C_{ab}}{\epsilon} = \frac{\eta_0}{4\epsilon_r}\left(\frac{Y_{12}^2 - Y_1^2}{Y_1}\right)$$

Fig. 13—Spur-line stub resonator and equivalent section of basic filter.

lines were used. The prototype circuit was chosen to have maximally flat response and its element values are, as before, $g_0 = g_3 = 1$, $g_1 = g_2 = 1.414$, with $\omega_1' = 1$.

The basic filter element values were found to be $Z_1 = 119.3\ \Omega$ and $Z_2 = 69.3\ \Omega$ for the shunt stubs and $Z_{12} = 86.05\ \Omega$ for the connecting line. A quarter-wavelength section of 50-Ω line was then connected to one end of the filter so that it could be properly partitioned. The two possible circuits thus obtained and their methods of partitioning are shown in Fig. 14 a) and b). Trial paper designs based on those circuits yielded coupled rectangular bar lines of such proportions that it was feared that the discontinuities at the junctions of the two sections and main line would be large enough to badly degrade the performance. The circuit of Fig. 14 c) was found superior in this respect. This circuit, which was used in the design of the filter, was derived from the basic circuit [Fig. 2a)] as follows. First, the dual of the basic circuit was calculated by the equation

$$Z_D = \frac{Z_0^2}{Z} \tag{8}$$

where $Z_0^2 = Z_A Z_B = (50)^2$, Z is the characteristic impedance of a stub or connecting line section and Z_D is the impedance of the dual of that section of line. (The dual of a shunt stub is a series short-circuited stub and the dual of a connecting line is again a connecting line.) Then, a quarter-wavelength section of 50-Ω line was added to each end of the filter and Kuroda's identity[2,6] was used to move the two stubs outward, each to its end of the network, and at the same time transform the series stubs back to shunt stubs with the circuit of Fig. 14c) as a result. This circuit was then partitioned so that it could be converted to a filter structure with two spur-line sections separated by a length of line of impedance $Z_{23} = 29.1\ \Omega$. The finished strip-line filter is shown in Fig. 15. (Note that the two spurs point toward each other and the spur-to-line junctions are separated by $3\lambda_0/4$. An alternative realization, in which the spurs point away from each other and their connecting points are separated $\lambda_0/4$, could be obtained from the original circuit [Fig. 14a)] by adding a section of 50-Ω line on the right.) A sketch of the spur-line type filter with essential dimensions is given in Fig. 16. The coupled lines were designed by the method described in Section IV.

The lengths of the spurs were individually adjusted to resonate at stop-band center as follows. The gap be-

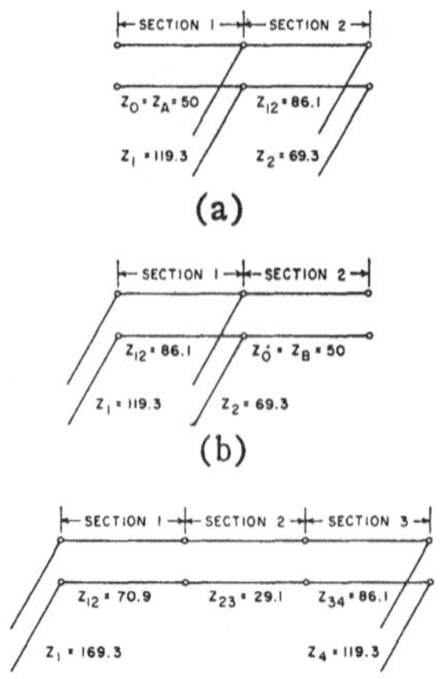

Fig. 14—Three ways of partitioning equivalent basic filters to form a spur-line filter.

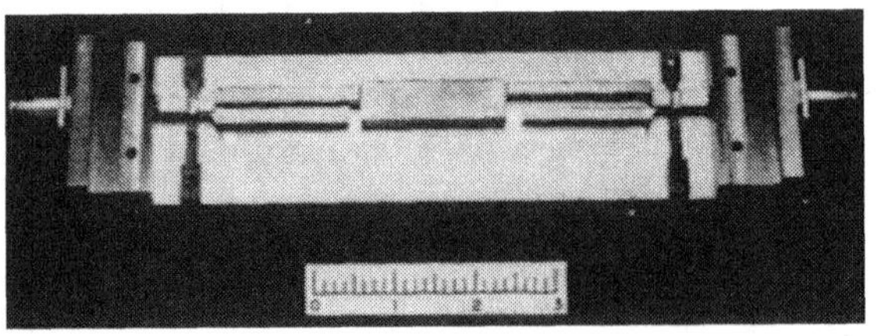

Fig. 15—A spur-line band-stop filter with cover plate removed.

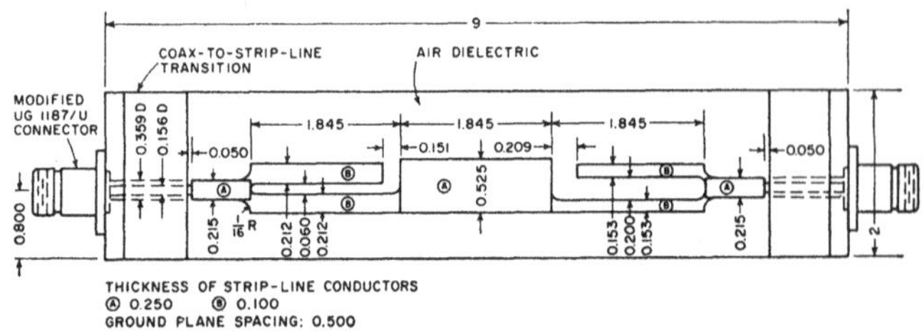

Fig. 16—Sketch of a spur-line band-stop filter with cover plate removed, showing dimensions in inches.

tween the end of one spur and the center section of the main line was bridged with adhesive aluminum foil, thereby making that spur-line section nonresonant. The frequency of maximum attenuation of the remaining resonant spur line was then measured and the unbridged spur was reduced in length about half the amount calculated to bring it to resonance at f_0, when considered to be a simple open-circuited shunt stub. (Half the calculated reduction in length was used as a safety precaution, because the capacitance of the spur end to main line is reduced at the same time and this latter value is not included in the calculation.) A few attempts brought the spur to resonance at the desired frequency and the same process was repeated for the other spur.

The measured values of attenuation loss and VSWR are shown to agree fairly well with the computed values in Fig. 17. The anomalous departure of measured values from computed values that does occur, mainly near the upper edge of the first stop-band and the lower edge of the second stop-band, may be due to the remaining discontinuities between sections, which could not be eliminated entirely.

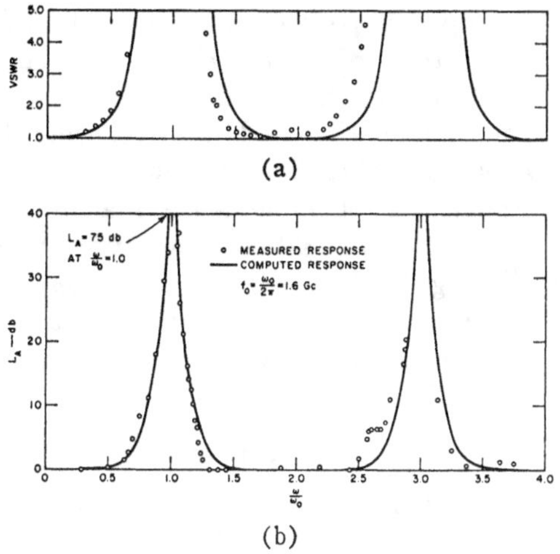

Fig. 17—Measured and computed VSWR and attenuation of filter of Fig. 15.

VII. Derivation of the Design Equations

The design procedure of this section hinges on Kuroda's identity.[2,4] This identity is realized in transmission line as shown in Fig. 18.[6] Note that this identity says that a circuit consisting of an open-circuited shunt stub and a connecting line of the same length has an exact equivalent circuit consisting of a short-circuited series stub with a connecting line at the opposite side.

Fig. 19 traces how Kuroda's identity in the transmission line form (Fig. 18) is used to relate a band-stop filter of the type in Fig. 2 a) to a low-pass prototype filter. Figure 19 a) shows a low-pass prototype filter for the case of $n=3$ reactive elements. Applying the mapping equation of Table I to the shunt susceptances and series reactances of this filter gives

$$\omega' C_j' = C_j' \omega_1' a \tan\left(\frac{\pi}{2}\frac{\omega}{\omega_0}\right), \tag{9}$$

$$\omega' L_j' = L_j' \omega_1' a \tan\left(\frac{\pi}{2}\frac{\omega}{\omega_0}\right). \tag{10}$$

Note that the right side of (9) corresponds to the susceptance of an open-circuited stub having a characteristic admittance.

$$Y_j = C_j' \omega_1' a, \tag{11}$$

the stub being $\lambda_0/4$ long at ω_0. Similarly, the right side of (10) corresponds to the reactance of a short-circuited stub of characteristic impedance

$$Z_j = L_j' \omega_1' a \tag{12}$$

when the stub is $\lambda_0/4$ long at frequency ω_0. Thus, the shunt capacitors in the low-pass prototype become open-circuited shunt stubs in the mapped filter, while the series inductance in the prototype becomes a short circuited series stub in the mapped filter.

Note that in the mapped filter in Fig. 19 b), the terminations seen by the reactive part of the filter are still R_A on the left and R_B on the right, even though two

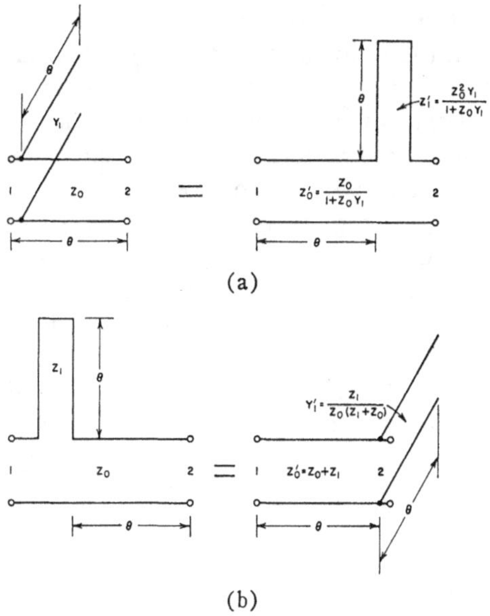

Fig. 18—Kuroda's identity in transmission-line form.

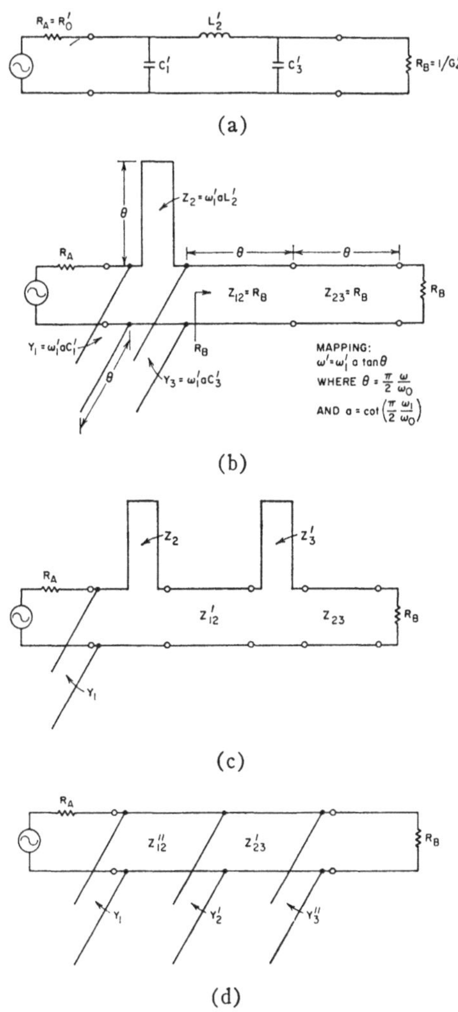

Fig. 19—Stages in the transformation of a low-pass prototype filter into a band-stop transmission-line filter. (a) Prototype. (b) Mapped prototype. (c) After applying Kuroda's identitity to Y_3 and Z_{12} in (b). (d) After applying Kuroda's identity to Z_2 and $Z_{12'}$ and to $Z_{3'}$ and Z_{23} in (c).

line sections of impedance $Z_{12}=Z_{23}=R_B$ have been added on the right to allow the use of Kuroda's identity in the next step. Since their characteristic impedance matches that of the termination, they do not affect the attenuation characteristic of the circuit; their only effect on the response is to give some added phase shift. The circuit in Fig. 19 b) then has a response which is the desired exact mapping of the low-pass prototype response. The only trouble with the filter in Fig. 19 b) is that it contains a series stub which is difficult to construct in a shielded TEM-mode microwave structure.

The series stub in Fig. 19 b) can be eliminated by application of Kuroda's identity (Fig. 18). Applying Kuroda's identity to stub Y_3 and line Z_{12} in Fig. 19 b) gives the circuit in Fig. 19 c). Then applying Kuroda's identity simultaneously to stub Z_2 and line Z_{12}' and stub Z_3' and line Z_{23} in Fig. 19 c) gives the circuit in Fig. 19 d). Note that the circuit in Fig. 19 d) has exactly the same input impedance and over-all transmission properties as the circuit in Fig. 19 b), while the circuit in Fig. 19 d) has no series stubs.

The equations in Tables II and III were derived by use of repeated applications of the procedures described above. For reasons of convenience, the equations in the tables use a somewhat different notation than does the example in Fig. 19; however, the principles used are the same. The equations in Tables II and III also provide for a shift in impedance level from that of the low-pass prototype.

Acknowledgment

R. B. Larrick and P. R. Reznek very ably performed the measurements and adjustments on the experimental filters. The assistance of Dr. Leo Young and P. H. Omlor in adapting an existing program for computing the filter responses is most gratefully acknowledged.

ELLIPTIC FUNCTION FILTERS VI

Elliptic-function filters have equal-ripple response in **both** the pass band and the stop band, and therein lies their usefulness, that is, they may be designed to almost exactly meet specifications without need for overdesign. Such overdesign often results in other ill effects, for example excessive stopband attenuation or stopband width is linked with unnecessarily large passband dissipation loss. Thus, for a given stopband attenuation, elliptic-function filters are preferred over Chebyshev filters if the passband loss is important. On the other hand, elliptic-function filters are more complicated to design and may lead to more complicated structures as well. Therefore for many routine applications the more conventional Chebyshev filters will continue to be used widely.

We are likely to see more papers and design data published on elliptic function filters in the future. A few additional references are given.[17-19] The reader is also referred to the short review paper by Rhodes (p. 239) reprinted in Section VIII.

SEPTEMBER

…ormation for Dis-

…using Richards' trans- …re used to separate the …ould all be located at the …d lengths of transmission … of this communication is …ormations are special cases … of a general … additional useful and interesting equivalent circu… …s may be derived.

Consider a network whose driving point impedance is a function of the complex variable t, where

$$t = \tanh \frac{ls}{c} \tag{1}$$

s being the complex frequency variable, and l/c is the ratio of the length of the basic commensurate transmission line element of the network to the phase velocity of the wave in such a line element. Equation (1) is Richards' transformation.[1] Richards showed that it is always possible to remove a unit element from any physical (positive real) driving point impedance $Z(t)$ leaving a remainder impedance which is positive real. The simplest proof is that given by Grayzel.[3] Since the unit element may be removed from either end of the network, it follows that it is always possible to reverse a unit element followed by a network in cascade to give a network followed by a unit element in cascade. The equivalence is shown in Fig. 1.

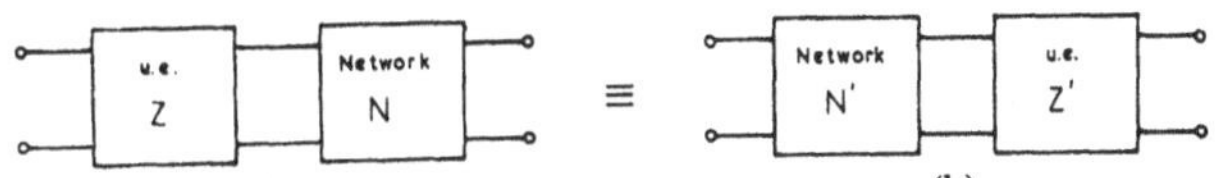

(a) (b)

Fig. 1. Generalized transformation.

The quantitative relationships between the network and its transform are best established by comparison of the transfer matrices. Thus, the transfer matrix of the network N in Fig. 1(a) is

$$\begin{bmatrix} A(t) & B(t) \\ C(t) & D(t) \end{bmatrix} \tag{2}$$

where the matrix elements are all positive real functions of t. The overall transfer matrix for Fig. 1(a) is

$$\frac{1}{\sqrt{1-t^2}}\begin{bmatrix} 1 & Zt \\ \frac{t}{Z} & 1 \end{bmatrix}\begin{bmatrix} A & B \\ C & D \end{bmatrix} = \frac{1}{\sqrt{1-t^2}}\begin{bmatrix} A+CZt & B+DZt \\ C+\frac{A}{Z}t & D+\frac{B}{Z}t \end{bmatrix}. \tag{3}$$

If this is to represent the transfer matrix of Fig. 1(b), then the unit element Z_5' may be removed from the rightby post-multi plication of (3) by the inverse matrix

$$\frac{1}{\sqrt{1-t^2}}\begin{bmatrix} 1 & -Z't \\ -\frac{t}{Z'} & 1 \end{bmatrix}$$

giving the transfer matrix of the remainder network N' as

$$\frac{1}{1-t^2}\begin{bmatrix} A+CZt & B+DZt \\ C+\frac{A}{Z}t & D+\frac{B}{Z}t \end{bmatrix}\begin{bmatrix} 1 & -Z't \\ -\frac{t}{Z'} & 1 \end{bmatrix}$$

$$= \frac{1}{1-t^2}\begin{bmatrix} A+CZt-\frac{B}{Z'}t-\frac{DZ}{Z'}t^2 & -AZ't-CZZ't^2+B+DZt \\ C+\frac{A}{Z}t-\frac{D}{Z'}t-\frac{Bt^2}{ZZ'} & -CZ't-\frac{AZ'}{Z}t^2+D+\frac{B}{Z}t \end{bmatrix}. \tag{4}$$

Since the $(1-t^2)$ factor must divide into all terms of the matrix in (4), this leads to the general formula for Z' as

$$Z' = \frac{B(1)+ZD(1)}{A(1)+ZC(1)}. \tag{5}$$

It may be checked that the four Kuroda transformations may be derived using (4) and (5).

As an example of the application of this general transformation, take the case of a unit element followed by a network N consisting of a distributed series resonant circuit in shunt, as shown in Fig. 2(a). The transfer matrix of N is

$$\begin{bmatrix} 1 & 0 \\ \frac{Ct}{1+LCt^2} & 1 \end{bmatrix} \tag{6}$$

Equation (4) gives the transfer matrix of the network N' as

$$\begin{bmatrix} 1+\frac{LC(Z-Z')t^2}{Z'(1+LCt^2)} & \frac{(Z-Z')}{(1+LCt^2)}t \\ \frac{LC(Z-Z')t}{ZZ'(1+LCt^2)} & 1+\frac{LC(Z'-Z)t^2}{Z(1+LCt^2)} \end{bmatrix} \tag{7}$$

while Z' is given by (5) as

Manuscript received December 16, 1964; revised May 11, 1965.

[1] P. I. Richards, "Resistor-transmission-line circuits," *Proc. IRE*, vol. 36, pp. 217–220, February 1948.

[2] H. Ozaki and J. Ishii, "Synthesis of a class of strip-line filters," *IRE Trans. on Circuit Theory*, vol. CT-5, pp. 104–109; June 1958.

[3] A. I. Grayzel, "A synthesis procedure for transmission line networks," *IRE Trans. on Circuit Theory*, vol. CT-5, pp. 172–181; September, 1958.

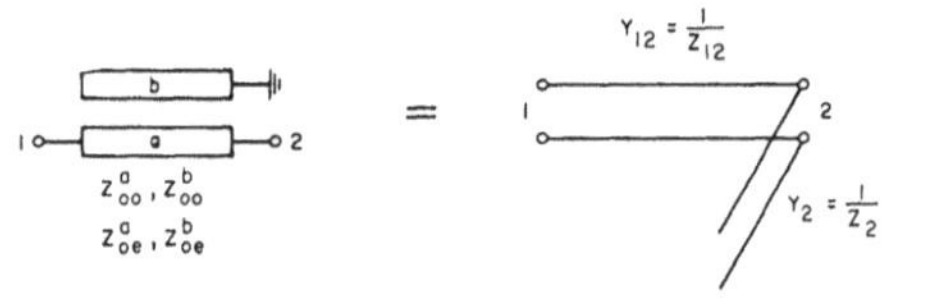

Unsymmetrical Case:

$$\frac{C_a}{\epsilon} = \frac{\eta_0}{\sqrt{\epsilon_r}}(Y_{12}+Y_2) - \frac{C_{ab}}{\epsilon}$$

$$\frac{C_b}{\epsilon} = \frac{\sqrt{\epsilon_r}}{Y_2\eta_0}\left(\frac{C_{ab}}{\epsilon}\right)^2 - \frac{C_{ab}}{\epsilon}$$

with C_{ab} arbitrary (see text)

Symmetrical Case:

$$\frac{C_a}{\epsilon} = \frac{C_b}{\epsilon} = \frac{\eta_0}{\sqrt{\epsilon_r}}\left[Y_{12}+Y_2 - \sqrt{Y_2(Y_{12}+Y_2)}\right]$$

$$\frac{C_{ab}}{\epsilon} = \frac{\eta_0}{\sqrt{\epsilon_r}}\sqrt{Y_2(Y_{12}+Y_2)}$$

Fig. 7—Parallel-coupled transmission-line section and equivalent section of basic filter.

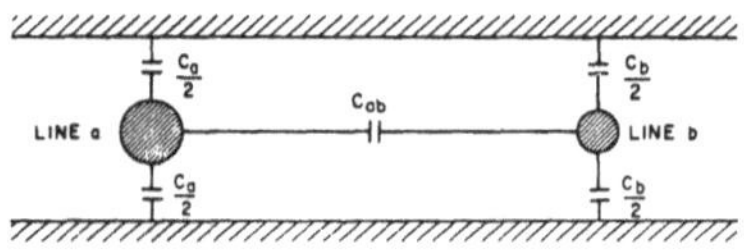

Fig. 8—An unsymmetrical pair of parallel-coupled lines; C_a, C_{ab} and C_b are line capacitances per unit length.

lines.[8,9] The formulas give distributed capacitances directly because these quantities are most useful in designing coupled-rectangular-bar strip lines. The coupled lines may be symmetrical ($C_a = C_b$) or asymmetrical ($C_a \neq C_b$).

If it is desired that the coupled lines be symmetrical, the right set of equations of Fig. 7 is used. If the coupled lines are made unsymmetrical, the value of C_{ab} is arbi-

trarily chosen (subject to the condition that positive values are obtained for C_a and C_b) and the left set of equations is used.

The relationship of the capacitances of the π-network of Fig. 8 to the component capacitances (*i.e.*, parallel-plate and fringing capacitances) of parallel-coupled rectangular bars is easily seen in Fig. 9. Since the capacitances C_a and C_b are not necessarily equal, the two coupled lines of each section need not be identical in cross section. However, as explained below, by restricting the heights so that the two rectangular bars are equal and further limiting the strip-line proportions, the coupled lines of both the symmetrical and asymmetrical type can be easily designed with the aid of Getsinger's graphs of fringing capacitances.[15]

Getsinger[15] has derived equations for the fringing capacitances C_{fe}', C_{fo}' and C_f' of rectangular bars shown in Fig. 9 and has prepared convenient charts which relate C_{fe}'/ϵ, C_{fo}'/ϵ, and C_f'/ϵ to rectangular-bar strip-line dimensions. Here ϵ is the dielectric constant of the medium of propagation, so that the above ratios are dimensionless and of moderate size. Getsinger gives equations for the design of symmetrical, parallel-coupled, rectangular strip-lines and here we adapt his

[15] W. J. Getsinger, "Coupled rectangular bars between parallel plates," IRE TRANS. ON MICROWAVE THEORY AND TECHNIQUES, vol. 10, pp. 65–72; January, 1962.

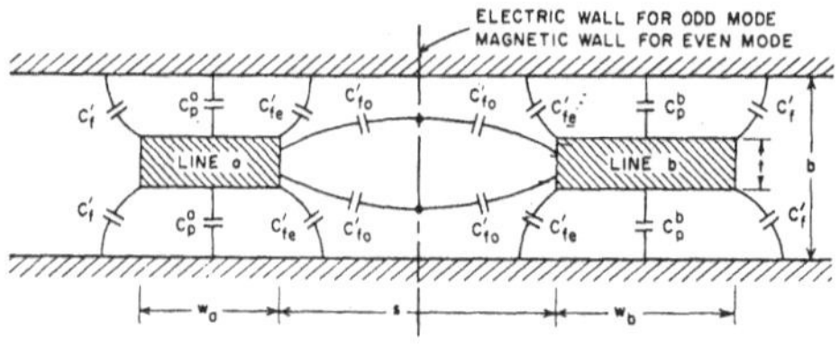

Fig. 9—Cross section of unsymmetrical, rectangular-bar parallel-coupled lines.

equations to fit the unsymmetrical, as well as the symmetrical, case.

Note that the shape of the strip lines in Fig. 9 is specified in terms of the dimensions t, b, s, w_a and w_b. To design a pair of lines such as those in Fig. 9, to have odd- and even-mode admittances or impedances as determined implicitly by the calculated values of C_a/ϵ, C_{ab}/ϵ and C_b/ϵ, first select a convenient value for t/b. Then, noting that

$$\frac{\Delta C}{\epsilon} = \frac{C_{ab}}{\epsilon}, \tag{4}$$

use Getsinger's chart[15] of $\Delta C/\epsilon$ and C_{fe}' vs s/b to determine s/b and also C_{fe}'/ϵ. Using t/b and Getsinger's chart of C_f' vs t/b, determine C_f'/ϵ and then compute

$$\frac{w_a}{b} = \frac{1}{2}\left(1 - \frac{t}{b}\right)\left[\frac{1}{2}\left(\frac{C_a}{\epsilon}\right) - \frac{C_{fe}'}{\epsilon} - \frac{C_f'}{\epsilon}\right], \tag{5}$$

$$\frac{w_b}{b} = \frac{1}{2}\left(1 - \frac{t}{b}\right)\left[\frac{1}{2}\left(\frac{C_b}{\epsilon}\right) - \frac{C_{fe}'}{\epsilon} - \frac{C_f'}{\epsilon}\right]. \tag{6}$$

When the ground plane spacing b is specified, the required bar widths w_a and w_b are determined. This procedure also works for the thin-strip case where $t/b = 0$. If either w_a/b or w_b/b is less than $0.35(1 - t/b)$, the width of the bar should be corrected using the approximate formula

$$\frac{w'}{b} = \frac{0.07\left(1 - \frac{t}{b}\right) + \frac{w}{b}}{1.20}, \tag{7}$$

providing that $0.1 < (w'/b)/(1 - t/b) < 0.35$. In (5), (6) and (7), w is the uncorrected bar width and w' is the corrected width. The need for this correction arises because of the interaction of the fringing fields at opposite sides of a bar, which will occur when a bar is relatively narrow.

The strip-line filter shown in Fig. 10 was designed by the methods outlined here. The stop-band center frequency is 1.6 Gc. The prototype filter has two elements and a maximally flat response. The prototype element values are $g_0 = g_3 = 1$, $g_1 = g_2 = 1.414$ and $\omega_1' = 1$. The characteristic impedance values of the basic two-stub filter, designed with the aid of the formulas in Table II for a 5 per cent width of the stop-band (between 3-db points) and 50-Ω terminations, are $Z_1 = 949.9$ Ω and $Z_2 = 899.9$ Ω for the shunt-stub characteristic impedances and $Z_{12} = 52.8$ Ω for the connecting-line imped-

Fig. 2. Case of the series resonant circuit in shunt.

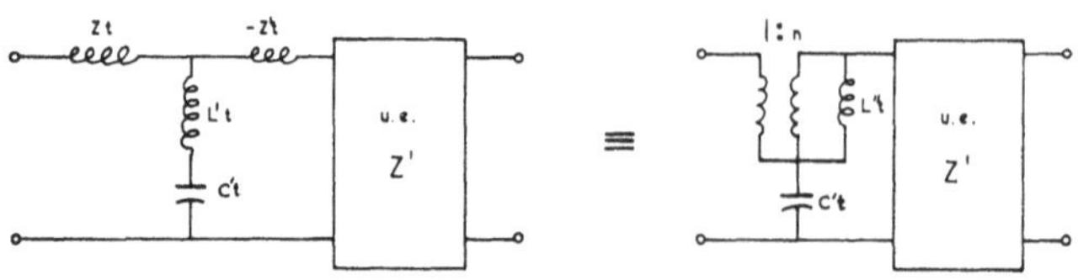

(a) (b)

Fig. 3. Physical realization of the transformation of Fig. 2.

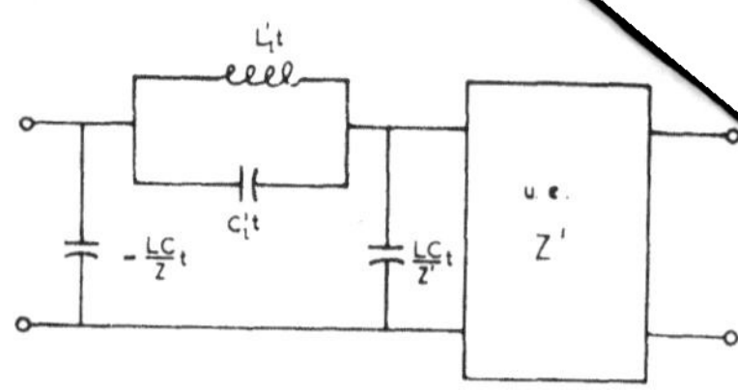

Fig. 4. Alternative realization of the transformation

$$L_1' = Z - Z' = \frac{CZ^2}{1 + LC + ZC}$$

$$C_1' = \frac{LC}{L_1'} = \frac{L(1 + LC + ZC)}{Z^2}.$$

$$\text{(a)} \quad L' = L + \frac{1}{C}$$

$$C' = \frac{LC^2}{1 + LC}$$

$$Z' = \frac{(1 + LC)Z}{1 + LC + ZC}.$$

$$\text{(b)} \quad L'' = L' - Z'$$

$$= \frac{(1 + LC)^2}{C(1 + LC + ZC)}$$

$$n = \frac{1 + LC}{1 + LC + ZC}.$$

$$Z' = \frac{1 + LC}{1 + LC + ZC} \cdot Z. \tag{8}$$

The matrix (7) may be synthesised as a Tee or Pi network with one negative element, which may then be eliminated either by conversion into a Brune type-C section, as shown in the case of the Tee-network realization in Fig. 3(b), or by absorption of the negative element into an adjacent positive element. The Brune type-C section plus the unit element of Fig. 3(b) may be realized in distributed form directly by means of one of a number of 3-wire lines, as described by Matsumoto.[4] The Pi-network realization is shown in Fig. 4. This too may be converted into a type-C section plus the unit element, but it is usually more convenient to absorb the negative shunt capacitance into an adjacent shunt capacitance derived from the previous part of the network. It is not possible to do this for a general network, but in the case of distributed elliptic function filters it appears that the absorption may usually be accomplished.

It is hoped to publish shortly a more complete account of the application of the generalized transformation to the realization of practical distributed networks, and to distributed elliptic function filters in particular.

Acknowledgment

The author is indebted to J. O. Scanlan for valuable discussions during the course of this work.

R. Levy
Elec. Engrg. Dept.
University of Leeds
Yorkshire, England

[4] A. Matsumoto, "Network synthesis with multiwire lines," Mono. Ser. of the Research Institute of Applied Electricity, no. 11, Hokkaido University, Sapporo, Japan, 1963.

Synthesis of Distributed Elliptic-Function Filters from Lumped-Constant Prototypes

R. LEVY, SENIOR MEMBER, IEEE, AND I. WHITELEY, STUDENT MEMBER, IEEE

***Abstract*—A synthesis procedure for distributed elliptic-function filters utilizing published tables of lumped-constant elliptic-function filters has been devised. It is dependent upon the application of a new generalized transformation for distributed networks. The filters often require the realization of distributed Brune sections, and several novel practical realizations are given both for general and degenerate Brune (i.e., Foster) sections. Examples are given of elliptic function stop band and pass band filters having both wide and narrow bandwidths, and experimental results show good agreement with theory.**

I. Introduction

THE FILTERS described in this paper are based on the elliptic-function or Cauer-parameter lumped-element prototype [1], [2]. A typical insertion loss vs. frequency characteristic for a low-pass prototype is shown in Fig. 1, which represents a filter having two transmission zeros at finite frequencies, and one at $\omega = \infty$. The ripple level in the pass band is A_D dB, and the minimum stop band attenuation is A_S dB, where the pass band edge is normalized to $\omega = 1$ and the stop band edge is at $\omega = \omega_s$. The main advantage of this type of characteristic is that the cutoff slope, which may be defined as

$$S_{\text{dB}} = \frac{A_S - A_D}{\omega_s - 1}, \tag{1}$$

may be made very large with a reasonably small number of circuit elements. This may be contrasted with Butterworth or Chebyshev filters, where very often a rather large number of circuit elements are required. A second advantage, of considerable practical importance, is that extensive design tables for elliptic-function filters are available [2].

Although elliptic-function filters have been developed and are employed extensively at low frequencies using lumped components, there has been little application of the technique in the microwave band, using distributed circuit elements. Ozaki and Ishii [3] made an early reference to a design procedure but their method consists of a direct synthesis from the network function, and appears to be quite complicated. An interesting procedure based on coupled transmission lines was introduced by Saito [4]. It is subject to limitations due to extreme impedance levels. A method (which also suffers from restrictions) similar to that described in Section VI of the present paper has been published independently by Blanc and Soldi [5], and a very compact design for broadband elliptic-function filters has been given by Horton and Wenzel [6]. A multiharmonic rejection filter designed from an elliptic-function prototype has been described by Schiffman [7], but is limited to prototypes having not more than one transmission zero at a finite frequency. A more empirical approach to obtaining high rates of cutoff at the band edges has also been described by Torgow and Lubell [8].

Although direct synthesis of distributed elliptic-function filters is certainly possible [3], a great deal of work is involved in the preparation of suitable computer programs. As an alternative it was decided to take advantage of the considerable effort which has led to the publication of comprehensive tables for lumped-element elliptic-function filters [2]. The distributed filters described in this paper use the low-pass prototypes of these tables.

II. Lumped to Distributed Transformations

The extension of network synthesis to distributed networks was demonstrated by P. I. Richards in 1948 [9]. It was shown that distributed networks composed of lumped resistors and equal length or commensurate lossless transmission lines (i.e., lines of an integral multiple of a given electrical length) can be treated by methods similar to those for lumped networks by using a transformation from the complex frequency variable s to a new complex variable

$$t = \tanh \frac{ls}{c} \tag{2}$$

where l is the free-space length of the basic transmission line element, and c is the velocity of EM waves. As presented, (2) applies to TEM transmission lines, but a similar transformation may be used in the case of waveguide circuits.

The mapping of the s-plane on to the t-plane is not one to one, but is periodic, corresponding to the periodic nature of the impedance function of the distributed network. The imaginary ($j\Omega$) axis of the t-plane is related to the imaginary axis (real frequency axis) of the s-plane by the equation

$$\Omega = \tan \phi \tag{3}$$

Manuscript received December 27, 1965; revised August 2, 1966. This work was supported by the Ministry of Aviation under Contract PD/62/022. The work of I. Whiteley was supported by a maintenance grant from the Science Research Council.

The authors are with the Department of Electrical and Electronic Engineering, The University of Leeds, Yorkshire, England.

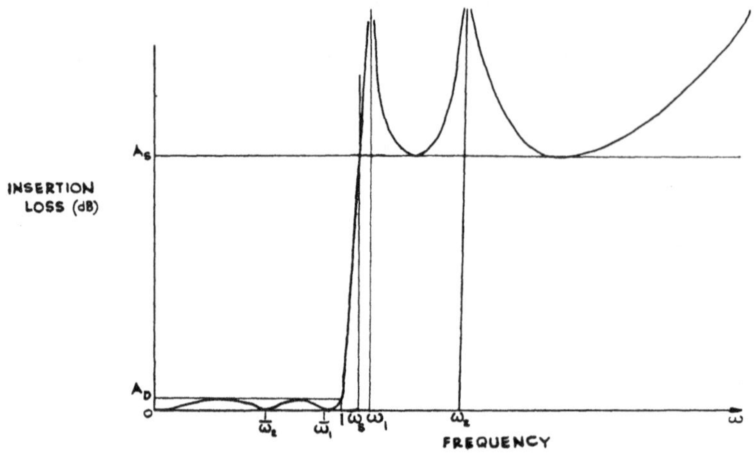

Fig. 1. Lumped-constant low-pass elliptic-function filter characteristic.

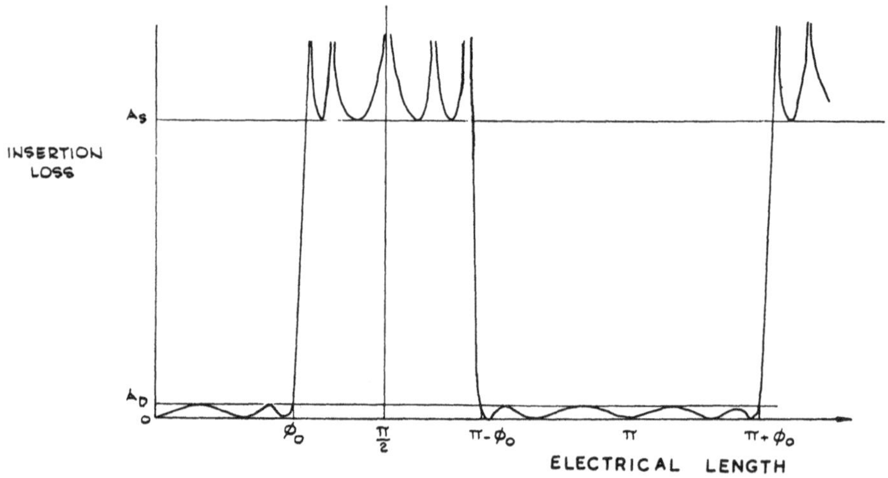

Fig. 2. Attenuation characteristic of a distributed-element elliptic-function filter.

where ϕ is the basic electrical length of the commensurate transmission lines. When the lumped network having the characteristic shown in Fig. 1 is transformed into a distributed network by Richards' transformation, the real-frequency behavior of the resulting network is shown in Fig. 2. In practice, this transformation is carried out by replacing the reactive lumped elements of the prototype by suitable transmission line stubs. Thus, an inductance of reactance $L'\omega_0$ at a frequency ω_0 is replaced by a short-circuited stub of input reactance $Z' \tan \phi_0$, and a capacitance of susceptance $C'\omega_0$ is replaced by an open-circuited stub of input susceptance $Y' \tan \phi_0$. ϕ_0 may be chosen arbitrarily, and in the example illustrated by Figs. 1 and 2 corresponds to the band edge of the low-pass filter. The choice of ϕ_0 controls the ratio of the cutoff frequencies of the second and first pass bands $(\pi-\phi_0)/\phi_0$, but too small a value of ϕ_0 may lead to extreme values for the characteristic immittances Z', Y' of the stubs.

In the transformed distributed network all the stubs are located at the same physical point, and as yet there is no provision for separation of the stubs by lengths of transmission line for ease of manufacture, and indeed, for physical realizability in the practical sense. Such a transmission line cascaded between two filter elements has the basic commensurate electrical length, and is termed a *unit element*. One of the main problems in the distributed filter design technique described in this paper is to introduce unit elements into the network without altering the insertion loss characteristic.

III. Generalized Kuroda Transformations

A method of stub separation which is used frequently in the case of certain restricted classes of ladder networks is the application of Kuroda's transformations [3], [10]. These establish an exact equivalence between a unit element in cascade with a series or shunt open- or short-circuited stub on the one hand, and the reversed cascade consisting of a stub followed by a unit element on the other. Since it is possible to incorporate any number of unit elements at either port of the ladder network, each of characteristic impedance equal to the terminating resistance at the respective ports without altering the insertion loss, then it is possible to transform these unit elements into the network by successive applications of the Kuroda transformation in such a way as to separate the stubs. This method is restrictive in that the simple Kuroda transformations apply only to single stubs,

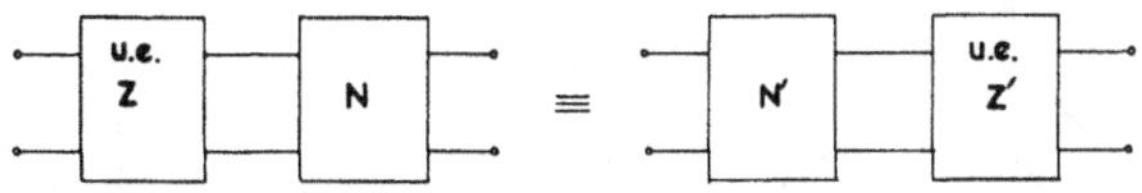

Fig. 3. Generalized Kuroda transformation.

(a) (b)

Fig. 4. Special case of a shunt Foster:

$$Z' = \frac{Z(1 + LC)}{1 + C(Z + L)}$$
$$C' = \frac{LC^2}{1 + LC}$$
$$M = \frac{1 + LC}{C}$$
$$L_p = \frac{1 + C(Z + L)}{C}$$
$$L_s = \frac{(1 + LC)^2}{C[1 + C(Z + L)]}.$$

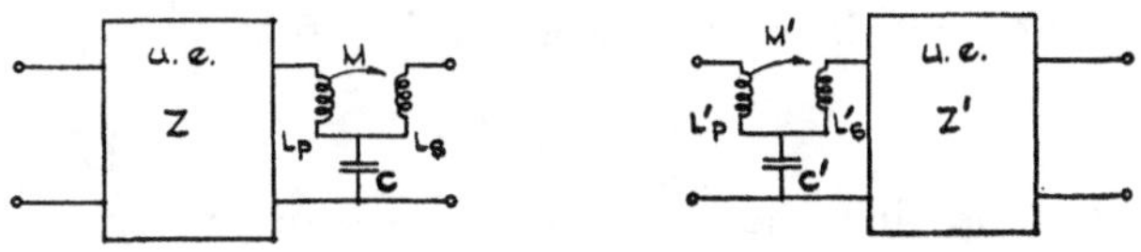

Fig. 5. The generalized Kuroda transformation applied to a Brune section:

$$Z' = \frac{Z(L_p + M^2C) + (L_p - M)^2}{L_p[1 + C(Z + L_p)]}$$
$$C' = C + \frac{1}{Z} - \frac{1}{Z'}$$
$$M'C' = MC$$
$$L_p' = \frac{ZCM^2}{Z'C'L_p}$$
$$L_s' = \frac{Z'C}{ZC'} \cdot L_p$$
$$M' - L_p' = \frac{MCZ(M - L_p - Z)}{MCZ + M - L_p}.$$

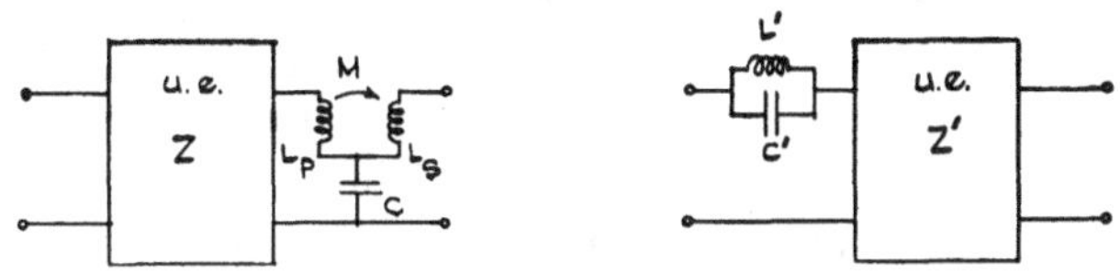

Fig. 6. Transformation of an important degenerate case for

$$Z = \frac{L_p - M}{MC}$$
$$Z' = \frac{Z}{CZ + 1}$$
$$L' = \frac{Z^2C(1 + MC)}{CZ + 1}$$
$$L'C' = MC.$$

equivalent to a series or shunt capacitor or inductor of the lumped element prototype. However, it has now been shown [11] that the simple Kuroda transformations are special cases of a more general transformation, depicted in Fig. 3, and which may be stated as the following theorem:

> A distributed network consisting of a unit element in cascade with a physically realizable two-port may always be replaced by an equivalent circuit consisting of a physically realizable two-port in cascade with a unit element.

Thus, if the network N of Fig. 3 is defined by its transfer (or chain) matrix

$$\begin{bmatrix} A(t) & B(t) \\ C(t) & D(t) \end{bmatrix} \tag{4}$$

where t is defined in (2), then in the equivalent circuit the network N' has the transfer matrix

$$\frac{1}{1-t^2}\begin{bmatrix} A + CZt - \frac{B}{Z'}t - \frac{DZ}{Z'}t^2 & -AZ't - CZZ't^2 + B + DZt \\ C + \frac{A}{Z}t - \frac{D}{Z'}t - \frac{Bt^2}{ZZ'} & -CZ't - \frac{AZ'}{Z}t^2 + D + \frac{B}{Z}t \end{bmatrix} \tag{5}$$

and the unit element Z' is given by

$$Z' = \frac{B(1) + ZD(1)}{A(1) + ZC(1)}. \tag{6}$$

The reader is referred to the original paper [11] for the proof of (5) and (6).

In the application of this generalized transformation to elliptic-function filters, it is frequently necessary to transform a unit element across a distributed series resonant circuit in shunt (i.e., a shunt Foster), as shown in Fig. 4(a). The result of the transformation is a microwave Brune section in cascade with a unit element [5], [11], as shown in Fig. 4(b).[1] The parameters of the Brune section are given in Fig. 4, using the notation of Youla [12].

If a further transformation of a unit element is required, this will be across the microwave Brune section, which is again accomplished by means of the generalized transformation. It is easy to show that the transformation of a unit element across a Brune section results in another Brune section, as shown in Fig. 5. It is evident that this depicts a more general case than that shown in Fig. 4, where the shunt Foster of Fig. 4(a) is to be regarded as a degenerate Brune section. The equations in Fig. 4 may be derived from those of Fig. 5 by substitut-

[1] This particular transformation was first formulated by Kuroda, but published only in an internal report of a Japanese university.

ing the conditions for degeneracy, i.e., that

$$L_p = L_s = M. \tag{7}$$

In applying the generalized transformation to the Brune section the resulting Brune section is frequently degenerate, i.e., the reverse of Fig. 4 may occur. However, if the condition

$$L_p = M(CZ + 1) \tag{8}$$

occurs, then inspection of the equations in Fig. 5 shows that $C' \to 0$, while M', L_p', and $L_s' \to \infty$ in such a way that the product $M'C' = MC$. This is because the Brune section degenerates not to a shunt Foster but to a series Foster, as shown in Fig. 6, where the equations to be applied in this case are presented.

IV. Realizations of Microwave Brune Sections

A considerable amount of work on the physical realization of microwave C, Brune, and D sections has been accomplished by several workers in Japan [13], [16]. Only Brune sections are required in the realization of elliptic-function filters, since the latter has only real-frequency transmission zeros [12]. At present, no realization exists for the microwave Brune section alone, but there are several realizations for the microwave Brune section in cascade with a unit element, i.e., for the network of Fig. 4(b). The earliest such realization appears to be the Ikeno loop [13], which is shown in Fig. 7. However, this is restricted to the cases where,

$$L_s > L_p, \text{ (or } M > L_p) \tag{9}$$

$$Z > M - L_p. \tag{10}$$

Even when these conditions are satisfied, it may not be
162 possible to construct the loop consisting of four sections of differing impedances because of impossible impedance levels.

A realization which may be more suitable than that of Ikeno was given by Saito [14], as shown in Fig. 8. Again, the impedance levels are likely to be awkward in many instances, and the restrictive conditions which apply are identical to those of the Ikeno loop, [(9), (10)].

The restrictions on the physical realization of Brune sections, as given by (9) and (10), were released to some extent by Matsumoto [14], who has given six types of realizations using three-wire lines. The formulas given by Matsumoto contain minor errors, which have been corrected by Rhodes and Scanlan [17]. In fact, it is not possible to realize a Brune section in cascade with a unit element of arbitrary impedance by a choice of one of Matsumoto's configurations. In practice it is found that some of his networks are difficult to construct, while the three-wire configuration is difficult to calculate as far as the self- and mutual-capacitance values of the wires are concerned. It has been found more convenient, therefore, to use new physical realizations as described by Rhodes, Scanlan, and Levy [18].

The first of these to be discovered was the series

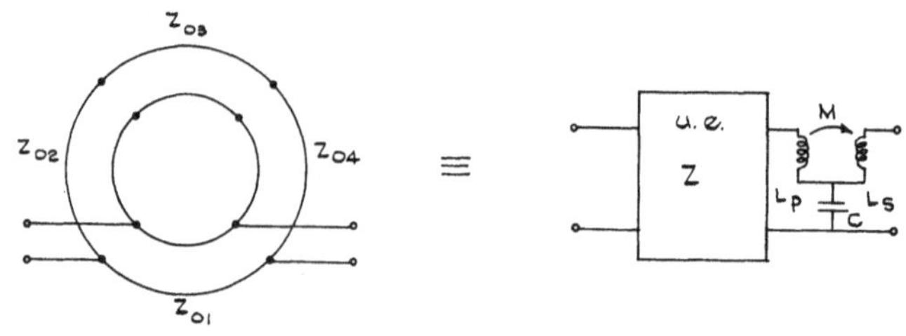

Fig. 7. Ikeno Loop. Restrictive Conditions: $M > L_p$ and $Z > M - L_p$

$$Z_{01} = \frac{(M - L_p)^2 + Z(L_p + M^2C)}{L_p(1 + MC)}$$

$$Z_{02} = \frac{\lfloor Z(M - L_p)^2 + Z(L_p + M^2C)\rfloor}{(M - L_p)(M - L_p + MCZ)}$$

$$Z_{03} = \frac{(M - L_p)^2 + Z(L_p + M^2C)}{MC(M - L_p + MCZ)}$$

$$Z_{04} = \frac{(M - L_p)^2 + Z(L_p + M^2C)}{L_pC(Z - M + L_p)}.$$

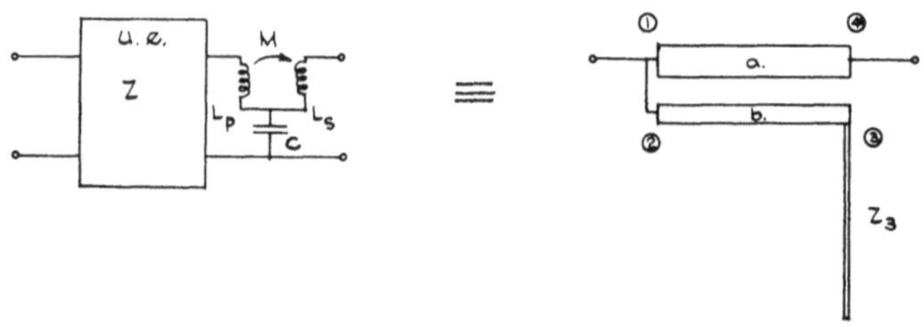

Fig. 8. Saito section. Restrictive Conditions: $M > L_p$ and $Z > M - L_p$

$$vC_a = \frac{L_p}{MZ} \qquad vC_b = \frac{M - L_p}{MZ}$$

$$vC_{ab} = \frac{L_p(Z - M + L_p)}{M^2 Z} \qquad Z_3 = \frac{1}{vC}$$

where the even and odd mode impedances are given in terms of the self and mutual capacitances of the coupled bars by the equations.

$$Z_{oe}^a = \frac{C_b + 2C_{ab}}{vF} \qquad Z_{oo}^a = \frac{C_b}{vF}$$

$$Z_{oe}^b = \frac{C_a + 2C_{ab}}{vF} \qquad Z_{oo}^b = \frac{C_a}{vF}$$

$$F = \begin{vmatrix} C_a + C_{ab} & -C_{ab} \\ -C_{ab} & C_b + C_{ab} \end{vmatrix}.$$

realization, shown in Fig. 9. The case with short-circuited stubs, Fig. 9(a), applies for $M > L_p$, and has restrictive conditons identical to those of the Ikeno loop or the Saito section. With open-circuited stubs, Fig. 9(b), cases for which $M < L_p$ may be realized, subject to a restrictive condition on Z_0 which may not be larger than $(L_p - M)/MC$. These series realizations were derived by forming the impedance matrix of the unit element/Brune section cascade, extracting the series line Z by one of two possible methods, and then synthesizing the remaining Z matrix to give the Pi network formed by Z_2, Z_3, and Z_4. It will be found that the circuits of Fig. 9 are particularly useful for filters with broad bandwidths where tight coupling of the Pi network to the series line is required. However, in cases of narrow bandwidths, the impedances Z_2 and Z_4 take extreme values. A more useful realization then is to use coupled bars and stubs,

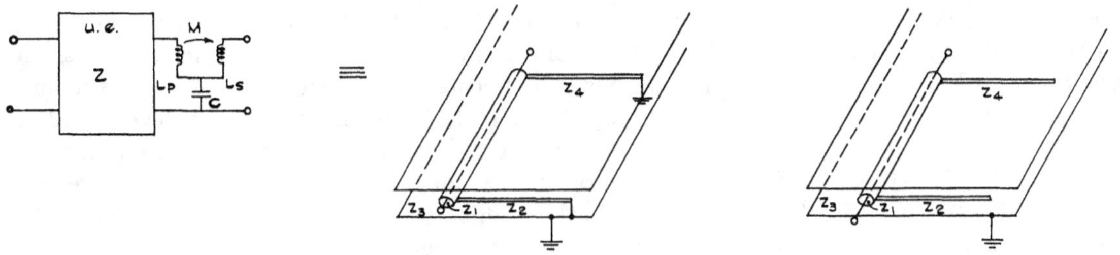

For $M > L_p \quad Z > M - L_p$

$$Z_1 = \frac{Z}{(1 + ZC)}$$

$$Z_2 = \frac{Z^2C(ZMC + M - L_p)}{(CZ + 1)^2(M - L_p)}$$

$$Z_3 = \frac{Z(ZMC + M - L_p)}{L_p(CZ + 1)}$$

$$Z_4 = \frac{Z(ZMC + M - L_p)^2}{L_p(1 + CZ)(L_p + Z + M)}$$

(a)

For $M < L_p \quad Z < (L_p - M)/MC$

$$Z_1 = \frac{MZ}{L_p}$$

$$Z_2 = \frac{(L_p - M + Z)Z}{L_p}$$

$$Z_3 = \frac{(L_p - M)(L_p - M + Z)}{L_p}$$

$$Z_4 = \frac{(L_p - M)(L_p - M - MCZ)}{L_p^2C}$$

(b)

Fig. 9. Series realization of a cascade of a unit element and a Brune section.

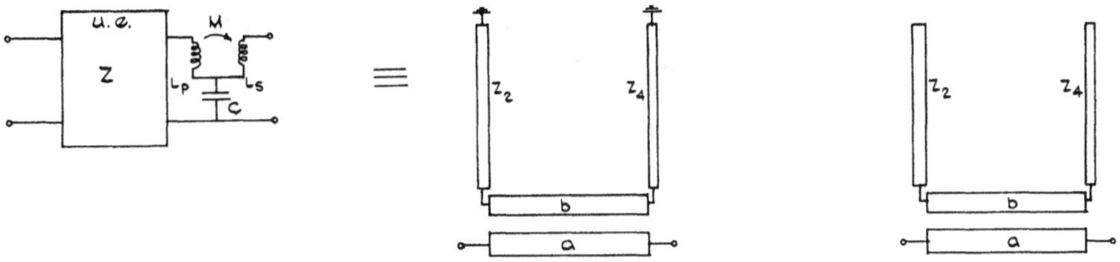

$M > L_p. \quad Z > M - L_p.$ $\qquad$ $M < L_p. \quad Z < \dfrac{L_p - M}{MC}$

$$Z_b = \frac{Z_{oe}{}^b + Z_{oo}{}^b}{2}, \text{ is a parameter}$$

(a):

$$\frac{Z_{oe}{}^a + Z_{oo}{}^a}{2} = \frac{MZ}{L_p}$$

$$\frac{Z_{oe}{}^a - Z_{oo}{}^a}{2} = \frac{Z_{oe}{}^b - Z_{oo}{}^b}{2} = \sqrt{\frac{(MCZ + M - L_p)ZZ_b}{L_p(CZ + 1)}}$$

$$Z_2 = \frac{CL_pZ}{(M - L_p)(CZ + 1)} \cdot Z_b$$

$$Z_4 = \frac{(MCZ + M - L_p)}{(Z + L_p - M)(CZ + 1)} \cdot Z_b$$

(a)

(b):

$$\frac{Z_{oe}{}^a + Z_{oo}{}^a}{2} = Z + \frac{(L_p - M)}{L_p}$$

$$\frac{Z_{oe}{}^a - Z_{oo}{}^a}{2} = \frac{Z_{oe}{}^b - Z_{oe}{}^b}{2} = \sqrt{\frac{(L_p - M)(Z + L_p - M)Z_b}{L_p}}$$

$$Z_2 = \frac{Z}{(L_p - M)} \cdot Z_b$$

$$Z_4 = \frac{(L_p - M - MCZ)}{L_pC(L_p - M + Z)} \cdot Z_b$$

(b)

Fig. 10. Parallel-coupled line realization of a unit element/Brune section cascade.

as shown in the two circuits of Fig. 10. Originally these were derived independently by analysis, but it is readily shown that the series circuits of Fig. 9 are special cases of those of Fig. 10. Thus Fig. 9(a) is derived by making the following substitution for Z_b in Fig. 10(a):

$$Z_b = \frac{MCZ + M - L_p}{L_p(CZ + 1)} Z. \tag{11}$$

Similarly, Fig. 10(b) reduces to Fig. 9(b) with the substitution

$$Z_b = \frac{(L_p - M)(Z + L_p - M)}{L_p} \cdot \tag{12}$$

From the restrictive conditions on Z it may appear that these coupled-line circuits are of limited use. However, in practice the restrictions are not serious, since Scanlan and Rhodes [17] have shown that, given a prescribed physically realizable driving point impedance with transmission zeros all at real frequencies, it is always possible to find a physical realization consisting of a cascade of unit elements and Brune sections satisfying these restrictions (which appear to be basic). Thus, in a synthesis procedure it may not be possible to extract a Brune section at some stages, but by first extracting one or more unit elements the eventual extraction of the Brune section for the particular transmission zero is guaranteed. In practice it is rarely necessary to extract more than one or two unit elements before the Brune section may be realized.

Examples showing how these unit element/Brune section structures are used in the design of elliptic-function microwave filters will now be given. The most important requirement is that the impedances of all lines within the filters have values which can be readily made in practice, i.e., lying in the range 10Ω to 200Ω. In this context it should be noted that physical realizability criteria in the strict circuit theory sense are not sufficient for actual practical realizations because they do not exclude structures with impossible impedance levels.

V. Examples of Elliptic-Function Microwave Filter Design Using Brune Sections

A. Band-Stop Filter from Single-Pole Prototype Having Approximately 6 Percent Bandwidth

The low-pass prototype taken from Saal's tables [2] is type number C04 15c for a modular angle $\theta = 31°$, having a low-pass ripple level corresponding to a reflection coefficient of 0.15, and a minimum stop band attenuation of 40.2 dB, as shown in Fig. 11(a), the prototype circuit being shown in Fig. 11(b). If Richards' transformation is applied directly to a low-pass prototype, then the resulting distributed filter is of band-stop character, with the stop bands repeating at harmonic frequencies. However if a narrow stop bandwidth is required, then it has been found necessary to transform the low-pass prototype into a lumped-element band-stop filter before applying Richards' transformation. This prevents the occurrence of lines with extreme values of impedance. The characteristic of the band-stop filter is shown in Fig. 11(c), and the circuit is shown in Fig. 11(d), having applied the frequency transformation

$$\omega \rightarrow \frac{1}{\delta\left(\omega - \dfrac{1}{\omega}\right)}$$

where

$$\delta = \left| \frac{\omega_2}{\sqrt{\omega_1\omega_2}} - \frac{\sqrt{\omega_1\omega_2}}{\omega_2} \right|, \tag{13}$$

ω_1, ω_2 being the band-stop edge frequencies shown in Fig. 11(c) and, in the present instance, $\delta = 0.1$. It will be noted that the band-stop filter is a cascade of shunt and series Foster sections, i.e., of degenerate Brune sections. When Richards' transformation is applied to this lumped-element filter, taking the electrical length at the center of the stop band as $\phi_0 = 45°$, [Fig. 11(e)], it is seen that the stop bandwidth of the first harmonic stop band, defined to the pass band ripple level is slightly greater than 6 percent. It is now necessary to transform unit elements from source and load ends of the filter to give a practical realization. The fact that the filter consists of a cascade of degenerate Brune sections facilitates this process. The distributed circuit resulting when one unit element is transformed from the left-hand side and two unit elements from the right-hand side of Fig. 11(d) are shown in Fig. 11(f). The transformations depicted in Figs. 4 and 6 are used in this example. The realization in the coupled-line form with stubs is shown in Fig. 11(g), which is obtained from Fig. 11(f) using the equivalence of Fig. 10(a), and also the equivalence of Fig. 12, which is used to realize the series Foster/unit element combination at the right-hand end of the network. Note that unit elements of value $Z_0 = 1$ external to the filter at both source and load ends were used in the transformations. The normalized impedances of all stubs lie in the range 0.896 to 1.09, the impedances of the coupled lines are also approximately unity, and the coupling coefficients of the parallel-coupled line sections are easily realizable. This filter is illustrated in Fig. 13, and the measured response in Fig. 14.

Using this technique of resonating the prototype filter, stop band filters for any bandwidth from 0.1 percent or less to 100 percent may be realized (Section VII).

B. An Octave-Bandwidth Band-Pass Filter Derived From a Double-Pole Prototype

The low-pass prototype is shown in Fig. 15(a), and has the response given in Fig. 15(b). In order to form a band-pass response in the distributed case, this must be converted to a high-pass structure, and to give a bandwidth of one octave, it is necessary to arrange the cutoff frequency to occur at an electrical length $\phi_0 = 60°$. The transformation to be applied to Fig. 15(a) is thus

$$s = \frac{\tan \phi_0}{t} \tag{14}$$

which gives the pseudo high-pass circuit of Fig. 15(c) in the t-variable. The response in the distributed domain is an octave bandwidth band-pass filter with harmonics, as shown in Fig. 15(d).

A convenient way to apply the general transformation to a case of this type is to combine the three shunt inductors using the transformation shown in Fig. 16. This gives an exact equivalent of a shunt Foster in cascade with an inductance and an inductance in cascade with a Brune section. This is a degenerate case of the interchange of a Brune section with a shunt inductance, shown in Fig. 17, which may be of use in some cases. When the result of Fig. 16 is applied to the circuit of Fig. 15(c), Fig. 15(e) results. Unit elements must now be transferred into the network across the Brune sections, and the result of transferring one unit element from the left and two from the right is given in Fig. 15(f). It is necessary to transform two unit elements over the right-hand Brune section because it is not possible to realize the Brune section in cascade with a unit element at the right (having $Z = 1$) after only one transfer. The next problem is to eliminate the ideal transformer at the left, and to obtain realizable impedances. This is achieved by dividing the inductance in a suitable optimum way, and transforming the various component parts across the available unit elements. A convenient way of carrying out this process is by use of the capacitance matrix transformation described by Wenzel [19] which is simpler than the repeated application of the Kuroda identities, particularly when several unit elements are involved. One result of this process is given by the circuit of Fig. 15(g). This could be realized in practice with some difficulty; the difficulty arising because the inductance of value 0.1958 is rather low. To eliminate this

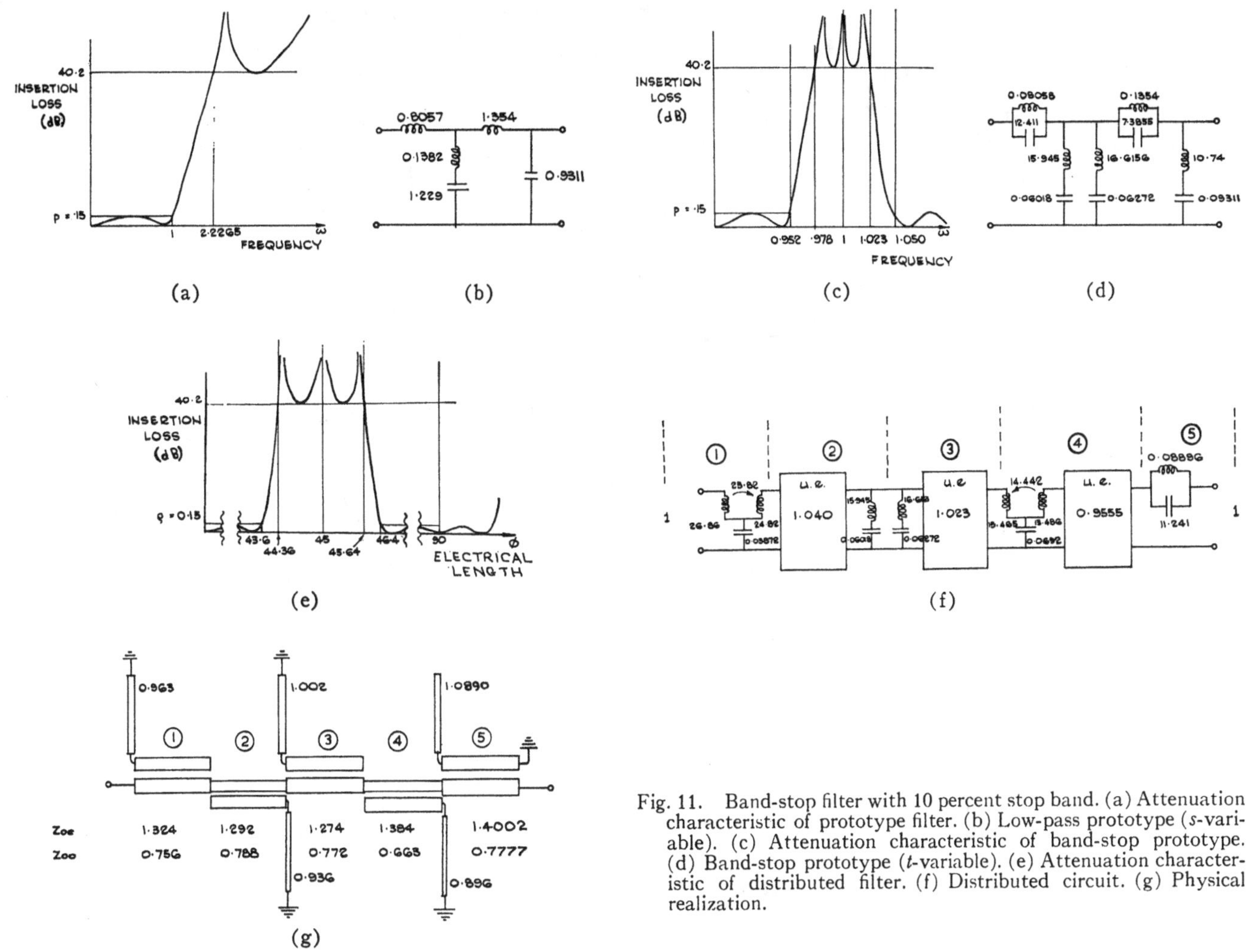

Fig. 11. Band-stop filter with 10 percent stop band. (a) Attenuation characteristic of prototype filter. (b) Low-pass prototype (s-variable). (c) Attenuation characteristic of band-stop prototype. (d) Band-stop prototype (t-variable). (e) Attenuation characteristic of distributed filter. (f) Distributed circuit. (g) Physical realization.

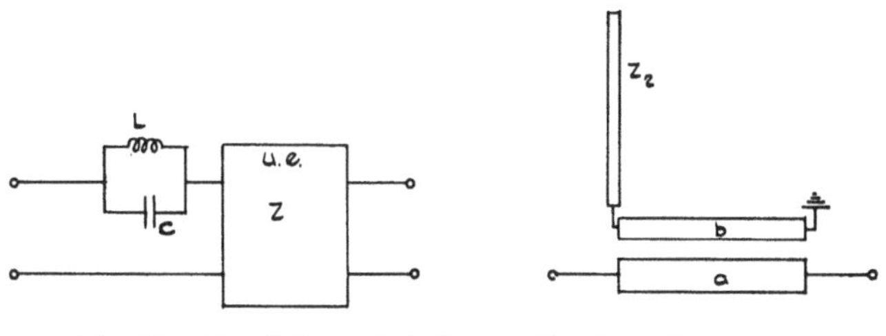

Fig. 12. Parallel-coupled line realization of a series Foster/unit element cascade.

$$Z_b = \frac{Z_{oe}^b + Z_{oo}^b}{2} \text{ is a parameter}$$

$$Z_a = \frac{Z_{oe}^a + Z_{oo}^a}{2} = Z + \frac{L^2C}{1 + LC}$$

$$\frac{Z_{oe}^a - Z_{oo}^a}{2} = \frac{Z_{oe}^b - Z_{oo}^b}{2} = \sqrt{L\left(\frac{Z}{L + Z\,(1 + LC)} + \frac{LC}{1 + LC}\right)\cdot Z_b}$$

$$Z_2 = \frac{(1 + LC)Z\,Z_b}{LC[L + Z\,(1 + LC)]}$$

Fig. 13. Band-stop filter with one ground plane removed.

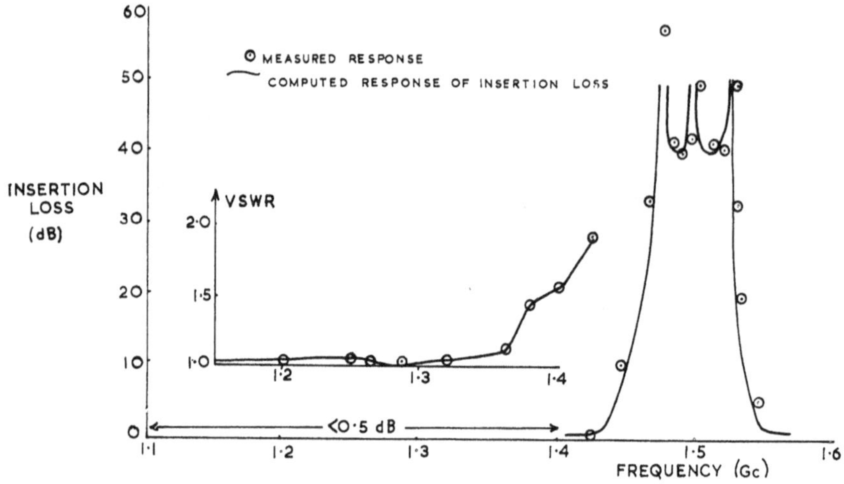

Fig. 14. Measured and computed characteristics of the band-stop filter shown in Figs. (11) and (13).

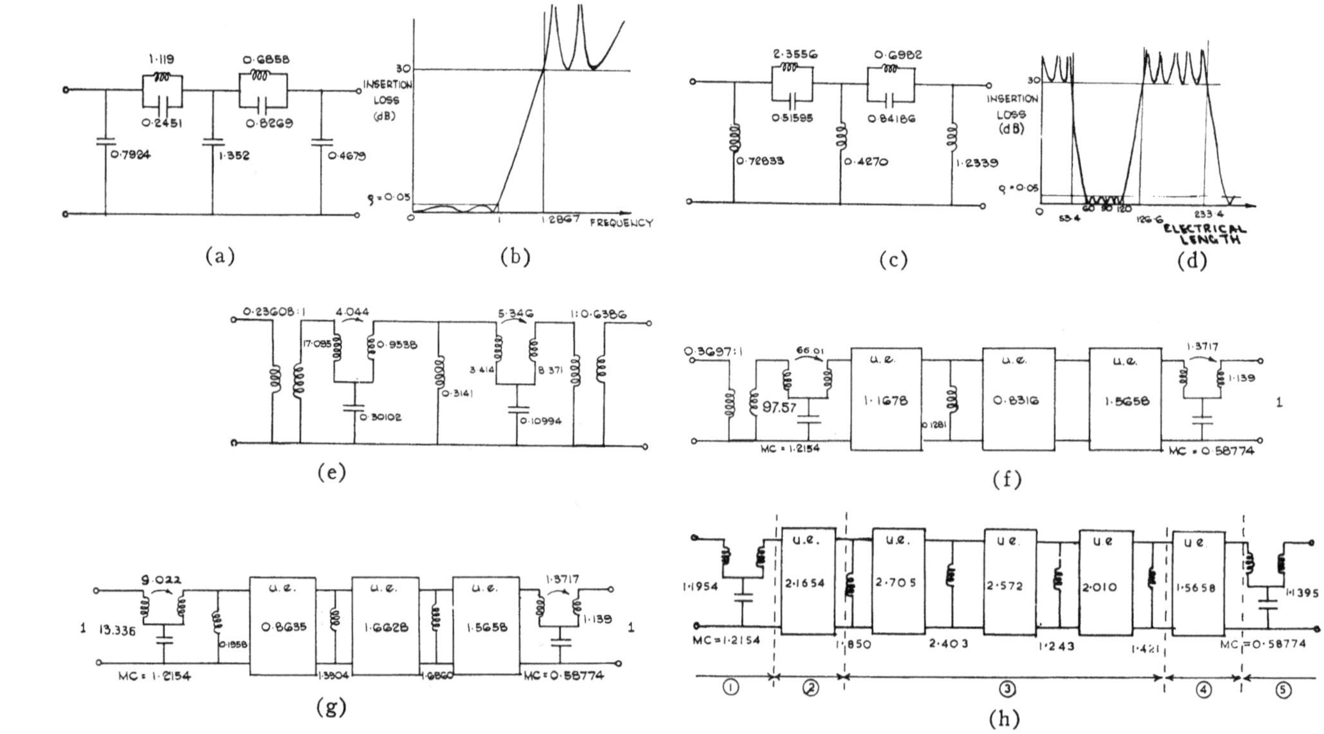

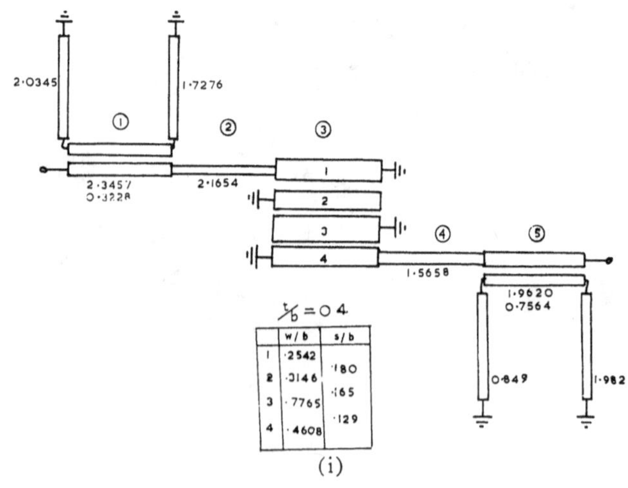

Fig. 15. Derivation of two-pole octave bandwidth filter. (a) Prototype low-pass filter (*s*-variable). (b) Attenuation characteristic of low-pass prototype. (c) Band-pass circuit (*t*-variable). (d) Attenuation characteristic of band-pass circuit. (e) Transference of inductors into center. (f) Introduction of 3 unit elements. (g) Removal of ideal transformers. (h) Introduction of two unit elements from LHS, and re-arrangement of inductors. (i) Physical realization of (h).

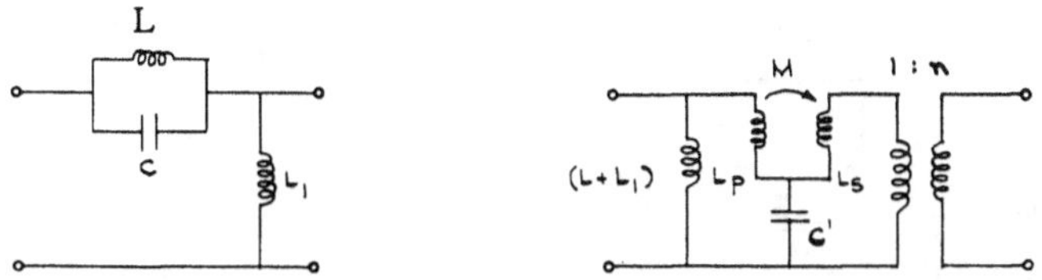

Fig. 16. Interchange of series Foster and an inductance:

$$M = \frac{(L + L_1)^2}{L}$$
$$L_p = \frac{L_1(L + L_1)}{L}$$
$$MC' = LC$$
$$n = \frac{L_1}{L + L_1}.$$

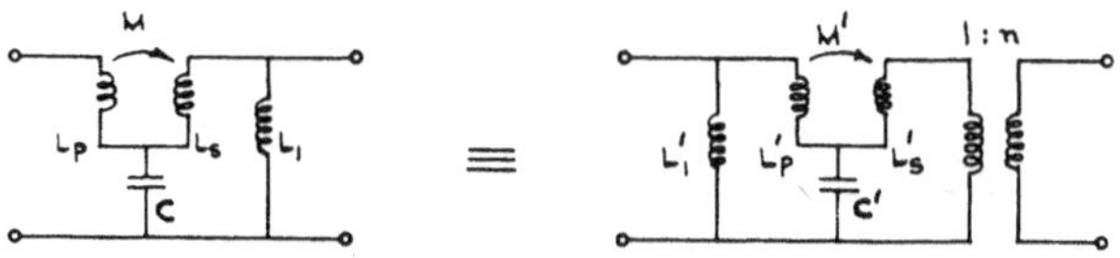

Fig. 17. Interchange of a Brune section and an inductance:

$$L_1' = L_1 + \frac{(L_p - M)^2}{L_p}$$
$$C' = \left(\frac{L_1'(L_1 + L_s)}{L_1^2} - \frac{L_p}{L_1}\right) C$$
$$M'C' = MC$$
$$n = L_1/L_1'$$
$$L_p'C' = nCL_p \qquad L_s'C' = CL_1'/n.$$

difficulty it was decided to introduce a certain degree of redundancy by transforming two additional unit elements across the left-hand Brune section. When the inductances had been redistributed the network of Fig. 15(h) resulted, which may be realized very easily as shown in Fig. 15(i). Note that the parallel-coupled line, section one, may be realized conveniently using the re-entrant line technique described by Cohn [20]. The values of ω/b and s/b for the interdigital section three are derived using Wenzel's results [19], the configuration used being rectangular coupled bars which are treated accurately by Getsinger [21].

The final configuration is somewhat bulky and inconvenient, and with further development might eventually be realized in more compact form, e.g., as an interdigital filter with stubs. Alternatively, the method of Horton and Wenzel [6], which is limited to broadband filters, would give a compact structure in the present instance.

VI. Alternative Procedure Using the Pi-Equivalent of a Brune-Section

As an alternative to the direct realization of the distributed section, it is occasionally possible to convert the Brune section into its Pi equivalent shown in Fig. 18, which consists of a series Foster and two shunt capacitors, one of the latter being negative. The negative capacitor may often be absorbed into an adjacent shunt capacitor of greater value.

As an example, a distributed pseudo low-pass filter with a stop band attenuation A_s of greater than 27.1 dB was designed. This rather low value of attenuation was chosen to simplify measurements on the filter. (A filter with $A_s > 60$ dB was also designed, and achieved the predicted rate of cutoff, but it was more difficult to establish the maxima and minima of stop band attenuation in this case.) The pass band ripple reflection coefficient was 0.20. The low-pass prototype is shown in Fig. 19(a), with the response characteristic of Fig. 19(b). The low-pass cutoff frequency for the distributed filter is shown at an electrical length $\phi_0 = 41°$, giving the characteristic shown in Fig. 19(c). Fig. 19(d) indicates the transformation of a unit element into the center of the filter, and the signs of the resulting shunt capacitors are indicated. Elimination of the negative capacitor C_2 is possible if C_1 is of greater magnitude, which is true for the prototype and cutoff electrical length ϕ_0 chosen. (It is usually true if the value of ϕ_0 is chosen to be near 45°). This process is repeated until five unit elements have been transformed, giving the circuit of Fig. 19(e). Note that the admittance of the first shunt stub is extremely low and, in practice, may be disregarded.

The physical realization using slab-line [22] is shown in Fig. 20, where the shunt Fosters are realized directly using the equivalence given in Fig. 21. The comparison between theory and experiment is shown in Fig. 22. No attempt was made to allow for fringing capacitances at the ends of the open-circuit stubs and no tuning was attempted, so that the results may be considered quite encouraging.

VII. Bandwidth Limitations of the Various Filter Designs

Investigations of the range of bandwidths which result in realizable filter designs have been carried out for some cases using a computer, and the results are presented in Table I. This shows that band-stop filters of the single-pole type are realizable over a wide range, and double-pole filters over a more restricted range of 167
bandwidths. Note that Table I does not present a complete picture even for these cases since it considers only particular forms of realization, i.e., with only the coupled bar type of unit element/Brune sections, and with a specified number of unit elements. By considering other Brune section realizations and by introduction of further unit elements, it is often possible to extend the range of realizable bandwidths. In fact, it would be difficult to present a complete picture because of the large number of circuits which exist.

Results are not yet available for band-pass filters of the type presented in Section V-B. However, it appears that the method described in that section, i.e., using an unresonated prototype, is suitable only for broadband filters for which, as stated previously, the method of Horton and Wenzel [6] is possibly more appropriate. It is expected that filters based on a resonated prototype would be realizable for narrow bandwidths as in the case of band-stop filters.

The double-pole resonated band-stop filter design of Table I, i.e., case (b), requires application of an approximation to a nearly degenerate Brune section. This

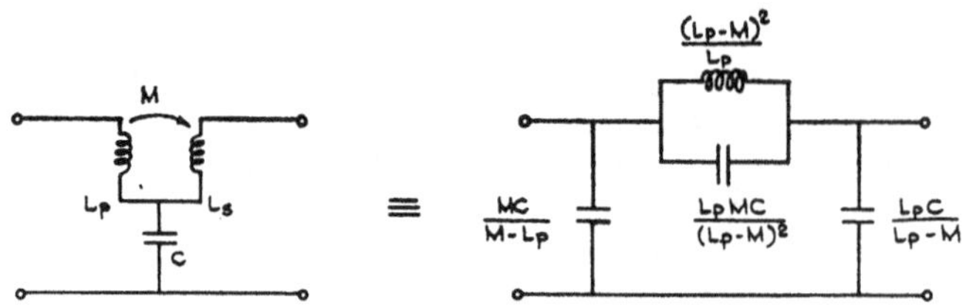

Fig. 18. Pi equivalent of a Brune section.

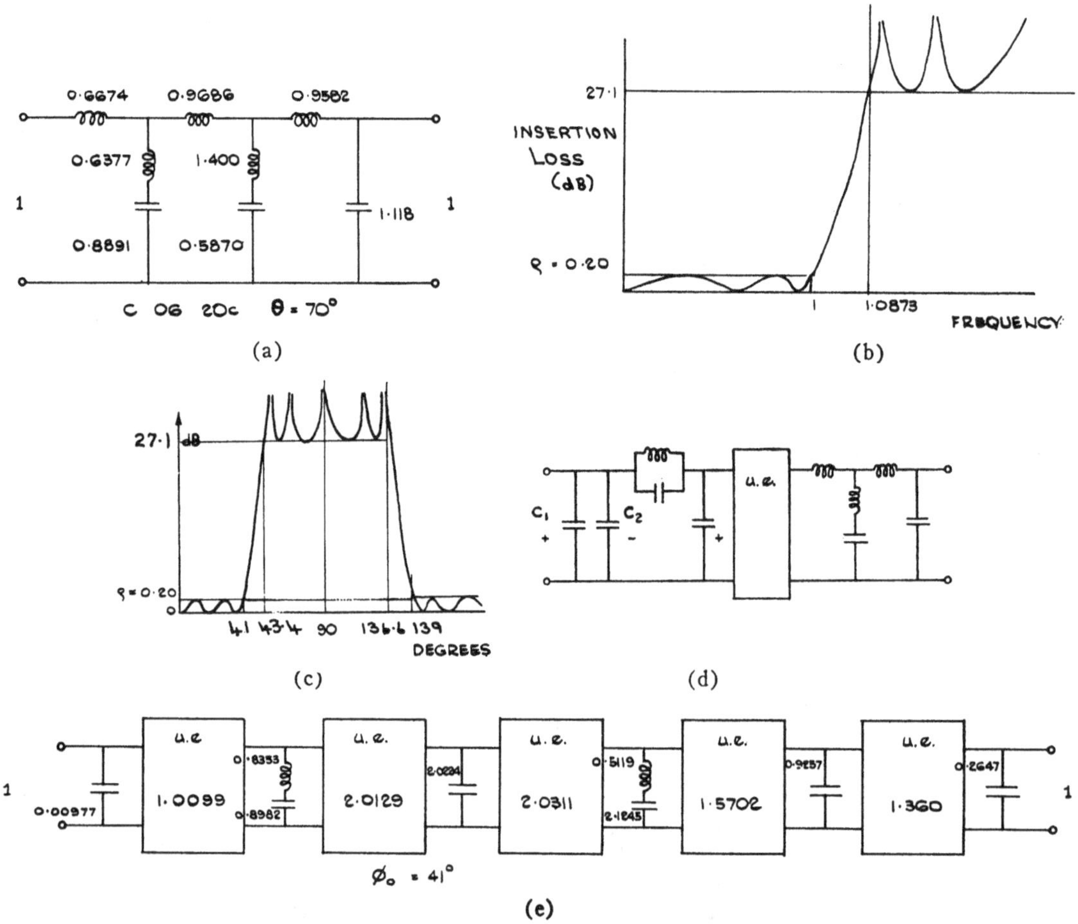

Fig. 19. Derivation of low-pass filter without Brune sections. (a) Low-pass prototype. (b) Attenuation characteristic of low-pass prototype (c) Attenuation characteristic of distributed circuit. (d) Transformation of one unit element. (e) Transformation of five unit elements.

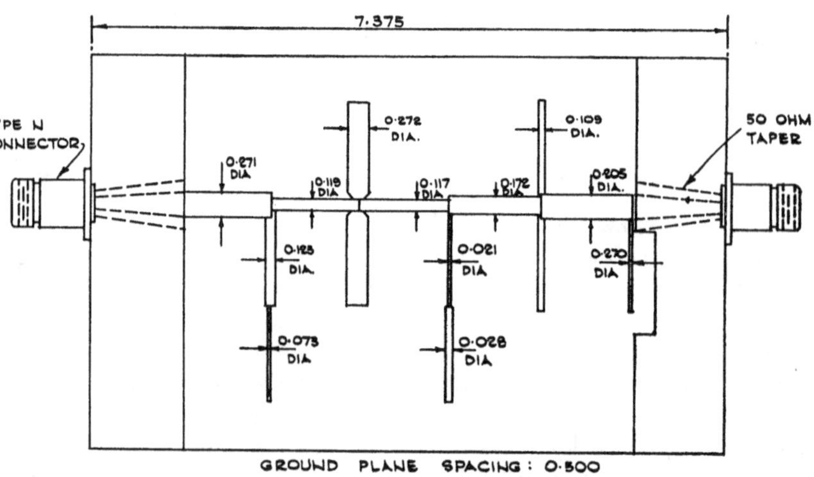

Fig. 20. Sketch of slab-line realization of Fig. 19(c).

168

Fig. 21. Equivalence of a distributed shunt Foster.

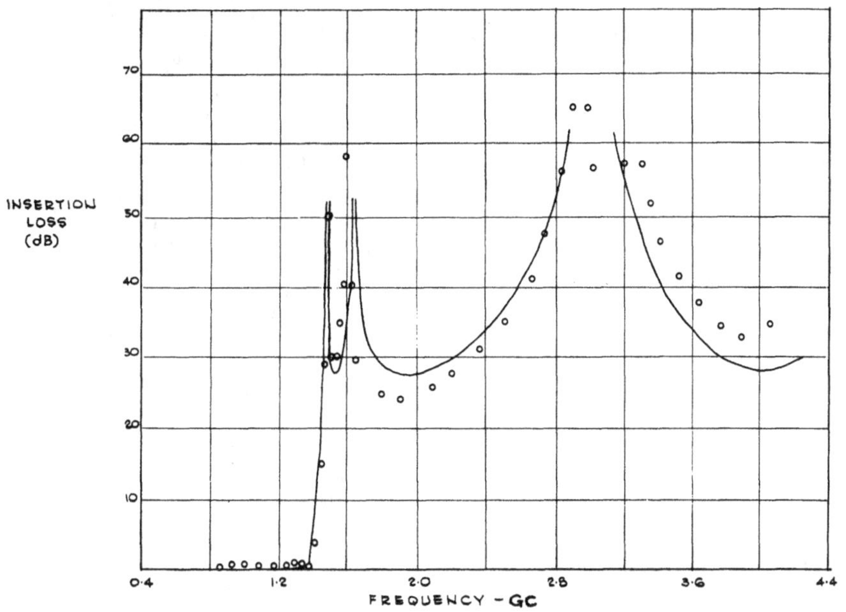

Fig. 22. Measured and computed attenuation characteristic of filter of Fig. 20.

169

TABLE I

REALIZABLE BAND-STOP FILTERS WITH COUPLED-BAR TYPE BRUNE SECTIONS (FIG. 10)

Type	Stop-band attenuation level, A_s (dB)	Pass-band reflection coefficient, ρ	Approximations required	Range of impedances occurring within the filter (Ω)	Range of bandwidths realizable
(a) Single-pole resonated prototype band-stop. See Fig. 11.	30–75	0.01–0.25	None	12–150	0–100%
(b) Double-pole resonated prototype band-stop—similar to Fig. 11 but having seven sections in cascade.	75	0.01	Nearly-degenerate Brune sections with one very low impedance stub occur, and are approximated by degenerate sections as described in the text.	Always approximately 50 Ω	0–15%
	30	0.01			0–5%
	60	0.15			0–6%
	30	0.15			0–4%

applies when (8) is very nearly satisfied, i.e., when $(L_p - M - MCZ_0)$ is positive but near zero. The stub impedance Z_4 of Fig. 10(b) is then very low, so that the Brune section is nearly degenerate and may be approximated by the realization shown in Fig. 12. In the exact Brune section M, L_p, and L_s are large and have values which are approximately equal, and C is small. MC has a value close to unity, i.e.,

$$MC \simeq L_pC \simeq L_sC \simeq 1. \tag{15}$$

The transfer matrix of a Brune section is

$$\frac{1}{1 + MCt^2}\begin{bmatrix} 1 + L_pCt^2 & \dfrac{(L_p - M)^2}{L_p}t \\ Ct & 1 + L_sCt^2 \end{bmatrix}. \tag{16}$$

In the nearly degenerate case this may be well-approximated by the matrix

$$\frac{1}{1 + L'C't^2}\begin{bmatrix} 1 + L'C't^2 & L't \\ 0 & 1 + L'C't^2 \end{bmatrix} \tag{17}$$

where

$$L' = \frac{(L_p - M)^2}{L_p}$$
$$L'C' = MC \tag{18}$$

giving a transmission zero at the same frequency. Matrix (17) is realized by the degenerate Brune section of Fig. 12.

VIII. Conclusions

Previously, most designs for microwave filters have been based on Butterworth or Chebyshev insertion loss functions. It has been shown how to design microwave filters with elliptic-function response, which has an advantage of giving poles of attenuation near the cutoff frequencies, and hence faster rates of cutoff than Chebyshev filters of equal degree and equal band-pass insertion loss. The design procedure commences from a lumped-constant prototype filter for which comprehensive tables of element values have been previously published [2]. This is standard practice in many existing microwave filter design methods. A generalization of the well-known Kuroda's identities is used to introduce unit elements (lengths of transmission-line in cascade) into the filter in order to separate the various stubs and other elements of the filter.

In general, it is found that distributed forms of the Brune section are required in order to realize the filters, and several such realizations have been presented. It has been shown that practical configurations result for both narrow and broadband filters of pseudo low-pass or pseudo high-pass character.

Reasonably good agreement with theory has been obtained for practical filters. In addition to the special case of elliptic function types, the method would be applicable to general filter characteristics.

Acknowledgment

The authors are grateful to J. O. Scanlan and J. D. Rhodes for valuable discussions in the course of this work.

References

[1] S. Darlington, "Synthesis of reactance 4-poles which produce prescribed insertion loss characteristics," *J. Math. Phys.*, vol. 18, pp. 257–353, September 1939.
[2] R. Saal, *Der Entwurf von filtern mit Hikfe des Kataloges Normieter Tiefpässe.* Backnang/Würtemberg, W. Germany: Telefunken, G.M.B.H.
[3] H. Ozaki and J. Ishii, "Synthesis of transmission-line networks and the design of UHF filters," *IRE Trans. on Circuit Theory*, vol. CT-2, pp. 325–336, December 1955.
[4] N. Saito, "Coupled transmission-line filter," *Sci. Repts. Res. Insts. Tohoku Univ.*, Sendai, Japan, Ser. B-(Elec. Comm.), vol. 14, no. 1, pp. 9–19, 1962.
[5] M. Blanc and M. Soldi, "Nota sul calcolo di filtri con parametri distribuiti," *Alta Frequenza*, vol. 32, pp. 649–657, September 1963.
[6] M. C. Horton and R. J. Wenzel, "Realization of microwave filters with equal ripple response in both pass and stop bands," presented at the 1966 Internat'l Symp. on Generalized Networks, Polytechnic Institute of Brooklyn, Brooklyn, N. Y.
[7] B. M. Schiffman, "A multiharmonic rejection filter designed by an exact method," *IEEE Trans. on Microwave Theory and Techniques*, vol. MTT-12, pp. 512–516, September 1964.
[8] E. N. Torgow and P. D. Lubell, "Band-pass filters with steep skirt selectivity," *IEEE Trans. on Microwave Theory and Techniques*, vol. MTT-13, pp. 124–126, January 1965.
[9] P. I. Richards, "Resistor-transmission-line circuits," *Proc. IRE*, vol. 36, pp. 217–220, February 1948.
[10] H. Ozaki and J. Ishii, "Synthesis of a class of strip-line filters," *IRE Trans. on Circuit Theory*, vol. CT-5, pp. 104–109, June 1958.
[11] R. Levy, "A general equivalent circuit transformation for distributed networks," *IEEE Trans. on Circuit Theory* (*Correspondence*), vol. CT-12, pp. 457–458, September 1965.
[12] D. C. Youla, "A new theory of cascade synthesis," *IRE Trans. on Circuit Theory*, vol. CT-9, pp. 244–260, September 1961.
[13] H. Ozaki and J. Ishii, "On the design of distributed constant networks," *IEEE Trans. on Circuit Theory* (*Reviews of Current Literature*), vol. CT-4, pp. 22–23, March 1957.
[14] *Network Synthesis With Multiwire Lines*, A. Matsumoto, Ed., Monograph series, Research Institute of Applied Electricity, Hokkaido University, Sapporo, Japan, no. 11, 1963.
[15] *Developments in Multiwire Networks*, A. Matsumoto, Ed., Monograph series, Research Institute of Applied Electricity, Hokkaido University, Sapporo, Japan, no. 12, 1964.
[16] G. I. Zysman and A. Matsumoto, "Properties of microwave C-sections," *IEEE Trans. on Circuit Theory*, vol. CT-12, pp. 74–82, March 1965.
[17] J. O. Scanlan and J. D. Rhodes, "Cascade synthesis of distributed networks," presented at the 1966 Internat'l Symp. on Generalized Networks, Polytechnic Institute of Brooklyn, Brooklyn, N. Y.
[18] J. D. Rhodes, J. O. Scanlan and R. Levy, "Realization of microwave Brune sections," *Elec. Lett.*, vol. 1, pp. 268–269, November 1965.
[19] R. J. Wenzel, "Exact theory of interdigital band-pass filters and related coupled structures," *IEEE Trans. on Microwave Theory and Techniques*, vol. MTT-13, pp. 559–575, September 1965.
[20] S. B. Cohn, "The re-entrant cross-section and wide-band 3 db hybrid couplers," *IEEE Trans. on Microwave Theory and Techniques*, vol. MTT-11, pp. 254–258, July 1963.
[21] W. J. Getsinger, "Coupled rectangular bars between parallel plates," *IRE Trans. on Microwave Theory and Techniques*, vol. MTT-10, pp. 65–72, January 1962.
[22] R. M. Chisholm, "The characteristic impedance of trough and slab lines," *IRE Trans. on Microwave Theory and Techniques*, vol. MTT-4, pp. 166–172, July 1956.

The Digital Elliptic Filter—A Compact Sharp-Cutoff Design for Wide Bandstop or Bandpass Requirements

M. C. HORTON, MEMBER, IEEE, AND R. J. WENZEL, MEMBER, IEEE

Abstract—Detailed design procedures are presented for a practical elliptic-function filter form capable of achieving high selectivity in a very compact configuration. This filter form, called the digital elliptic because of its digital construction and elliptic function response, can provide either bandpass or bandstop characteristics. Examples are given to illustrate typical design procedures for both bandpass and bandstop applications. Experimental results are presented for an octave bandstop design.

Introduction

RECENTLY, the application of modern network techniques to distributed lines [1]–[5] has permitted the exact design of microwave TEM filters by using methods employed in the exact synthesis of lumped element filters. The relationship between a lumped element prototype filter and its distributed element counterpart is accomplished by employing Richards' [6] tangent transformation (1). Using this transform, a shorted quarter-wavelength stub is represented as an inductor L and an open stub is represented as a capacitor C. The "unit element," a quarter-wavelength of line, can also be represented and used in the theory but it has no lumped element counterpart.

Several previous papers have covered in detail the design of distributed filters having equal-ripple response, or maximally flat response, in the passband only [1]–[5]. Such lumped element filter prototypes are usually referred to as Chebyshev or Butterworth, respectively. A lumped element prototype which has sharper cutoff characteristics for a given number of sections is the well-known Cauer parameter or elliptic-function filter which exhibits equal-ripple response in both passband and stopband. Tables of element values are available for the simple design of such lumped element filters. This paper presents a simple exact design procedure for the construction of an elliptic-function distributed-line filter using the available tables. The procedure is a specific application of the general theory presented in a previous paper [7]. Because of the finger-like appearance of the structure, the new filter has been termed the digital elliptic. For wide bandwidth designs (greater than 30 percent), either distributed bandpass or bandstop, the digital elliptic filter is an extremely compact microwave structure for attaining sharp cutoff characteristics.

Manuscript received September 6, 1966; revised December 22, 1966. This work was supported by the U. S. Army Electronics Laboratory, Fort Monmouth, N. J., under Contract DA 28-043 AMC-01869(E).

The authors are with The Bendix Corporation, Research Laboratories Division, Southfield, Mich.

The remainder of this paper presents the simple design procedure used in converting from the lumped element low-pass (or high-pass) prototype to the distributed element bandstop (or bandpass) realization, and gives example designs for both bandstop and bandpass filters. Test results are presented for an experimental model of an octave bandstop design. The paper concludes with a summary of the practical advantages and limitations of the digital elliptic filter.

Design Procedure

The low-pass and high-pass lumped element prototypes of an elliptic-function filter are shown in Fig. 1(a) and (b), respectively. Parallel resonant tank circuits in the series branches create transmission zeros, or attenuation peaks, in the stopband of each prototype at finite frequencies as shown in the response characteristics. The approximation problem for filters of this type which give equal-ripple response in both passband and stopband has been solved and the synthesis of appropriate element values may be found in published tables [8]–[11]. Also presented in these tables are the finite frequencies of zero transmission and the ratio of cutoff frequencies between regions of high and low transmission.

Conversion from the lumped element frequency variable Ω to the distributed element frequency variable $\omega=2\pi f$ is accomplished through Richards' transformation,

$$\Omega = \tan \frac{\pi\omega}{2\omega_0} = \tan \frac{\pi f}{2f_0} \tag{1}$$

where $\omega_0=2\pi f_0$ is the frequency for which all distributed TEM lines are one quarter-wave in length. This transformation involves only a ratio of distributed frequencies, i.e., $\omega/\omega_0=f/f_0$. Since most desired response characteristics are specified in terms of f rather than ω, the subsequent design of digital elliptic filters, including characteristics and equations, will be given in terms of f. The entire positive frequency range in Ω is mapped by the transformation into the positive range in f from zero to f_0. Because of the repetitive nature of the tangent function, the response characteristic then repeats on an interval of $2f_0$ producing response "windows" as shown in Fig. 2(a) and (b).

Physically, the conversion using (1) implies that each inductor in the prototype be replaced by a shorted quarter-

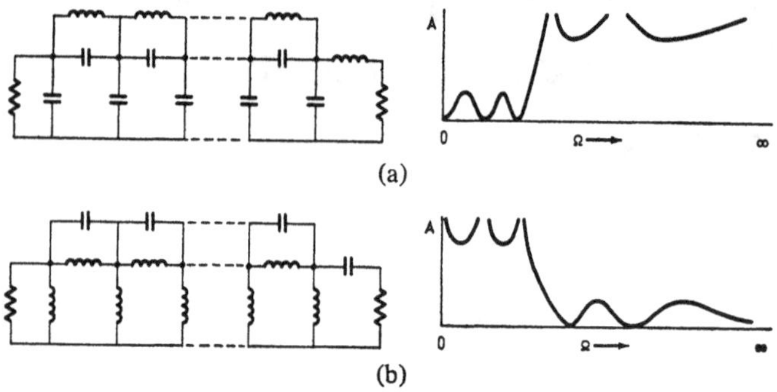

Fig. 1. Elliptic function prototype filter networks and attenuation characteristics: (a) prototype low-pass, and (b) prototype high-pass.

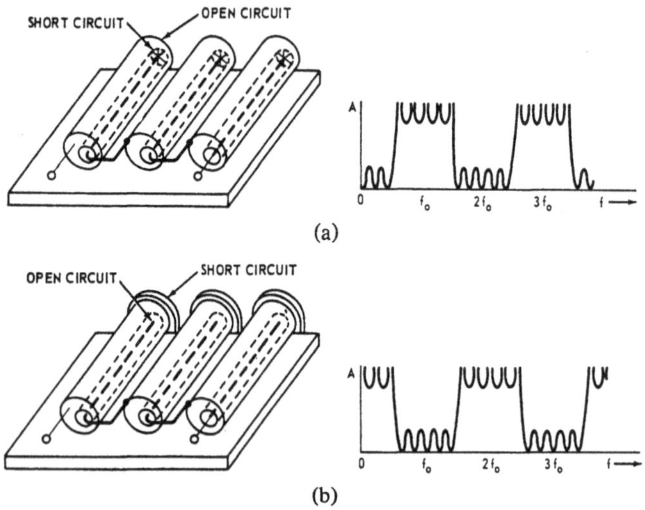

Fig. 2. Digital elliptic microwave filter configurations and responses for (a) distributed bandstop designs, and (b) distributed bandpass designs.

172

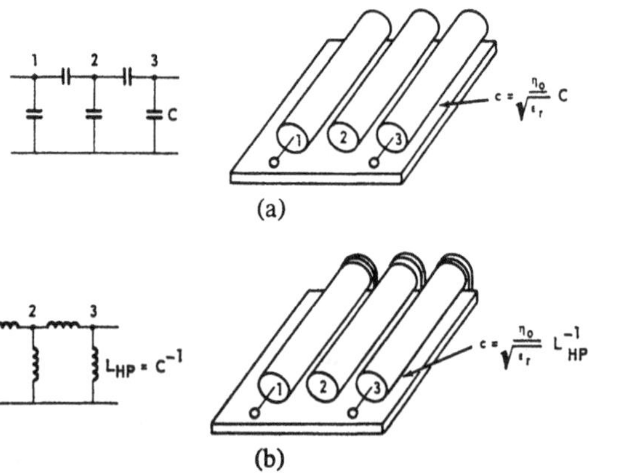

Fig. 3. Parallel conductor distributed representations of transformed lumped element prototype networks: (a) capacitive, and (b) inductive.

wave stub of normalized characteristic impedance given by the inductive prototype value, and that each capacitor be replaced by an open quarter-wave stub of normalized characteristic admittance given by the capacitive prototype value. Concentrating first on the low-pass prototype of Fig. 1(a), it has been pointed out [12]–[14] that the capacitive portion of this network is the lumped element representation of a parallel array of quarter-wavelength conductors above a ground plane, one conductor representing each ungrounded node in the prototype, as shown for a three-node network in Fig. 3(a). Further, all lumped element capacitance values, e.g., C in Fig. 3(a), are proportional to the respective "static capacitance" values c of the distributed line[1] such that $c=(\eta/\sqrt{\epsilon_r})C$, where $\eta=376.7/Z_0$ for a filter terminated in Z_0 ohms and ϵ_r is the relative dielectric constant of the medium. Again, glancing at Fig. 1(a), the inductances bridging each node and in series with the output port can be represented in the distributed parallel line system of Fig. 3(a) by incorporating shorted coaxial stubs within each line and attaching each stub center conductor to the next adjacent parallel line, as shown in the digital elliptic bandstop configuration of Fig. 2(a). The repetitious response of this filter with frequency f is the same as the prototype network response mapped by (1).

In similar manner, the inductive portion of the high-pass prototype network of Fig. 1(b) is represented in distributed form by a parallel array of conductors shorted at the far end as in Fig. 3(b). All distributed "static capacitance" values c are now related to the respective prototype inductance values L_{HP} by the proportional relation $c=(\eta/\sqrt{\epsilon_r})L_{HP}^{-1}$. The bridging series capacitors are incorporated internally within the parallel conductors as open stubs. The resulting configuration and response of the digital elliptic bandpass filter is given in Fig. 2(b).

A. Prototype L-P (Microwave Bandstop)

The procedure for obtaining bandstop digital elliptic filter dimensions will be described briefly. Convenient charts [15]–[16] are used to design the prototype capacitive portion, i.e., the parallel conductor array. The bandwidth-normalized values of self and mutual "static capacitance" required for use with the referenced charts are obtained by multiplying the low-pass prototype capacitance values C by $(\eta/\sqrt{\epsilon_r})$, thus

$$c_{\text{chart}} = \frac{376.7}{Z_0\sqrt{\epsilon_r}} C. \tag{2}$$

The characteristic impedances of the shorted series stubs are given in terms of the low-pass prototype inductance values L by

$$\sqrt{\epsilon_r}\, Z_{\text{series stub}} = \sqrt{\epsilon_r}\, Z_0 L, \tag{3}$$

and physical dimensions can be obtained from nomographs [17].

[1] Note that c as defined here is dimensionless, and is actually the ratio of static capacitance between conductors per unit length to the dielectric constant of the medium (this ratio is independent of the dielectric medium and depends only on cross-sectional geometry) [13].

B. Prototype H-P (Microwave Bandpass)

The normalized self and mutual static capacitance values for the prototype inductive portion of the bandpass digital elliptic filter are obtained by dividing $\eta/\sqrt{\epsilon_r}$ by the high-pass prototype inductance values L_{HP} to give

$$C_{\text{chart}} = \frac{376.7}{Z_0\sqrt{\epsilon_r}} L_{HP}^{-1}, \tag{4}$$

and the characteristic impedances of the series open stubs are given in terms of the high-pass prototype capacitance values C_{HP} by

$$\sqrt{\epsilon_r}\, Z_{\text{series stub}} = \sqrt{\epsilon_r}\, Z_0 C_{HP}^{-1}. \tag{5}$$

All dimensions are obtained as in the *L-P* prototype by using the referenced charts and nomographs.

C. Bandwidth Scaling

The tables of prototype elliptic-function filter element values given in Saal and Ulbrich [8], and Saal [9] are normalized to a cutoff frequency of unity, $\Omega_c=1$. Therefore, when a digital elliptic filter is constructed as described above, it will provide a stop or pass bandwidth of 100 percent (3 to 1) centered about the quarter-wave frequency f_0. Frequency scaling to other than 100 percent bandwidth can be achieved by dividing all prototype element values in (2)–(5) by a scaling constant Ω_c', which may be either greater or less than unity. This is equivalent to changing to a new bandwidth-scaled frequency variable Ω' given by

$$\Omega' = \Omega_c'\Omega = \tan\frac{\pi f'}{2f_0}\cdot \tag{6}$$

Values of $\Omega_c'>1$ give rise to a reduced bandwidth in the case of either a bandstop or bandpass filter, and $\Omega_c'<1$ corresponds to increased bandwidth.[2] The scaling constant, from (1), is given by

$$\Omega_c' = \tan\frac{\pi f_{c1}'}{2f_0} = \tan\frac{\pi f_{c1}'}{f_{c1}'+f_{c2}'},$$

where f_{c1}' is the lower, and f_{c2}' is the higher desired bandedge cutoff frequency for either bandstop or bandpass digital elliptic filters. The resulting percentage bandwidth is given by [3]:

$$\%\text{BW} = 200\left(1-\frac{2}{\pi}\tan^{-1}\Omega_c'\right).$$

The elliptic-function filter tables of Saal and Ulbrich [8] and Saal [9] list only low-pass prototype element values, C and L. The normalized-bandwidth high-pass prototype values, $L_{HP}=C^{-1}$ and $C_{HP}=L^{-1}$, are reciprocals of the low-pass prototype values. Thus the normalized-bandwidth high-pass prototype equations (4) and (5) become identical to those for the low-pass prototype equations (2) and (3). When bandwidth scaling is performed, the factor Ω_c' defined before, divides the low-pass prototype element values of (2) and (3) in a low-pass prototype design, and multiplies them in a high-pass prototype design.

The complete set of design equations for digital elliptic filters is given in Table I for handy reference. Equations are included for calculating bandwidth scaling factor Ω_c', and bandwidth-normalized prototype skirt selectivity factor Ω_s. The selectivity factor symbol (Ω_s), the low-pass capacitance and inductance symbols (C and L), and the maximum passband attenuation (A_D), and minimum stopband attenuation (A_s) symbols used in Table I are identical to those used in the tables of Saal and Ulbrich [8], and Saal [9].

[2] In all cases, percentage bandwidth (%BW) is bandwidth about f_0 and may be either a stopband or passband.

Design Examples

A. Example I: Octave Bandstop Design

Assume that a bandstop filter is to be designed to meet the following specifications:

a) Passband cutoff frequencies of $f_{c1}'=1.0$ GHz, $f_{c2}'=2.0$ GHz.
b) Passband VSWR $\leq 1.6{:}1$.
c) Attenuation (A_s) to be ≥ 30 dB for 1.050 GHz $\leq f' \leq 1.950$ GHz; therefore $f_s'=1.05$ GHz.

Referring to Table I, from (T-1),

$$\Omega_c' = \tan\frac{\pi f_{c1}'}{f_{c1}'+f_{c2}'} = \tan\frac{\pi}{3} = \sqrt{3}, \tag{7}$$

and from (T-3),

$$\Omega_s = \frac{\tan\dfrac{\pi f_s'}{f_{c1}'+f_{c2}'}}{\Omega_c'} = \frac{\tan 0.35\pi}{\sqrt{3}} = 1.13. \tag{8}$$

Note from specification c) and (8) that the tangent transformation increases the skirt selectivity factor of the distributed network relative to that of the lumped element prototype. A selectivity factor of 1.13 for the lumped element prototype corresponds to a selectivity factor of 1.05 for the distributed filter.

From the tables of Saal [9], a filter-type numbered *C-06-20c*, with modular angle $\theta=66°$, provides the following characteristics which somewhat exceed the given specifications:

Maximum passband reflection coefficient $|\rho|=0.20$ ($\text{VSWR}_{\max}=1.5{:}1$).
Skirt selectivity factor $\Omega_s=1.124$.
Minimum stopband attenuation $A_s=31.5$ dB.
Frequencies of zero transmission $\Omega_4=1.143$; $\Omega_2=1.405$; $\Omega_\infty=\infty$.

The prototype circuit is shown in Fig. 4(a). From (6) normalized bandwidth-scaled real frequencies $(f'/f_0)=(2/\pi)$

TABLE I

DESIGN EQUATIONS FOR DIGITAL ELLIPTIC FILTERS

	Prototype Low-Pass Microwave Bandstop	Prototype High-Pass Microwave Bandpass
	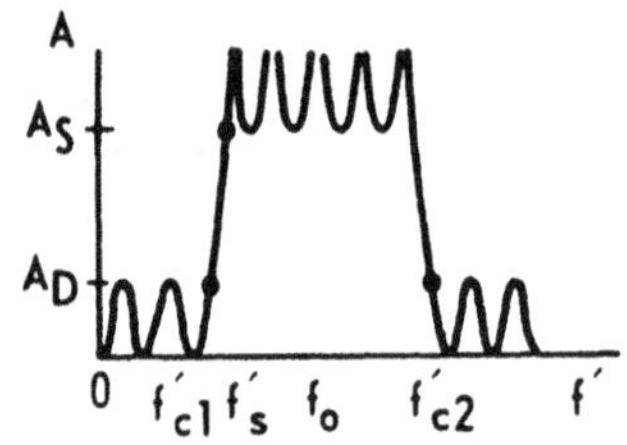	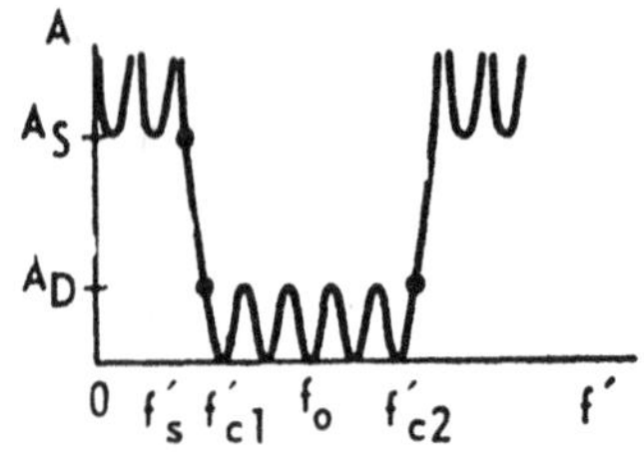
Bandwidth Scaling Factor, Ω_c'	$\Omega_c' = \tan\frac{\pi f_{c1}'}{2f_0} = \tan\frac{\pi f_{c1}'}{f_{c1}'+f_{c2}'}$ (T-1) $\% \text{ BW} = 200\left(1 - \frac{2}{\pi}\tan^{-1}\Omega_c'\right)$ (T-2)	
Skirt Selectivity Factor, Ω_s	$\Omega_s = \frac{\Omega_s'}{\Omega_c'} = \frac{\tan\frac{\pi f_s'}{2f_0}}{\Omega_c'}$ (T-3)	$\Omega_s = \frac{\Omega_c'}{\Omega_s'} = \frac{\Omega_c'}{\tan\frac{\pi f_s'}{2f_0}}$ (T-4)
Parallel Conductor Array	$c_{\text{chart}} = \frac{376.7}{\sqrt{\epsilon_r}\,Z_0}\frac{C}{\Omega_c'}$ (T-5)	$c_{\text{chart}} = \frac{376.7}{\sqrt{\epsilon_r}\,Z_0}C\Omega_c'$ (T-6)
Internal Series St bs	$\sqrt{\epsilon_r}Z_{\text{ser. stub}} = \sqrt{\epsilon_r}Z_0\frac{L}{\Omega_c'}$ (T-7)	$\sqrt{\epsilon_r}Z_{\text{ser. stub}} = \sqrt{\epsilon_r}Z_0L\Omega_c'$ (T-8)

A_D = maximum passband attenuation level in dB.
A_S = Minimum stopband attenuation level in dB.
f_s' = Lower-edge frequency for which the attenuation is A_S.
f_{c1}' = Lower-edge passband cutoff frequency.
f_{c2}' = Higher-edge passband cutoff frequency.
$f_0 = (f_{c1}' + f_{c2}')/2$ = frequency for which the lines are quarter-wavelength.
Ω_s = Skirt selectivity parameter listed in Saal and Ulbrich [8] and Saal [9].
L, C = Low-pass prototype element values listed in Saal and Ulbrich [8] and Saal [9].
Z_0 = Characteristic terminating impedance (ohms).
ϵ_r = Relative dielectric constant of medium.
c_{chart} = "Static capacitance" to be found in charts [15], [16].
$\sqrt{\epsilon_r}Z_{\text{ser stub}}$ = Characteristic impedance to be found in chart [17].

$\tan^{-1}\Omega_c'\Omega$ corresponding to important response points are given by

$$\frac{f_{c1}'}{f_0} = \frac{2}{\pi}\tan^{-1}\sqrt{3}\,(1) = 0.667$$

$$\frac{f_s'}{f_0} = \frac{2}{\pi}\tan^{-1}\sqrt{3}\,(1.124) = 0.698$$

$$\frac{f_2'}{f_0} = \frac{2}{\pi}\tan^{-1}\sqrt{3}\,(1.143) = 0.703$$

$$\frac{f_4'}{f_0} = \frac{2}{\pi}\tan^{-1}\sqrt{3}\,(1.405) = 0.752$$

$$\frac{f_\infty'}{f_0} = \frac{2}{\pi}\tan^{-1}\infty = 1.000.$$

The octave bandwidth theoretical response characteristic is shown in Fig. 4(b). The actual distributed filter selectivity factor is

$$\frac{f_s'}{f_{c1}'} = \frac{0.698}{0.667} = 1.045. \tag{9}$$

For a 50-ohm realization, the self and mutual static capacitance values are obtained by direct substitution in (T-5) and the series stub impedance values likewise from (T-7). The appropriate static capacitance network is shown in Fig. 4(c), along with the stub impedance values. To enable the practical construction of the series stubs, a dielectric constant larger than that used for the coupled lines is desirable. With air-loaded coupled lines, Teflon ($\sqrt{\epsilon_r} = 1.44$) provides a convenient dielectric.

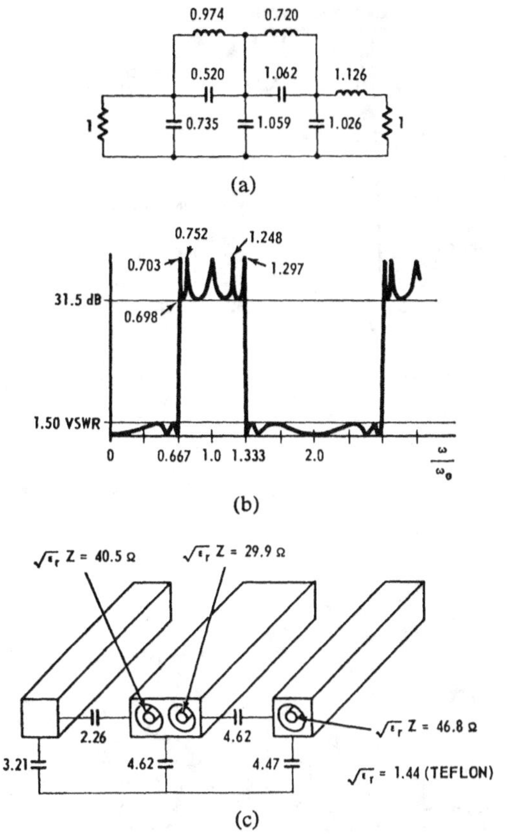

Fig. 4. Octave bandstop digital elliptic design. (a) Low-pass prototype element values. (b) Theoretical response characteristic. (c) Static capacitance values and series stub impedances.

Dimensions for the coupled-line portion of the filter are determined as for interdigital filters as described by Matthaei [18]. Thus, for a rectangular bar realization, $(s_{k,k+1})/b$ is determined directly from Fig. 3 of Getsinger [15] using $c_{m(k,k+1)}$, and

$$\frac{w_k}{b} = \frac{1}{2}\left(1 - \frac{t}{b}\right)\left[\frac{1}{2} c_{sk} - (c_{fe})_{k-1,k} - (c_{fe})_{k,k+1}\right] \quad (10)$$

is computed using the referenced charts. In the previous calculations

b = ground plane spacing,
t = line thickness,
$s_{k,k+1}$ = spacing between lines k and $k+1$,
w_k = width of the kth line,
c_{sk} = self capacitance of the kth line,
$c_{m(k,k+1)}$ = mutual capacitance between lines k and $k+1$,
$(c_{fe})_{k,k+1}$ = even-mode fringing capacitance between lines k and $k+1$ as determined from Fig. 3 of Getsinger [15].

Note that as $s/b \to \infty$, the corresponding $c_{fe} \to c_f$ where c_f is the fringing capacitance of an isolated line (Fig. 5 of Getsinger [15]). Thus for an N-line network

$$\frac{w_1}{b} = \frac{1}{2}\left(1 - \frac{t}{b}\right)\left[\frac{c_{s1}}{2} - c_f - (c_{fe})_{12}\right]$$

and

$$\frac{w_N}{b} = \frac{1}{2}\left(1 - \frac{t}{b}\right)\left[\frac{c_{sN}}{2} - (c_{fe})_{N-1,N} - c_f\right]. \quad (11)$$

Using the preceding equations, line dimensions for the bandstop filter of Fig. 4 were determined. The effect of the side walls was taken into account by using $c_{f0}(s_w/b)$ in place of c_f for the end lines as shown in Fig. 6 of Getsinger [15], where $s_w/2$ is the spacing between the end lines and the walls. A line drawing showing cross-sectional dimensions for $b = 0.500$ inch and $t = 0.200$ inch is given in Fig. 5(a). The diameter ratios for the series stubs were determined from the referenced nomograph [17]. Note that each w_k/b should satisfy the inequality

$$\frac{\dfrac{w_k}{b}}{1 - \dfrac{t}{b}} < 0.35 \quad (12)$$

as discussed by Getsinger [15]. If this is not satisfied (as was the case with the first line in the above example), the correction formula [15]

$$\frac{w_k'}{b} = \frac{0.07\left(1 - \dfrac{t}{b}\right) + \dfrac{w_k}{b}}{1.2} \quad (13)$$

can be used providing $0.1 < (w_k'/b)/[1 - (t/b)] < 0.35$.

Using the cross-sectional dimensions of Fig. 5(a), an experimental model was constructed using quarter-wave-length lines at 1.5 GHz. Assembly sketches of the filter structure are shown in Fig. 5(b) and (c). Several junctions were mitered to reduce discontinuity effects. Line spacings were maintained by use of Teflon fiberglasss pacers as shown. Movable shorting slugs were provided to allow adjustment of the series stub lengths.

The measured response of the filter is shown in Fig. 6, and a photograph of the experimental model is shown in Fig. 7. The experimental results shown were obtained with no physical alterations other than adjustment of the series shorting slugs and matching of the connectors. The dielectric spacers provided a high degree of mechanical stability and did not appear to affect the electrical response. Junction effects and the nonzero width of the structure did not influence significantly the electrical performance.

B. Example II: Wide Bandpass Design (Pseudo High Pass)

Assume a wide bandpass filter is to be designed to satisfy the following specifications:

a) Passband cutoff frequencies $f_{c1}' = 1.0$ GHz, $f_{c2}' = 4.0$ GHz.
b) Passband VSWR less than 1.5:1.
c) Attenuation (A_s) to be ≥ 50 dB for $f' \leq 0.95$ GHz ($f_s' = 0.95$ GHz).

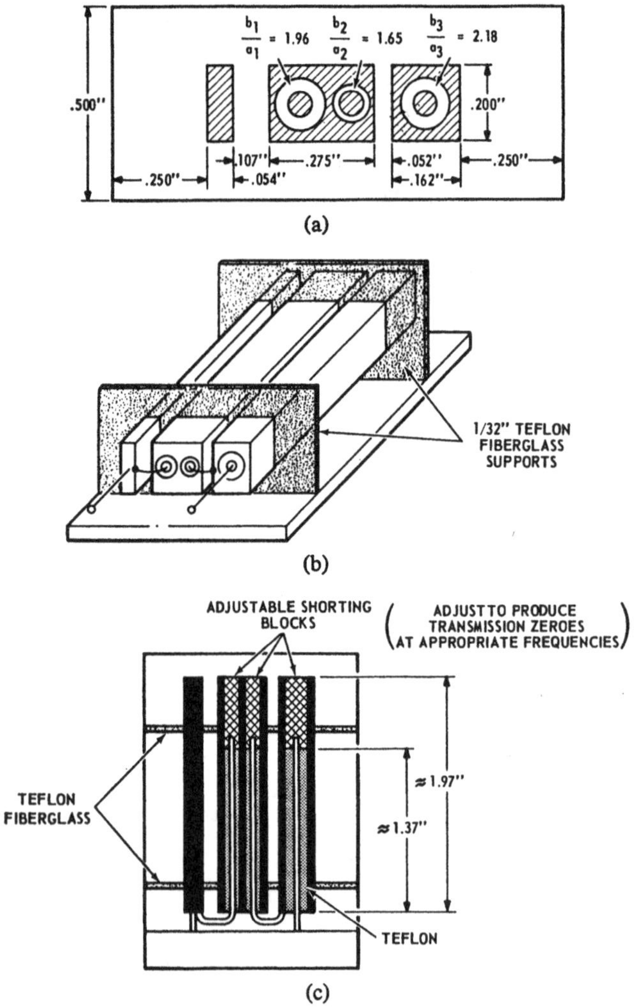

Fig. 5. Construction details of octave bandwidth digital elliptic filter. (a) Cross section. (b) Perspective view. (c) Plan view.

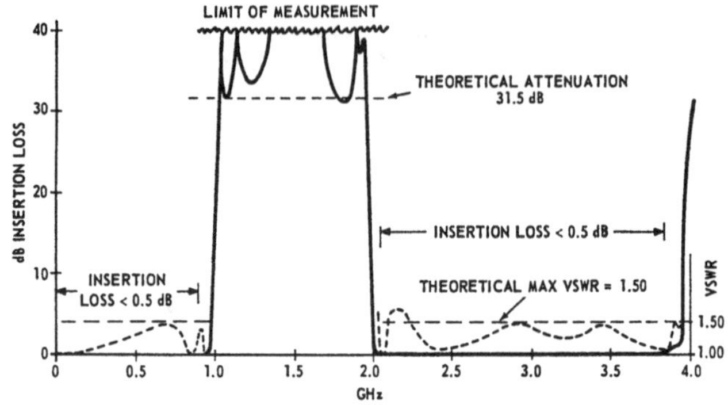

Fig. 6. Octave band digital elliptic measured response.

Fig. 7. Experimental octave bandstop digital elliptic filter.

176

From Table I, (T-1),

$$\Omega_c' = \tan\frac{\pi}{5} = 0.727 \tag{14}$$

and from (T-4),

$$\Omega_s = \frac{\Omega_c'}{\tan\dfrac{\pi(0.950)}{5}} = \frac{0.727}{0.680} = 1.07. \tag{15}$$

From Saal's tables [9] a filter-type numbered *C-09-15*, with modular angle $\theta = 70°$, provides the following satisfactory characteristics:

Maximum passband reflection coefficient $|\rho| = 0.15$ ($\text{VSWR}_{\max} = 1.36:1$).
Skirt selectivity factor $\Omega_s = 1.064$.
Minimum stopband attenuation $A_s = 51.0$ dB.
Frequencies of zero transmission $\Omega_8 = 1.302$, $\Omega_6 = 1.069$, $\Omega_4 = 1.119$, $\Omega_2 = 2.094$, $\Omega_\infty = \infty$.

The low-pass prototype element values are shown in Fig. 8(a). The high-pass response characteristics are obtained by substituting $1/\bar{\Omega}$ for Ω as in the usual low-pass to high-pass transformation [19]. Therefore,

$$\bar{\Omega}_c = 1.0,\ \bar{\Omega}_s = 0.940,\ \bar{\Omega}_8 = 0.768,\ \bar{\Omega}_6 = 0.936,$$
$$\bar{\Omega}_4 = 0.895,\ \bar{\Omega}_2 = 0.478,\ \bar{\Omega}_\infty = 0.$$

Bandwidth-scaled distributed network frequencies, corresponding to important response points, are obtained by applying (6) and using the high-pass barred frequencies given by

$$\frac{f'}{f_0} = \frac{2}{\pi}\tan^{-1}\Omega_c'\bar{\Omega}.$$

With $f_0 = 2.5$ GHz, the theoretical response characteristic is that shown in Fig. 8(b). The actual distributed filter selectivity factor is $\bar{f}_c/\bar{f}_s' = 1.047$. Note that in this example, because the tangent function is more nearly linear for small angles, the difference between the lumped element prototype selec-

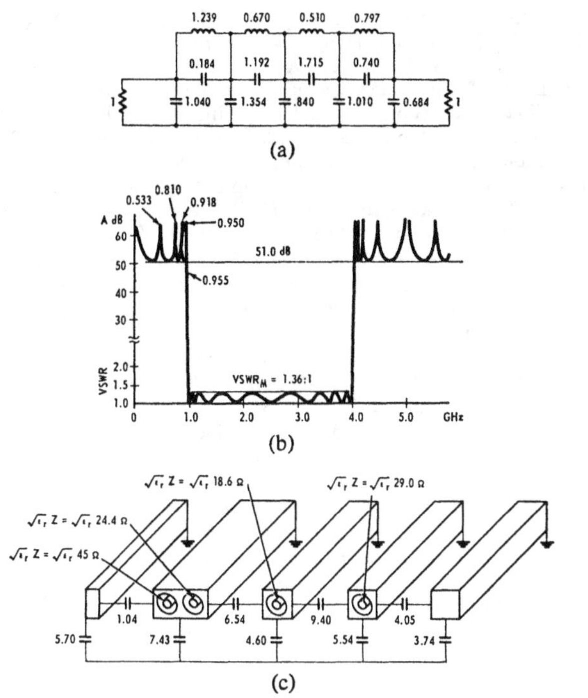

Fig. 8. Wide bandpass digital elliptic design. (a) Low-pass prototype. (b) Theoretical response characteristic. (c) Static capacitance values and series stub impedances.

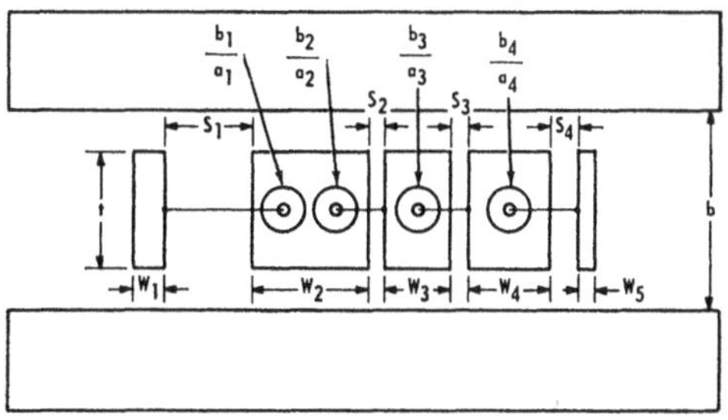

DIMENSIONS (INCHES)		
b = .500, t = .300		
S_1 = .237	W_1 = .085	b_1/a_1 = 2.95
S_2 = .050	W_2 = .269	b_2/a_2 = 1.80
S_3 = .033	W_3 = .172	b_3/a_3 = 1.56
S_4 = .078	W_4 = .228	b_4/a_4 = 2.00
	W_5 = .039	WITH $\sqrt{\epsilon_r}$ = 1.44 (TEFLON)

Fig. 9. Sample cross-sectional dimensions for wide bandpass design.

tivity factor and the distributed filter selectivity factor is very slight.

For a 50-ohm realization, the self and mutual static capacitance values are obtained by substitution of the low-pass prototype element values in (T-6) and the series stub impedance values are obtained likewise from (T-8). The appropriate static capacitance network and stub impedance values are shown in Fig. 8(c). Dimensions for the coupled-line portion of the filter are obtained by use of (10) through (13) of Example I, and the stub impedances are obtained from the referenced nomograph [17]. A line drawing showing computed cross-sectional dimensions for $b=0.500$ inch and $t=0.300$ inch is shown in Fig. 9. The final realization takes the form shown in Fig. 2(b).

CONCLUSIONS

The digital elliptic distributed TEM filter is an extremely compact microwave structure capable of very sharp cutoff response in wideband applications throughout the frequency range up to X-band. The filter is finger-like in appearance, and realizes equal-ripple response in both passband and stopband as do elliptic-function lumped element filters; thus the name digital elliptic. The practical advantages and limitations of the digital elliptic filter are listed below.

Advantages

1) Existing tables [8]–[11] may be used to obtain prototype inductance and capacitance element values, thus circumventing tedious approximation and synthesis procedures.
2) The digital elliptic filter can be realized in a physical form identical to its lumped element prototype; thus no redundant elements nor equivalent circuit transformations are necessary.
3) Bandwidth scaling is accomplished by applying a multiplying factor to the tabulated element values; this simple recipe cannot be used in the exact design of most wideband microwave filters which employ quarter-wavelength elements.
4) The physical form is extremely compact, e.g., it is often four or five times smaller than an interdigital bandpass filter of comparable selectivity. It is also much more compact than other microwave realizations for elliptic-function response [7], [20].
5) The same design procedure holds equally well for either bandstop or bandpass designs, and the filter structural form is identical.
6) The digital elliptic filter has been found to give measured performance close to theoretical with relatively few adjustments.

Limitations

1) The structure is inherently wideband (>30 percent) and is not suited to narrowband designs due to physical limitations.
2) The structure requires series internal stubs and therefore more machined parts than some filter forms.
3) At X-band frequencies and above, practical ground plane spacings will result in junction effects which are difficult to compensate.

The practical design procedure for digital elliptic filters is extremely simple. All necessary equations are listed for handy reference in Table I. Two design examples, a bandstop and a bandpass, are given to demonstrate each step in the design.

References

[1] H. Ozaki and J. Ishii, "Synthesis of a class of strip-line filters," *IRE Trans. on Circuit Theory*, vol. CT-5, pp. 104–109, June 1958.

[2] A. I. Grayzel, "A synthesis procedure for transmission line networks," *IRE Trans. on Circuit Theory*, vol. CT-5, pp. 172–181, September 1958.

[3] R. J. Wenzel, "Exact design of TEM microwave networks using quarter-wave lines," *IEEE Trans. on Microwave Theory and Techniques*, vol. MTT-12, pp. 94–111, January 1964.

[4] B. M. Schiffman and G. L. Matthaei, "Exact design of band-stop microwave filters," *IEEE Trans. on Microwave Theory and Techniques*, vol. MTT-12, pp. 6–15, January 1964.

[5] M. C. Horton and R. J. Wenzel, "General theory and design of optimum quarter-wave TEM filters," *IEEE Trans. on Microwave Theory and Techniques*, vol. MTT-13, pp. 316–327, May 1965.

[6] P. I. Richards, "Resistor transmission-line circuits," *Proc. IRE*, vol. 36, pp. 217–220, February 1948.

[7] M. C. Horton and R. J. Wenzel, "Realization of microwave filters with equal ripple response in both pass and stop bands," presented at the Internat'l Symp. on Generalized Networks, Polytechnic Institute of Brooklyn, Brooklyn, N. Y., April 12–14, 1966.

[8] R. Saal and E. Ulbrich, "On the design of filters by synthesis," *IRE Trans. on Circuit Theory*, vol. CT-5, pp. 284–327, December 1958.

[9] R. Saal, "Der Entwurf von Filtern mit Hilfe des Kataloges normierter Tiefpässe," (The design of filters using the catalogue of normalized low-pass filters), Telefunken, G.M.B.H., Backnang/Württemberg, Western Germany, 1961.

[10] D. C. Pawsey, "Element coefficients for symmetrical two-section filters having Tchebycheff response in both pass and stop bands," *IRE Trans. on Circuit Theory*, vol. CT-7, pp. 19–31, March 1960.

[11] J. K. Skwirzynski, *Design Theory and Data for Electric Filters*. London: Van Nostrand, 1965.

[12] E. Ott, "A network approach to the design of multi-line 2N-port directional couplers," Polytechnic Institute of Brooklyn, Brooklyn, N. Y., Rept. PIBMRI-1236-64; Contract AF 30(602)-2868, Tech. Rept. RADC-TR-65-41, April 1965.

[13] R. J. Wenzel, "Exact theory of interdigital band-pass filters and related coupled structures," *IEEE Trans. on Microwave Theory and Techniques*, vol. MTT-13, pp. 559–575, September 1965.

[14] ——, "Theoretical and practical applications of capacitance matrix transformations to TEM network design," *IEEE Trans. on Microwave Theory and Techniques*, vol. MTT-14, pp. 635–647, December 1966.

[15] W. J. Getsinger, "Coupled rectangular bars between parallel plates," *IRE Trans. on Microwave Theory and Techniques*, vol. MTT-10, pp. 65–72, January 1962.

[16] E. G. Cristal, "Coupled circular cylindrical rods between parallel ground planes," *IEEE Trans. on Microwave Theory and Techniques*, vol. MTT-12, pp. 428–439, July 1964.

[17] *The Microwave Engineers' Handbook and Buyers' Guide*. Brookline, Mass.: Horizon House-Microwave, Inc., 1966, p. 91.

[18] G. L. Matthaei, "Interdigital band-pass filters," *IEEE Trans. on Microwave Theory and Techniques*, vol. MTT-10, pp. 479–491, November 1962.

[19] E. A. Guillemin, *Synthesis of Passive Networks*. New York: Wiley, 1957, p. 602.

[20] R. Levy, "Synthesis of distributed elliptic-function filters from lumped-constant prototypes," *IEEE Trans. on Microwave Theory and Techniques*, vol. MTT-14, pp. 506–517, November 1966.

Reprinted from IEEE TRANSACTIONS
ON *MICROWAVE THEORY AND TECHNIQUES*
Volume MTT-15, Number 5, May, 1967
pp. 307-314

The Stepped Digital Elliptic Filter

JOHN DAVID RHODES, MEMBER, IEEE

Abstract—**The design and synthesis of various types of microwave elliptic function filters has been accomplished by a number of authors. However, one problem in this field which remains is the realization of compact narrow-band bandpass elliptic function filters. In this paper, a procedure is presented which enables this class of filters to be constructed in a compact digital form.**

Since the physical realization is in the form of an *n*-wire line, one-quarter of a wavelength long at the center frequency of the passband, where the impedance levels are stepped along the center of the coupled lines, the filter has been termed the stepped digital elliptic filter. The absence of awkward interconnections in the filter due to the stepped digital structure inherently implies that reasonable insertion loss characteristics may be achieved in the *X*-band region and above, and also simplifies the mechanical construction.

It is shown that the resonant elements in the filter, due to the design procedure adopted, are relatively insensitive to the absolute bandwidth of the filter, and consequently fractional bandwidths of approximately 30 percent and below may be readily achieved while the normalized impedance values of the elements in the network remain of the order of unity. This latter result is similar to that obtainable from conventional interdigital filters but in the case of narrow bandwidths the stepped digital filter is considerably smaller in physical size.

A systematic procedure is also formulated for the inclusion of the parasitic lumped end effect capacitances into the overall design procedure in order to maintain the equiripple passband and stopband responses.

Experimental results are presented for a five-element, 11 percent bandwidth filter and are shown to be in good agreement with theoretical predictions.

Introduction

THE KNOWN theoretical advantages which arise from designing a microwave filter to exhibit an elliptic function response in preference to a Chebyshev response have recently led to the publication of a number of contributions on this topic. These papers have in common the fact that the same electrical response is realized but they vary according to the forms of physical networks chosen to realize this elliptic response.

Saito [1] utilized the symmetrical nature of the odd-ordered elliptic filter in his realization in the form of a cascade of symmetrical two-wire coupled lines terminated in either an open- or short-circuited stub. There exist theoretical limitations upon the realizability of this structure, and the second-order case has been discussed by Matsumoto [2]. A more severe limitation which arises is due to the inability to synthesize physically realizable impedance values except in the case of bandwidths of the order of an octave.

This problem of achieving physically realizable impedance values has been common to most of the subsequent design procedures. Schiffman and Young [3] have published design tables for bandstop elliptic function filters of degree 5 using redundant unit elements. The synthesis technique used was based upon the conventional partial pole extraction procedure, thus allowing the poles of attenuation to be realized by simple and double shunt stubs. It was found that for most bandwidths practical techniques for overcoming high-impedance values had to be incorporated, but this approach fails in the case of narrow bandwidths. Levy and Whiteley [4], by using coupled-line structures, found that a systematic procedure could be formulated for the introduction of redundant unit elements into a distributed network based upon the lumped prototype, while still preserving reasonable element values. In the case of the narrow-band stopband filter, it was found that by initially resonating the low-pass prototype before incorporating the unit elements, normalized impedance values of the order of unity could be maintained. Successful narrow-band bandpass filters using this technique were not reported, however.

The design procedures described in [3] and [4] rely upon the tables published by Saal [5] for lumped low-pass prototype elliptic function filters, thus eliminating the task of constructing bounded real reflection coefficients from the prescribed insertion loss function. However, it has been shown by Horton and Wenzel [6] that the unit elements which are necessary in these realization procedures may contribute to the insertion loss characteristic, and therefore a superior characteristic may be numerically computed thus making Saal's tables inapplicable in this optimum design case.

A compact realization of wide-band elliptic function filters has also been reported by Horton and Wenzel [6], [7]. This direct interdigital realization of the basic lumped form of elliptic function filter has been termed the "digital elliptic filter." In this case no unit elements were incorporated into the network resulting in a canonic realization. This particular network possesses the fundamental advantage of direct conversion of the tabulated element values of the low-pass prototype into physical dimensions of the interdigital line. Unfortunately, this realization exhibits a number of disadvantages, the two most important being that 1) it is impossible to construct narrow-band filters and 2) difficulties are encountered when the operating range is extended into *X*-band and above, due to the awkward interconnection of series stubs.

Hence, it may be stated that it is difficult, if not impossible, to construct narrow-band bandpass elliptic function filters using any of the realization techniques described. A qualitative explanation may be developed, as follows, to show why these procedures are inadequate when it is required to realize narrow-band elliptic function filters.

The resonant circuits which provide the poles of attenuation in the elliptic function filter possess elements which

Manuscript received June 6, 1968; revised November 14, 1968.
The author is with Microwave Development Laboratories, Inc., Natick, Mass. 01760.

are proportional to the bandwidth scaling factor and others which are inversely proportional. For small (or large) scaling factors, one set of elements become physically unrealizable if a direct digital elliptic realization is used. Similarly, in Saito's direct synthesis procedure, the parameters of the adjacently extracted two-wire lines are approximately proportional to and inversely proportional to the scaling factor, respectively. In the cases where redundant unit elements are introduced into the networks, the situation tends to deteriorate since the impedances of the unit elements are invariant to the bandwidth scaling factor, unless of course, the resonating technique of Levy and Whiteley [4] is applied in the bandstop case. However, with their technique for realizing resonant sections in cascade with a unit element, a successful solution to the narrow-band bandpass case does not necessarily follow. Similarly, other forms of resonant section realizations proposed by Levy [8] are inherently broadband.

The basic problem to be solved follows directly from this qualitative reasoning. A realization must be sought whereby the resonant sections which provide the poles of attenuation must be relatively insensitive to the bandwidth scaling factor. Such a realization is presented in this paper.

Theoretical Design Procedure

One of the two basic forms of the odd-order low-pass prototype elliptic function filter is shown in Fig. 1, and it is this cascade of pi sections which is to be realized by the stepped digital elliptic filter. Initially, we shall consider the realization of a typical section with shunt capacitances C_{r-1} and C_{r+1} and with a transmission zero at $\Omega_r = 1/\sqrt{L_r C_r}$.

The normal procedure for the conversion into a microwave bandpass filter is to apply the high-pass transformation

$$\omega \rightarrow \frac{\tan \omega_c}{\tan \omega} \tag{1}$$

where the normalized band-center frequency is at $\omega = \pi/2$ and the band-edge frequencies are situated at $\omega = \omega_c$ and $\omega = \pi - \omega_c$ in the fundamental period. However, this transformation implies that the impedance values of three of the elements are inversely proportional to $\tan \omega_c$ and one proportional to $\tan \omega_c$. For narrow bandwidths $\tan \omega_c \gg 1$; this results in at least one extreme impedance value even if the entire impedance level of the network is adjusted by means of transformer action at the input and output ports, as in case of conventional interdigital networks [9], [10].

The solution to this problem is to apply the conventional bandpass transformation

$$\omega \rightarrow a\left(\tan \omega - \frac{1}{\tan \omega}\right) \tag{2}$$

where

$$a = \frac{\tan 2\omega_c}{2}. \tag{3}$$

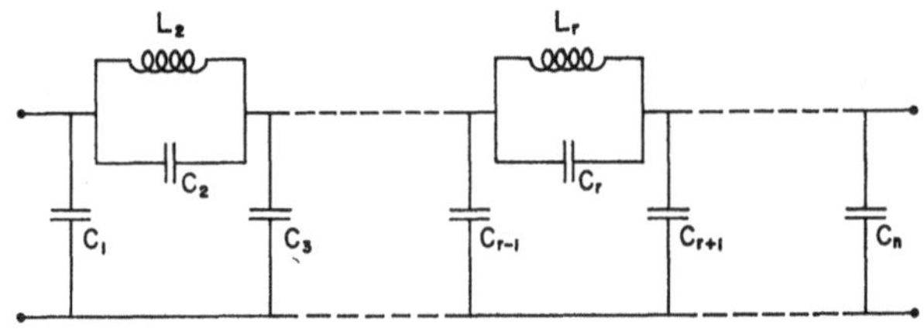

Fig. 1. The odd-order low-pass prototype elliptic function filters.

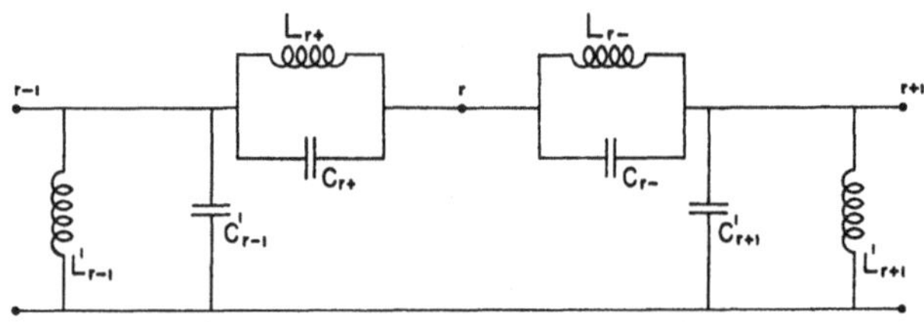

Fig. 2. A typical resonated pi section.

In this case the band-center frequency is normalized to $\omega = \pi/4$ and the band-edge frequencies $\omega = \omega_c$ and $\omega = (\pi/2) - \omega_c (\omega_c < \pi/4)$. This procedure has, of course, theoretically doubled the degree of the network but it will be shown that by using the proposed stepped digital realization the number of nonredundant coupled lines necessary to realize any given elliptic filter is equal to the degree of the low-pass prototype. Throughout this paper the degree of the network is assumed to be the degree of the resulting reflection coefficient.

Applying this resonating technique to the basic section under consideration, we generate the section shown in Fig. 2 where the pertinent element values are the following:

$$\begin{aligned} C'_{r-1} &= aC_{r-1}, \qquad C'_{r+1} = aC_{r+1} \\ L'_{r-1} &= \frac{1}{C'_{r-1}}, \qquad L'_{r+1} = \frac{1}{C'_{r+1}} \\ C_{r+} &= \frac{1}{L_{r-}} = aC_r(1 + \lambda_{r-}{}^2) \\ C_{r-} &= \frac{1}{L_{r+}} = aC_r(1 + \lambda_{r+}{}^2) \end{aligned} \tag{4}$$

where

$$\lambda_{r\pm} = \sqrt{\left(\frac{\Omega r}{2a}\right)^2 + 1} \pm \frac{\Omega r}{2a}. \tag{5}$$

For a narrow-band filter we have $a \gg 1$, and for most practical filter specifications this implies that $a \gg \Omega_r$ for all r. It follows immediately from (5) that λ_{r+} and λ_{r-} are of the order of unity, and hence, from (4), the admittance levels of all of the elements in the network are approximately directly proportional to the bandwidth scaling factor a. It is apparent therefore, that if the coupling into the network can be made through transformer action, then the resonant sections will become relatively invariant to bandwidth scaling. Before considering the method for achieving this transformer action, the coupled-line realization of the basic network will be discussed.

From Fig. 2, the obvious direct realization of the resonated section is in the form of a pair of 3-wire line interdigital networks connected in parallel. One network possesses lines which are short circuited to ground at one end, whereas the other is terminated in open circuits. The physical configuration is depicted in Fig. 3 where the nodes $r-1$, r, $r+1$ refer to the corresponding nodes in Fig. 2. The characteristic admittance matrices of networks 1 and 2 may therefore be identified as follows.

The interconnection of the basic sections must now be accomplished.

It may readily be shown [9], [10] that if any pair of conductors in a uniform n-wire line are at the same potential at both ends, then at any intermediate point they are at the same potential and may consequently be replaced by a single line. For the overall filter, the interconnection of consecutive sections requires that the node at the junction of the outer pair of lines of one section be at the same potential as the node at the junction of the corresponding pair of lines of the adjacent section. Since the opposite ends of the lines in network 1 are short circuited to ground, and therefore at the same zero potential, the appropriate pair of lines from adjacent sections in network 1 may be condensed into a single line [10].

Network 1

$$\begin{array}{ccc} (r-1) & (r) & (r+1) \\ \left[\begin{array}{ccc} a[C_{r-1} + C_r(1+\lambda_{r+}^2)] & -aC_r(1+\lambda_{r+}^2) & 0 \\ -aC_r(1+\lambda_{r+}^2) & aC_r[2+\lambda_{r+}^2+\lambda_{r-}^2] & -aC_r(1+\lambda_{r-}^2) \\ 0 & -aC_r(1+\lambda_{r-}^2) & a[C_{r+1} + C_r(1+\lambda_{r-}^2)] \end{array}\right] \end{array} \quad (6)$$

Network

$$\begin{array}{ccc} (r-1) & (r) & (r+1) \\ \left[\begin{array}{ccc} a[C_{r-1} + C_r(1+\lambda_{r-}^2)] & -aC_r(1+\lambda_{r-}^2) & 0 \\ -aC_r(1+\lambda_{r-}^2) & aC_r(2+\lambda_{r+}^2+\lambda_{r-}^2) & -aC_r(1+\lambda_{r+}^2) \\ 0 & -aC_r(1+\lambda_{r+}^2) & a[C_{r+1} + C_r(1+\lambda_{r+}^2)] \end{array}\right] \end{array} \quad (7)$$

The characteristic admittance of line r to ground may be made finite by scaling the admittance level of the center rows and columns in the above matrices by a factor greater than unity, as in the case of conventional interdigital networks [9], [10], [11]. This operation does not alter the external performance of the section between nodes $r-1$ and $r+1$ if the same scaling factor is used on both networks. One additional condition which is necessary for physical realization other than the normal hyperdominancy condition, is that the distance between adjacent lines of one network must approximately equal the distance between corresponding lines on the other network, in order to obtain simple physical interconnections.

From the matrices (6) and (7) it may be seen that the admittance to ground of lines r on both networks 1 and 2 will be the same after the necessary admittance scaling. Furthermore, since the sum of the coupling element values between lines $r-1$ and r, r and $r+1$ are the same for both networks, the distance between lines $r-1$ and $r+1$ must inherently be approximately equal. In addition, since λ_{r+} and λ_{r-} are of the order of unity, the physical separation in a transverse direction of lines r on both networks must necessarily be small. These points will be discussed further after the construction of the entire network has been achieved.

We have now shown how each of the basic resonated sections in the bandpass microwave elliptic function filter may be constructed utilizing a pair of 3-wire interdigital lines.

Since the opposite ends of the lines in network 2 are open circuited, and therefore not necessarily at the same potential, this simple principle may not be applied directly. However, it may be shown that a sufficient condition for the condensation of open-circuited lines connected together at one end is that neither line couples directly nor indirectly to any other line which is not also open circuited. This property is of fundamental importance when consideration is given to the method of transformer coupling into the network at the input and output ports.

As an example of the line condensation procedure for a two-section (fifth-degree) filter the overall network reduces to a pair of five-wire lines as shown in Fig. 4. The characteristic admittance matrices of these two lines are

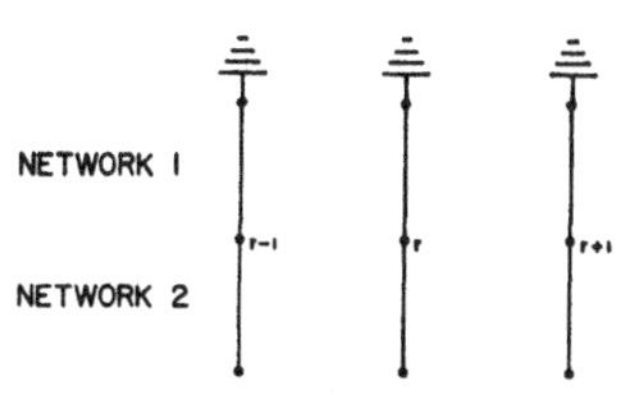

Fig. 3. The coupled-line realization of the basic resonated section.

Network 1

$$\begin{bmatrix} a[C_1 + C_2(1+\lambda_{2+}^2)] & -aC_2(1+\lambda_{2+}^2) & 0 & 0 & 0 \\ -aC_2(1+\lambda_{2+}^2) & aC_2(2+\lambda_{2+}^2+\lambda_{2-}^2) & -aC_2(1+\lambda_{2-}^2) & 0 & 0 \\ 0 & -aC_2(1+\lambda_{2-}^2) & a[C_3 + C_2(1+\lambda_{2-}^2) + C_4(1+\lambda_{4+}^2)] & -aC_4(1+\lambda_{4+}^2) & 0 \\ 0 & 0 & -aC_4(1+\lambda_{4+}^2) & aC_4(2+\lambda_{4+}^2+\lambda_{4-}^2) & -aC_4(1+\lambda_{4-}^2) \\ 0 & 0 & 0 & -aC_4(1+\lambda_{4-}^2) & a[C_5 + C_4(1+\lambda_{4-}^2)] \end{bmatrix} \quad (8)$$

Network 2

$$\begin{bmatrix} a[C_1 + C_2(1+\lambda_{2-}^2)] & -aC_2(1+\lambda_{2-}^2) & 0 & 0 & 0 \\ -aC_2(1+\lambda_{2-}^2) & aC_2(2+\lambda_{2+}^2+\lambda_{2-}^2) & -aC_2(1+\lambda_{2+}^2) & 0 & 0 \\ 0 & -aC_2(1+\lambda_{2+}^2) & a[C_3 + C_2(1+\lambda_{2+}^2) + C_4(1+\lambda_{4-}^2)] & -aC_4(1+\lambda_{4-}^2) & 0 \\ 0 & 0 & -aC_4(1+\lambda_{4-}^2) & aC_4(2+\lambda_{4+}^2+\lambda_{4-}^2) & -aC_4(1+\lambda_{4+}^2) \\ 0 & 0 & 0 & -aC_4(1+\lambda_{4+}^2) & a[C_5 + C_4(1+\lambda_{4+}^2)] \end{bmatrix} \quad (9)$$

For an elliptic function filter of degree n in the low-pass prototype version, the complete network degenerates into a pair of uniform digital n-wire lines which are one eighth of a wavelength long at the band-center frequency. Due to the shunt connection of this pair of digital networks, the complete network may be viewed as consisting of a single stepped digital line which is one quarter of a wavelength long at the band-center frequency. However, it must be remembered that the input and output terminals are situated at the centers of the first and last lines. The direct connection of these points to the external ports would yield a filter with the required electrical response, but the internal elements would be of low impedance. As mentioned previously, to overcome this problem coupling from the ports to the centers of these lines must be made using transformer action.

To accomplish this, unit elements are introduced into the network, one at each end, of unity characteristic impedance, and one eighth of a wavelength long at the band-center frequency. These unit elements may then be realized in the conventional interdigital form as shown in Fig. 5 for the case of a fifth-degree network. The number of lines in network 1 has now been increased to $n+2$ and its resulting augmented characteristic admittance matrix becomes

$$\begin{bmatrix} 1 & -1 & 0 & \cdots \\ -1 & 1 + a[C_1 + C_2(1+\lambda_{2+}^2)] & -aC_2(1+\lambda_{2+}^2) & \cdots \\ 0 & -aC_2(1+\lambda_{2+}^2) & aC_2(1+\lambda_{2+}^2+\lambda_{2-}^2) & \cdots \\ \cdot & \cdot & \cdot & \cdot \end{bmatrix} \quad (10)$$

while the characteristic admittance matrix of network 2 remains unchanged. A dual coupling system might appear to be the coupling into the open-circuited ends of the input and output lines through open-circuited coupled lines. This procedure, however, is incorrect since it changes the properties of the complete network due to the implied sufficiency condition for the line condensation of open-circuited lines. The justification for using the former method of coupling follows from the condition for line condensation of short-circuited lines [10].

The only factor which now remains to be considered is that of the admittance scaling of the entire digital line. Initially, every line in the stepped digital filter, except the lines on which the input and output ports are situated, may be scaled by the factor $1/\sqrt{a}$. The characteristic admittance matrices

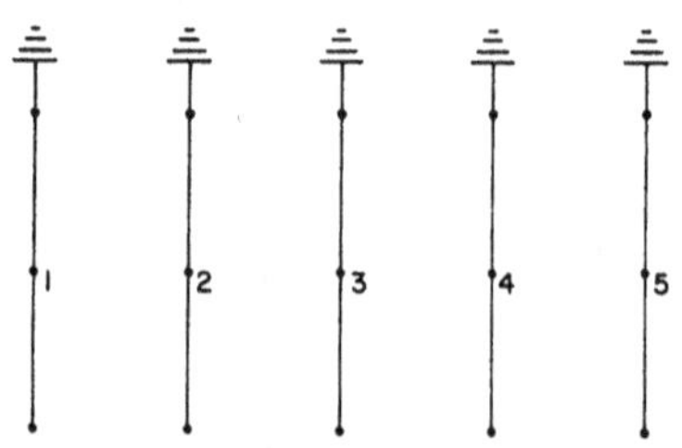

Fig. 4. Degenerate realization of a two-section filter.

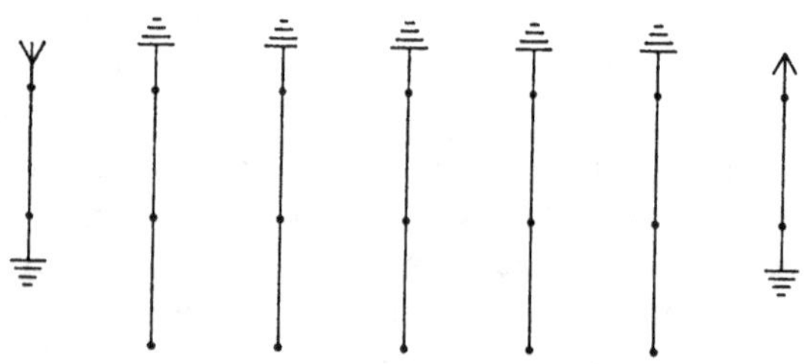

Fig. 5. Realization of fifth-order filter with transformer coupling elements.

of networks 1 and 2 then become

$$\begin{bmatrix} 1 & -\frac{1}{\sqrt{a}} & 0 & \\ -\frac{1}{\sqrt{a}} & \frac{1}{a}+C_1+C_2(1+\lambda_{2+}^2) & -C_2(1+\lambda_{2+}^2) & \\ 0 & -C_2(1+\lambda_{2+}^2) & C_2(2+\lambda_{2+}^2+\lambda_{2-}^2) & -C_2(1+\lambda_{2-}^2) \\ 0 & 0 & -C_2(1+\lambda_{2-}^2) & C_3+C_2(1+\lambda_{2-}^2)+C_4(1+\lambda_{4+}^2) \\ \vdots & \vdots & \vdots & \vdots \end{bmatrix} \tag{11}$$

for network 1, and

$$\begin{bmatrix} C_1+C_2(1+\lambda_{2-}^2) & -C_2(1+\lambda_{2-}^2) & 0 \\ -C_2(1+\lambda_{2-}^2) & C_2(2+\lambda_{2+}^2+\lambda_{2-}^2) & -C_2(1+\lambda_{2+}^2) \\ 0 & -C_2(1+\lambda_{2+}^2) & C_3+C_2(1+\lambda_{2+}^2)+C_4(1+\lambda_{4-}^2) \\ \vdots & \vdots & \vdots \end{bmatrix} \tag{12}$$

for network 2.

It follows immediately that all of the lines forming the resonant circuits are relatively insensitive to the bandwidth scaling factor since the only variation is with respect to $\lambda r\pm$ which, for narrow bandwidths, are all of the order of unity. Thus, the only significant change in the network due to a variation in bandwidth is in the values of the coupling elements at the input and output lines.

Admittance scaling may now be applied to the even-numbered lines in order to produce finite values of capacitance to ground, and to the odd-numbered lines to produce additional flexibility in the physical realization. In conventional interdigital filters, the scaling factors are normally chosen such that the capacitances to ground of all of the lines, except possibly the input and output lines, are equal. Physically, this results in bars of approximately the same width when a rectangular bar configuration is used [12]. In the present case, however, additional points must be considered.

As mentioned previously in the case of a single section, it is necessary for the complementary lines in networks 1 and 2 to be separated by approximately the same physical distance. This constraint has been shown to be inherently satisfied due to the resonated realization, and the scaling factors should therefore be chosen such that there is a minimum amount of variation in the ground and coupling capacitances throughout both networks 1 and 2. To illustrate the general procedure of admittance scaling, an example of a five-element stepped digital elliptic filter is presented.

Numerical Example

The element values for the fifth degree low-pass prototype filter to be considered may be found in [5, p. 82). The filter with a stopband attenuation greater than 60.1 dB and a passband VSWR less than 1.22 has been chosen with an 11-percent bandwidth. The element values of this low-pass prototype are

$$C_1 = 0.9265 \qquad C_2 = 0.05866 \qquad C_3 = 1.666$$

$$C_4 = 0.1607 \qquad C_5 = 0.8363$$

with

$$\Omega_2 = 3.6119 \quad \text{and} \quad \Omega_4 = 2.3038.$$

If the center of the passband is normalized to a frequency $\pi/4=0.7854$ radians, then the band-edge frequency ω_c will be 0.7422 for an 11-percent bandwidth filter. From (3) we therefore have

$$a = \frac{11.54}{2} = 5.77$$

Utilizing (5) we then have

$$\lambda_{2+} = 1.361$$
$$\lambda_{2-} = 0.735$$
$$\lambda_{4+} = 1.219$$
$$\lambda_{4-} = 0.820$$

These values may now be used to determine the characteristic admittance matrices (9) and (10) for networks 2 and 1, respectively. Due to the number of algebraic operations necessary to derive these matrices, the entire procedure was programmed on a digital computer. After these matrices had been obtained, it was then necessary to apply admittance scaling to obtain a realizable network.

Initially, lines 1 and 5 on both networks were scaled by the factor $1/\sqrt{a}$. The remaining lines were scaled such that the corresponding main diagonal entries in the capacitance matrix representing network 1 were respectively equal to, and a factor 1.2 greater than, the main diagonal value of line 1 up to the center line from the input, and line 5 up to and including the center line from the output. This simple procedure appears to result in or near optimum relationship between the coupling and ground capacitances of both networks 1 and 2.

After scaling the entire matrices by the factor 7.534 in order to convert to capacitance values in a 50-ohm system, the resulting capacitance matrices for the particular numerical example under consideration were the following.

The dimensions of the bars in network 1 may be calculated from the capacitance matrix (13). However, before the dimensions of the bars in network 2 can be obtained, due account must be taken of the parasitic lumped end effect capacitances.

The following approximate compensation procedure may be adopted which neglects the interbar fringing capacitances.

Network 1

$$\begin{bmatrix} (0) & (1) & (2) & (3) & (4) & (5) & (6) \\ 7.534 & -3.136 & 0 & 0 & 0 & 0 & 0 \\ -3.136 & 9.546 & -3.062 & 0 & 0 & 0 & 0 \\ 0 & -3.062 & 11.456 & -1.395 & 0 & 0 & 0 \\ 0 & 0 & -1.395 & 11.558 & -3.848 & 0 & 0 \\ 0 & 0 & 0 & -3.848 & 11.558 & -3.068 & 0 \\ 0 & 0 & 0 & 0 & -3.068 & 9.631 & -3.136 \\ 0 & 0 & 0 & 0 & 0 & -3.136 & 7.534 \end{bmatrix} \tag{13}$$

Network 2

$$\begin{bmatrix} (1) & (2) & (3) & (4) & (5) \\ 7.661 & -1.653 & 0 & 0 & 0 \\ -1.653 & 11.456 & -2.583 & 0 & 0 \\ 0 & -2.583 & 11.269 & -2.588 & 0 \\ 0 & 0 & -2.588 & 11.558 & -4.561 \\ 0 & 0 & 0 & -4.561 & 9.312 \end{bmatrix} \tag{14}$$

Using a rectangular bar configuration, Getsinger's graphs [12] may be used to calculate the physical dimensions of the filter. Using a bar thickness to ground plate spacing of 0.6, which appears to be an ideal value for all realizable bandwidths, a typical filter of the form of the sketch shown in Fig. 6 may be obtained. In this sketch, it may be noted that the capacitance to ground of the first and last lines in network 2 are partially given by the capacitance to the grounded end walls. The introduction of these grounded walls enable the transformer elements to be readily introduced into the overall physical configuration.

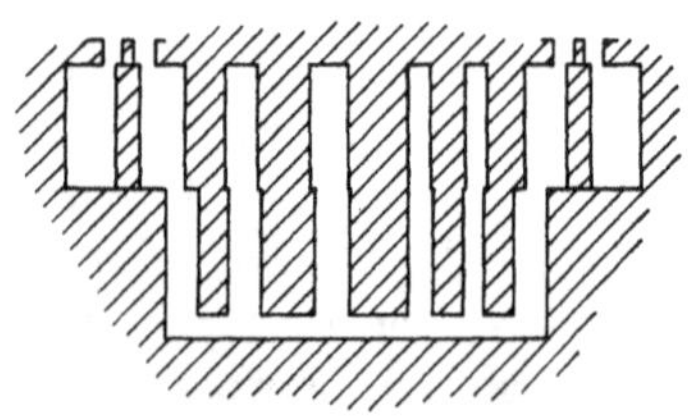

Fig. 6. Sketch of physical configuration of the designed filter.

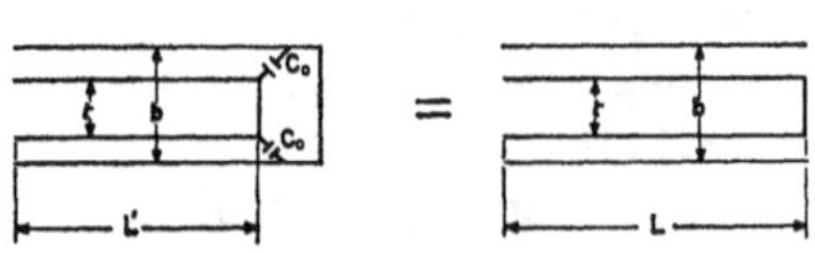

Fig. 7. Compensation for end effect capacitances in network 2.

If l is the length corresponding to an eighth-wavelength at center frequency, then the required lengh l' of network 2 will be given by

$$l' = l - 2C_0 b \tag{15}$$

where C_0 and b are defined in Fig. 7.

This equation has been obtained from the assumption that the distribution of lumped capacitance along the end of any bar in network 2 is proportional to the distribution of capacitance to the ground plates across the width of the corresponding bar, and that the distance l-l' is considerably shorter than a quarter wavelength.

In order to maintain the correct location of the poles of infinite attenuation, due to the reduction in length of network 2, the coupling admittances in network 2 must be augmented according to the equation

$$Y_{r\pm}' = \frac{Y_{r\pm}\lambda_{r\pm}}{\tan\left(\frac{l'}{l}\tan^{-1}\lambda_{r\pm}\right)} \tag{16}$$

where $Y_{r\pm}$ is the theoretically derived coupling admittance forming part of the resonant circuit which produces the pole of infinite attenuation at the normalized frequency $\tan^{-1}\lambda_{r\pm}$. This correction must also be applied at the center frequency to the admittances between the end lines and the grounded end walls.

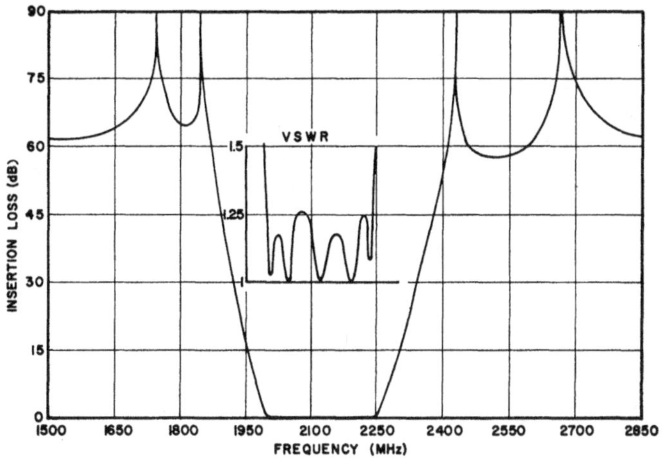

Fig. 8. Insertion loss and VSWR characteristics of experimental filter.

An experimental filter has been constructed using the values obtained from the numerical example with the above compensation for the end effect capacitances. The center frequency was 2150 MHz, from which we obtain $l=0.686$ inches. A ground plate spacing of 0.625 was chosen with a bar thickness to ground plate spacing of 0.6.

From (15) if C_0 is assumed to be the fringing capacitance for an isolated bar [12], we have

$$l' = 0.686 - 2 \times 1.2 \times 0.625 = 0.536.$$

Applying (16) to the coupling admittance given in the matrix (14) results in

$r^{\pm}$	$Y_{r\pm}$	$Y_{r\pm}'$
0±	2.400	3.41
2+	1.653	2.50
2−	2.583	3.51
4+	2.588	3.81
4−	4.561	6.27

where $Y_{0\pm}$ is assumed to be the isolated bar fringing capacitance of the end lines and is used in the calculation of the width of these lines.

The measured values of this experimental filter are shown in Fig. 8. Apart from a slight shift in the center frequency and poles of infinite attenuation, this response is in good agreement with the theoretical response.

Conclusions

In this paper a method has been presented whereby compact narrow-band bandpass elliptic function filters may be constructed. The realization is in the form of a pair of digital lines and has been called the stepped digital elliptic filter due to its basic physical configuration. Fractional bandwidths of approximately 30 percent and below may readily be obtained, but in the case of very narrow bandwidths (2 percent and below); difficulties are encountered due to the sensitivity of the response with respect to fractional percentage changes in the locations of the poles of infinite attenuation. However, due to the unique independent realization of these poles above and below the passband by the interbar coupling regions, every pole location may be individually tuned.

The main practical advantage of this filter in comparison to a conventional interdigital filter which exhibts a similar electrical response is the significant reduction in physical size. This comparison is based upon the realization of the same stopband attenuation level, the same passband VSWR and similar passband dissipation loss. The main reason for the similar dissipation loss in this reduced size filter may be attributed to the low midband group delay which in the experimental filter was less than 4 ns, resulting in a loss of 0.3 dB in the unplated brass structure.

Acknowledgment

The author is grateful to E. Curley for his assistance in the construction and measurement of the experimental filter.

References

[1] N. Saito, "Coupled transmission-line filter," *Sci. Repts. Res. Inst. Tohoku Univ., Sendai, Japan*, ser. B, vol. 14, no. 1, pp. 9–19. 1962.

[2] A. Matsumoto, "Network synthesis with multiwire lines," Research Institute of Applied Electricity, Hokkaido University, Sapporo, Japan, monograph 12, 1964.

[3] B. M. Schiffman and L. Young, "Design tables for an elliptic-function band-stop filter ($N=5$)," *IEEE Trans. Microwave Theory and Techniques*, vol. MTT-14, pp. 474–482, October 1966.

[4] R. Levy and I. Whiteley, "Synthesis of distributed elliptic-function filters from lumped-constant prototypes," *IEEE Trans. Microwave Theory and Techniques*, vol. MTT-14, pp. 506–517, November 1966.

[5] R. Saal, "Der Entwurf von Filtern mit Hilfe des Kataloges Normierter Tiefpasse," Germany: Telefunken, Backnay/Württemberg, W., GmbH, 1964.

[6] M. C. Horton and R. J. Wenzel, "Realization of microwave filters with equal ripple response in both pass and stop bands," *Proc. Symp. on Generalized Networks*, vol. 16. Brooklyn, N. Y.: Polytechnic Institute of Brooklyn Press, 1966, pp. 257–287.

[7] ——, "The digital elliptic filter—A compact sharp-cutoff design for wide bandstop or bandpass requirements," *IEEE Trans. Microwave Theory and Techniques*, vol. MTT-15, pp. 307–314, May 1967.

[8] R. Levy, "Three-wire-line interdigital filters of Chebyshev and elliptic-function characteristic for broad bandwidths," *IEE Electronics Letters*, vol. 2, pp. 455–456, December 1966.

[9] R. J. Wenzel, "Exact theory of interdigital bandpass filters and related coupled structures," *IEEE Trans. Microwave Theory and Techniques*, vol. MTT-13, pp. 559–575, September 1965.

[10] J. D. Rhodes, "The theory of generalized interdigital networks" *IEEE Trans. Circuit Theory* (to be published).

[11] R. J. Wenzel, "Theoretical and practical applications of capacitance matrix transformations to TEM network design," *IEEE Trans. Microwave Theory and Techniques*, vol. MTT-14, pp. 635–647, December, 1966.

[12] W. J. Getsinger, "Coupled rectangular bars between parallel plates," *IRE Trans. Microwave Theory and Techniques*, vol. MTT-10, pp. 65–72, January 1962.

MULTIPLEXERS VII

Multiplexers are used widely in communications, electronic countermeasures, in the design of efficient frequency multipliers, and for other applications. The design principles laid down in the five papers reprinted here apply to microwave multiplexers generally. For additional reading, we refer the reader to the chapter by Matthaei and Cristal.[4]

A Technique for the Design of Multiplexers Having Contiguous Channels*

E. G. CRISTAL†, MEMBER, IEEE, AND G. L. MATTHAEI†, MEMBER, IEEE

Summary—**A general procedure for the design of multiplexers having contiguous channel frequency bands is presented. Using this procedure, the individual channel band-pass filters are designed from low-pass prototype filters having a resistive termination at one end only. The use of parallel-connected band-pass filters designed in this fashion, along with a susceptance-annulling network, is shown to be capable of giving a nearly constant input conductance across the operating band of a multiplexer. A three-channel design example using comb-line band-pass filters was worked out and its input admittance and attenuation characteristics were computed. This design was also constructed and tested. The computer and experimental results demonstrated the validity of the theory.**

Introduction

MULTIPLEXERS are often needed in order to split a single channel carrying many frequencies into a number of separate channels carrying narrower bands of frequencies. They are also often used for the inverse process of summing a number of channels carrying different bands of frequencies, so that all of the frequencies can be put in a single broad-band channel without the loss of energy which would otherwise occur due to leakage of energy from any one of the input channels into the other input channels.

It might at first appear that the design of multiplexers could easily be accomplished by simply designing several filters using any of various band-pass filter design procedures [1] and then connecting the filters in parallel or in series, as the case may be. However, useful as these filter design procedures are for multiplexer design, they must be accompanied by special techniques in order to avoid undesirable interaction between the filters, which could result in very poor performance.

A conceptually simple way of solving the multiplexer problem is to use directional filters [2]. Filters of this sort, which are all designed for the same terminating resistance, can be cascaded to form a multiplexer that in theory completely avoids interaction between filters. In many cases, it is a very practical way of dealing with multiplexing problems, even though it is by no means always the most practical way. Each filter will generally have some parasitic VSWR which will affect the system significantly, if many filters are to be cascaded. However, probably the greatest practical drawback of directional filters is that each resonator of each filter has two different orthogonal modes and, if more than one or two resonators are required per filter, the tuning of the filters may be difficult. Also, if multiplexers with contiguous pass-bands are to be designed, directional filters are not very suitable, since it would be difficult to give them the 3-db cross-overs usually desired. (For example, if at the desired cross-over frequency the first directional filter removed half of the signal power, and if the second filter were also at a 3-db point, then the second filter would remove only half of the remaining power.)

Other design methods are available, if the channels of a multiplexer are quite narrow (say, of the order of 1 per cent bandwidth or less) and if the channels are separated by guard bands that are several times the pass-band width of the individual filters (or more). Then, relatively simple decoupling techniques should work quite well for preventing harmful interaction between filters [3]–[5].

When the channels of the multiplexer are contiguous, so that adjacent channels have attenuation characteristics that typically cross over at their 3-db points, decoupling techniques are not appropriate. In this case, the multiplexer should be designed as an integral unit. One design method—the method used in the synthesis of the multiplexer described in this paper—can be explained in terms of the circuit in Fig. 1. The schematic drawing represents an N-channel multiplexer consisting of specially designed band-pass filters. The filters are connected in parallel and a shunt, susceptance-annulling network is added at the input of the multiplexer to help provide a nearly constant total input admittance Y_{TN}. When properly designed, Y_{TN} will be nearly real and will approximate the generator conductance G_B across the operating band of the multiplexer. The band-pass filters that constitute the channels of the multiplexer are designed from a singly terminated, low-pass prototype filter rather than a doubly terminated prototype filter (*i.e.*, the prototype filter has a resistor termination at one end only). The reason is that, for the singly terminated prototype filter, the real part of the filter input admittance has the same frequency dependence as the filter transmission characteristic. This may be seen by observing the singly

* Received July 11, 1963; revised manuscript received September 13, 1963. This work was sponsored by the U. S. Army Electronics Research and Development Lab., Fort Monmouth, N. J., Contract DA 36-039 SC-87398.

† Stanford Research Institute, Menlo Park, Calif.

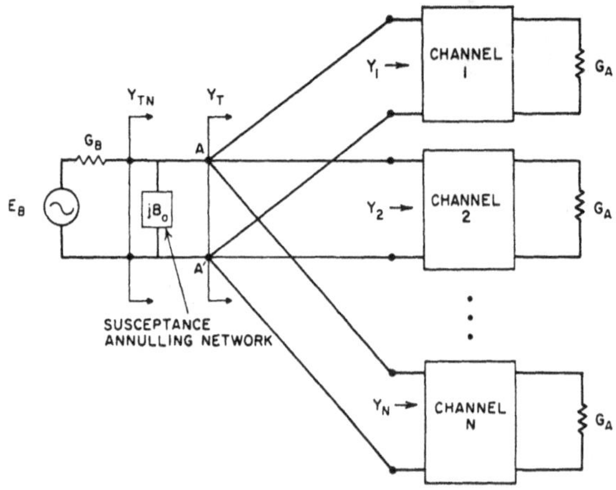

Fig. 1—A parallel-connected multiplexer.

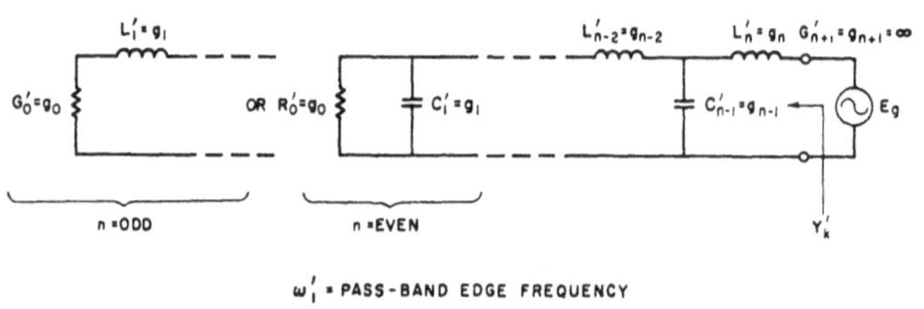

Fig. 2—Low-pass lumped element singly terminated prototype filter driven by a zero-impedance generator.

terminated low-pass prototype filter shown in Fig. 2.[1] This filter is driven by a zero-impedance generator at the right end and has a resistor termination at the left end. Because the filter is lossless, the power delivered to the load on the left in this circuit is given by

$$P = |E_g|^2 \operatorname{Re} Y_k' \tag{1}$$

where Y_k' is the admittance seen from the generator. Thus, if the filter has a Chebyshev transmission characteristic, for example, the Re Y_k' must also have a Chebyshev characteristic.

Fig. 3 shows typical Re Y_k' characteristics for Chebyshev, low-pass, prototype filters designed to be driven by zero-impedance generators. The input admittance Y_i of the channel filters in Fig. 1 are band-pass mappings of such low-pass, prototype, admittance characteristics. Notice that in Fig. 3 a low-pass prototype filter parameter g_{n+1}'' is defined. This parameter will be referred to in the discussion of the details of multiplexer design. However, it should be noted at this point that either g_{n+1}'' or its reciprocal corresponds to the geometric mean between the value of Re Y_k' at the top of the ripples and Re Y_k' at the bottom of the ripples. This admittance level for the low-pass prototype is analogous to the driving source admittance G_B in Fig. 1. That is, the multiplexer described in this paper is designed to have a Chebyshev Re Y_{TN} characteristic with G_B equal to the mean value of the ripples.

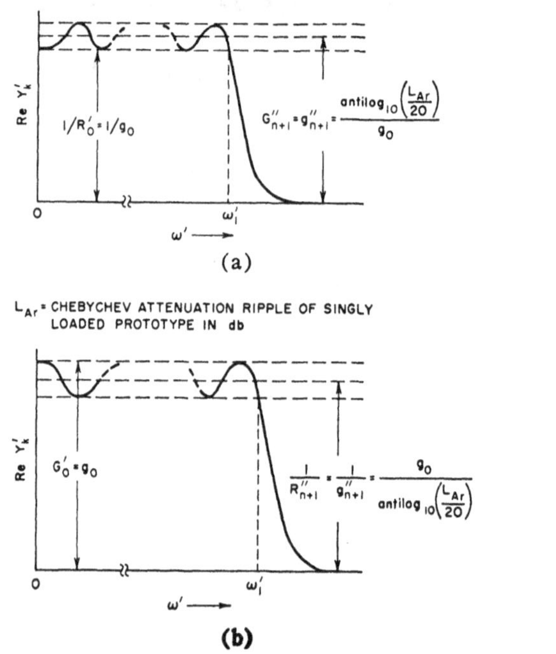

Fig. 3—Re Y'_k response of a low-pass prototype filter and definition of the quantity g''_{n+1}. (a) Case of n even. (b) Case of n odd.

Comb-Line Filter Description[2]

The multiplexer design techniques described in this paper are applicable for use with many different types of filter structures. However, to illustrate one application of these multiplexer design techniques, the use of comb-line filter structures will be assumed herein.

Fig. 4 shows a comb-line band-pass filter in strip-line form. The resonators in this type of filter consist of TEM-mode transmission-line elements that are short-circuited at one end and have a lumped capacitance C_k^s between the other end of each resonator line element and ground. In Fig. 4, lines 1 to n and their associated capacitances, C_1^s to C_n^s, constitute resonators, while lines 0 and $n+1$ are not resonators but simply part of impedance-transforming sections at the ends of the filter. Coupling between resonators is achieved in this type of filter by way of fringing fields between resonator lines. With the capacitors C_k^s present, the resonator lines will be less than $\lambda_0/4$ long at resonance (where λ_0 is the wavelength in the medium of propagation at midband), and the coupling between resonators is predominantly magnetic. Interestingly enough, if the capacitors C_k^s were not present, the resonator lines would be a full $\lambda_0/4$ long at resonance, *and the structure would have no pass-band* [8]. Without some kind of reactive loading at the ends of the resonator line elements, the magnetic and electric coupling effects would cancel each other, and the comb-line structure would become an all-stop structure.

For reasons described above, it is usually desirable to

[1] Tabulated element values for such filters will be found in Matthaei [1] and Weinberg [6].

[2] The design equations for comb-line filters and their derivation are given by Matthaei [7].

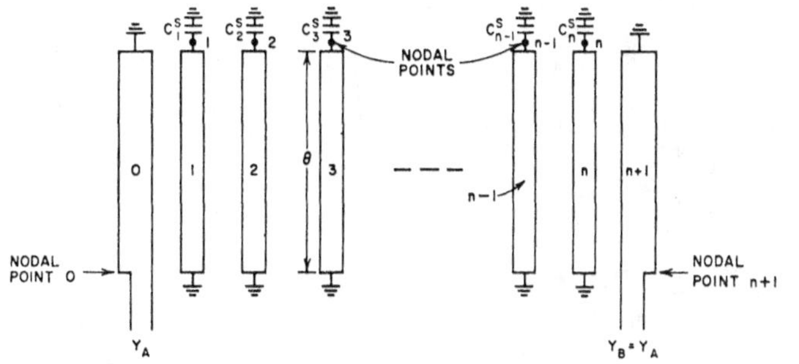

Fig. 4—A comb-line band-pass filter.

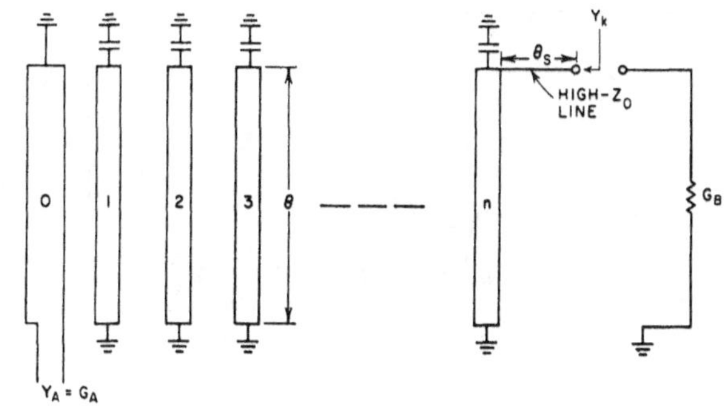

Fig. 5—A possible comb-line filter configuration for use in multiplexers.

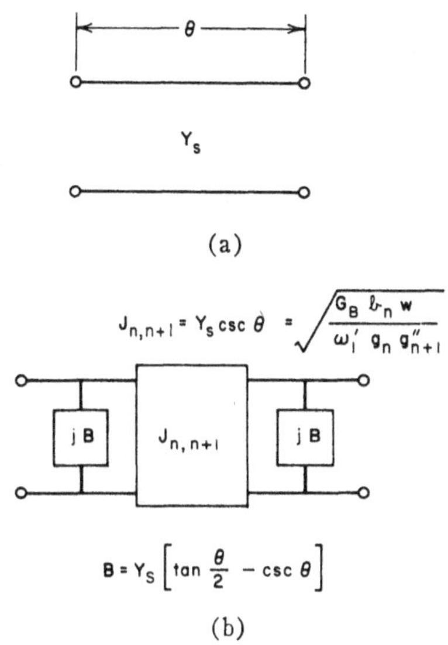

Fig. 6—Use of a high-impedance wire as a J-inverter.

make the capacitances C_k^s in this type of filter sufficiently large so that the resonator lines will be $\lambda_0/8$ or less long at resonance. Besides permitting efficient coupling between resonators (with sizeable spacings between adjacent resonator lines), the resulting filter will be relatively small. In this type of filter, the second pass-band occurs when the resonator line elements are somewhat over one-half wavelength long, so that, if the resonator lines are $\lambda_0/8$ long at the primary pass-band, the second pass-band will have its center at slightly over four times the frequency of the center of the first pass-band. If the resonator line elements are made to be less than $\lambda_0/8$ long at the primary pass-band, the second pass-band will be even further removed from the primary pass-band.

To adapt the comb-line filter design [7] for use in multiplexers, it was necessary to modify the input end of the filter in order to provide for coupling several filters to a common input junction. The modification used is shown schematically in Fig. 5. The comb-line filter in this figure is seen to include a fine-wire, high-impedance line at the input end.

Fig. 6 shows a high-impedance wire and its equivalent circuit when used as an admittance inverter $J_{n,n+1}$. (An ideal admittance inverter is defined as a device which operates like a quarter-wavelength line of characteristic admittance $J_{n,n+1}$ mhos at all frequencies [7].[3]) The The comb-line-filter design equations [7] of Matthaei, which utilize the admittance inverter concept, were modified to use this high-impedance-wire type of inverter at one end of the filter. The susceptance at one side of the inverter in Fig. 6 was compensated for by absorbing it into resonator n, while the susceptance B on the other side of the inverter was effectively absorbed into the susceptance-annulling network. In this case, the high-impedance wire type of coupling was suggested because it would help prevent the common junction from being crowded. However, it is desirable to keep the high-impedance wire short in order to avoid unwanted resonances, while not having it so small in diameter that it would increase the losses.

The design equations for the modified comb-line filter having a high impedance line in the input section are given in Getsinger [10]. These equations are equivalent to those of Matthaei [7] except for the input section. Hence, by use of the design data in Matthaei [7] along with the information in Fig. 6, the desired comb-line filter designs can be obtained.

A Computed Example Using Comb-Line Filters

Using the equations for the modified comb-line filter, a four-resonator, 10-per cent bandwidth filter design was worked out using a $n=4$, singly terminated, Chebyshev, prototype filter having 1-db ripple. The filter design was expressed analytically in an approximate form and its input admittance Y_k was computed using a digital computer. The results are shown in Fig. 7, normalized with respect to G_A. Note that Re Y_k/G_A is very nearly perfectly Chebyshev, while in the pass-band the slope of Im Y_k/G_A is negative on the average.

If band-pass filters with input admittance characteristics such as that in Fig. 7 are designed to cover contiguous bands, and if they are designed so that the Re Y_k/G_A characteristics of adjacent filters overlap at approximately their Re $Y_k/G_A = 0.5$ point[4] or slightly

[3] Cohn [9] describes impedance inverters. An admittance inverter is an admittance representation of an impedance inverter.

[4] By (1), it can be shown that the Re Y_k/G_A characteristic for a singly terminated Chebyshev filter can be computed using standard Chebyshev transfer functions. Equations for doing this are given in Getsinger [10].

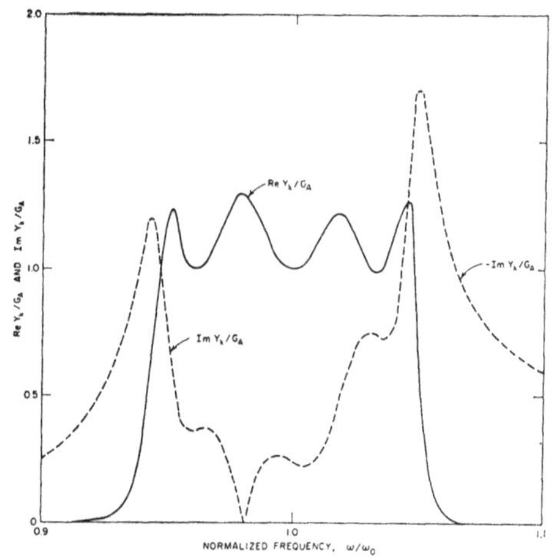

Fig. 7—Computed normalized Y_k of a modified comb-line filter.

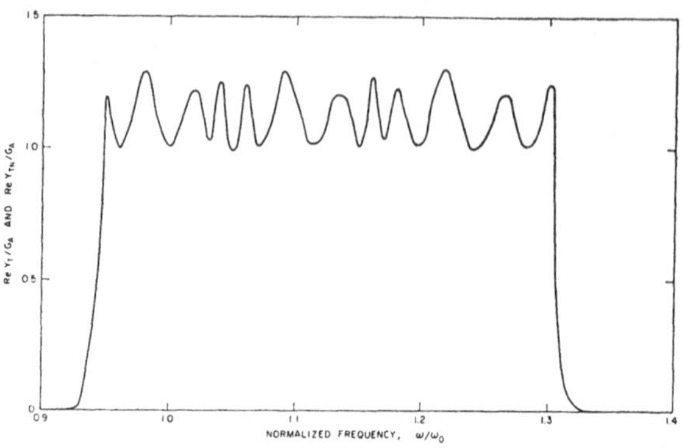

Fig. 8—Computed normalized Re Y_T of a 3-channel multiplexer using modified comb-line filters.

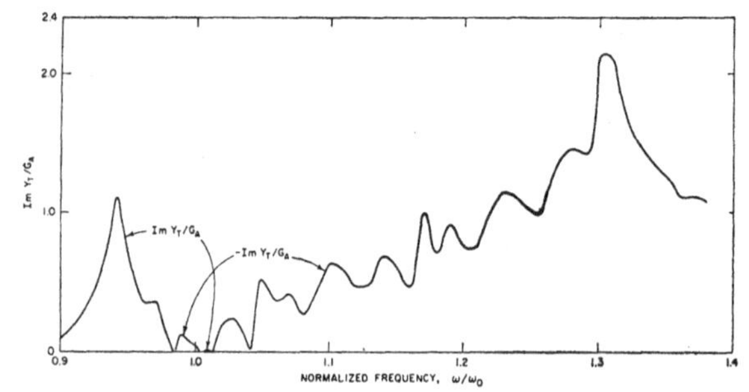

Fig. 9—Computed Im Y_T of a 3-channel multiplexer using modified comb-line filters.

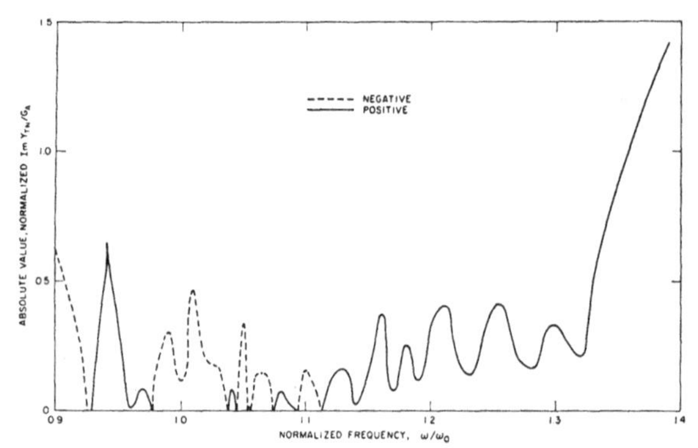

Fig. 10—Computed normalized Im Y_{TN} of a 3-channel multiplexer using modified comb-line filters and appropriate annulling network.

below, then their total input admittance $Y_T = Y_1 + Y_2 + Y_3 + \cdots + Y_N$, when they are paralleled, will also have an approximately Chebyshev real-part characteristic. This was done in the example in Fig. 8. This figure shows the computed real-part characteristic for three, paralleled, comb-line filters with contiguous pass-bands, where the individual filters have input admittance characteristics as shown in Fig. 7 except for a shift in frequency. Fig. 9 shows the corresponding Im Y_T/G_A characteristic for this three-channel design.

Note that the Im Y_T/G_A characteristic in Fig. 9 is negative on the right side of the figure and that, on the average, the slope of the curve is negative throughout the operating band of the multiplexer. Since, by Foster's reactance theorem, the susceptance slope of a lossless network is always positive, the operating-band susceptance (Fig. 9) with its average negative slope can be largely cancelled by adding an appropriate lossless shunt branch (which will have a positive susceptance slope). In this case, a susceptance-annulling branch could consist of a short-circuited stub of such a length as to give resonance at the normalized frequency $\omega/\omega_0 = 1.02$, where the Im Y_T/G_A curve in Fig. 9 is approximately zero. Estimates indicate that, if the stub
192
had a normalized characteristic admittance of $Y_s/G_A = 3.85$, the susceptance slope of the annulling network should be about right. Fig. 10 shows the normalized susceptance Im Y_{TN}/G_A after the susceptance-annulling network has been added. Note that although Im Y_{TN} has not been completely eliminated, it has been greatly reduced.

Since Re Y_{TN}/G_A = Re Y_T/G_A, the total input admittance Y_{TN} with the annulling network in place is given by Figs. 8 and 10. To the extent that Y_{TN} approximates a constant conductance across the operating band, G_B and Y_{TN} in Fig. 8 will act as a resistive voltage divider, and the voltage developed across terminals A–A' of Fig. 1 will be constant with frequency. Under these conditions, the individual filters would have exactly the same response as if they were driven by zero-impedance generators (as they were designed to be driven). However, since Y_{TN} only approximates a constant conductance, the performance will be altered somewhat from this idealized performance.

The driving generator conductance G_B for the trial multiplexer design was given a normalized value of $G_B/G_A = 1.15$, which makes the generator conductance equal to the mean value of the ripples in Fig. 8.[5] Fig. 11 shows the computed response of the multiplexer, while Fig. 12 shows the details of the pass-band response in enlarged scale. Note that the attenuation characteristics cross over at about the 3-db points, and that, although the filters were designed to have 1-db Chebyshev ripple when driven by a zero-impedance generator, the pass-

[5] This makes G_B correspond to $G_{n+1}'' = g_{n+1}''$ for the prototype as indicated in Fig. 3(a).

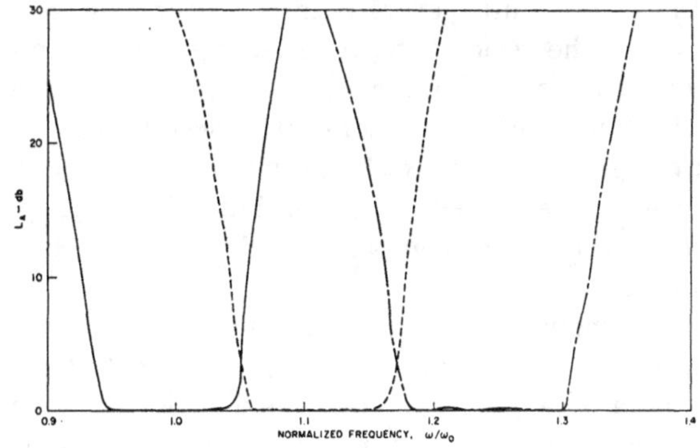

Fig. 11—Computed attenuation characteristics of the multiplexer channels.

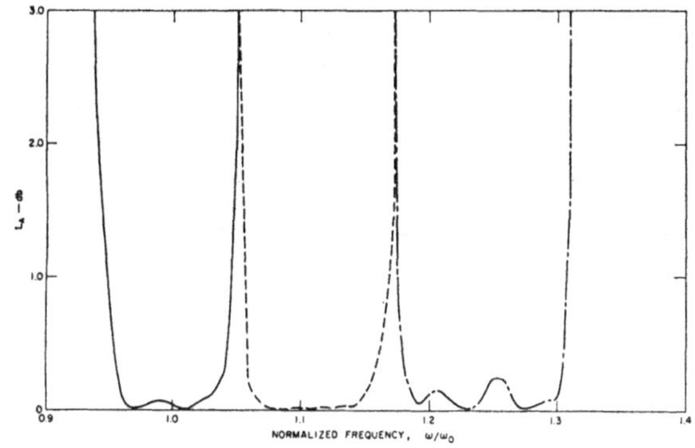

Fig. 12—Computed attenuation characteristics of the multiplexer channels in the pass-band.

band attenuation is much less than that in the completed multiplexer. This is largely due to the fact that adding the generator internal conductance G_B tends to mask out the variations in Re Y_{TN}. Also, choosing G_B to be equal to the mean value of Re Y_{TN} in the operating band tends to reduce the amount of mismatch that will occur.

Experimental Work

Using the design theory and equations previously discussed and referenced, a microwave multiplexer based on the preceding trial design was constructed. A sketch of the multiplexer is given in Fig. 13. The view in Fig. 13 is that seen when looking at the top of the multiplexer. The susceptance-annulling network of the multiplexer cannot be seen in this particular view. It consists of a specially constructed, low-impedance, coaxial line that is connected to the common junction and extends through the lower ground plane opposite the common input of the triplexer. The electrical length of the coaxial line was varied by sliding a short-circuit block.[6]

The multiplexer attenuation, after final tuning and adjusting of the annulling network for an optimum response, is given in Fig. 14. Fig. 15 gives the multiplexer attenuation in the pass-band using an expanded ordinate. The attenuation in the pass-band of the lower- and middle-frequency channels is typically 0.5 db. The

[6] For a more detailed description of the construction of the triplexer, see Matthaei [11].

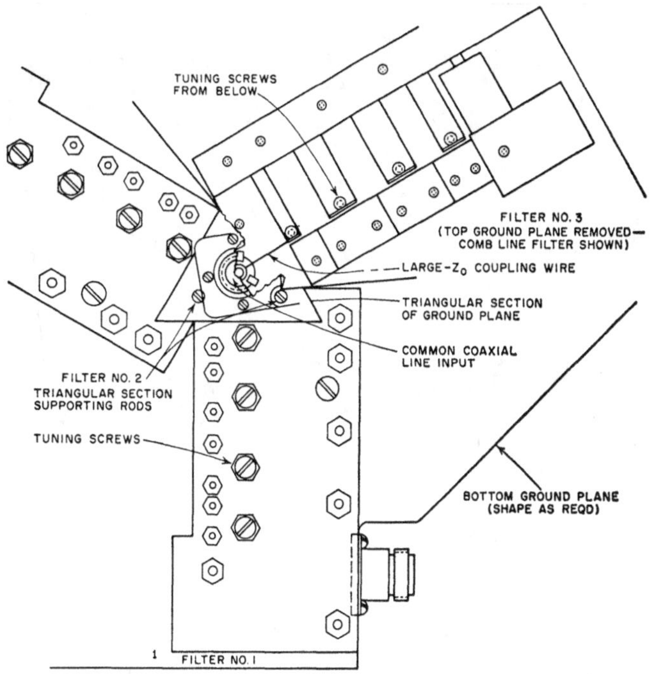

Fig. 13—Sketch of the top view of the 3-channel multiplexer showing one of the comb-line filters with its cover plate removed.

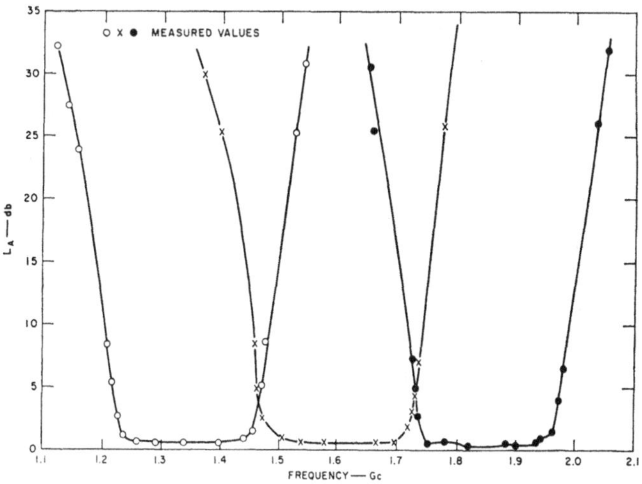

Fig. 14—Measured attenuation of the 3-channel multiplexer.

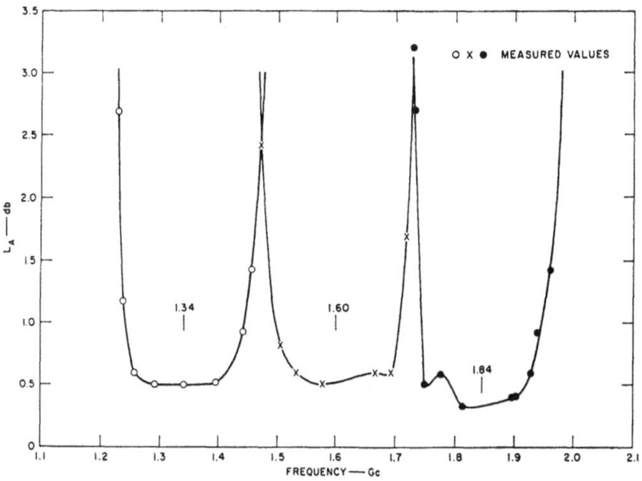

Fig. 15—Measured attenuation in the pass-band of the 3-channel multiplexer.

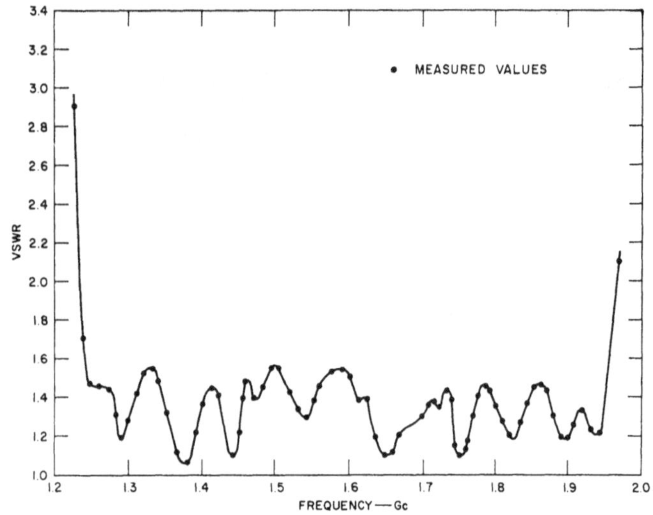

Fig. 16—Measured VSWR in the pass-band of the 3-channel multiplexer.

attenuation in the pass-band of the highest-frequency channel is approximately 0.4 db. The VSWR of the multiplexer (measured at the common input) is shown in Fig. 16 and is less than 1.6 throughout the pass-band except at the multiplexer band edges.

The 3-db percentage bandwidths of each channel were determined. These were defined as the difference of the frequencies at which the attenuation is 3 db, divided by the arithmetic mean of those frequencies. The bandwidths were found to be 17, 16, and 13 per cent for the lowest-, middle-, and highest-frequency channels, respectively. These percentages are larger than the corresponding values for the triplexer example previously discussed, whose channel 3-db percentage bandwidths were 11 per cent each. "Bandwidth spreading" has been noted in comb-line filters before [7]. However, in the case of a previous experimental comb-line filter having a design bandwidth of 10 per cent [7], the bandwidth spreading was less than in the present cases. Bandwidth spreading is believed to be due to coupling effects beyond nearest neighbor line elements, which were neglected in the derivation of the design equations for comb-line filters. It can be compensated for by increasing the spacings between the resonators slightly.

194

Conclusions

A design theory for multiplexers having contiguous channels was presented. The examples discussed were cases of multiplexers using filters connected in parallel. However, the same approach can be carried out on a dual basis for multiplexers with the filters connected in series. (In the series case, a series reactance-annulling network is required instead of a shunt, susceptance-annulling network.) The computed performance of the trial design showed that the proposed technique works very well. The measured performance of the triplexer based on the design compared well with the computed performance.

One of the main difficulties of this approach for some applications is that the filters are assumed to be interconnected at a point. Space considerations may make it difficult to interconnect a very large number of filters, although in many cases three or four can be interconnected very easily. The series-inductance coupling proposed for the case of multiplexing comb-line filters provides one way of interconnecting the filters while keeping them separated from each other as much as possible.

Acknowledgment

The experimental three-channel multiplexer was constructed by R. Pierce and C. A. Knight. P. R. Rezneck took the experimental data. The computed example of the three-channel multiplexer design was programmed for the Burroughs 220 computer by J. R. Herndon.

References

[1] G. L. Matthaei, L. Young, and E. M. T. Jones, "Design of Microwave Filters, Impedance-Matching Networks, and Coupling Structures," Stanford Research Inst., Menlo Park, Calif., SRI Project 3527, Contract DA 36-039 SC-87398, vols. I and II; January, 1963.

[2] S. B. Cohn and F. S. Coale, "Directional channel separation filters," Proc. IRE, vol. 44, pp. 1018–1024; August, 1956.

[3] G. C. Southworth, "Principles and Applications of Waveguide Transmission," D. Van Nostrand Co., Inc., New York, N. Y.; 1950.

[4] D. Alstadter and E. O. Houseman, Jr., "Some notes on strip transmission line and waveguide multiplexers," 1958 IRE Wescon Convention Record, pt. 1, pp. 54–69.

[5] G. L. Ragan, "Microwave Transmission Circuits," MIT Rad. Lab. Series, McGraw-Hill Book Co., Inc., New York, N. Y.; vol. 9, pp. 708–709; 1948.

[6] L. Weinberg, "Additional tables for design of optimum ladder networks," *J. Franklin Inst.*, vol. 264, pts. I and II, pp. 7–23 and 127–138; July and August, 1957.

[7] G. L. Matthaei, "Comb-line band-pass filters of narrow or moderate bandwidth," *Microwave J.*, vol. 6, pp. 82–91; August, 1963.

[8] J. T. Bolljahn and G. L. Matthaei, "A study of the phase and filter properties of arrays of parallel conductors between ground planes," Proc. IRE, vol. 50, pp. 299–311; March, 1962.

[9] S. B. Cohn, "Direct-coupled resonator filters," Proc. IRE, vol. 45, pp. 187–196; February, 1957.

[10] W. J. Getsinger, E. G. Cristal, and G. L. Matthaei, "Microwave Filters and Coupling Structures," Stanford Research Inst., Menlo Park, Calif., SRI Project 3527, Contract DA 36-039 SC-87398, QPR 6, pp. 85–90; August, 1962.

[11] G. L. Matthaei, B. M. Schiffman, E. G. Cristal, and L. A. Robinson, "Microwave Filters and Coupling Structures," Stanford Research Inst., Menlo Park, Calif., SEI Project 3527, Contract DA 36-039 SC-87398, Final Rept.; February, 1963.

Application of Exact Synthesis Methods to Multichannel Filter Design

R. J. WENZEL, MEMBER IEEE

Abstract—An exact design procedure is presented for contiguous-band multichannel filters. The procedure is theoretically valid at all frequencies for filters which employ TEM-mode (transverse-electromagnetic-mode), quarter-wave line networks. The design of diplexing networks is considered in detail with multichannel filters being realized as a cascade of suitable diplexers. Microwave complementary component filters are shown to fulfill the requirement for the design of diplexing and multichannel filters with theoretically perfect match at all frequencies.

A design procedure using component networks with Chebyshev characteristics is given. This allows a narrow crossover region and high isolation to be obtained for a given network complexity at the cost of a slight increase in input-port VSWR (Voltage Standing-Wave Ratio). The design procedure makes use of published tables and does not require a detailed knowledge of synthesis techniques.

Experimental results for two diplexing networks, one a complementary pair and the other a Chebyshev pair, are given in verification of the theory presented.

I. Introduction

THE DESIGN of passive multichannel filters requires a network that will separate a given frequency band into N channels with minimum insertion loss and low-input-port VSWR. The complexity of the network is determined by the allowable width of contiguous channel crossover regions and the required isolation between bands.

The design approach reported in this paper uses a number of two-channel filters (diplexers) as basic building blocks in obtaining the N-channel filter as shown in Fig. 1. A design technique for multiplexers using a different network configuration is given in Cristal and Matthaei [1]. By using exact synthesis techniques, a three-port network with the desired band separation and a low-input VSWR over a broad frequency range is obtained. Networks of this type can then be used, in cascade, to realize multichannel filters.

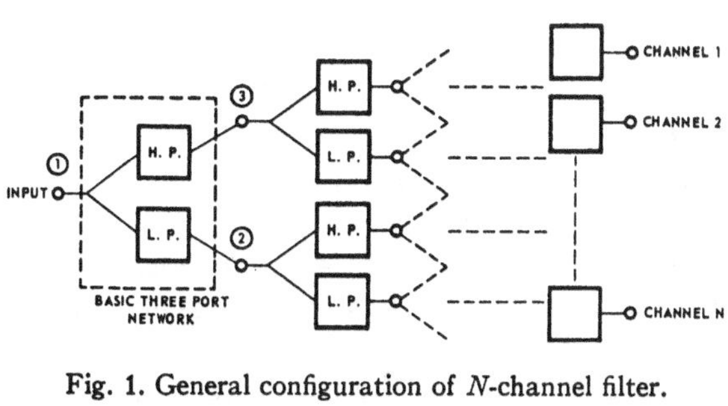

Fig. 1. General configuration of N-channel filter.

Manuscript received April 13, 1964; revised June 25, 1964.
The author is with the Research Labs. Div., Bendix Corporation, Southfield, Mich.

II. The "Ideal" Diplexer

An "ideal" diplexer is defined as a lossless, resistively terminated three-port network that separates the frequency spectrum into two isolated bands with a perfect match at the input port.

The network to be considered for the ideal diplexer is that of two parallel-connected, resistively terminated, lossless filters, one of which is low-pass, and the other high-pass[1] as indicated schematically in Fig. 2.

A simple parallel connection of low-pass and high-pass filters does not insure that the resulting network will approximate the ideal diplexer characteristic. In most single-filter design procedures, only the transmission response is considered and little attention is given to the input admittance characteristics. In the stop-band, the input admittance of the filter will be a susceptance that varies with frequency, and thus, one that will seriously affect the performance of another filter connected in parallel. For narrow-band channels separated by guard bands, a good match only may be required over the narrow pass-bands involved, and conventional design procedures can be employed [3]. Where the bandwidths involved are moderate to wide and the channels are contiguous, special techniques are required.

A parallel connection of networks of the type shown in Fig. 2 provides a perfect match at the input port if the component filters are defined as complementary.[2] This requires that the input admittances of the low-pass and of the high-pass filters obey the normalized relationship

$$Y_{\text{in}} = Y_{\text{inLP}} + Y_{\text{inHP}} = 1. \tag{1}$$

Consequently, the sum of real parts of the input admittances must be constant and the reactive parts must be equal and of opposite sign. Thus

$$\begin{aligned} \text{Re}\,[Y_{\text{inLP}}] + \text{Re}\,[Y_{\text{inHP}}] &= 1 \\ \text{Im}\,[Y_{\text{inLP}}] + \text{Im}\,[Y_{\text{inHP}}] &= 0. \end{aligned} \tag{2}$$

[1] The analysis for the dual case of series-connected networks follows in an analogous manner. The admittance case is, in general, more applicable for implementation at microwave frequencies. Because of the repetitive nature of the microwave networks to be considered, the high-pass filter becomes a band-pass filter, and the low-pass filter becomes a stop-band filter [2]. Thus, only low-pass and high-pass characteristics need be investigated.

[2] A thorough discussion of complementary filters can be found in Guillemin [4] and Weinberg [5]. Examples of microwave complements are given by Wenzel [2].

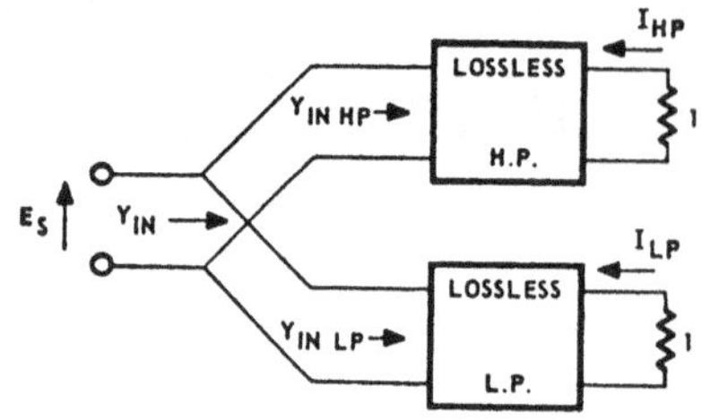

$$Y_{IN} = Y_{IN\,LP} + Y_{IN\,HP} = \mathrm{Re}\left[Y_{IN\,LP}\right] + j\,\mathrm{Im}\left[Y_{IN\,LP}\right] + \mathrm{Re}\left[Y_{IN\,HP}\right] + j\,\mathrm{Im}\left[Y_{IN\,HP}\right]$$

$$\left|Y_{12}\right|^2_{HP} = \left|\frac{I_{HP}}{E_S}\right|^2$$

$$\left|Y_{12}\right|^2_{LP} = \left|\frac{I_{LP}}{E_S}\right|^2$$

Fig. 2. Basic diplexer configuration and network definitions.

For lossless networks terminated in a unit resistance

$$\mathrm{Re}\,[Y_{\mathrm{in}}] = |\,Y_{12}\,|^2 \tag{3}$$

where Y_{12} is the ratio of output current to input voltage. (See Guillemin [4], p. 446.) Therefore, (1), (2), and (3) require that

$$|\,Y_{12}\,|^2_{\mathrm{LP}} + |\,Y_{12}\,|^2_{\mathrm{HP}} = 1. \tag{4}$$

To design a complementary filter pair, it is necessary to designate the $|\,Y_{12}|^2$ in (4) with suitable functions and then to synthesize networks that realize these functions. Details of methods used in function designation, sometimes referred to as the "approximation problem," can be found in Guillemin [4] and Weinberg [5]. In most instances, this lengthly process can be avoided by the use of tables [5], [6]. To obtain low-input VSWR at all frequencies, it is necessary that the input admittances of the component filters be realized exactly. Exact syntnesis procedures for microwave TEM-mode networks consisting of quarter-wave lines can be found in Ozaki and Ishii [7], [8], Grayzel [9], Shiffman and Matthaei [10], and also in Wenzel [2]. The microwave networks can be designed by applying lumped element

196 techniques, and using the transformation

$$S = j\Omega = j\tan\frac{\pi\omega}{2\omega_0}\,.$$

A straightforward application of Kuroda's identities [2], [8] and/or Richard's theorem [11] then leads to practical microwave structures with characteristics that are known theoretically at all frequencies.

Complementary filter design, based on transfer admittance prototype networks, requires the use of singly terminated networks rather than the doubly terminated networks more commonly used at microwave frequencies. Thus, each component network is designed as though it were being driven by an ideal voltage source rather than with a matched generator. When the two complementary networks are connected in parallel, the input admittance becomes purely conductive. Thus, if the network is driven by a matched source, the input voltage remains constant as was assumed in the design of the component filters. The characteristics of the network pairs are determined by both the magnitude and phase of the input admittance rather than by a magnitude function alone as in conventional, doubly terminated microwave filter design. To achieve the desired performance, techniques must be used that preserve both the magnitude and phase characteristics of the input admittance. Exact synthesis procedures are ideally suited to meeting these requirements.

A commonly used low-pass filter characteristic, suitable for use in designing a complementary filter pair is the nth order Butterworth function given by

$$|\,Y_{12}\,|^2_{\mathrm{LP}} = \frac{1}{1+\Omega^{2n}} \tag{5}$$

where

$$\Omega = \tan\frac{\pi\omega}{2\omega_0}\,.$$

The complementary characteristic is given by

$$|\,Y_{12}\,|^2_{\mathrm{HP}} = \frac{\Omega^{2n}}{1+\Omega^{2n}}\,. \tag{6}$$

The element values in the ladder prototype for the low-pass filter can be obtained from tables [5], [6]. The high-pass network has the same geometry as that of the low-pass network with every element being replaced by its dual and having a reciprocal element value [2].

Prototype networks for the third-order filter pair are shown in Fig. 3(a).[3] To obtain a realizable microwave structure, unit elements of characteristic im-

[3] The capacitor and inductor symbols in Fig. 3 are used to represent open-circuited and short-circuited quarter-wave sections of transmission line, respectively [2], [12].

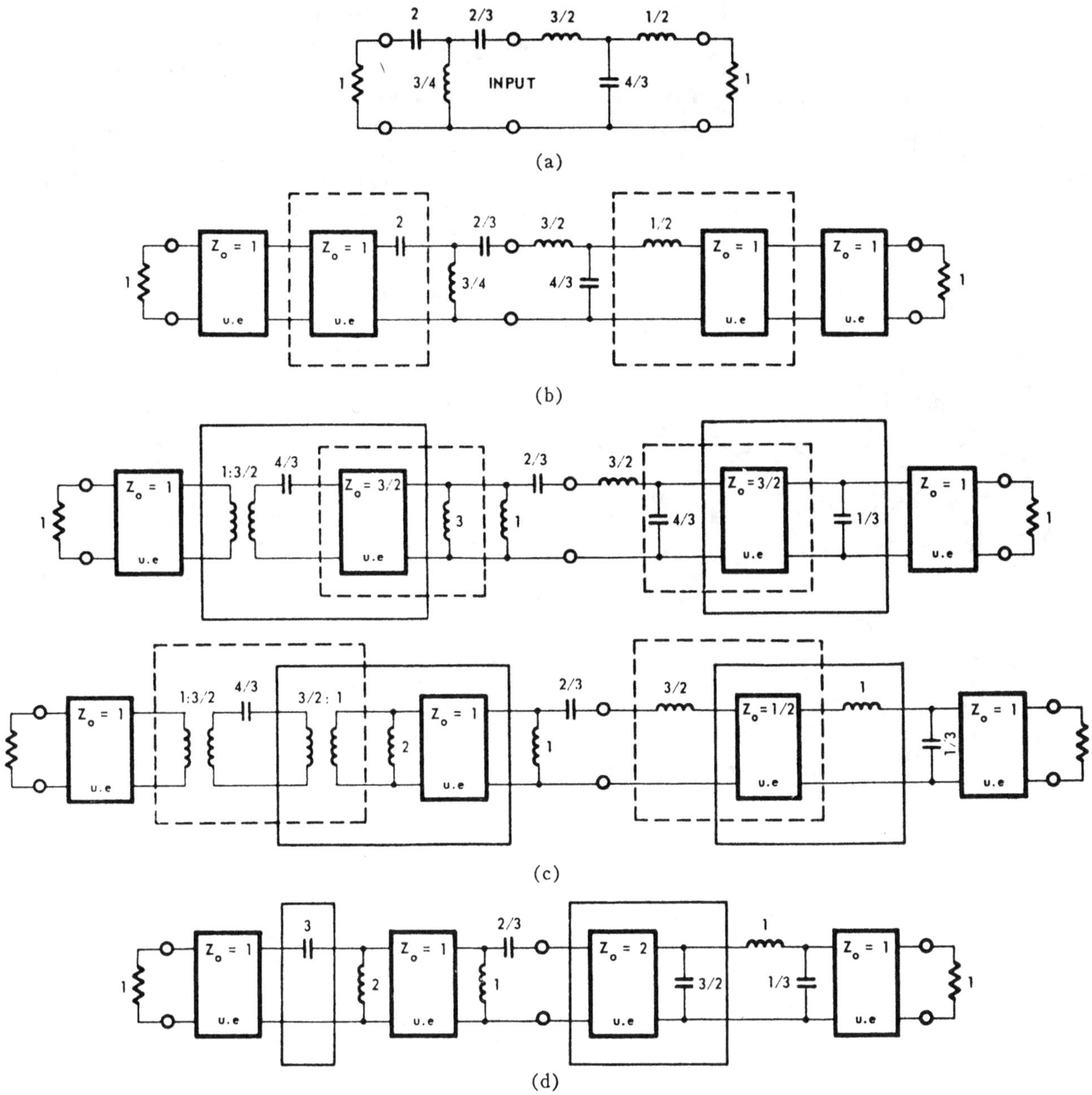

Fig. 3. (a) Third-order butterworth complementary filter pair prototype. (b) Insertion of unit elements (μ.e.). Note: The dotted lines indicate the section that will be transformed in the next diagram. The solid lines indicate the results of the previous transformation. (c) Application of Kuroda's identities. (d) Final circuit.

pedance $Z_0=1$ are inserted between each network and its load as shown in Fig. 3(b). This changes the phase of the transfer admittances, but maintains the input admittance unchanged. Kuroda's identities now can be applied to obtain a practical realization. One method of accomplishing this is demonstrated in Fig. 3(c). As is evident, the physical form of the filter pairs is not unique. The identities can be further applied to the final circuit, Fig. 3(d), and many other realizations can be obtained. The manner in which the identities are applied is dictated by the practical realizability of resulting element values. For the series capacitors and shunt inductors, care must be taken in applying the transformations so that ideal transformers are not generated. In the foregoing example, this is accomplished by splitting the shunt inductor in Fig. 3(c) in such a way that the resulting transformer has the same turns ratio, but of opposite sense, as that generated by the series capacitor transformation [2].

A practical microwave realization[4] of the above diplexer is shown in Fig. 4 and an experimental model is shown in Fig. 5. The experimental input VSWR and transfer characteristics in the frequency range 0 to 8 Gc/s are given in Fig. 6. The repetitive nature of the distributed lines results in multiple crossover points. If only one crossover is of interest—as is often the case—maximum experimental VSWRs of less than 1.2 are easily

[4] The manner in which the numbers associated with the prototype are converted to transmission line dimensions is discussed thoroughly in [2], [7], [9], and [12].

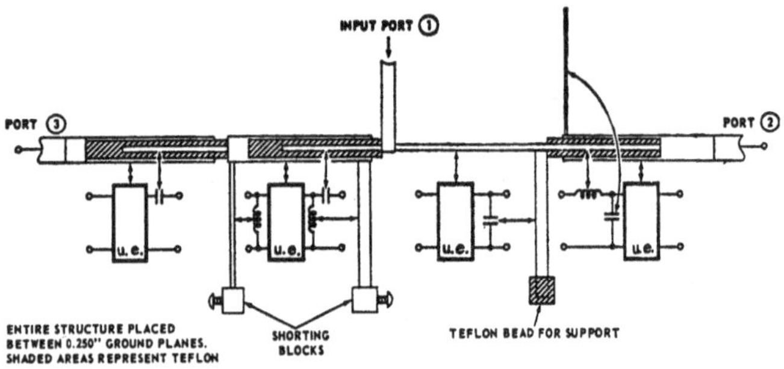

Fig. 4. Diagram of TEM circuit that realizes a third-order Butterworth complementary filter pair.

Fig. 5. Third-order Butterworth complementary filters.

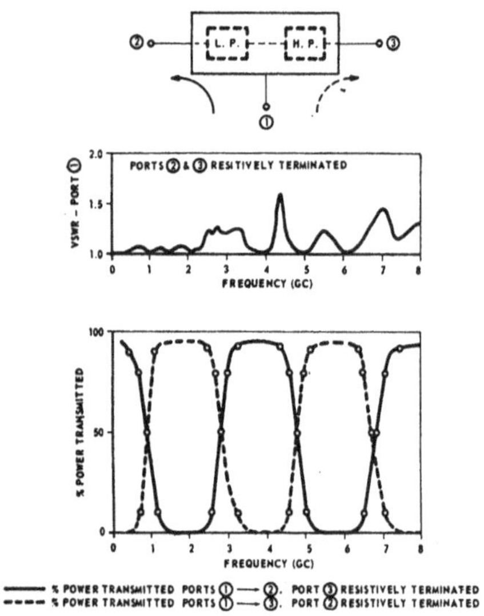

Fig. 6. Measured insertion loss and VSWR characteristics of third-order Butterworth complementary filters.

obtainable. Sharper crossover characteristics can be obtained by using filters of a higher order. Any desirable bandwidth can be obtained by frequency scaling the prototype networks [2].

The above diplexer is "ideal" in the sense that it has an input match that is theoretically perfect. It is "non-ideal" in that only for an infinite number of filter elements does the crossover region become zero in width. Steeper filter skirts, and consequently a narrower crossover region, can be obtained for a given number of filter elements if a Chebyshev approximation to the desired characteristic is employed. From the viewpoint of input VSWR, the optimum condition occurs when the filters are complementary. However, imposing this condition on Chebyshev filter pairs leads to poor isolation. This effect can be understood by considering a 0.1 dB ripple filter. When the power response of this filter is down 0.1 dB or approximately two per cent, the response of a true complement must pass the two per cent power resulting in only 17 dB isolation. However, this amount of isolation is insufficient in many practical situations. Thus, requiring the input VSWR to be 1.0 puts a severe restriction on the width of the crossover region and the realizable isolation. Intuitively, one expects that it should be possible to improve the isolation and crossover characteristics by relaxing the input VSWR requirement.

III. Narrower Crossover Regions Obtained by Input-Port Mismatch

Insight into the manner in which the VSWR requirement might be relaxed can be obtained by considering the mathematical form of the pertinent filter characteristics. A significant fact is that the input admittances associated with parallel-connected complementary filters are minimum susceptive, i.e., they do not contain j-axis poles. Under this condition, the admittances are analytic in a one-half plane and have real and imaginary parts that are Hilbert transforms. Thus

$$\operatorname{Im}\left[Y_{\text{in}}(\Omega)\right] = -\frac{1}{\pi}\int_{-\infty}^{+\infty} \frac{\operatorname{Re}\left[Y_{\text{in}}(x)\right]dx}{\Omega - x}. \tag{7}$$

198

For the functions under consideration, one can differentiate both real and imaginary parts of Y_{in} a number of times, carry out the indicated integration, and then integrate the left-hand side the same number of times it had been originally differentiated to obtain $\operatorname{Im}\left[Y_{\text{in}}(\Omega)\right]$. That is

$$\left.\begin{aligned} \frac{d^n \operatorname{Im}\left[Y_{\text{in}}(\Omega)\right]}{d\Omega^n} &= -\frac{1}{\pi}\int_{-\infty}^{+\infty} \frac{\dfrac{d^n \operatorname{Re}\left[Y_{\text{in}}(x)\right]}{dx^n}}{\Omega - x}\,dx \\ &= -\frac{1}{\pi}\int_{-\infty}^{+\infty} \frac{\dfrac{d^n\left[\,|Y_{12}(x)|^2{}_{LP} + |Y_{12}(x)|^2{}_{HP}\right]}{dx^n}}{\Omega - x}\,dx \end{aligned}\right\} \tag{8}$$

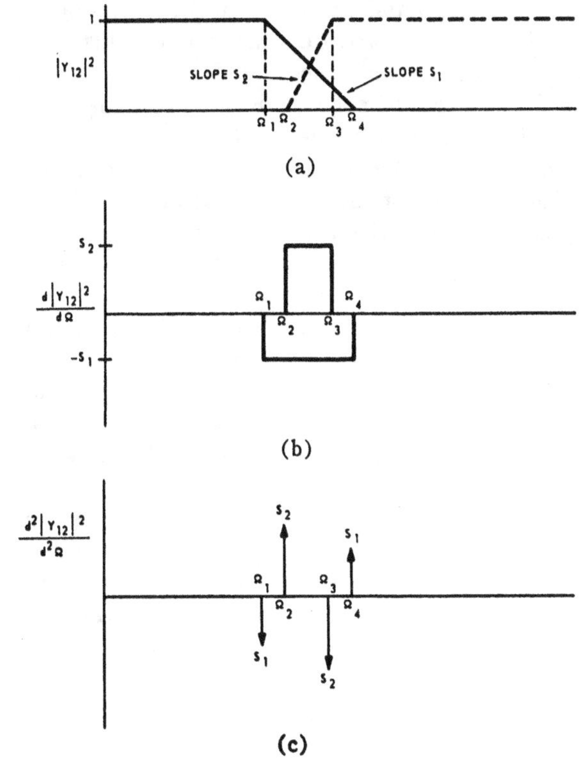

Fig. 7.—(a) Piecewise linear low-pass high-pass filter characteristics $|Y_{12}|^2 = \text{Re}\,(Y_{in}^{\circ})$. (b) First derivative of filter characteristics. (c) Second derivative of filter characteristics.

Equation (8) provides a convenient method of calculating—numerically and graphically—the input VSWR of a filter pair and demonstrates the factors which most influence the obtainable characteristics.[5] Using criteria derived from this relation, it is possible to build filter pairs with low-input VSWR and crossover characteristics that are significantly better than the Butterworth characteristics of comparable order. These filters are synthesized with the use of an exact theory, and thus have characteristics that can be specified theoretically at all frequencies.

As an example of applying the above relation, consider the piecewise linear, low-pass high-pass filter pair characteristics shown in Fig. 7(a). First and second derivatives of the characteristics are shown, respectively, in Fig. 7(b) and 7(c). Since the second derivative functions have been reduced to a set of impulses, the integral in (8) can be evaluated by inspection and the imaginary part can be determined by two integrations of the resulting function. A good design criterion can be obtained from Fig. 7(c). Location of the impulses is determined by the filter break points, and their magnitudes are determined by the slope of the filter characteristics. Thus, if the break points are forced to occur at the same frequencies and the slopes of the filter characteristics are made equal and opposite, the impulses will cancel, and result in a zero imaginary part. Under these conditions, the real parts must add to unity as in the case of true complements.

Computations similar to the above example show that any filter pair that has an input admittance with a real part of approximately unity at all frequencies, and that is devoid of extremely rapid variations in its characteristic will have an imaginary part that is quite small and will exhibit good diplexing properties. Therefore, any low-pass high-pass filter pair for which $|Y_{12}|^2_{LP} + |Y_{12}|^2_{HP} \approx 1$ will be suitable.

One method of obtaining input admittances that add to approximately unity is to set the 3 dB points of the filter characteristics equal to each other, and to set the slope of the high-pass filter equal to the negative of the low-pass slope at the 3 dB point. As a starting point, a low-pass Chebyshev prototype network is chosen where

$$|Y_{12}(\Omega)|^2_{LP} = \frac{1}{1 + \epsilon^2 T_n^2(\Omega)}. \tag{9}$$

This function has a pass-band ripple determined by ϵ and is monotonic in the stop-band. For reasons previously discussed, the true complement should not be used for the high-pass network. A suitable functional characteristic can be obtained by applying the low-pass to high-pass transformation (by substituting $1/\Omega$ for Ω) to (9) giving

$$|Y_{12}(\Omega)|^2_{HP} = \frac{1}{1 + \epsilon^2 T_n^2\left(\dfrac{1}{\Omega}\right)}. \tag{10}$$

[5] For examples of numerical calculations see Guillemin [4], pp. 296–308.

This characteristic has a monotonic stop-band, and pass-band ripple determined by ϵ. It does not provide the desired characteristic in that it "crosses over" the chosen low-pass function of (9) at $\Omega=1$, or at the specified ripple value rather than the 3 dB point. This problem can be eliminated by frequency scaling the high-pass network which is accomplished by substituting Ω/k for Ω in (10). The 3 dB frequency for the low-pass filter [13] is given by

$$\Omega_{3\,\text{dB}} = \cosh\left[\frac{\cosh^{-1}\left(\frac{1}{\epsilon}\right)}{n}\right] \tag{11}$$

where n is the number of filter sections. Therefore, k is given by solution of

$$T_n{}^2(\Omega_{3\,\text{dB}}) = T_n{}^2\left(\frac{k}{\Omega_{3\,\text{dB}}}\right) \tag{12}$$

or

$$k = \Omega^2{}_{3\,\text{dB}} = \cosh^2\left[\frac{\cosh^{-1}\left(\frac{1}{\epsilon}\right)}{n}\right]. \tag{13}$$

Thus, the requirement that the 3 dB points for each characteristic occur at the same frequency has been satisfied.

The second requirement, that the characteristics have slopes that are equal and opposite at the 3 dB frequency, is also satisfied by the above transformation and renormalization as shown in the Appendix. The desired network characteristics thus are given by

$$|Y_{12}|^2{}_{LP} = \frac{1}{1+\epsilon^2 T_n{}^2(\Omega)} \tag{14}$$

and

$$|Y_{12}|^2{}_{HP} = \frac{1}{1+\epsilon^2 T_n{}^2\left(\frac{k}{\Omega}\right)}.$$

200

An important aspect of the above procedure is that networks with quite high ripples (on the order of 1 dB) in their $|Y_{12}|^2$ functions, with consequent steep band-edge skirts, can be used without obtaining significant ripple in insertion loss. Thus, one obtains the advantage of skirt selectivity associated with a high-ripple filter together with the pass-band insertion loss characteristic of a low-ripple filter. At first thought, this result may appear to contradict the known optimum responses obtainable using a given number of filter elements. For example, it is known that a low-pass filter containing N elements has an established optimum response in terms of skirt slope and pass-band ripple. An improvement over this optimum response for the diplexing network containing two N-element networks is possible since the high-pass filter stop-band input admittance aids the pass-band response of the low-pass filter, and vice versa. In effect, the pass-band insertion loss response of each filter in the pair is determined by two N-element filters and thus can provide a response better than the optimum for an N-element filter.

The fact that low-pass-band ripple can be obtained from a filter pair having large ripple in its transfer admittance characteristics also can be seen by considering the input admittance of each filter in the pair. Since $\text{Re}[Y_{in}]=|Y_{12}|^2$, each filter has an input admittance whose real part exhibits relatively large ripples. If the real parts add to approximately unity (within 20 per cent, for example) and do not vary rapidly, the imaginary parts will be approximately equal and opposite and will tend to cancel. Under these conditions, the component networks appear as two conductances in parallel, one that varies as $|Y_{12}|^2{}_{LP}$ and the other as $|Y_{12}|^2{}_{HP}$. Ripple in the $|Y_{12}|^2$ functions thus is presented at the input port as a resistive ripple. $A|Y_{12}|^2$ ripple of 1 dB corresponds to a resistive variation of approximately 1.25 to 1 and, therefore, to an input VSWR variation of 1.25 to 1. This results in a very small transmission ripple. In general, for ripple values for 1 dB or less, the maximum input VSWR is approximately $\text{antilog}_{10}(\alpha/10)$, where α is the ripple in dB. In practice, the input VSWR is slightly higher than this value since the input suscep' ance is not exactly zero.

Element values for the network that realizes the low-pass characteristic of (14) can be obtained from tables in Weinberg [5], [6]. The high-pass network is obtained by inspection using a procedure similar to that for the Butterworth networks. The low-pass to high-pass transformation results in a high-pass network having the same geometry as the low-pass network with every element having been replaced by its dual with a reciprocal value. Frequency scaling by the factor

$$k = \cosh^2\left[\frac{\cosh^{-1}\left(\frac{1}{\epsilon}\right)}{n}\right]$$

is then accomplished by dividing all of the element values in the high-pass network by k.

The following design example demonstrates the technique described above. For simplicity, a two-section design will be described. Extension to higher order filter pairs follows in a straightforward manner.

The design procedure is initiated by choosing the low-pass filter which, in the present example, is a two-section 1-dB ripple Chebyshev transfer admittance prototype shown in Fig. 8(a) (page 12). The low-pass to high-pass transformation gives the network of Fig. 8(b).

ore, from (13), $k=1.48$. :he network of Fig. 8(b) shown in Fig. 8(c). To ents are inserted adjacent rth example and Kuroda's se of the simple nature of without actually applying introduction of a one-unit is sufficient. However, this requ... inductor and capacitor be placed on opp... of the line to insure that there is no interaction betw...n their fields. The input port then has to be connected at right angles to the plane of the filter structure. This can be a suitable configuration at low frequencies but might easily lead to radiation in C band or X band. To alleviate this problem, the high-pass prototype was left in its original form and Kuroda's identities were applied to the low-pass structure to move the capacitor away from the filter junction. One application of Kuroda's identity yields the circuit of Fig. 8(d). In order to make the construction simple, the input inductor was separated into a series connection of two inductors, and Kuroda's identity was applied to yield a design in which all the unit elements had a value equal to the characteristic impedance of the system, and all the stub impedances were easily realizable. The final prototype network is shown in Fig. 8(e). A diagram of an experimental circuit that realizes the network of Fig. 8(e) is shown in Fig. 9(a). The performance of the filter pair can be obtained by direct analysis. Referring to Fig. 8(c), the normalized output powers are $4|V_3(j\Omega)|^2$ and $4|V_2(j\Omega)|^2$ for the low-pass and high-pass channels, respectively. Both theoretical and experimental power transmission curves as well as input VSWR are shown in Fig. 9(b).

The computed characteristics of a five-section 1-dB ripple Chebyshev transfer admittance characteristic are shown in Fig. 10. The improvement in crossover width relative to that of a Butterworth characteristic of the same order is significant. The cost is only a small increase in input VSWR.

IV. Multichannel Filters

The exact design procedure for the diplexer now may be extended to handle contiguous multiple channels. The theoretically perfect match of each diplexer can be maintained at the input of the N-channel filter by using complementary filter pairs. If a slight input-port mismatch can be tolerated, Chebyshev filter characteristics can be used resulting in a decrease in network complexity for a given crossover width and isolation requirement. As an example, assume that it is desired to separate the frequency band 1 to 12 Gc/s into four bands—1 to 2, 2 to 4, 4 to 8, and 8 to 12 Gc/s, respectively. One method of accomplishing this separation is indicated in Fig. 11(a). The first diplexer has a crossover at 4 Gc/s and is constructed of transmission lines having a quarter-wavelength at a frequency, f_0, of 8 Gc/s. The resulting 1 to 4 Gc/s and 4 to 12 Gc/s bands are then split by the use of two more diplexing networks.

An alternate method that uses the repeating nature of the exactly designed networks is shown in Fig. 11(b). The first diplexer is constructed about a center frequency of 6 Gc/s and divides the 1 to 12 Gc/s band into three bands: 1 to 4, 8 to 12, and 4 to 8 Gc/s. The 1 to 4 Gc/s and 8 to 12 Gc/s bands are widely separated, and can be split with a relatively simple diplexer (three sections may be adequate). The 1 to 4 Gc/s band is then split as in the first method.

The second approach to multiplexing the contiguous bands in the above example has distinct advantages over the first. Only two of the component diplexers need have narrow crossover regions. Furthermore, the 1 to 12 Gc/s diplexer which, in general, might be the most difficult to construct because it must work over a very wide band, now is constructed by using 6 Gc/s quarter-wave elements rather than 8 Gc/s elements and, therefore, is less susceptable to undesirable junction effects.

V. Alternate Forms of Complementary Pairs

The diplexing networks discussed so far have had L–C[6] ladder prototypes, and have led to physical realizations of the type shown in Fig. 4 and 9(a). Use can be made of a number of circuit equivalents [2] to obtain different practical realizations. Another important class of microwave filters are those that do not have a prototype of this form. These include:

1) Parallel coupled bars with open ends
2) Parallel coupled bars with shorted ends
3) Uniform line with quarter-wave spaced series open stubs
4) Uniform line with quarter-wave spaced shunt-shorted stubs
5) Stepped quarter-wave sections with one series open stub
6) Stepped quarter-wave sections with one shunt-shorted stub

Diplexers also can be designed in these forms using the same methods outlined in Sections II and III. An exact synthesis procedure for networks of this type is given by Wenzel [2]. Because these networks have but a single pole in their transfer response, they are most useful in relatively narrow-band designs. The existence of only one singularity degrades the realizable skirt characteristics for wide bandwidths. On the other hand, for narrow-band designs, the networks are often easier to construct in this form.

[6] Again, the use of the symbols L and C are taken to mean the short-circuited and open-circuited distributed quarter-wave line, respectively.

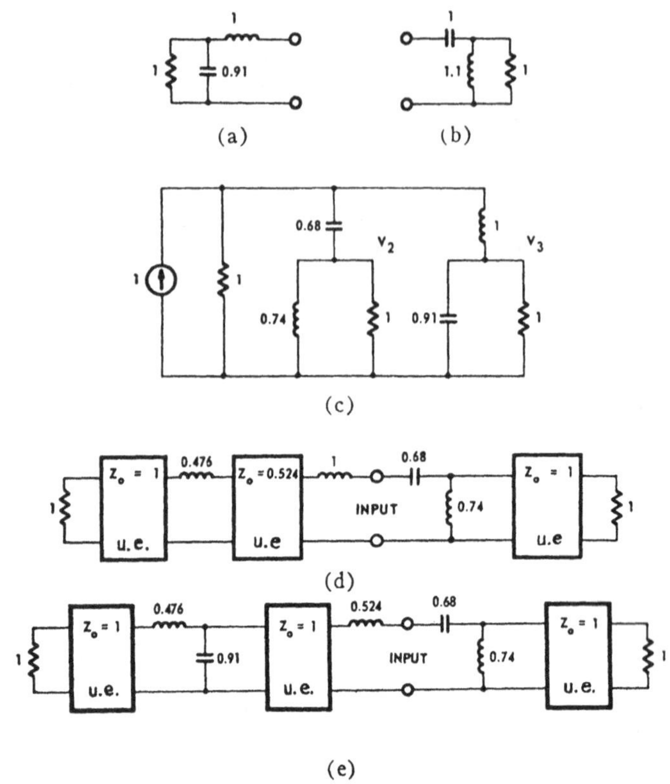

Fig. 8.—(a) Low-pass 1-dB ripple $|Y_{12}|^2$ prototype. (b) High-pass filter prototype. (c) Final prototype circuits from which performance characteristics can be obtained by direct analysis. (d) Prototype after one application of Kuroda's identity. (e) Practical prototype after insertion of unit elements and application of Kuroda's identities.

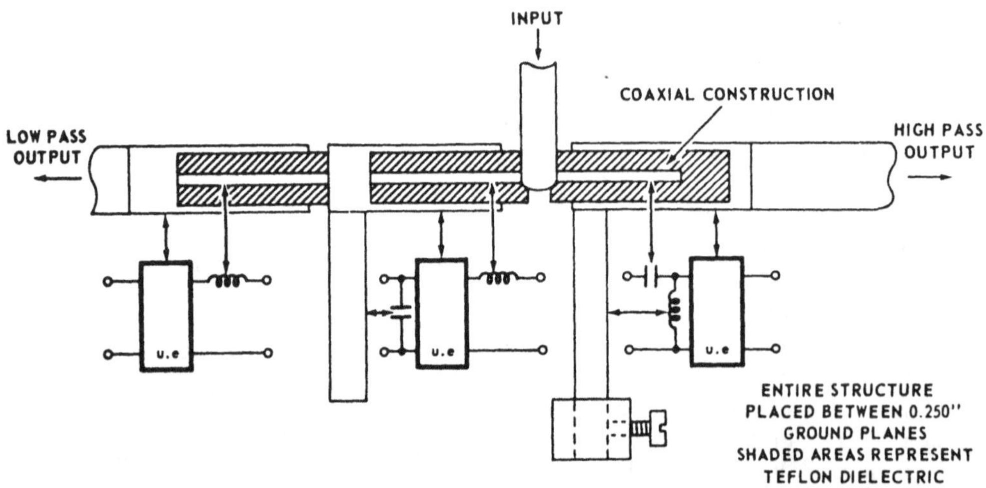

Fig. 9. (a) Experimental circuit that realizes a two-section 1-dB ripple Chebyshev filter pair. (*cont'd.*)

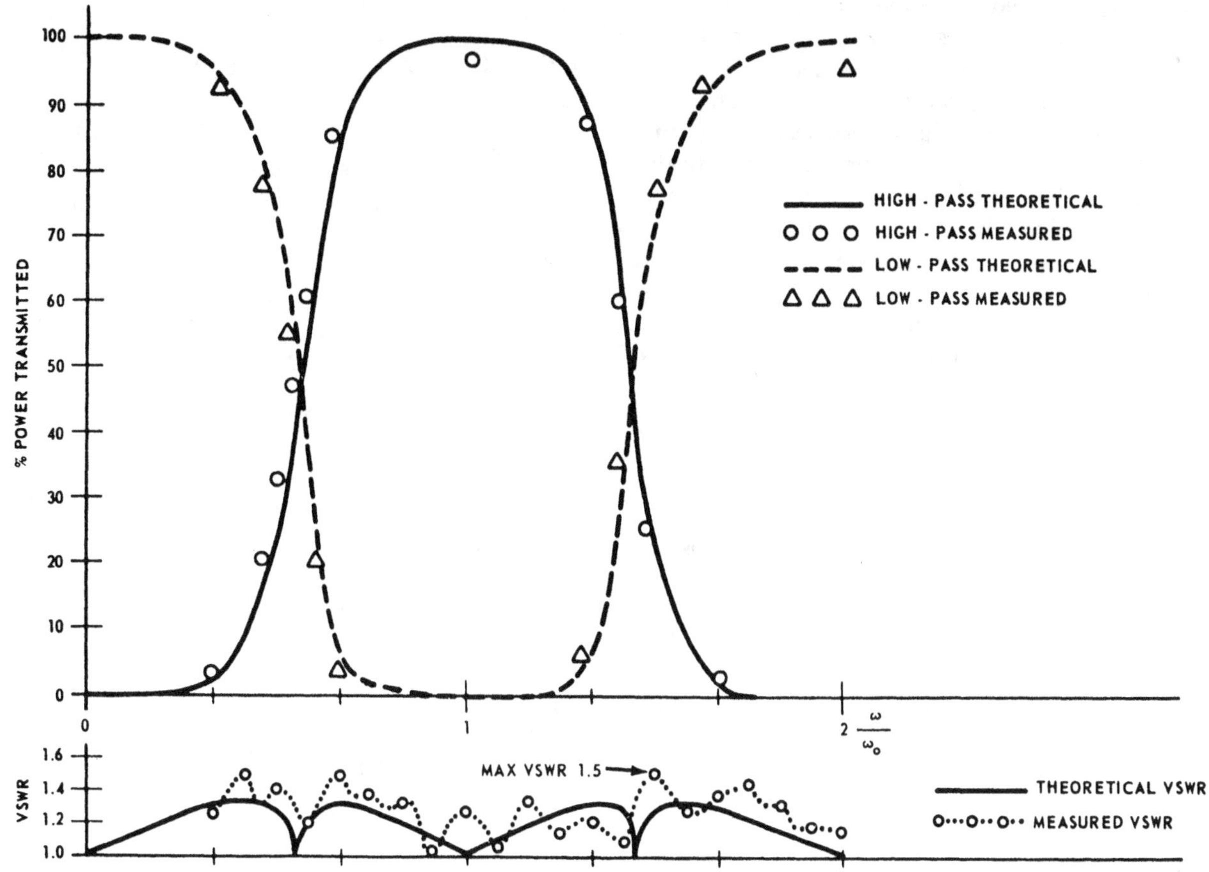

Fig. 9 (b) Theoretical and experiment characteristics of the filter pair.

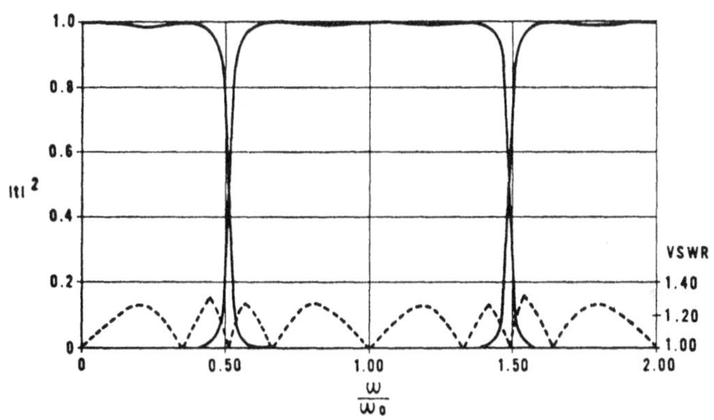

Fig. 10.—Computed characteristics of a five-section filter pair that uses 1-dB ripple transfer admittance prototypes.

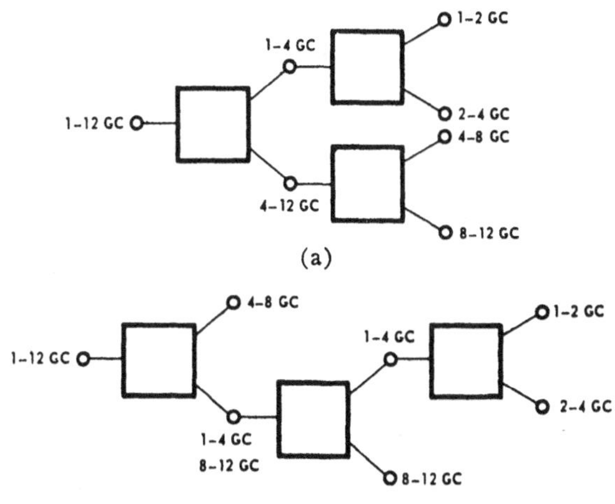

Fig. 11—Wide-band multichannel filters. (a) Network requiring narrow crossover characteristics for each diplexer. (b) Network requiring only two diplexers with narrow crossover regions.

203

VI. Conclusions

Microwave complementary component filters, based on exact design procedures, have been shown to fulfill the requirement for the design of diplexing and multichannel filters with theoretically perfect match at all frequencies. By allowing the input to be slightly mismatched, steeper skirt characteristics, and consequently, narrower crossover regions, can be obtained for a given number of sections. In general, there are five factors of major importance in a multiplexing network—input VSWR, crossover width, isolation, loss, and network complexity. An optimum network might be defined in different ways depending on which of these five factors is of most importance. If an extremely good input match is required, the filters should be made complementary. Losses in multichannel filters of this type will tend to be minimal due to the good match and absence of internal reflections. If slight input-port mismatch can be tolerated, use of Chebyshev charactertistics provides better isolation and narrower crossover width for a given filter complexity. In multichannel applications requiring a large number of cascaded filter pairs, the use of Chebyshev characteristics with large ripple values may result in a high overall input VSWR due to an accumulation of the relatively small VSWRs of each component filter pair. The overall loss of a structure of this type will increase due to the input mismatch and dissipation loss caused by internal reflections. The designer must consider which of the performance factors are of greatest importance, and then must select the network characteristics to obtain the best results.

The procedure given for designing networks with input mismatch leads to networks with good performance characteristics and, in general, is quite simple to apply. In verification of the theory presented, the measured performance of two experimental diplexing networks closely followed the predicted performance over an extremely wide frequency range. The computed characteristics of a five-section filter pair designed from 1-dB ripple Chebyshev transfer admittance prototypes demonstrated that, for moderate ripple values, the resulting maximum input VSWR is approximately that which would be caused by a resistive mismatch of antilog$_{10}$ $(\alpha/10)$, where α is the prototype ripple in dB.

Appendix

Given the characteristics

$$| Y_{12}(\Omega) |^2_{LP} = \frac{1}{1+\epsilon^2 T_n^2(\Omega)} \tag{1}$$

and

$$| Y_{12}(\Omega) |^2_{HP} = \frac{1}{1+\epsilon^2 T_n^2\left(\dfrac{k}{\Omega}\right)}$$

where

$$k = (\Omega_{3\text{ dB}})^2 = \cosh^2\left[\frac{\cosh^{-1}\left(\dfrac{1}{\epsilon}\right)}{n}\right] \tag{2}$$

and n = the number of filter sections, then

$$\left.\frac{d\,| Y_{12}(\Omega) |^2_{LP}}{d\Omega}\right|_{\Omega_{3\text{ dB}}} = \left.\frac{-2\epsilon^2 T_n(\Omega)\dfrac{d}{d\Omega}T_n(\Omega)}{[1+\epsilon^2 T_n^2(\Omega)]^2}\right|_{\Omega_{3\text{ dB}}} \tag{3}$$

and

$$\left.\frac{d\,| Y_{12}(\Omega) |^2_{HP}}{d\Omega}\right|_{\Omega_{3\text{ dB}}} = \left.\frac{-2\epsilon^2 T_n\left(\dfrac{k}{\Omega}\right)\dfrac{d}{d\Omega}T_n\left(\dfrac{k}{\Omega}\right)}{\left[1+\epsilon^2 T_n\left(\dfrac{k}{\Omega}\right)\right]^2}\right|_{\Omega_{3\text{ dB}}}.$$

Use of (2) shows that the only factors that need be considered are

$$\left.T_n(\Omega)\frac{d}{d\Omega}T_n(\Omega)\right|_{\Omega_{3\text{ dB}}} \tag{4}$$

and

$$\left.T_n\left(\frac{k}{\Omega}\right)\frac{d}{d\Omega}T_n\left(\frac{k}{\Omega}\right)\right|_{\Omega_{3\text{ dB}}}.$$

Now

$$T_n(X) = \cos\,(n \cos^{-1} X).$$

Therefore,

$$\left.T_n(\Omega)\frac{d}{d\Omega}T_n(\Omega)\right|_{\Omega_{3\text{ dB}}} = \left.\frac{nT_n(\Omega)\sin\,(n\cos^{-1}\Omega)}{[1-\Omega^2]^{1/2}}\right|_{\Omega_{3\text{ dB}}} \tag{5}$$

and

$$\left.T_n\left(\frac{k}{\Omega}\right)\frac{d}{d\Omega}T_n\left(\frac{k}{\Omega}\right)\right|_{13\text{ dB}} = \left.\frac{-nkT_n\left(\dfrac{k}{\Omega}\right)\sin\left(n\cos^{-1}\dfrac{k}{\Omega}\right)}{\Omega^2[1-\Omega^2]^{1/2}}\right|_{\Omega_{3\text{ dB}}}$$

Substitution of (2) shows the slopes to be equal and opposite.

Acknowledgment

The author is indebted to M. C. Horton for proofreading the original manuscript and for his many helpful comments and suggestions. Appreciation is also expressed for discussions with P. C. Goodman which were very helpful in formulating the approach taken; to A. Gianetti and T. Gordon for their work in constructing the experimental diplexers.

References

[1] Cristal, E. G., and G. L. Matthaei, A technique for the design of multiplexers having contiguous channels, *IEEE Trans. on Microwave Theory and Techniques*, vol MTT-12, Jan 1964, pp 88–93.

[2] Wenzel, R. J., Exact design of TEM microwave network using quarter-wave lines, *ibid*, pp 94–111.

[3] Ragan, G. L., *Microwave Transmission Circuits*, MIT Rad Lab Series, New York: McGraw-Hill, vol 9, 1948, pp 708–9.

[4] Guillemin, E. A., *Synthesis of Passive Networks*, New York: John Wiley, 1957.

[5] Weinberg, L., *Network Analysis and Synthesis*, New York: McGraw-Hill, 1962.

[6] Weinberg, L., Network design by use of modern synthesis techniques and tables, *Proc. NEC*, vol 12, 1956, pp 704–817.

[7] Ozaki, H., and J. Ishii, Synthesis of transmission-line networks and the design of UHF filters, *IRE Trans. on Circuit Theory*, vol CT-3, Dec 1956, pp 325–336.

[8] Ozaki, H., and J. Ishii, Synthesis of a class of strip-line filters, *IRE Trans. on Circuit Theory*, vol CT-5, Jun 1958, pp 104–109.

[9] Grayzel, A. I., A synthesis procedure for transmission-line networks, *IRE Trans. on Circuit Theory*, vol CT-5, Sep 1958, pp 172–181.

[10] Schiffman, B. M., and G. L. Matthaei, Exact design of band-stop microwave filters, *IEEE Trans. on Microwave Theory and Techniques*, vol MTT—12, Jan 1964, pp 6–15.

[11] Richards, P. I., Resistor transmission-line circuits, *Proc. IRE*, vol 36, Feb 1948, pp 217–220.

[12] Horton, M. C., and R. J. Wenzel, Distributed filter design using exact techniques, *Microwaves*, Apr 1964, pp 16–21.

[13] Magèr, H., The Chebyshev normalized low-pass 3 dB frequency —a bone from the technical graveyard? *IEEE Trans. on Circuit Theory (Correspondence)*, vol CT-10, Jun 1963, pp 287–288.

Multiplexer Channel-Separating Units Using Interdigital and Parallel-Coupled Filters

G. L. MATTHAEI, FELLOW, IEEE, AND E. G. CRISTAL, MEMBER, IEEE

Abstract—**Design theory is presented for strip-line multiplexer channel-separating units which have constant-resistance input impedances so that many units can be cascaded without reflection effects. Each channel unit consists of an interdigital band-pass filter and a band-stop filter which uses parallel-coupled resonators. In order to obtain the desired constant-resistance input impedance, both filters are designed from singly loaded, maximally flat, low-pass prototype filters. A trial design was worked out and constructed so as to have five per cent bandwidth for the separated channel. Excellent agreement between theory and experiment was obtained.**

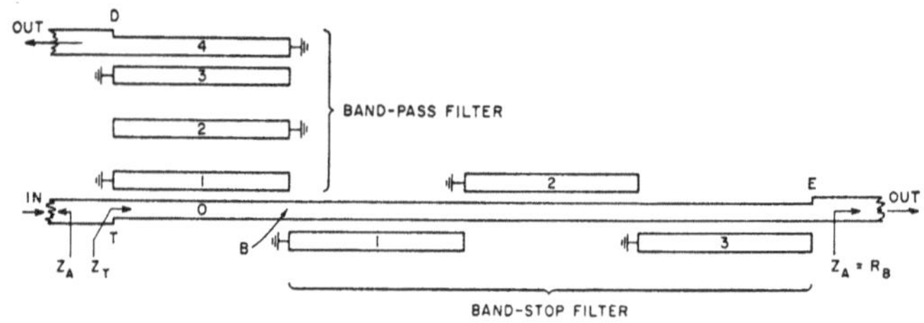

Fig. 1. A strip-line multiplexer channel unit.

I. Introduction

THE DESIGN OF multiplexers, which separate out frequencies in certain ranges from a spectrum of signals covering a large range of frequencies, is a problem frequently encountered in microwave engineering. The separation of the desired frequency bands can be accomplished by use of band-pass filters, but special design procedures are required so that the filters can be interconnected without undesirable interaction effects. In this paper a design technique is described that makes use of channel-separating units, each of which consists of a band-pass filter along with a band-stop filter. In theory, these channel units present a constant-resistance input impedance, and as a result a number of units designed to separate out a number of frequency bands can be cascaded without harmful interaction effects [1], [2], [3]. The filter structures used have the additional advantage of being relatively easy to fabricate.

A sketch of the strip-line multiplexer channel unit is given in Fig. 1. It consists of an interdigital band-pass filter and a parallel-coupled-resonator band-stop filter. In Fig. 1, Z_A represents the generator and terminating resistance. The input impedance Z_T is, in theory, exactly equal to Z_A at all frequencies. Thus the load impedance Z_A may be replaced by another channel unit, and a number of channel units may be cascaded.

The basic method for designing a reflectionless multiplexer channel unit uses a singly loaded (i.e., with a resistor termination at one end only) low-pass prototype filter in the design of both the band-pass filter and the band-stop filter. The low-pass prototypes for both filters should have the same number of elements and should ideally be maximally flat in the pass band,[1] in order for the derived band-pass and band-stop filter to be complementary filters (that is, filters having input impedances whose imaginary parts cancel and whose real parts add to give a constant). Herein the two filters are connected in series, giving an input impedance for the channel unit which is a constant resistance equal to the generator resistance. A proof for this constant-resistance input-impedance property is given in the Appendix.

II. Synthesis Method for the Band-Stop Filter Used in the Channel Unit

The synthesis procedure for the realization of the band-stop filter of the channel unit is basically the exact synthesis method of Ozaki and Ishii [4], and of Schiffman and Matthaei [5]. However, it is fundamental to the design theory presented in this paper that an alternate method of applying Kuroda's Identity to that previously given [5], [6], be used in the synthesis of the band-stop filter of the channel unit.

The reasons for using an alternate method of applying Kuroda's Identity are as follows: The band-stop filter design equations of Schiffman and Matthaei [5] were derived by starting from a lumped-element low-pass prototype filter, then converting this low-pass filter into a band-stop filter by use of a tangent-function mapping. The mapping function converted shunt capacitors into shunt, open-circuited stubs, and series inductances into series, short-circuited stubs. The input impedance of this filter had the properties desired for use in multiplexer channel units, but the series stubs in the structure were impractical to build. Kuroda's Identity [4], [5], [6] was then used to feed sections of line into the structure from both ends, so as to yield a modified structure having the form of shunt stubs separated

Manuscript received October 19, 1964; revised January 25, 1965. The research in this paper was supported by the U. S. Army Electronics Research and Development Lab., Ft. Monmouth, N. J., under Contract DA 36-039-AMC-00084(E).

G. L. Matthaei is with the Dept. of Electrical Engineering, University of California, Santa Barbara, Calif. He was formerly with the Stanford Research Inst., Menlo Park, Calif.

E. G. Cristal is with the Stanford Research Inst., Menlo Park, Calif.

[1] It would also be satisfactory for many applications to use singly loaded filters with small Chebyshev pass band ripples. This would, however, cause the input impedance to deviate from the desired constant resistance.

by sections of lines (the stubs and lines all being a quarter-wavelength long at mid-stop-band). Feeding the extra sections of line in from both ends of the filter had the effect of altering the input impedance characteristic at both ends of the filter. In the case of a band-stop filter having a sizeable stop-band width, these altered input impedances would have added series- and shunt-resonance effects that would disrupt the performance of the band-pass filter in the multiplexer channel unit.

In order to eliminate this problem, an alternate method of applying Kuroda's Identity was proposed. In the alternate method, the Identity is applied from the load end only (i.e., from the end away from the band-pass filter). In this way the input impedance Z_s of the band-stop filter is a simple tangent-function mapping of the input impedance of the prototype filter. That is, if $Z(\omega')$ is the input impedance of the prototype filter, then $Z_s = Z[\Lambda \tan (\pi\omega/2\omega_0)]$ is the input impedance of the derived band-stop filter as indicated in Fig. 2. The symbols used as the argument of the impedance function are defined below:

- ω' is the frequency variable of the low-pass prototype filter.
- ω_1' is the cutoff frequency of the low-pass prototype filter.
- ω is the frequency variable of the band-stop filter.
- ω_0 is the center frequency of the band-stop filter.
- Λ is a scaling parameter that is defined by

$$\Lambda = \omega_1' \tan [(\pi/4)w]. \tag{1}$$

The symbol w in (1) is the fractional stop-band width of the band-stop filter. It is defined by

$$w = \frac{\omega_2 - \omega_1}{\omega_0} \tag{2}$$

where ω_2 and ω_1 are the upper and lower cutoff frequencies of the band-stop filter corresponding to ω_1' of the low-pass prototype filter.

By using this alternate method, both the band-pass and band-stop filter may be synthesized from singly loaded low-pass prototype filters, which should permit close realization of the desired theoretical performance. The alternate method of applying Kuroda's Identity is shown schematically in Fig. 2. The example in Fig. 2 uses a singly loaded prototype, since the primary use of the alternate method will probably be in its application to design of multiplexer channel units. However, this technique can also be used for doubly loaded prototypes. When the alternate method of applying Kuroda's Identity is used, the previously presented exact synthesis design formulas [5] and tables of h parameters [6] do not apply. However, the algorithm [6] for computing the h parameters of band-stop filters remains valid and can be easily adapted to include the case where Kuroda's Identity is applied from the load end only. Only two modifications to the algorithm are necessary in order to use this alternate design method:

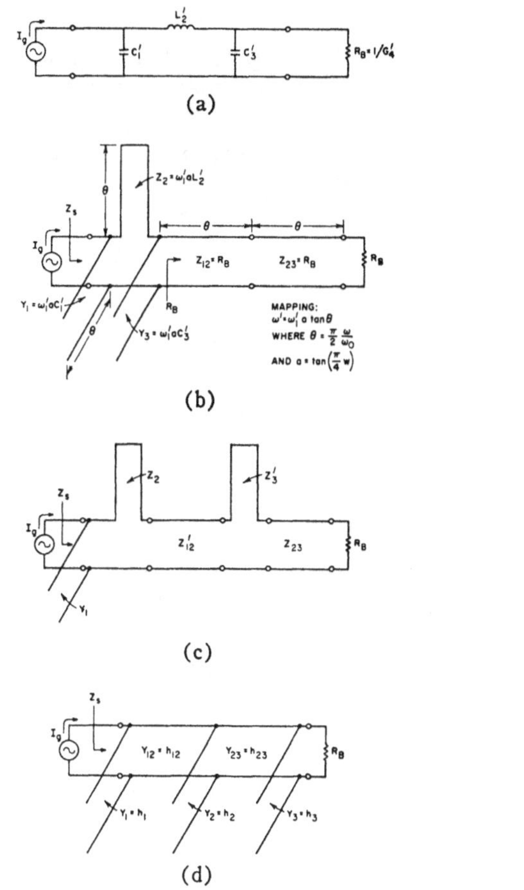

Fig. 2. Stages in the transformation of a singly loaded, low-pass, prototype filter into a band-stop, transmission-line filter. (a) Singly loaded prototype. (b) Mapped prototype. (c) After applying Kuroda's Identity to Y_3 and Z_{12} in (b). (d) After applying Kuroda's Identity to Z_2 and Z_{12}' and to Z_3' and Z_{23} in (c).

1) The determination of the parameter P is now given by[2]

$$P = n - 1 \tag{3}$$

for both n even and odd, where n is the number of reactive elements in the low-pass prototype filter.

2) $$h_1 = g_1'. \tag{4}$$

All other details of applying the algorithm are as previously described [6].

An example will now be given of the calculation of the h values of a three-resonator, 0.05-fractional-bandwidth, band-stop filter based on a maximally flat prototype [7], [8].[3] The results will be used in the synthesis of a trial channel unit in strip line which is described in Section IV.

[2] See Cristal [6], p 370, for the significance of the parameter P.

[3] The g_i values for singly loaded low-pass prototype filters given in [7] and [8] are numbered so that the resistive load element is given by g_0, and the adjacent reactive element is then denoted by g_1, and so on. In order to use these element values in the algorithm, and in order to make the notation conform to that in Fig. 2(a), the subscripts of the g's in [7] and [8] should be altered according to the rule

$$g_i = g_{n+1-i}$$

where n is the number of reactive elements in the low-pass prototype filter.

TABLE I

g_i AND g_i' VALUES FOR A MAXIMALLY FLAT SINGLY LOADED LOW-PASS PROTOTYPE FILTER [7], [8]

$g_0=\infty$	$g_1=1.500$	$g_2=1.333$	$g_3=0.500$	$g_4=(\text{load})=1.0$
$g_0'=\infty$	$g_1'=0.058935$	$g_2'=0.052386$	$g_3'=0.019645$	$g_4'=1.0$

2P + 1 COLUMNS

P ELEMENTS g_i', AND P ROWS	1	2	3	4	5
0	$g'_{n+1}=1$	0	$g'_{n+1}=1$	0	$g'_{n+1}=1$
$g'_n = g'_3 =$ 0.019645	1.0	0	1	0	1
$g'_{n-1} = g'_2 =$ 0.052386	0.98073	0.019266	1	0	1
X	0.967942	0.051703	0.981097	0.018902	1
	h_{12}	h_2	h_{23}	h_3	h_1

Fig. 3. Computation matrix for h parameters of a singly loaded band-stop filter.

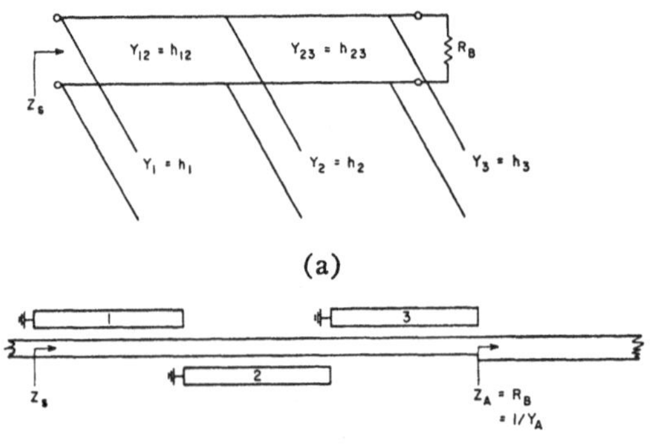

Fig. 4. (a) Open-wire-line, band-stop filter. (b) Its parallel-coupled strip-line equivalent.

The g and g' values used in the example calculations are given in Table I. Recall that g and g' are related by [6]

$$g_i' = g_i\omega_1' \tan\left(\frac{\pi}{4} w\right). \tag{5}$$

The parameter P, given by (3), is

$$P = n - 1 = 3 - 1 = 2. \tag{6}$$

The computation matrix [6] is shown in Fig. 3. The parameter h_1 is given according to (4) by g_1' and is therefore equal to 0.05894.

The transmission-line filter corresponding to the h values given in Fig. 3 is shown in Fig. 4(a), and its exact realization as a parallel-coupled-resonator strip-line filter is given in Fig. 4(b). Design equations that relate the admittances of the transmission lines and stubs of the filter shown in Fig. 4(a) to the normalized self and mutual distributed capacitances of the coupled resonators in Fig. 4(b) are given in [5]. From these equations the band-stop filter may be constructed [5].

208

III. THE BAND-PASS FILTER AND THE INTERCONNECTION

Figure 5(a) shows an interdigital band-pass filter such as is used in the multiplexer channel unit in Fig. 1. Equations for the design of filters of this type from low-pass prototype filters are presented in Matthaei [9]. Note that the interdigital filter structure in Fig. 5(a) is of the type that uses short-circuited input and output lines, which act as impedance transforming sections (not as resonators) [9]. Thus, the structure shown is a three-resonator filter (any number of resonators may be used, of course), the resonators being formed from Lines 1, 2, and 3. This form of interdigital band-pass filter, which uses short-circuited input and output lines, is most practical for narrow to moderate bandwidths [9].

Figure 5(b) shows an equivalent circuit for the filter in Fig. 5(a). Though this equivalent circuit involves some approximations [9] it has been found to give very good accuracy in representing the performance of interdigital filters of the form shown in Fig. 5(a). Two parts of this equivalent circuit are not evident in Matthaei [9]. These are the lengths of transmission line of length θ and characteristic admittance Y_A at each end of the filter. The exact equivalent circuit in Fig. 21 of Matthaei [9] should have included such a length of line between the ideal transformer and the load conductance Y_A at the left in the open-wire circuit, in order for the open-wire equivalent circuit to be exactly equivalent to the parallel-coupled-strip circuit in all respects. However, the discussion in Matthaei [9] was only concerned with the *input admittance* seen looking in from the right, so the presence of this section of matched transmission line was of no importance for that analysis. But for the present situation, the presence of this length of line is of importance in understanding the operation of the multiplexer channel unit. [Note that points T and D in Fig. 5(a) correspond to points T and D indicated in Fig. 5(b).]

In order to use a filter of the form in Fig. 5(a) in a band-pass-plus-band-stop filter connection such as that in Fig. 1, it is necessary to break the ground connection on line 0 at point B indicated in Fig. 5(a). By study of various parallel-coupled line configurations and their known, exact, open-wire equivalent circuits [10], it was determined that at least to an excellent approximation (and probably as an exact equivalence), breaking the circuit in Fig. 5(a) at point B, and inserting an added impedance between the end of line 0 and ground, has the effect of breaking the circuit in Fig. 5(b) at the indi-

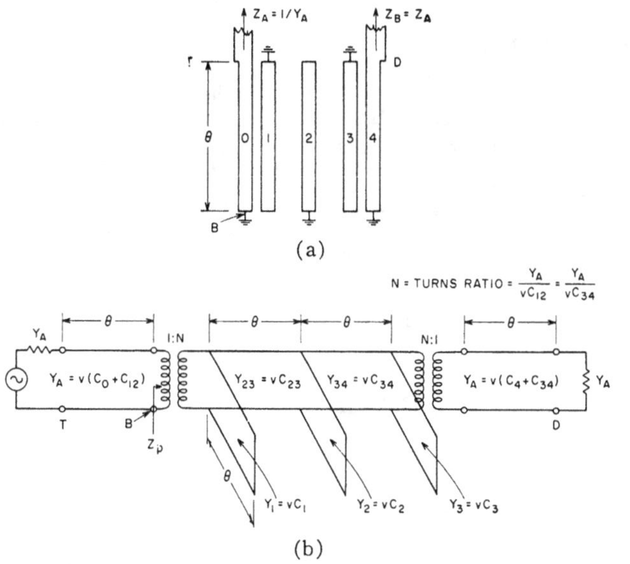

Fig. 5. An interdigital filter and its open-wire-line equivalent circuit. The capacitance C_i is the capacitance per unit length between strip i and ground, and capacitance $C_{i,i+1}$ is the capacitance per unit length between strips i and $i+1$. The transformers shown are ideal transformers, and v is the velocity of propagation.

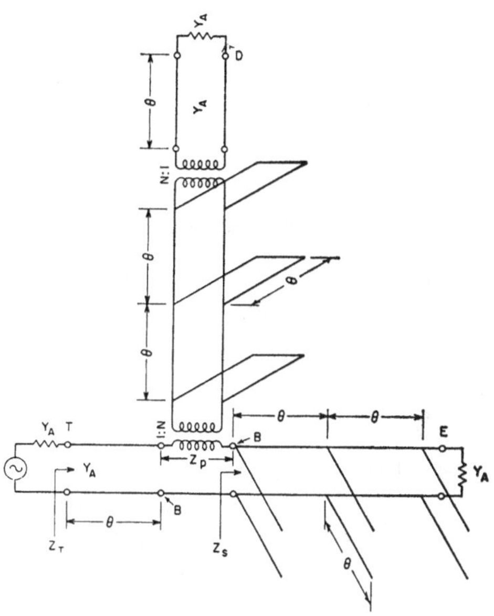

Fig. 6. Open-wire equivalent circuit for the multiplexer channel unit in Fig. 1.

cated point B, and inserting the added impedance in series with the ideal transformer.

Making use of the equivalent circuits in Figs. 4(a) and 5(b), along with the foregoing observation, we see that the multiplexer channel unit in Fig. 1 has the open-wire equivalent circuit shown in Fig. 6. Note that in this circuit the connection at point B in Fig. 5(b) has been broken, and the input terminals of the band-stop filter circuit have been connected so as to reclose the circuit. Note that looking left from the left half of the bisected Point B, a constant-resistance Z_A will be seen since the transmission line of characteristic impedance Z_A is terminated in a matching generator impedance Z_A. The ideal transformer at the input of the band-pass filter has an impedance-level-transforming effect, but otherwise has no effect on the character of the input impedance to the band-pass filter. Thus, the impedance Z_p associated with the band-pass filter is of the form typical of a band-pass filter that starts out with a shunt, band-pass resonator. The band-stop filter starts out with a shunt, band-stop resonator, which makes the two filters of the proper form for series interconnection. (Note that if the band-stop filter started out with a series band-stop resonator branch, it would present an open circuit at resonance, which would disrupt the performance of the band-pass filter.)

If the band-stop filter is designed from a singly loaded, maximally flat low-pass prototype filter using the methods described in Section II, and if the band-pass filter is designed using the same low-pass prototype and for the same 3-dB bandwidth as the band-stop filter, then the input impedances Z_p and Z_s will be complementary. This means that the imaginary parts of these impedances will cancel, while their real parts will add up to a constant such that

$$Z_p + Z_s = Z_A. \tag{7}$$

Thus, as has been previously mentioned, it is possible (at least to a good approximation) for the input impedance of multiplexer units of the form in Fig. 1 to have a perfect impedance match with respect to the line impedance Z_A at all frequencies.

The perfect impedance match described above is contingent upon the use of singly loaded, maximally flat prototype filters. However, in some cases it may be desirable to use singly loaded, Chebyshev low-pass prototype filters instead, as a basis for design. In this case the input VSWR of the multiplexer channel unit would not be as good (though it would be acceptable for most applications), but the filters would have the advantage of more uniform pass-band attenuation and sharper rate of cutoff, as is typical of Chebyshev filters in comparison to corresponding maximally flat filters. If Chebyshev filters are used, they should also be designed so that the band-stop and band-pass filters have the same 3-dB bandwidth.

The design equations in Table II of Matthaei [9] can be used without any changes, for the design of band-pass interdigital filters of the form in Fig. 1. However, some additional comments are required with respect to the use of the singly loaded low-pass prototype element values in Matthaei, et al. [7], (and similar comments apply with respect to those in Weinberg [8]). First, the element values in Matthaei, et al. [7], are numbered, so that the resistive load element is $g_0 = 1$, the adjacent reactive element is g_1, and the element values are numbered consecutively up to g_n, which is the reactive element adjacent to the generator; and $g_{n+1} = \infty$ is the internal impedance (or, for the dual case, internal admittance) of the generator. To correspond to the notation in Fig. 5 and in Matthaei [9], we wish to reverse the order of numbering all these element values so that $g_0 = \infty$, and so on consecutively to $g_{n+1} = 1$. Now for

TABLE II
DIMENSIONING OF FILTERS IN CHANNEL UNIT

(a) Band-pass filter rod diameters and center to center spacing in inches

k	Diameter of Rod k	k	Center to Center Spacing ($c_{k,k+1}$) of Rods k and $k+1$
0	0.331	0	0.486
1	0.217	1	0.792
2	0.224		
3	0.203	2	0.674
4	0.318	3	0.384

(b) Band-stop filter rod diameters and center to center spacing in inches

k	Diameter of Rod k: k_a Rod	Diameter of Rod k: k_b Rod	k	Center to Center Spacing ($c_{k,k}$) of k_a Rod and k_b Rod
1	0.332	0.215	1	0.663
2	0.336	0.219	2	0.686
3	0.344	0.237	3	0.846

Ground Plane Spacing 0.625

computing an interdigital filter design such as that in Fig. 5, we want the impedance Z_p looking into the filter to be that of a filter designed so that it would give a prescribed maximally flat (or Chebyshev) response if driven by an infinite-internal-impedance current generator. However, in this application we are not actually going to drive the filter from a current generator, but instead, from a source of impedance Z_A. For this reason we must change the prototype driving source impedance from $g_0=\infty$ to $g_0=g_{n+1}=1$. After reversing the order of numbering of the element values as described, and then setting $g_0=g_{n+1}=1$, the prototype element values can be used directly in Table II of Matthaei [9], and the band-pass filter design worked out as described in that paper. As is indicated in Fig. 1, line 0 in Fig. 5(a) becomes part of the main transmission line in the completed multiplexer channel unit.

IV. Trial Multiplexer Channel Unit in Strip Line

Based on the theory and design techniques presented in Sections II and III a multiplexer channel unit was constructed in strip line. The filters of the channel unit were based on a three-resonator, singly loaded, maximally flat low-pass prototype. The design center frequency was 1.5 Gc, and the design fractional bandwidth was 0.05. The interdigital band-pass filter and the parallel-coupled-resonator band-stop filter were both constructed using round rods between parallel ground planes [11].

A drawing of the channel unit, showing important dimensions of the filters, is given in Fig. 7, while Table II gives additional dimensions. A photograph of the constructed channel unit with its top ground plane removed is shown in Fig. 8.

The band-stop filter was tuned by individually resonating each of the band-stop filter resonators while the remaining two were detuned. Resonance was determined by tuning for minimum transmission from Port 1 to Port 3. Initial tuning of the band-pass filter was accomplished using the alternating short-circuit and open-circuit procedure [12], [13]. Fine tuning adjustments were made by observing the reflected wave of the channel unit using an electronically swept frequency source, a reflectometer, and an oscilloscope.

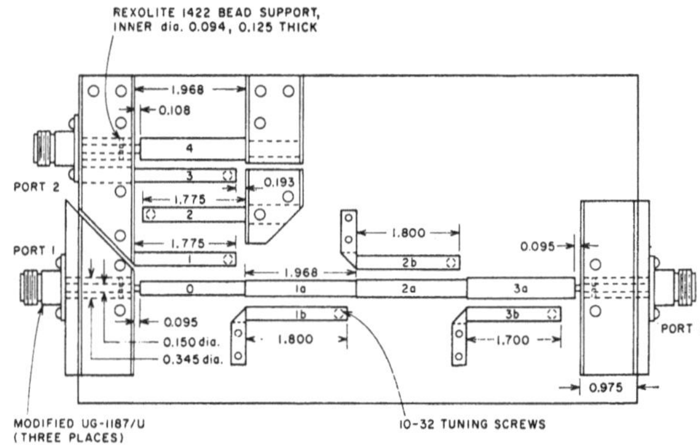

Fig. 7. Drawing of the trial multiplexer channel unit in strip line.

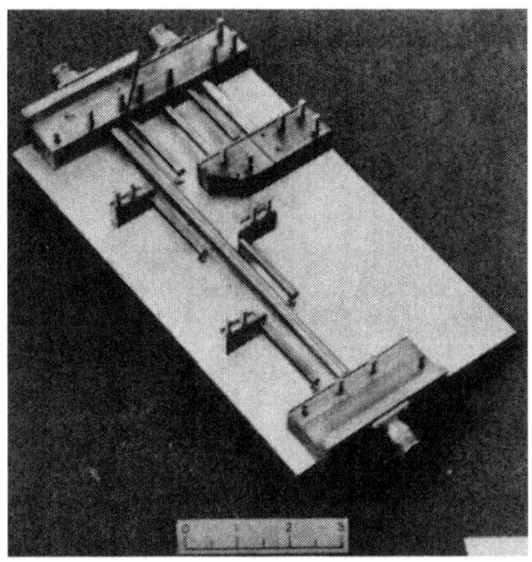

Fig. 8. Photograph of the trial multiplexer channel unit with its cover plate removed.

The VSWR of the channel unit (at Port 1) from 1 to 4 Gc is given in Fig. 9. An expanded frequency scale is also shown in this figure, giving the VSWR in the vicinity of the pass band of the interdigital filter. It is seen to be better than 1.25 everywhere in the 1-to-4-Gc interval. The sinusoidal variation of VSWR seen in Fig. 9 is due primarily to the junction effect of the transitions associated with the type-N connector at Ports 1 and 3, although there is some contribution to the VSWR from the connectors themselves as well as from the termination at Port 3. The maximum magnitude of the sinusoidal variation of VSWR suggests that the VSWR of the transition from coaxial line to the round conductor

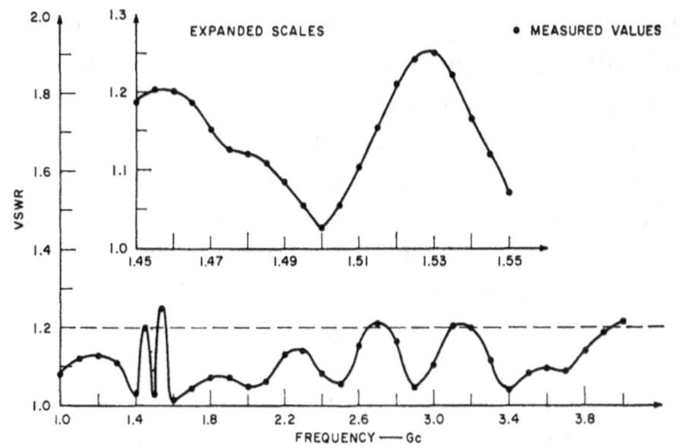

Fig. 9. Measured VSWR of a trial channel unit in strip line.

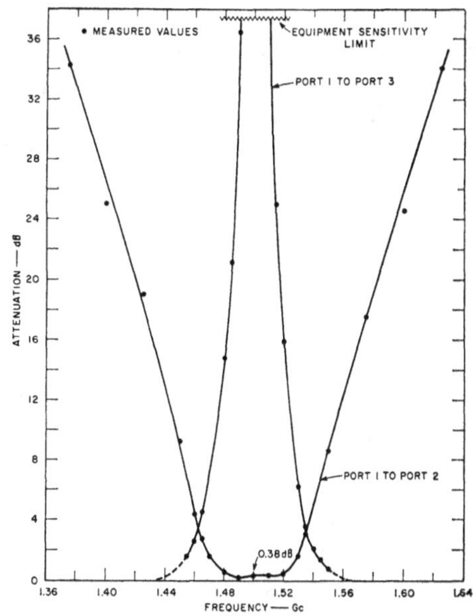

Fig. 10. Measured attenuation characteristics of a trial channel unit in strip line.

between parallel ground planes is approximately 1.1 per transition. This could be reduced by further refining the transitions. However, the measured results adequately demonstrate the design principles that were to be shown.

The measured attenuations from Port 1 to Port 3 and from Port 1 to Port 2 in the vicinity of the pass band of the band-pass filter are given in Fig. 10. The attenuation at the band-center frequency, 1.5 Gc, is 0.38 dB. This value compares favorably with that given by a first-order formula for the attenuation of the band-pass filter of a channel unit at center frequency:

$$(L_A)_0 \approx \frac{4.34\omega_1'}{w}\left[\left(\sum_{i=1}^{n}\frac{g_i}{Q_{u_i}}\right)+\frac{1}{\hat{Q}_{u_1}\hat{g}_1}\right] \text{dB} \qquad (8)$$

where

- $(L_A)_0$ is the attenuation in dB at the center frequency
- w is the fractional bandwidth
- ω_1' is the cutoff frequency of the low-pass prototype filter
- Q_{u_i} is the unloaded Q of the ith resonator of the band-pass filter
- $\hat{Q}_{u_i}$ is the Q of the first resonator of the band-stop filter
- g_i is the ith element value of the low-pass prototype filter used for designing the band-pass filter
- $\hat{g}_1$ is the g_1 element value of the low-pass prototype filter used for designing the band-stop filter.

Using the same unloaded Q of 1000[4] for each of the resonators and (8) gives a calculated value of 0.347 dB for L_{A_0}.

From Fig. 10, it is seen that the crossover frequencies occur virtually at the 3-dB points, and that the 3-dB fractional bandwidth is

$$2\left(\frac{1.536 - 1.463}{1.536 + 1.463}\right) = 0.0488,$$

which is in good agreement with the design value of 0.05.

V. Conclusions

The multiplexer channel unit in strip line was found to have a VSWR and an attenuation response that corresponded excellently with that predicted by the theory. The off-band VSWR was sufficiently small (and could be made smaller by improving the design of the transitions from the type-N connectors to the round conductor between parallel ground planes) to enable several channel units to be cascaded without undue interaction effects. Use of the new method of designing the band-stop filter of the channel unit resulted in a measured bandwidth of both filters that was accurately predictable from the theory. Aligning of the channel unit was straightforward and is not made appreciably more complicated by adding more resonators.

Appendix

Proof that Series-Connected Maximally Flat, Singly Loaded Filters Can Give a Constant-Resistance Input Impedance

Let us first consider the case of lumped-element LC low-pass and high-pass filters such as might be used in a lumped-element diplexer. A singly terminated, maximally flat, low-pass filter having a one-ohm resistor at one end and an infinite-internal-impedance current generator (of current I_g) at the other end would deliver power P to the resistor in accordance with

$$P = |I_g|^2 \text{ Re } Z_L \qquad (8)$$

where

$$\text{Re } Z_L = \frac{1}{1+\omega^{2n}}, \qquad (9)$$

[4] A Q_u of 1000 represents a typical value of unloaded Q for strip-line resonators of the type used.

Z_L is the input impedance seen from the current generator, ω is the radian frequency, and n is the number of reactive elements in the filter. Thus, we see that Re Z_L gives the power transfer function, and the 3-dB point occurs where $\omega = 1$ (and Re $Z_L = 0.5$). To obtain Re Z_H for a corresponding high-pass filter having a 3-dB point at $\omega = 1$, we map (9) by replacing ω by $1/\omega$ which gives

$$\text{Re } Z_H = \frac{\omega^{2n}}{1 + \omega^{2n}}. \qquad (10)$$

If these two filters are connected in series, the real part of the total input impedance Z_T will be

$$\text{Re } Z_T = \frac{1}{1 + \omega^{2n}} + \frac{\omega^{2n}}{1 + \omega^{2n}} = 1. \qquad (11)$$

Thus we see that the real parts add to one at all frequencies.

The next point, which we shall briefly prove, is that if the real part of a minimum-reactance impedance (i.e., an impedance having no poles on the $j\omega$ axis) is a constant vs. frequency, then its imaginary part is zero. For reasons indicated later in the paper, Z_L and Z_H are minimum-reactance impedances, so Z_T is also a minimum-reactance impedance. Thus in view of (11),

$$Z_T = Z_L + Z_H = 1 \qquad (12)$$

at all frequencies ω, and the filters are said to be complementary.

The poles of Z_L are the natural frequencies of vibration when Z_L is driven by a current generator, and for a maximally flat filter of the type under discussion the natural frequencies of vibration are well known to lie on a semicircle in the left half of the complex-frequency plane. Because there are no poles of Z_L on the $j\omega$ axis, Z_L is a minimum-reactance impedance. Z_H is also a minimum-reactance impedance for similar reasons.

To prove that Im $Z_T = 0$ for all ω if Re $Z_T =$ (a constant), we invoke equation (8.05-3) of Tuttle [14], which in our notation is

$$\text{Im } Z_T(\omega) = \frac{1}{\pi} \int_{-\infty}^{\infty} \frac{\text{Re } Z_T(\lambda) - \text{Re } Z_T(\omega)}{\lambda - \omega} d\lambda. \qquad (13)$$

(This relation applies only to minimum-reactance impedances.) Since Re $Z_T(\lambda) =$ Re $Z_T(\omega) = 1$ for all λ and ω, the integral must equal zero.

If a low-pass-to-band-pass transformation is applied to the filters previously discussed, the low-pass filter will become a band-pass filter, and the high-pass filter will become a band-stop filter. This mapping process would distort the frequency scale of Z_T from what it was before being mapped, but since $Z_T = 1$, the mapped Z_T will still be a constant resistance. Thus singly loaded, maximally flat band-pass and band-stop filters connected in series can also be complementary and give a constant-resistance input. (Also, a similar proof on the dual basis applies to the parallel connection of filters.)

In the case of the filters in Fig. 6, the input impedance Z_s of the band-stop filter can be obtained exactly from a *high-pass*, lumped-element prototype by use of the mapping $\omega = -A \cot \theta$, where A is a constant. The band-pass filter is not obtainable *exactly* from a conventional lumped-element prototype, but if the band-pass filter is designed by the method of Matthaei [9], Z_p will be a close approximation to the impedance obtained by applying the same $\omega = -A \cot \theta$ mapping to a conventional *low-pass*, lumped-element-filter prototype. Thus if the prototypes are complementary, the mapped filters will also be complementary (at least to an excellent approximation).

Acknowledgment

The authors wish to thank R. Pierce who fabricated the channel unit, and E. Fernandes who made the laboratory tests.

References

[1] Matthaei, G. L., et al., Novel microwave filter design techniques, Quarterly Progress Rept 1, SRI Project 4344, Contract DA 36-039-AMC-00084(E), Stanford Research Institute, Menlo Park, Calif., Apr 1963.

[2] Cristal, E. G., et al., Novel microwave filter design techniques, sec 3, Quaterly Progress Rept 3, SRI Project 4344, Contract DA 36-039-AMC-00084(E), Stanford Research Institute, Menlo Park, Calif., Sep 1963. Also on same contract, Quarterly Progress Rept 4, sec 4, Jan 1964.

[3] The procedures described by Wenzel are closely related to those discussed herein. However, he has made use of different forms of filter structures. See: Wenzel, R. J., Application of exact synthesis methods to multichannel filter design, *IEEE Trans. on Microwave Theory and Techniques*, vol MTT-13, Jan 1965, pp 5–15.

[4] Ozaki, H., and J. Ishii, Synthesis of a class of strip-line filters, *IRE Trans. on Circuit Theory*, vol CT-5, Jun 1958, pp 104–109.

[5] Schiffman, B. M., and G. L. Matthaei, Exact design of band-stop microwave filters, *IEEE Trans. on Microwave Theory and Techniques*, vol MTT-12, Jan 1964, pp 6–15.

[6] Cristal, E. G., Addendum to "An exact synthesis method for microwave band-stop filters," *IEEE Trans. on Microwave Theory and Techniques (Correspondence)*, vol MTT-12, May 1964, pp 369–382.

[7] Matthaei, G. L., L. Young, and E. M. T. Jones, *Microwave Filters, Impedance-Matching Networks, and Coupling Structures*, sec 4.06, New York: McGraw-Hill, 1964.

[8] Weinberg, L., *Network Analysis and Synthesis*. New York: McGraw-Hill, 1962, ch 13.

[9] Matthaei, G. L., Interdigital band-pass filters, *IRE Trans. on Microwave Theory and Techniques*, vol MTT-10, Nov 1962, pp 479–491.

[10] Matthaei, G. L., L. Young, and E. M. T. Jones, [7], sec 5.09.

[11] Cristal, E. G., Coupled circular rods between parallel ground planes, *IEEE Trans. on Microwave Theory and Techniques*, vol MTT-12, Jul 1964, pp 428–439.

[12] Dishal, M., Alignment and adjustment of synchronously tuned multiple-resonator-circuit filters, *Proc. IRE*, vol 39, Nov 1951, pp 1448–1455.

[13] Matthaei, G. L., L. Young, and E. M. T. Jones, [7], pp 668–673.

[14] Tuttle, D. F., Jr., *Network Synthesis*, vol 1. New York: Wiley, 1958, p 388.

Wideband High-Selectivity Diplexers Utilizing Digital Elliptic Filters

ROBERT J. WENZEL, MEMBER, IEEE

Abstract—**Design techniques and element value tables are presented for the construction of compact high-selectivity diplexers using digital elliptic component filters. A brief review of an applicable diplexer theory is given and the derivation and use of the element value tables are discussed. The tables provide values for component filters with an even number of branches n for $n=4$ through $n=12$ with a prototype ripple value corresponding to maximum diplexer input VSWR of approximately 1.26 to 1. For $n=8$, 10, and 12 branch filters, additional tables with permuted transmission zero orders are given.**

Test results for an $n=6$ branch experimental diplexer of 2.25 to 1 bandwidth are presented. The design and construction of this diplexer is described in detail and serves to illustrate the use of the element value tables and design procedures presented. Many practical construction and alignment suggestions are given which are useful in obtaining designs with good response characteristics. The experimental diplexer has crossover frequencies at 1.5 and 3.4 GHz and provides greater than a 50 dB isolation at frequencies 0.2 GHz from crossover. The diplexer has an input VSWR ≤1.5 to 1 from dc to 6.0 GHz, and has package dimensions of approximately 2.0 by 2.0 by 0.75 inches.

I. Introduction

A PROBLEM often encountered at microwave frequencies is the separation of contiguous frequency bands using passive filters. Detailed design techniques have been presented[1]–[4] for achieving suitable multiplexers; however, the structures used to realize the component filters often have been large and bulky, especially for high-selectivity wideband requirements. The high-selectivity of digital elliptic filters[5]–[7] makes them attractive for use in wideband diplexers, and their compact size results in structures that are substantially smaller than previous designs. This paper presents design techniques and element value tables that allow the simple design of compact diplexers and multiplexers using digital elliptic component filters. The tables presented give element values for filters of 4, 6, 8, 10, and 12 branches. These tables provide sufficient complexity to satisfy most wideband requirements.

This paper is presented in the following manner. In Section II, a brief review of diplexer filter theory relevant to digital elliptic filters is given. General criteria for multiplexer design are stated and related to elliptic function type responses. In Section III, design techniques and element value tables are presented for the design of wideband high-selectivity diplexers. Both the derivation and use of the element value tables are discussed. In Section IV, experimental results for a wideband diplexer model are presented. The design is described in detail and serves to illustrate the application of the tables to a typical diplexer requirement. Practical construction techniques and alignment suggestions are also presented. Conclusions and an overall evaluation of digital elliptic diplexers are given in Section V.

Manuscript received May 5, 1967. The work reported here was supported by the USAECOM, Fort Monmouth, N. J., under Contract DA 28-043 AMC-01869(E).

The author is with the Bendix Research Laboratories, Southfield, Mich. 48075

II. Digital Elliptic Diplexing Filters

The theory of diplexers has been described in several papers[1]–[4] where it is shown that a perfect match at the input port of a diplexer requires the component filters to be complementary. Filters with equal-ripple response in both passband and stopband can be designed to be complementary; however, it has been pointed out[2] that this places an undesirable restriction on the isolation characteristics. The use of "pseudo-complementary filters"[1],[2] allows the achievement of equal-ripple designs with high-isolation characteristics at the cost of a slight increase in the input VSWR. The performance of such filters is described in detail in Wenzel[2] where it is shown that for proper operation the component filters should:

1) be designed on a singly terminated or transfer immittance basis ($|Z_{12}|^2$ for series connection and $|Y_{12}|^2$ for parallel connection),

2) have attenuation characteristics that "cross over" at the 3 dB level,

3) have component attenuation characteristics whose slopes are equal and of opposite sign at the crossover frequency, and

4) have a total real part input immitance that is approximately constant (to within 20 percent, for example) and devoid of extremely rapid variations.

Under these conditions, the maximum input VSWR ($VSWR_M$) is given to a good approximation by[2]

$$VSWR_M \approx 10^{\alpha/10} \tag{1}$$

where α is the transfer immittance prototype ripple value in decibels. Although the transfer immittance function can have relatively large ripples, the actual power transmission ripple is determined by the resulting input VSWR. For example, a 1 dB $|Y_{12}|^2$ design gives from (1) a $VSWR_M$

$=1.26$, and a corresponding power transmission ripple of approximately 0.06 dB.

In general, conditions 1) through 4) can be satisfied by: a) synthesizing a singly terminated lowpass prototype filter, b) bandwidth scaling this design to have the 3 dB frequency occur at the desired crossover frequency, and c) using a low to highpass transformation to achieve the complementary or pseudo-complementary highpass prototype.

Digital elliptic filter prototypes contain only distributed L-C type elements (i.e., no unit elements) and a suitable diplexer prototype is readily obtained by bandwidth scaling the appropriate lowpass transfer immittance values in the conventional manner[8],[9] to produce the required 3 dB level crossover. Element values for singly terminated elliptic function filters of three through seven branches (one to three finite transmission zeros) are given in Skwirzynski.[10] The particular filter configuration to be described uses a parallel connection of digital elliptic component filters and thus requires the input admittances to be minimum susceptive.[2],[8],[9] This, in turn, requires the lowpass prototype to begin with a series inductance and the highpass prototype to begin with a series capacitance at the junction of the two component filters. Consequently, the prototype filters are limited to those having an even number of branches (i.e., 4, 6, etc.).[1] Although filters of four or six branches are of sufficient complexity to satisfy many requirements, higher-order filters are sometimes needed. To satisfy high-selectivity requirements and to obtain a consistent set of tables applicable to microwave digital elliptic realizations, element values have been computed for prototypes with an even number of branches n for $n=4$ through $n=12$. The derivation and use of these tables is described in Section III.

III. Element Value Tables for Digital Elliptic Diplexers

A. Derivation of Element Value Tables

The element value tables were derived using the detailed theory and synthesis procedures described by Skwirzynski (in particular Chapters 2 through 6).[10] Because of the tedious nature and length of these calculations, only those aspects of general interest will be discussed. In essence, the utilization of elliptic functions allows the closed form factorization of the transfer admittance function. With suitable manipulation, these roots are used to generate an appropriate reactance function from which the desired network can be obtained using well-known synthesis techniques.[8]–[10] The use of elliptic functions to obtain closed form solutions is very important in obtaining accurate element values for complex filters. Using the parameter methods described in Skwirzynski, reactance functions were obtained whose coefficients were very accurately known without resorting to specialized transformation techniques.[11]

Initially, the lowpass prototype was synthesized with the prototype ripple value[2] occurring at $\Omega_C=1$. This lowpass prototype was then bandwidth scaled to obtain $\Omega_{3\,\mathrm{dB}}=1$. As a check on the accuracy of the element values and the maximum input VSWR, the response of each diplexer derived from the tables was computed by direct analysis. The uniformity of the resultant maximum VSWR for all tables is good indication that the element values are highly accurate. Those element values which could be checked against the tables in Skwirzynski[10] ($n=4$ and 6) showed agreement to five significant figures (the maximum given in Skwirzynski). The maximum VSWR was in close agreement with (1) for all cases computed.

An important practical aspect of the synthesis procedure is the physical order in which transmission zeros are produced. That is, for K finite transmission zeros, there are $K!$ possible physical networks, each possessing resonant circuits that realize the K transmission zeros in a different order. For example, with the $n=12$ branch filter, there are five (5) transmission zeros or $5!=120$ possible network realizations. Consequently, a decision had to be made as to the transmission zero order to utilize in the tables. Generally speaking, the order of transmission zero removal influences the number of negative element values that will be encountered as well as the numerical spread of the element values. For certain transmission zero ordering, negative element values occur for extreme values of passband ripple and selectivity. For other orders, the numerical spread of element values is larger than necessary. Previous work on doubly terminated filters has indicated that best results are achieved if transmission zeros closest to the cutoff frequency are realized physically in the center of the filter, and transmission zeros further from the band edge are realized physically at the ends of the filter.[12]–[14] To investigate these realization possibilities, element values for several permutations of the transmission zero locations were computed for each value of n, including choosing the transmission zeros in monotonically increasing order. The results of this investigation showed that both the monotonic order and the order that realized transmission zeros closest to cutoff in the center of the filter resulted in acceptable element values for singly terminated designs. Other permutations gave element values that were "different," but no improvement over the aforementioned orders was noted. In the tables to be presented, for $n=8$, 10, and 12, two transmission zero orders are given for each value of n. In these cases, the first table gives the monotonic order, and the second gives the permuted order with zeros closest to cutoff realized in the center of the filter. This last order is that utilized by Saal in his extensive tables of doubly terminated element values.[15]

[1] Elliptic function prototype filters with an odd number of branches can be utilized in diplexer design if a physical configuration and interconnection other than a parallel connection of digital elliptic component filters is considered.

[2] The prototype frequency variable is taken to be Ω, where $S=j\Omega=j\tan(\pi f/2f_0)$ is Richards' transformation.

214

B. Content of the Element Value Tables

The complete prototype and typical response characteristics for digital elliptic diplexers are shown in Fig. 1(a). The mapped response using Richards' transformation $S=j\Omega=j\tan(\pi f/2f_0)$ is shown in Fig. 1(b). The structure and response shown are for $n=12$ branches. The prototype for other values of n is identical in form where C_n and all other element values with subscripts greater than n are taken to be zero. For lower values of n there is also a corresponding reduction in the number of finite transmission zeros. The element values in the tables are normalized such that the crossover frequency occurs at $\Omega_{3\text{ dB}}=1$ (i.e., $f_{3\text{ dB}}/f_0=\frac{1}{2}$) resulting in a 3 to 1 bandwidth for an unscaled distributed realization.

Element values are given in Tables I through VIII for $n=4$ through 12 branches with a transfer admittance prototype ripple value of 1 dB. This ripple value results in a maximum diplexer input VSWR of 1.26 to 1 and a power transmission ripple value of 0.06 dB. For $n=8$, 10, and 12, Tables IV, VI, and VIII, respectively, having permuted transmission zero locations (permuted tables are designed with a P) are included. The eight tables presented are adequate in satisfying most practical requirements.

The following symbols are utilized in the tables:

k = selectivity parameter; ratio of the cutoff frequency to the frequency at which the attenuation first reaches A_s in decibels. For high selectivity, the cutoff frequency is essentially the crossover frequency

C_i (i odd) = shunt capacitor

C_i (i even) = series bridging capacitor

L_i (i even) = series bridging inductor

$\Omega_i=1/\sqrt{L_iC_i}$ (i even) = finite transmission zero frequencies

A_s = minimum stopband attenuation level in decibels

VSWR_M = theoretical maximum input VSWR of diplexer

Ω_{A_s} = frequency at which the attenuation first reaches A_s in decibels.

In all cases, the tables have been reduced in size by limiting selectivity and attenuation levels to those values that are likely to be encountered in a practical requirement. For small n, the selectivity is given in steps of 0.05, while for large n it is given in steps of 0.02.

It is interesting to note that the normalized element values in the tables are relatively close to unity for all values of n and for a wide range in selectivity factor. Some small values of series bridging capacitors do occur, but these are realized by a parallel coupling of lines and are usually convenient. The uniformity of element values indicates that digital elliptic diplexer designs are practical to construct.

C. Use of Element Value Tables

The practical design of digital elliptic diplexing filters is identical to that of doubly terminated digital elliptic filters and all previous design information given in Wenzel and Horton[6],[7] is applicable. The only difference between singly and doubly terminated designs is the numerical value of the elements. However, there are several aspects of diplexer design that are not relevant to a single filter design and these will be discussed.

In designing a digital elliptic diplexing filter, one must decide whether a single or double crossover filter is desired. This consideration determines the center frequency f_0 that must be chosen. Because of the repeating response of distributed quarterwave filters [see Fig. 1(b)], the design may be either pseudo highpass-lowpass, or bandpass-bandstop. If only one crossover is desired in the frequency band of interest, f_0 can be varied to obtain a design with most convenient element values. If a bandpass-bandstop combination is desired, f_0 is constrained to be $(f_{C1}'+f_{C2}')/2$, where f_{C1}' and f_{C2}' are the crossover frequencies. The advantages of using the repeating response of distributed filters in multiplexers is discussed in Wenzel.[2]

Another important consideration in distributed filter design is the effect of the tangent mapping function on selectivity. The effect of the tangent mapping is to increase selectivity of the distributed filter in comparison with that of the prototype filter, the amount increasing as the bandwidth is reduced. Thus, it is important to check the mapped response to insure that the filter has not been overdesigned. This check is automatically included in the design procedures to be presented.

Complete design data for digital elliptic diplexing filters is given in Table IX. For filters with two crossovers, the crossover frequencies are f_{C1}' and f_{C2}'. For a single crossover, f_{C2}' is arbitrary and is chosen to yield a desired f_0. Table IX gives expressions for computing the required selectivity factor k and converting the lowpass prototype element values into static capacitances and stub impedances for both bandpass and bandstop component filters. These static capacitances and stub impedances can be utilized directly with available design data[16],[17] to obtain dimensions. The applicable techniques and detailed design examples are given in Wenzel and Horton.[6],[7] Many of the above considerations are described in the design example of Section IV.

The practical limitations of digital elliptic diplexers are the same as those listed in Wenzel and Horton. Several designs have been worked out in detail and practical dimensions have resulted for cases in which the component filters are wideband. These trial designs have indicated that bandwidths in the range of 2 to 1 (octave) to 4 to 1 usually result in practical structures.

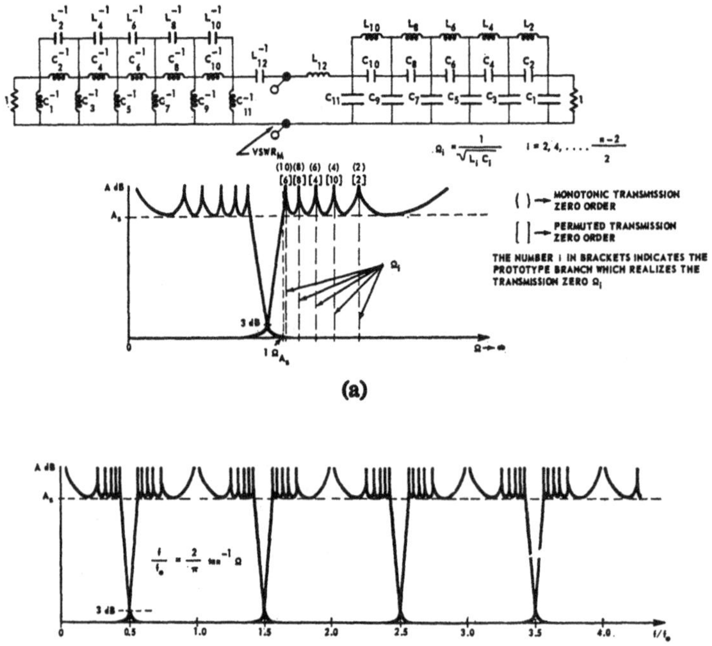

Fig. 1. Digital elliptic prototype and response characteristics. (a) L-C prototype and response characteristics. (b) Mapped response using Richards' transformation $S=j\Omega=j\tan(\pi f/2f_0)$.

TABLE I

MODIFIED LOWPASS PROTOTYPE ELEMENT VALUES FOR NORMALIZED ($\Omega_{3\text{ dB}}=1$) DIGITAL ELLIPTIC PSEUDO-COMPLEMENTARY FILTER PAIRS
$n=4$ BRANCHES
PROTOTYPE PASSBAND RIPPLE = 1.00 dB
(0.06 dB TRANSMISSION RIPPLE)

k	C_1	C_2	L_2	C_3	L_4	Ω_2	A_s	$VSWR_M$	Ω_{A_s}
0.30	0.7749	0.0541	1.5562	1.6146	1.6606	3.4461	66	1.2645	3.1470
0.35	0.7533	0.0755	1.5185	1.6017	1.6619	2.9528	60	1.2647	2.6999
0.40	0.7276	0.1017	1.4740	1.5865	1.6634	2.5827	55	1.2648	2.3651
0.45	0.6974	0.1335	1.4223	1.5692	1.6651	2.2947	51	1.2649	2.1051
0.50	0.6623	0.1722	1.3627	1.5496	1.6670	2.0641	47	1.2650	1.8975
0.55	0.6215	0.2197	1.2945	1.5276	1.6691	1.8753	43	1.2650	1.7280
0.60	0.5741	0.2785	1.2167	1.5032	1.6714	1.7178	39	1.2652	1.5873
0.65	0.5188	0.3532	1.1279	1.4763	1.6736	1.5843	35	1.2651	1.4687
0.70	0.4536	0.4511	1.0264	1.4471	1.6757	1.4696	32	1.2646	1.3677
0.75	0.3755	0.5855	0.9100	1.4161	1.6771	1.3700	29	1.2634	1.2807
0.80	0.2793	0.7843	0.7753	1.3844	1.6766	1.2824	26	1.2599	1.2055
0.85	0.1551	1.1155	0.6178	1.3553	1.6716	1.2046	22	1.2527	1.1401

TABLE II

MODIFIED LOWPASS PROTOTYPE ELEMENT VALUES FOR NORMALIZED ($\Omega_{3\,dB}=1$)
DIGITAL ELLIPTIC PSEUDO-COMPLEMENTARY FILTER PAIRS
$n=6$ BRANCHES
PROTOTYPE PASSBAND RIPPLE = 1.00 dB
(0.06 dB TRANSMISSION RIPPLE)

k	C_1	C_2	L_2	C_3	C_4	L_4	C_5	L_6	Ω_2	Ω_4	A_s	$VSWR_M$	ΩA_s
0.30	0.8714	0.0283	1.5832	1.7021	0.0458	1.9167	1.6715	1.7093	4.7233	3.3736	112	1.2686	3.2553
0.35	0.8593	0.0393	1.5645	1.6779	0.0636	1.8803	1.6564	1.7097	4.0324	2.8912	104	1.2687	2.7914
0.40	0.8449	0.0526	1.5423	1.6494	0.0851	1.8375	1.6386	1.7101	3.5116	2.5292	96	1.2689	2.4437
0.45	0.8279	0.0685	1.5162	1.6162	0.1107	1.7878	1.6179	1.7106	3.1041	2.2477	90	1.2689	2.1734
0.50	0.8080	0.0874	1.4858	1.5781	0.1413	1.7306	1.5941	1.7111	2.7755	2.0223	83	1.2691	1.9573
0.55	0.7848	0.1099	1.4506	1.5344	0.1778	1.6652	1.5668	1.7117	2.5041	1.8379	77	1.2693	1.7808
0.60	0.7578	0.1370	1.4096	1.4846	0.2217	1.5907	1.5357	1.7124	2.2753	1.6841	72	1.2693	1.6338
0.65	0.7260	0.1699	1.3619	1.4278	0.2751	1.5056	1.5003	1.7131	2.0788	1.5539	67	1.2695	1.5097
0.70	0.6884	0.2105	1.3060	1.3626	0.3414	1.4084	1.4597	1.7138	1.9071	1.4422	61	1.2698	1.4035
0.75	0.6431	0.2621	1.2394	1.2876	0.4263	1.2962	1.4128	1.7146	1.7544	1.3452	56	1.2701	1.3117
0.80	0.5871	0.3304	1.1586	1.2000	0.5403	1.1653	1.3581	1.7153	1.6162	1.2602	51	1.2702	1.2318
0.85	0.5150	0.4272	1.0570	1.0957	0.7058	1.0090	1.2926	1.7159	1.4881	1.1850	46	1.2705	1.1616
0.90	0.4145	0.5828	0.9209	0.9670	0.9828	0.8142	1.2110	1.7159	1.3650	1.1179	41	1.2703	1.0998

TABLE III

MODIFIED LOWPASS PROTOTYPE ELEMENT VALUES FOR NORMALIZED ($\Omega_{3\,dB}=1$)
DIGITAL ELLIPTIC PSEUDO-COMPLEMENTARY FILTER PAIRS
$n=8$ BRANCHES
PROTOTYPE PASSBAND RIPPLE = 1.00 dB
(0.06 dB TRANSMISSION RIPPLE)

k	C_1	C_2	L_2	C_3	C_4	L_4	C_5	C_6	L_6	C_7	L_8	Ω_2	Ω_4	Ω_6	A_s	$VSWR_M$	ΩA_s
0.50	0.8844	0.0520	1.5259	1.6820	0.1013	1.7908	1.6062	0.1372	1.8013	1.5990	1.7286	3.5513	2.3480	2.0116	119	1.2694	1.9766
0.55	0.8695	0.0652	1.5053	1.6494	0.1270	1.7446	1.5565	0.1725	1.7344	1.5711	1.7289	3.1922	2.1246	1.8284	111	1.2697	1.7977
0.60	0.8520	0.0809	1.4812	1.6118	0.1576	1.6915	1.4996	0.2148	1.6580	1.5393	1.7291	2.8880	1.9371	1.6757	104	1.2693	1.6487
0.65	0.8313	0.0999	1.4527	1.5683	0.1943	1.6301	1.4346	0.2662	1.5710	1.5030	1.7293	2.6251	1.7769	1.5464	97	1.2699	1.5227
0.70	0.8066	0.1230	1.4188	1.5176	0.2391	1.5589	1.3598	0.3298	1.4715	1.4614	1.7295	2.3937	1.6380	1.4355	90	1.2701	1.4149
0.75	0.7767	0.1519	1.3778	1.4581	0.2931	1.4753	1.2733	0.4109	1.3568	1.4133	1.7297	2.1861	1.5156	1.3394	83	1.2696	1.3215
0.80	0.7393	0.1892	1.3269	1.3867	0.3677	1.3754	1.1717	0.5190	1.2230	1.3571	1.7298	1.9957	1.4063	1.2551	76	1.2704	1.2401
0.82	0.7214	0.2076	1.3027	1.3539	0.4034	1.3296	1.1258	0.5738	1.1628	1.3317	1.7299	1.9230	1.3654	1.2243	73	1.2700	1.2103
0.84	0.7014	0.2286	1.2757	1.3179	0.4445	1.2795	1.0761	0.6379	1.0979	1.3043	1.7298	1.8517	1.3260	1.1949	70	1.2691	1.1820
0.86	0.6787	0.2532	1.2451	1.2782	0.4926	1.2242	1.0219	0.7148	1.0275	1.2744	1.7298	1.7811	1.2877	1.1668	68	1.2707	1.1551
0.88	0.6525	0.2824	1.2099	1.2338	0.5501	1.1627	0.9623	0.8095	0.9505	1.2417	1.7297	1.7107	1.2504	1.1400	65	1.2708	1.1294
0.90	0.6216	0.3183	1.1686	1.1834	0.6210	1.0929	0.8961	0.9306	0.8653	1.2052	1.7295	1.6396	1.2138	1.1144	62	1.2680	1.1050
0.92	0.5838	0.3642	1.1187	1.1250	0.7124	1.0122	0.8210	1.0941	0.7695	1.1639	1.7291	1.5667	1.1776	1.0898	59	1.2710	1.0817
0.94	0.5355	0.4268	1.0556	1.0550	0.8386	0.9156	0.7336	1.3346	0.6591	1.1157	1.7284	1.4899	1.1412	1.0662	56	1.2691	1.0595
0.96	0.4683	0.5225	0.9691	0.9662	1.0351	0.7929	0.6271	1.7445	0.5264	1.0568	1.7272	1.4053	1.1038	1.0436	53	1.2705	1.0383
0.98	0.3553	0.7114	0.8289	0.8389	1.4367	0.6156	0.4840	2.7196	0.3523	0.9768	1.7242	1.3022	1.0633	1.0217	50	1.2592	1.0183

TABLE IV (P)

MODIFIED LOWPASS PROTOTYPE ELEMENT VALUES FOR NORMALIZED ($\Omega_{3\,dB}=1$)
DIGITAL ELLIPTIC PSEUDO-COMPLEMENTARY FILTER PAIRS
$n=8$ BRANCHES, PERMUTED TRANSMISSION ZEROS
PROTOTYPE PASSBAND RIPPLE = 1.00 dB
(0.06 dB TRANSMISSION RIPPLE)

k	C_1	C_2	L_2	C_3	C_4	L_4	C_5	C_6	L_6	C_7	L_8	Ω_2	Ω_4	Ω_6	A_s	$VSWR_M$	ΩA_s
0.50	0.8844	0.0520	1.5259	1.6444	0.1446	1.7090	1.6081	0.0963	1.8831	1.6346	1.7286	3.5513	2.0116	2.3480	119	1.2694	1.9766
0.55	0.8695	0.0652	1.5053	1.6036	0.1818	1.6450	1.5588	0.1208	1.8340	1.6146	1.7289	3.1922	1.8284	2.1246	111	1.2697	1.7977
0.60	0.8520	0.0809	1.4812	1.5568	0.2265	1.5721	1.5024	0.1499	1.7775	1.5914	1.7291	2.8880	1.6757	1.9371	104	1.2693	1.6487
0.65	0.8313	0.0999	1.4527	1.5032	0.2809	1.4888	1.4378	0.1850	1.7123	1.5648	1.7293	2.6251	1.5464	1.7769	97	1.2699	1.5227
0.70	0.8066	0.1230	1.4188	1.4414	0.3482	1.3936	1.3637	0.2277	1.6368	1.5338	1.7295	2.3937	1.4355	1.6380	90	1.2701	1.4149
0.75	0.7767	0.1519	1.3778	1.3695	0.4342	1.2838	1.2778	0.2812	1.5483	1.4974	1.7297	2.1861	1.3394	1.5156	83	1.2696	1.3215
0.80	0.7393	0.1892	1.3269	1.2847	0.5493	1.1556	1.1769	0.3505	1.4429	1.4540	1.7298	1.9957	1.2551	1.4063	76	1.2704	1.2401
0.82	0.7214	0.2076	1.3027	1.2461	0.6077	1.0978	1.1313	0.3846	1.3946	1.4340	1.7299	1.9230	1.2243	1.3654	73	1.2700	1.2103
0.84	0.7014	0.2286	1.2757	1.2041	0.6764	1.0355	1.0819	0.4238	1.3419	1.4122	1.7298	1.8517	1.1949	1.3260	70	1.2691	1.1820
0.86	0.6787	0.2532	1.2451	1.1582	0.7589	0.9679	1.0282	0.4697	1.2839	1.3881	1.7298	1.7811	1.1668	1.2877	68	1.2707	1.1551
0.88	0.6525	0.2824	1.2099	1.1075	0.8608	0.8938	0.9691	0.5245	1.2194	1.3613	1.7297	1.7107	1.1400	1.2504	65	1.2708	1.1294
0.90	0.6216	0.3183	1.1686	1.0505	0.9920	0.8118	0.9033	0.5920	1.1465	1.3309	1.7295	1.6396	1.1144	1.2138	62	1.2680	1.1050
0.92	0.5838	0.3642	1.1187	0.9854	1.1704	0.7194	0.8288	0.6788	1.0624	1.2956	1.7291	1.5667	1.0898	1.1776	59	1.2710	1.0817
0.94	0.5355	0.4268	1.0556	0.9086	1.4356	0.6127	0.7423	0.7982	0.9620	1.2534	1.7284	1.4899	1.0662	1.1412	56	1.2691	1.0595
0.96	0.4683	0.5225	0.9691	0.8131	1.8965	0.4842	0.6372	0.9828	0.8351	1.1999	1.7272	1.4053	1.0436	1.1038	53	1.2705	1.0383
0.98	0.3553	0.7114	0.8289	0.6800	3.0396	0.3152	0.4969	1.3550	0.6527	1.1228	1.7242	1.3022	1.0217	1.0633	50	1.2592	1.0183

TABLE V

MODIFIED LOWPASS PROTOTYPE ELEMENT VALUES FOR NORMALIZED ($\Omega_{3\ dB}=1$)
DIGITAL ELLIPTIC PSEUDO-COMPLEMENTARY FILTER PAIRS
$n=10$ BRANCHES
PROTOTYPE PASSBAND RIPPLE = 1.00 dB
(0.06 dB TRANSMISSION RIPPLE)

k	C_1	C_2	L_2	C_3	C_4	L_4	C_5	C_6	L_6	C_7	C_8	L_8
0.80	0.8265	0.1234	1.4102	1.5244	0.2604	1.5135	1.3246	0.4060	1.3729	1.1418	0.5117	1.2444
0.82	0.8136	0.1351	1.3939	1.4978	0.2850	1.4783	1.2840	0.4459	1.3216	1.0942	0.5654	1.1835
0.84	0.7990	0.1485	1.3757	1.4684	0.3129	1.4393	1.2398	0.4918	1.2658	1.0427	0.6284	1.1179
0.86	0.7824	0.1640	1.3548	1.4355	0.3453	1.3958	1.1910	0.5457	1.2047	0.9867	0.7037	1.0467
0.88	0.7632	0.1823	1.3306	1.3983	0.3836	1.3466	1.1367	0.6103	1.1369	0.9253	0.7963	0.9687
0.90	0.7403	0.2046	1.3018	1.3554	0.4301	1.2899	1.0753	0.6905	1.0609	0.8571	0.9146	0.8825
0.92	0.7122	0.2328	1.2665	1.3047	0.4889	1.2229	1.0043	0.7944	0.9737	0.7801	1.0740	0.7856
0.94	0.6759	0.2707	1.2209	1.2422	0.5678	1.1405	0.9195	0.9388	0.8707	0.6907	1.3077	0.6739
0.96	0.6249	0.3271	1.1568	1.1600	0.6857	1.0321	0.8123	1.1659	0.7423	0.5823	1.7044	0.5396
0.98	0.5384	0.4332	1.0484	1.0346	0.9087	0.8671	0.6590	1.6357	0.5624	0.4374	2.6389	0.3634

k	C_9	L_{10}	Ω_2	Ω_4	Ω_6	Ω_8	A_s	$VSWR_M$	Ω_{A_s}
0.80	1.3548	1.7378	2.3975	1.5928	1.3394	1.2531	101	1.2702	1.2437
0.82	1.3292	1.7377	2.3044	1.5408	1.3027	1.2224	97	1.2699	1.2137
0.84	1.3016	1.7376	2.2127	1.4900	1.2674	1.1931	94	1.2679	1.1851
0.86	1.2715	1.7374	2.1216	1.4404	1.2334	1.1652	90	1.2631	1.1579
0.88	1.2385	1.7372	2.0302	1.3913	1.2005	1.1386	86	1.2658	1.1320
0.90	1.2017	1.7370	1.9375	1.3425	1.1684	1.1131	83	1.2701	1.1073
0.92	1.1601	1.7365	1.8415	1.2933	1.1371	1.0887	79	1.2701	1.0836
0.94	1.1115	1.7359	1.7395	1.2426	1.1061	1.0653	75	1.2598	1.0611
0.96	1.0520	1.7348	1.6256	1.1887	1.0750	1.0428	72	1.2598	1.0396
0.98	0.9708	1.7326	1.4838	1.1266	1.0426	1.0211	68	1.2631	1.0191

TABLE VI (P)

MODIFIED LOWPASS PROTOTYPE ELEMENT VALUES FOR NORMALIZED ($\Omega_{3\ dB}=1$)
DIGITAL ELLIPTIC PSEUDO-COMPLEMENTARY FILTER PAIRS
$n=10$ BRANCHES, PERMUTED TRANSMISSION ZEROS
PROTOTYPE PASSBAND RIPPLE = 1.00 dB
(0.06 dB TRANSMISSION RIPPLE)

k	C_1	C_2	L_2	C_3	C_4	L_4	C_5	C_6	L_6	C_7	C_8	L_8
0.80	0.8265	0.1234	1.4102	1.4070	0.4397	1.2678	1.1625	0.5173	1.2310	1.2526	0.2415	1.6321
0.82	0.8136	0.1351	1.3939	1.3731	0.4840	1.2175	1.1130	0.5706	1.1727	1.2113	0.2644	1.5932
0.84	0.7990	0.1485	1.3757	1.3360	0.5354	1.1628	1.0595	0.6328	1.1100	1.1664	0.2905	1.5503
0.86	0.7824	0.1640	1.3548	1.2951	0.5961	1.1027	1.0013	0.7068	1.0421	1.1172	0.3208	1.5024
0.88	0.7632	0.1823	1.3306	1.2494	0.6697	1.0362	0.9375	0.7971	0.9678	1.0627	0.3567	1.4484
0.90	0.7403	0.2046	1.3018	1.1975	0.7619	0.9614	0.8665	0.9114	0.8856	1.0014	0.4002	1.3863
0.92	0.7122	0.2328	1.2665	1.1372	0.8832	0.8757	0.7864	1.0636	0.7933	0.9311	0.4553	1.3132
0.94	0.6759	0.2707	1.2209	1.0646	1.0554	0.7745	0.6935	1.2830	0.6868	0.8479	0.5292	1.2237
0.96	0.6249	0.3271	1.1568	0.9719	1.3344	0.6485	0.5808	1.6459	0.5587	0.7440	0.6395	1.1067
0.98	0.5384	0.4332	1.0484	0.8365	1.9428	0.4736	0.4307	2.4624	0.3895	0.5979	0.8473	0.9299

k	C_9	L_{10}	Ω_2	Ω_4	Ω_6	Ω_8	A_s	$VSWR_M$	Ω_{A_s}
0.80	1.5235	1.7378	2.3975	1.3394	1.2531	1.5928	101	1.2702	1.2437
0.82	1.5078	1.7377	2.3044	1.3027	1.2224	1.5408	97	1.2699	1.2137
0.84	1.4905	1.7376	2.2127	1.2674	1.1931	1.4900	94	1.2679	1.1851
0.86	1.4712	1.7374	2.1216	1.2334	1.1652	1.4404	90	1.2631	1.1579
0.88	1.4493	1.7372	2.0302	1.2005	1.1386	1.3913	86	1.2658	1.1320
0.90	1.4242	1.7370	1.9375	1.1684	1.1131	1.3425	83	1.2701	1.1073
0.92	1.3944	1.7365	1.8415	1.1371	1.0887	1.2933	79	1.2701	1.0836
0.94	1.3579	1.7359	1.7395	1.1061	1.0653	1.2426	75	1.2598	1.0611
0.96	1.3099	1.7348	1.6256	1.0750	1.0428	1.1887	72	1.2598	1.0396
0.98	1.2366	1.7326	1.4838	1.0426	1.0211	1.1266	68	1.2631	1.0191

TABLE VII

MODIFIED LOWPASS PROTOTYPE ELEMENT VALUES FOR NORMALIZED ($\Omega_{3\ dB}=1$)
DIGITAL ELLIPTIC PSEUDO-COMPLEMENTARY FILTER PAIRS
$n=12$ BRANCHES
PROTOTYPE PASSBAND RIPPLE = 1.00 dB
(0.06 dB TRANSMISSION RIPPLE)

k	C_1	C_2	L_2	C_3	C_4	L_4	C_5	C_6	L_6	C_7	C_8	L_8
0.80	0.8819	0.0869	1.4548	1.6231	0.1924	1.6038	1.4526	0.3175	1.4886	1.2607	0.4294	1.3600
0.82	0.8720	0.0951	1.4435	1.6015	0.2103	1.5768	1.4176	0.3477	1.4454	1.2167	0.4718	1.3060
0.84	0.8609	0.1045	1.4306	1.5773	0.2306	1.5468	1.3792	0.3823	1.3982	1.1689	0.5210	1.2469
0.86	0.8481	0.1153	1.4159	1.5502	0.2541	1.5129	1.3365	0.4225	1.3457	1.1164	0.5787	1.1828
0.88	0.8333	0.1280	1.3987	1.5192	0.2817	1.4741	1.2885	0.4702	1.2869	1.0582	0.6482	1.1119
0.90	0.8155	0.1435	1.3780	1.4831	0.3150	1.4289	1.2335	0.5285	1.2200	0.9927	0.7347	1.0325
0.92	0.7937	0.1629	1.3525	1.4398	0.3568	1.3745	1.1690	0.6027	1.1419	0.9176	0.8475	0.9421
0.94	0.7653	0.1889	1.3191	1.3856	0.4123	1.3063	1.0905	0.7032	1.0476	0.8288	1.0054	0.8359
0.96	0.7251	0.2272	1.2714	1.3126	0.4938	1.2140	0.9886	0.8556	0.9267	0.7179	1.2562	0.7049
0.98	0.6560	0.2979	1.1883	1.1971	0.6432	1.0674	0.8366	1.1505	0.7497	0.5627	1.7841	0.5241

k	C_9	C_{10}	L_{10}	C_{11}	L_{12}	Ω_2	Ω_4	Ω_6	Ω_8	Ω_{10}	A_s	$VSWR_M$	Ω_{A_s}
0.80	1.1217	0.5080	1.2556	1.3527	1.7426	2.8117	1.8002	1.4546	1.3086	1.2521	125	1.2710	1.2457
0.82	1.0733	0.5612	1.1944	1.3269	1.7426	2.6985	1.7366	1.4105	1.2739	1.2215	121	1.2699	1.2155
0.84	1.0211	0.6235	1.1282	1.2993	1.7424	2.5866	1.6743	1.3678	1.2407	1.1923	116	1.2691	1.1868
0.86	0.9644	0.6980	1.0565	1.2691	1.7422	2.4752	1.6129	1.3262	1.2088	1.1644	112	1.2638	1.1594
0.88	0.9023	0.7897	0.9781	1.2360	1.7420	2.3632	1.5519	1.2855	1.1780	1.1378	108	1.2600	1.1334
0.90	0.8335	0.9066	0.8913	1.1991	1.7417	2.2490	1.4905	1.2454	1.1482	1.1124	103	1.2600	1.1085
0.92	0.7558	1.0641	0.7937	1.1574	1.7413	2.1304	1.4280	1.2054	1.1192	1.0881	99	1.2600	1.0847
0.94	0.6661	1.2948	0.6812	1.1087	1.7407	2.0035	1.3627	1.1650	1.0908	1.0648	94	1.2700	1.0619
0.96	0.5575	1.6857	0.5460	1.0490	1.7397	1.8608	1.2916	1.1231	1.0627	1.0424	90	1.2628	1.0402
0.98	0.4134	2.6042	0.3685	0.9674	1.7377	1.6808	1.2069	1.0768	1.0341	1.0209	85	1.2599	1.0195

TABLE VIII (P)

MODIFIED LOWPASS PROTOTYPE ELEMENT VALUES FOR NORMALIZED ($\Omega_{3\ dB}=1$)
DIGITAL ELLIPTIC PSEUDO-COMPLEMENTARY FILTER PAIRS
$n=12$ BRANCHES, PERMUTED TRANSMISSION ZEROS
PROTOTYPE PASSBAND RIPPLE = 1.00 dB
(0.06 dB TRANSMISSION RIPPLE)

k	C_1	C_2	L_2	C_3	C_4	L_4	C_5	C_6	L_6	C_7	C_8	L_8
0.80	0.8819	0.0869	1.4548	1.5131	0.3429	1.3785	1.2354	0.5241	1.2171	1.1521	0.4328	1.3494
0.82	0.8720	0.0951	1.4435	1.4840	0.3762	1.3362	1.1877	0.5780	1.1596	1.1024	0.4758	1.2949
0.84	0.8609	0.1045	1.4306	1.4519	0.4144	1.2898	1.1361	0.6408	1.0978	1.0487	0.5258	1.2355
0.86	0.8481	0.1153	1.4159	1.4163	0.4591	1.2383	1.0797	0.7156	1.0306	0.9902	0.5845	1.1710
0.88	0.8333	0.1280	1.3987	1.3761	0.5126	1.1806	1.0175	0.8069	0.9572	0.9261	0.6552	1.0999
0.90	0.8155	0.1435	1.3780	1.3300	0.5784	1.1148	0.9479	0.9225	0.8759	0.8549	0.7434	1.0204
0.92	0.7937	0.1629	1.3525	1.2757	0.6630	1.0380	0.8687	1.0766	0.7846	0.7743	0.8584	0.9300
0.94	0.7653	0.1889	1.3191	1.2094	0.7794	0.9452	0.7759	1.2987	0.6791	0.6808	1.0194	0.8244
0.96	0.7251	0.2272	1.2714	1.1227	0.9597	0.8262	0.6615	1.6667	0.5522	0.5671	1.2746	0.6947
0.98	0.6560	0.2979	1.1883	0.9918	1.3226	0.6521	0.5047	2.4960	0.3844	0.4148	1.8062	0.5177

k	C_9	C_{10}	L_{10}	C_{11}	L_{12}	Ω_2	Ω_4	Ω_6	Ω_8	Ω_{10}	A_s	$VSWR_M$	Ω_{A_s}
0.80	1.3394	0.1750	1.7630	1.5709	1.7426	2.8117	1.4546	1.2521	1.3086	1.8002	125	1.2710	1.2457
0.82	1.3035	0.1915	1.7318	1.5584	1.7426	2.6985	1.4105	1.2215	1.2739	1.7366	121	1.2699	1.2155
0.84	1.2644	0.2102	1.6970	1.5448	1.7424	2.5866	1.3678	1.1923	1.2407	1.6743	116	1.2691	1.1868
0.86	1.2211	0.2319	1.6579	1.5293	1.7422	2.4752	1.3262	1.1644	1.2088	1.6129	112	1.2638	1.1594
0.88	1.1728	0.2574	1.6134	1.5117	1.7420	2.3632	1.2855	1.1378	1.1780	1.5519	108	1.2600	1.1334
0.90	1.1180	0.2882	1.5615	1.4911	1.7417	2.2490	1.2454	1.1124	1.1482	1.4905	103	1.2600	1.1085
0.92	1.0544	0.3270	1.4996	1.4665	1.7413	2.1304	1.2054	1.0881	1.1192	1.4280	99	1.2600	1.0847
0.94	0.9779	0.3787	1.4222	1.4357	1.7407	2.0035	1.1650	1.0648	1.0908	1.3627	94	1.2700	1.0619
0.96	0.8802	0.4547	1.3185	1.3941	1.7397	1.8608	1.1231	1.0424	1.0627	1.2916	90	1.2628	1.0402
0.98	0.7377	0.5942	1.1554	1.3282	1.7377	1.6808	1.0768	1.0209	1.0341	1.2069	85	1.2599	1.0195

TABLE IX

DESIGN EQUATIONS FOR DIGITAL ELLIPTIC DIPLEXERS

	PROTOTYPE LOWPASS MICROWAVE BANDSTOP	PROTOTYPE HIGHPASS MICROWAVE BANDPASS
	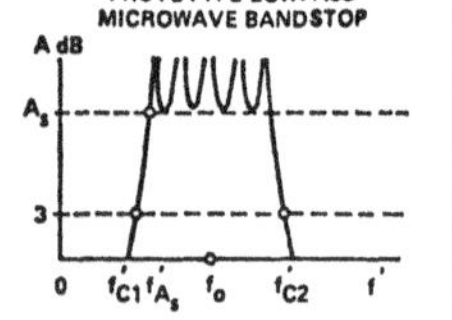	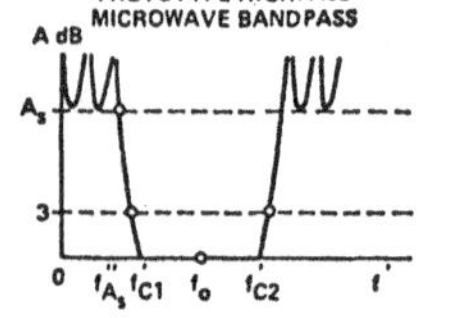
BANDWIDTH SCALING, FACTOR, Ω_c	$\Omega_c' = \tan \frac{\pi f_{C1}'}{2 f_0} = \tan \frac{\pi f_{C1}'}{f_{C1}' + f_{C2}'}$ (T-1) $\% BW = 200 \left(1 - \frac{2}{\pi} \tan^{-1} \Omega_c'\right)$ (T-2)	
SKIRT* SELECTIVITY FACTOR, k	$k_L = \frac{\Omega_c'}{\Omega_{A_s}'} = \frac{\Omega_c'}{\tan \frac{\pi f_{A_s}'}{2 f_0}}$ (T-3)	$k_H = \frac{\Omega_{A_s}''}{\Omega_c'} = \frac{\tan \frac{\pi f_{A_s}''}{2 f_0}}{\Omega_c'}$ (T-4)
PARALLEL CONDUCTOR ARRAY	$\epsilon_{chart} = \frac{376.7}{\sqrt{\epsilon_r} Z_0} \frac{C}{\Omega_c'}$ (T-5)	$\epsilon_{chart} = \frac{376.7}{\sqrt{\epsilon_r} Z_0} C \Omega_c'$ (T-6)
INTERNAL SERIES STUBS	$\sqrt{\epsilon_r} Z_{ser,stub} = \sqrt{\epsilon_r} Z_0 \frac{L}{\Omega_c'}$ (T-7)	$\sqrt{\epsilon_r} Z_{ser,stub} = \sqrt{\epsilon_r} Z_0 L \Omega_c'$ (T-8)

A_s – MINIMUM STOPBAND ATTENUATION LEVEL IN dB
f_{A_s}' – LOW PASS FILTER FREQUENCY FOR WHICH THE ATTENUATION IS A_s dB
f_{A_s}'' – HIGH PASS FILTER FREQUENCY FOR WHICH THE ATTENUATION IS A_s dB
f_{C1}' – LOWER 3 dB CROSSOVER FREQUENCY
f_{C2}' – HIGHER 3 dB CROSSOVER FREQUENCY
$f_0 = (f_{C1}' + f_{C2}')/2$ – FREQUENCY FOR WHICH THE LINES ARE QUARTER-WAVELENGTH
k – SKIRT SELECTIVITY PARAMETER LISTED IN ELEMENT VALUE TABLES
L, C – LOW PASS PROTOTYPE ELEMENT VALUES LISTED IN ELEMENT VALUE TABLES
Z_0 – CHARACTERISTIC TERMINATING IMPEDANCE (ohms)
ϵ_r – RELATIVE DIELECTRIC CONSTANT OF MEDIUM
ϵ_{chart} – "STATIC CAPACITANCE" TO BE FOUND IN CHARTS [16]
$\sqrt{\epsilon_r} Z_{ser,stub}$ – CHARACTERISTIC IMPEDANCE TO BE FOUND IN CHART [17]
*CHOOSE $k \geq k_H$ IF $\Omega_c' > 1$ AND $k \geq k_L$ IF $\Omega_c' < 1$

IV. EXPERIMENTAL RESULTS INCLUDING CONSTRUCTION AND ALIGNMENT SUGGESTIONS

To illustrate in detail the design of digital elliptic diplexers using the element value tables and design Table I, consider the following specifications typical of those likely to be encountered:

1) Bandpass-bandstop diplexer with crossover frequencies $f_{C1}' = 1.50$ GHz and $f_{C2}' = 3.375$ GHz (i.e., 2.25 to 1 bandwidth)
2) Isolation greater than 60 dB at $f_{C1}' \pm 0.30$ GHz (i.e., $f_{A_s}'' = 1.20$ GHz and $f_{A_s}' = 1.80$ GHz)
3) Maximum input VSWR$\leq$1.5 to 1.

From design Table IX one determines

$$f_0 = \frac{f_{C1}' + f_{C2}'}{2} = \frac{4.875}{2} \approx 2.44 \text{ GHz}. \quad (2)$$

Then from equation (T-1)

$$\Omega_c' = \tan \frac{\pi f_{C1}'}{2 f_0} = \tan 55.4° = 1.45 \quad (3)$$

and from equation (T-4)[3]

$$k \geq k_H = \frac{\tan \frac{\pi f_{A_s}''}{2 f_0}}{\Omega_c'} = 0.672. \quad (4)$$

The desired parameters are now completely specified. That is, one wishes to find in the tables a design with $k \geq 0.672$, $A_s \geq 60$ dB, and VSWR$_M \leq 1.5$ to 1. The calculations of (2) through (4) determine the actual prototype selectivity required. A simple ratio of the highpass A_s in decibels attenuation frequency to the crossover frequency gives $k = 1.20/1.50 = 0.80$. However, as explained in Section III-C, the effect of the tangent mapping function is to increase the selectivity of the distributed realization over the prototype selectivity and the calculations of (2) through (4) indicate that any $k \geq 0.672$ is suitable. The use of the direct ratio $k = 0.80$ would lead to a filter that was overdesigned. If one wishes to insure that a specification will be met, some overdesign may be desirable.

Inspection of the element value tables shows that an $n = 6$ branch filter in Table II has the following satisfactory specifications:

$$k = 0.70, \qquad A_s = 61 \text{ dB},$$
$$\text{VSWR}_M \approx 1.27, \qquad \Omega_{A_s} = 1.404$$

frequencies of zero transmission

$$\Omega_2 = 1.907, \qquad \Omega_4 = 1.442. \quad (5)$$

The normalized low and highpass prototype element values are shown in Fig. 2(a). Bandwidth scaled real frequencies suitable for obtaining the actual frequency response are given by[6],[7]

$$f' = \frac{2 f_0}{\pi} \tan^{-1} \Omega_c' \Omega. \quad (6)$$

For the highpass prototype, one substitutes Ω^{-1} for Ω in (6) to obtain the bandwidth scaled frequencies. Applying (6) to the design data gives the final theoretical mapped response of Fig. 2(b).

Static capacitance values and series stub impedance values are obtained by substituting the normalized lowpass prototype values of Fig. 2(a) in equations (T-5) through (T-8) of Table IX. The static capacitance network for the complete diplexer along with stub impedance values is shown in Fig. 2(c). Dimensions for the coupled-line portions of each filter are determined using Getsinger's data[16] as described in Matthaei *et al.*,[4] and Wenzel and Horton,[6],[7] and the series stub dimensions are obtained from the graph in *The Microwave Engineer's Handbook and Buyer's Guide.*[17]

[3] *Note:* In determining the required selectivity factor, choose $k \geq k_H$ when $\Omega_c' > 1$ and $k \geq k_L$ when $\Omega_c' < 1$. The choice of k is influenced by the tangent mapping function. The values of k_H and k_L do not differ significantly for bandwidths in the 2 to 1 to 4 to 1 range.

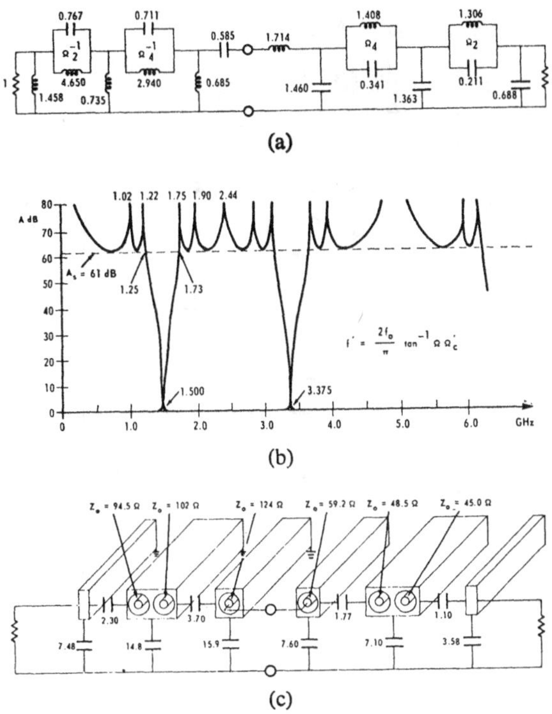

Fig. 2. Digital elliptic diplexer design example. (a) Diplexer prototype network obtained from element value Table II. (b) Mapped theoretical response characteristics. (c) Static capacitance network and series stub impedance values.

Inspection of the static capacitance values shows the shunt lines closest to the common junction to be quite wide. To obtain narrower width lines and to allow the side by side placement of the component filters, a wall was inserted between the component filters and the first shunt line of each filter was coupled to this wall. The widths of both first shunt lines were chosen to be $b/2$ (where b is the ground plane spacing) and dimensions were calculated using the procedures in Getsinger[16] to obtain the desired shunt static capacitance. Complete cross-sectional dimensions, as determined by the preceding procedures, and a plan view of the diplexer are given in Fig. 3. The ground plane spacing b was chosen to be 0.250 inch and the coupled-line thickness t was chosen to be 0.100 inch for the bandstop filter and 0.150 inch for the bandpass filter.

A perspective sketch of the complete diplexer showing the physical layout of the coupled lines is given in Fig. 4. In the design of the series stubs, the outer diameters were chosen such that the center conductors would have a standard diameter. Teflon loading was utilized for all stubs except the first series stub in the bandpass filter which was the highest impedance line (124 ohms). All coupled lines were initially 1.210 inches long ($\lambda/4$ at 2.44 GHz) with the teflon loaded lines being 1.210/1.44 inches = 0.850 inch long. The teflon loading of the series stubs was shortened by approximately 0.100 inch and the center conductors were extended to allow the inclusion of tuning slugs. The tuning slugs designated A in Fig. 4 were of teflon and provided tuning of the bandpass transmission zeros while the B slugs were of brass and provided tuning of the bandstop transmission zeros. In the initial design, the teflon tuning slugs were of sufficient length to extend beyond the back of the filter allowing adjustment with both ground planes intact. Likewise, the brass tuning slugs were threaded at the rear and nylon rods were inserted to provide external adjustment. The bandstop parallel-coupled lines were provided with small 0-80 set screws to firmly hold the brass shorting slugs in place after they were adjusted. The teflon tuning slugs were a tight fit in the shorting block and were cut off after final adjustment. It is important that the brass shorting slugs not extend beyond the end of the open-circuited parallel-coupled lines. In total, approximately 0.100 to 0.150 inch of adjustment in the lengths of the series stubs was obtained using the aforementioned procedures. This was found to be more than adequate in adjusting the filter. The slight air gaps that may occur due to these tuning adjustments were found to have no noticeable effects on the filter response. The tuning slugs are not necessary and could be eliminated in subsequent filters. They are, however, very convenient for obtaining the initial design dimensions.

In the model, most of the junctions were mitered to reduce discontinuities. Hardened berillium copper wire was used for the 0.020 inch center conductors and gold-plated tungsten wire was used for the 0.010 inch center conductors. The teflon cylinders with small center holes were realized using tandem sections one-half inch in length. The center conductor holes were made on a jeweler's lathe and little manufacturing difficulty was encountered. All coupled bars were machined from brass. The bandstop filter portion was supported by two $\frac{1}{32}$ inch machined fiberglass plates. A photograph of the complete model is shown in Fig. 5. The final dimensions of the entire structure are approximately 2 by 2 by 0.75 inch.

Before describing the final measured response characteristics, several practical alignment and construction techniques will be discussed. When the filter was initially tested, the attenuation response was fairly good, but the input VSWR was as large as 3 to 1 in the 2.0 to 3.0 GHz region. Adjustment of the tuning slugs further improved the attenuation characteristics but the input VSWR was lowered only slightly. Inspection of the structure (see Fig. 4) indicated that the junctions could be the cause of the high VSWR. The connecting lines between parallel lines and series stubs are assumed to be of zero length; however, at high frequencies this is unjustified. One must then answer the question as to what impedance a finite length of connecting line should be made when the design theory says it is of zero length. To answer this question, consider the bandpass filter in Fig. 4. In the passband, the series stubs are of low impedance and the shunt stubs are of high impedance. Therefore, between input and output ports there is a length of transmission line whose impedance is approximately the impedance of the connecting lines. Since these connecting lines were very small wires and consequently of high impedance, it was thought that increasing their size to achieve 50 ohm connecting lines would lower the VSWR. This was accomplished by increasing the size of the connecting lines and coupling them to the front wall as shown in Fig. 5. A brass block was added to achieve suitable coupling to the wall. Similar statements apply to the bandstop filter and the size of the connecting lines was increased accordingly. The results were quite gratifying in that the input VSWR was immediately reduced to

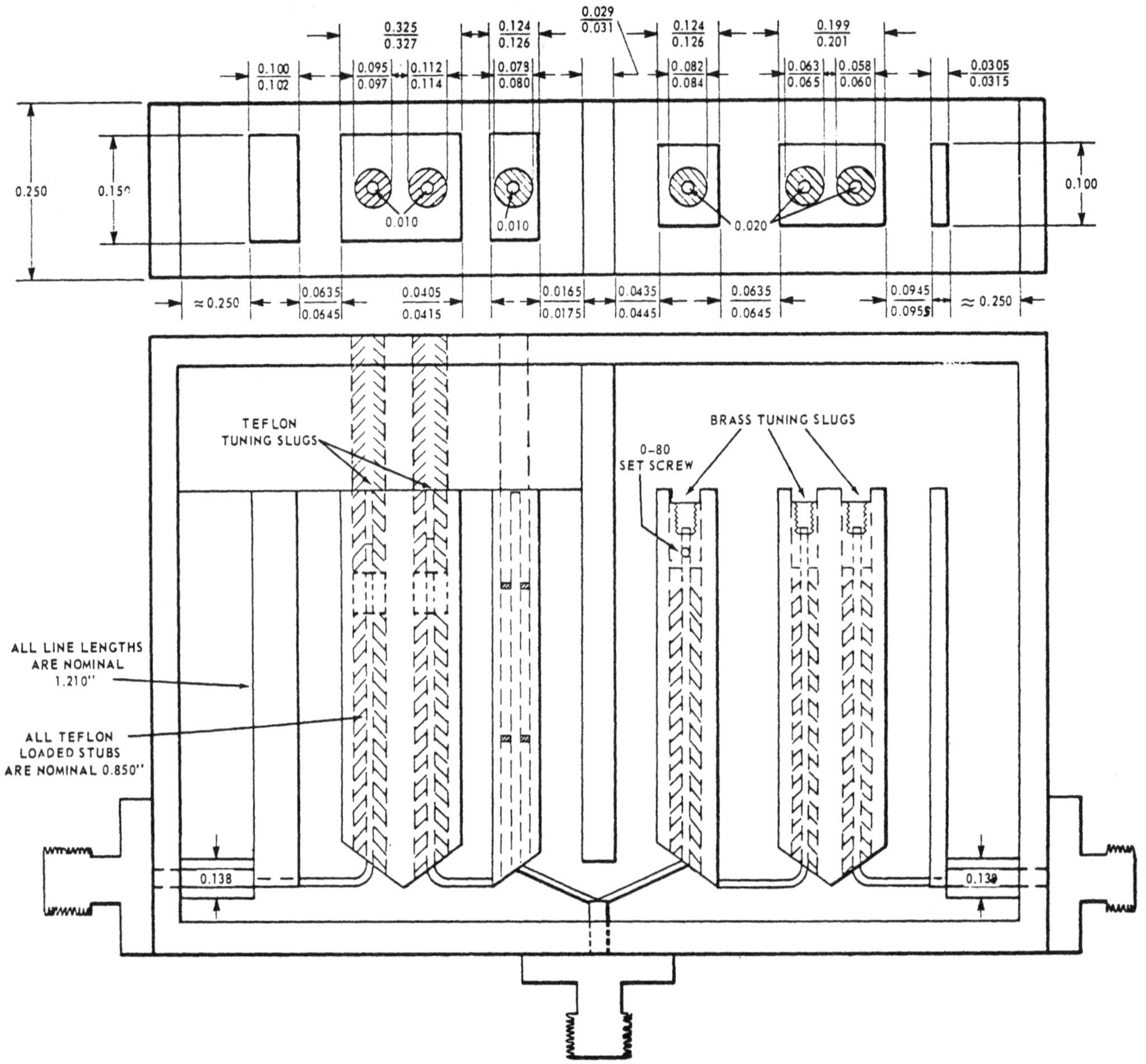

Fig. 3. Cross-sectional dimensions and plan view of experimental digital elliptic diplexer.

below 1.8 to 1 everywhere and below 1.5 to 1 over most of the dc to 6 GHz frequency band. It is suggested that for high-frequency filters the short connecting lines be made relatively large and the filter structures be placed sufficiently close to the front wall to achieve a near 50 ohm cross section. Filters for frequencies below 2.0 GHz should provide satisfactory performance without increasing the connecting line size.

In the filter design, the parallel-coupled lines are chosen to be one-quarter wavelength at the center frequency of the filter. Due to junction effects and finite widths, the exact lengths required cannot be determined analytically. Fortunately, good estimates of the lengths can be obtained by measurement. For example, consider the bandstop filter of Fig. 4 with the series shorting slugs removed. The series inductors are now capacitors and are practically shorts near $f=f_0$. One is then left with essentially the parallel capacitance array of Fig. 2(c). By shorting the ends of all but one of the coupled lines and by monitoring the insertion loss, a good estimate of the effective length of each conductor can be obtained. This length can then be adjusted to produce a transmission zero at $f=f_0$. A similar procedure can be utilized for the series stubs and also for all elements in the bandpass filter to obtain line lengths that are very close to those desired. In the experimental model, the line lengths in the bandstop filter were originally shortened to account for fringing effects. A measurement of the type described above then indicated that the center line was short and a tuning block was added as shown in Fig. 5 to achieve the desired effective length.

The measured response of the completed diplexer is shown in Fig. 6 and was obtained by adjusting all series

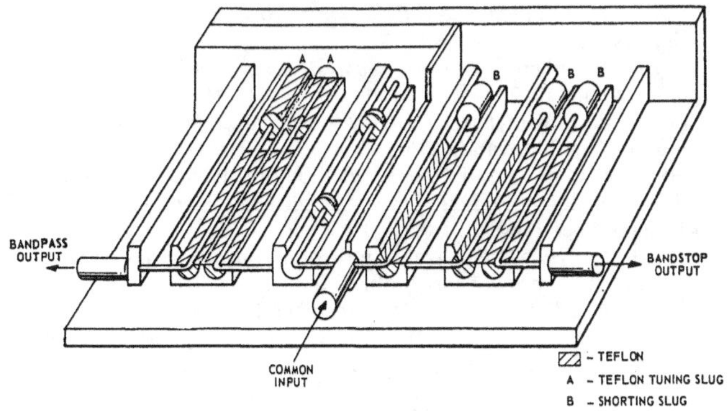

Fig. 4. Perspective sketch of experimental wideband digital elliptic diplexer.

Fig. 5. Compact broadband digital elliptic diplexer.

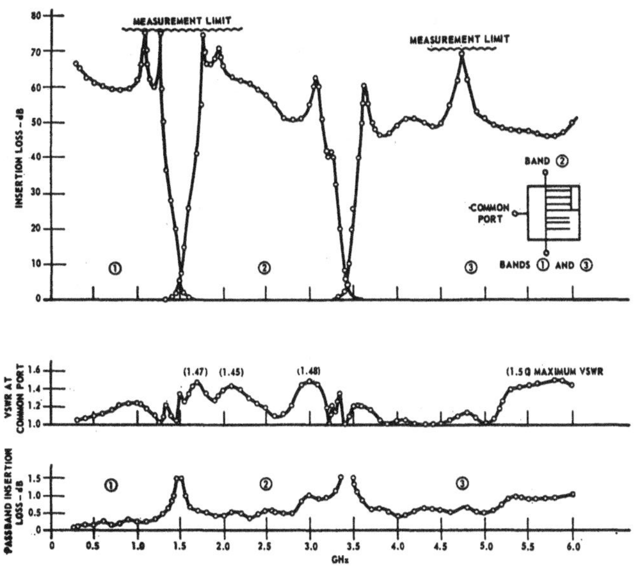

Fig. 6. Measured response of experimental digital elliptic diplexer.

shorting slugs after the connecting line sizes had been increased. This adjustment was carried out while monitoring input reflection only. The maximum input VSWR between dc and 6 GHz was 1.5 to 1. Inspection of the attenuation curves shows the lower crossover to be almost perfect, even exceeding the specifications at the high-frequency end of the crossover. The price paid for this increased attenuation is the lower than theoretical attenuation near the second crossover. The departure of the attenuation from theoretical is caused by slight misplacement of the bandstop transmission zeros as is evident in Fig. 6. There is also a tendency for dissipation loss and junction effects to obscure transmission zeros for higher repeating passbands. If the filter had been adjusted while monitoring the bandstop attenuation, some improvement could have been obtained, but the response was judged to be satisfactory as shown. A trial adjustment of transmission zeros using a one-watt amplifier and an oscilloscope enabled the setting of transmission zeros quite accurately. In performing an adjustment and measurement of this type, it is very important that the source be well filtered to eliminate harmonic outputs.

The measured insertion loss at both crossover frequencies was between 5.0 and 5.5 dB giving a dissipation loss of 2.0 to 2.5 dB. Passband loss was substantially lower as shown in Fig. 6. Since the filter had been handled quite extensively and no parts in the final structure were cleaned or plated, it is felt that the insertion loss could have been lowered. However, for the high selectivity and compactness of the structure, the measured values are quite encouraging. The crossover frequencies were very close to those predicted and the attenuation was good, especially at the first crossover.

Although the total multiplexer is a complex structure, the construction of the device was judged to be straightforward and no difficult problems were encountered.

V. Conclusions

Element value tables and design techniques for the construction of compact high-selectivity digital elliptic diplexers have been presented. The element values in the tables are very uniform and should result in practical filter structures particularly for bandwidths in the 2 to 1 to 4 to 1 range. Because the digital elliptic prototype is an L–C-type network, bandwidth scaling is simply accomplished and a large number of designs can be obtained from relatively few tables. The computed tables present element values heretofore unavailable, and these values are directly applicable to the design of lumped element diplexers as well as distributed digital elliptic realizations. The small size of the component filters will make them particularly desirable at UHF and low microwave frequencies; however, they should also find wide application at frequencies through *C* band.

By utilizing the repeating response of the distributed component filters, further size reduction in multichannel multiplexers can also be obtained. For example, the experimental diplexer described in Section IV can be made into a triplexer with the addition of a very simple and compact diplexer. This is possible because of the wide frequency separation between bands 1 and 3 shown in Fig. 6.

The test results for the experimental diplexer agreed well with theory and verified the design techniques presented. Although a multisection digital elliptic diplexer is a complex structure, little difficulty was encountered in constructing and aligning the experimental filter. The final experimental filter provided a combination of electrical and physical characteristics substantially better than those obtainable using other available techniques.

References

[1] R. G. Veltrop and R. B. Wilds, "Modified tables for the design of optimum diplexers," *Microwave J.*, vol. 7, pp. 76–80, June 1964.

[2] R. J. Wenzel, "Application of exact synthesis methods to multichannel filter design," *IEEE Trans. Microwave Theory and Techniques*, vol. MTT-13, pp. 5–15, January 1965.

[3] G. L. Matthaei and E. G. Cristal, "Multiplexer channel-separating units using interdigital and parallel-coupled filters," *IEEE Trans. Microwave Theory and Techniques*, vol. MTT-13, pp. 328–334, May 1965.

[4] G. L. Matthaei, L. Young, and E. M. T. Jones, *Microwave Filters, Impedance Matching Networks and Coupled Structures.* New York: McGraw-Hill, 1964.

[5] M. C. Horton and R. J. Wenzel, "Realization of microwave filters with equal ripple response in both pass and stop bands" in *Proceedings of the Symposium on Generalized Networks*, vol. 16. Brooklyn, N. Y.: Polytechnic Institute of Brooklyn Press, 1966, pp. 257–287.

[6] R. J. Wenzel and M. C. Horton, "Design techniques for TEM networks," USAEL Contract DA28-043 AMC-01869(E), Semi-Annual Rept., August 1966.

[7] M. C. Horton and R. J. Wenzel, "The digital elliptic filter—a compact sharp-cutoff design for wide bandstop or bandpass requirements," *IEEE Trans. Microwave Theory and Techniques*, vol. MTT-15, pp. 307–314, May 1967.

[8] E. A. Guillemin, *Synthesis of Passive Networks.* New York: Wiley, 1957.

[9] L. Weinberg, *Network Analysis and Synthesis.* New York: McGraw-Hill, 1962.

[10] J. K. Skwirzynski, *Design Theory and Data for Electric Filters.* London: Van Nostrand, 1965.

[11] J. A. C. Bingham, "A new method of solving the accuracy problem in filter design," *IEEE Trans. Circuit Theory*, vol. CT-11, pp. 327–341, September 1964.

[12] T. Fujisawa, "Realizability theorem on mid-series or mid-shunt low-pass ladders without mutual induction," *IRE Trans. Circuit Theory*, vol. CT-2, pp. 320–325, December 1955.

[13] R. Saal and E. Ulbrich, "On the design of filters by synthesis," *IRE Trans. Circuit Theory*, vol. CT-5, pp. 284–327, December 1958.

[14] J. Meinguet and V. Belevitch, "On the realizability of ladder filters," *IRE Trans. Circuit Theory*, vol. CT-5, pp. 253–255, December 1958.

[15] R. Saal, "Der Entwurf von Filtern mit Hilfe des Kataloges normierter Tiefpässe" (in German), Telefunken, Backnang, Germany, 1961.

[16] W. J. Getsinger, "Coupled rectangular bars between parallel plates," *IRE Trans. Microwave Theory and Techniques*, vol. MTT-10, pp. 65–72, January 1962.

[17] *The Microwave Engineer's Handbook and Buyer's Guide.* Brookline, Mass.: Horizon House-Microwave, Inc., 1966, p. 91.

Reprinted from IEEE TRANSACTIONS
ON *MICROWAVE THEORY AND TECHNIQUES*
Volume MTT-15, Number 12, December, 1967
pp. 669-680

TABLE I

MINIMUM STOPBAND ATTENUATION (A_s) FOR NORMALIZED ($\Omega_{3dB}=1$) DIGITAL ELLIPTIC PSEUDO-COMPLEMENTARY FILTER PAIRS $n=4, 6, 8, 10, 12$ BRANCHES
PROTOTYPE PASSBAND RIPPLE = 1.00 dB (0.06 dB TRANSMISSION RIPPLE)

k	$\eta=4$ Branches Table I	$\eta=6$ Branches Table II	$\eta=8$ Branches Tables III and IV(P)	$\eta=10$ Branches Tables V and VI(P)	$\eta=12$ Branches Tables VII and VIII(P)
$k<0.80$	Values Listed in Original Tables[1] are Correct to Within 1 dB				
0.80	25 dB	50 dB	75 dB	100 dB	124 dB
0.82	—	—	72 dB	46 dB	119 dB
0.84	—	—	69 dB	92 dB	115 dB
0.85	21 dB	45 dB	—	—	—
0.86	—	—	66 dB	88 dB	110 dB
0.88	—	—	62 dB	84 dB	105 dB
0.90	—	38 dB	59 dB	79 dB	99 dB
0.92	—	—	55 dB	74 dB	93 dB
0.94	—	—	50 dB	68 dB	86 dB
0.96	—	—	45 dB	61 dB	78 dB
0.98	—	—	37 dB	52 dB	66 dB

224

Wideband, High-Selectivity Diplexers Utilizing Digital-Elliptic Filters[1]

The author wishes to thank Dr. E. M. T. Jones of Technology for Communications International for pointing out a discrepancy between the minimum stopband attenuation values (A_s dB) listed in the element value tables and those obtained by direct analysis for high values of the selectivity parameter k. The minimum stopband attenuation values (A_s dB) have been recomputed and the correct values (to within 1 dB) are listed in Table I above.

The errors in the original tables are small for $k<0.90$, but become substantial as k approaches unity. The element values and maximum input standing-wave ratio ($VSWR_M$) are correct as given, and have been verified by direct analysis of each complete diplexer network.

R. J. WENZEL
Bendix Research Laboratories
Southfield, Mich. 48075

Manuscript received June 28, 1968.
[1] R. J. Wenzel, *IEEE Trans. Microwave Theory and Techniques,* vol. MTT-15, pp. 669–680, December 1967.

Reprinted from IEEE TRANSACTIONS
ON *MICROWAVE THEORY AND TECHNIQUES*
Volume MTT-16, Number 3, March, 1968
pp. 147-157

Printed-Circuit Complementary Filters for Narrow Bandwidth Multiplexers

ROBERT J. WENZEL, MEMBER, IEEE

Abstract—**Design techniques and interconnection equivalent circuits are presented for constructing low-cost printed-circuit narrowband complementary filters. The techniques and circuits described allow the achievement of contiguous band multiplexers using a single printed circuit board with no series or shorted stubs. Equivalent circuit transformations useful in obtaining practical designs and experimental results for a two-section stripline complementary filter pair are also given. While emphasis is placed on printed-circuit filters, the interconnection circuits and design techniques presented can be applied in the design of narrowband multiplexers using many other filter types.**

I. INTRODUCTION

THE use of complementary or pseudo-complementary filters in achieving well-matched contiguous-band multiplexers has been recently described in detail.[1]–[4] Specific network configurations and physical structures have been given for both wideband and narrowband applications. This paper is concerned with the interconnection, synthesis, and realization of low-cost printed-circuit stripline complementary filters.[1] A specific network to be described incorporates a singly terminated parallel-coupled bandpass filter and a bandstop complement applicable to printed-circuit construction. No series stubs nor shorted stubs are required and a multichannel filter can be obtained with a single printed circuit. Either exact or approximate design procedures can be used to obtain element values for the parallel-coupled and bandstop filters. Equivalent circuit transformations useful in investigating possible physical realizations are also described.

II. THE INTERCONNECTION PROBLEM

Narrowband-bandpass filters require the addition of redundant network elements to achieve practical impedance levels and physical dimensions. Therefore, special precautions must be taken to insure that the connection of component filters results in a complementary filter pair. In the case of doubly terminated narrowband parallel-coupled bandpass filters, a convenient impedance level for the interior sections of the filter is obtained by adding redundant unit elements of impedance Z_0 both after the source and before the load, where the load and source impedances are equal to Z_0. Application of Kuroda's identities then provides the desired impedance level change.[2] The redundant unit elements alter the transfer phase but do not change the transfer amplitude. In multiplexer design employing singly terminated filters, proper network performance is dependent on the input admittance (impedance) characteristics of the component filters, and these input characteristics are altered by the addition of a redundant unit element before the network. In wideband designs, a redundant unit element (ue) before the network is not required to obtain practical dimensions, and any number of unit elements may be used in the circuit realization by introducing them only from the back end of the network,[2],[3],[6] or by including them in the synthesis

Manuscript received July 14, 1967.
The author is with Research Laboratories Division, Bendix Corporation, Southfield, Mich. 48075.

[1] Henceforth, only complementary filters will be discussed. The design of narrowband pseudo-complementary filters follows in an analogous manner. Wideband pseudo-complementary filters whose prototypes contain unit elements require special consideration. See Wenzel.[4]

[2] For the case of approximate narrowband design procedures, the necessary redundant unit elements are accounted for in the design formulas. For example, see the discussion given by Matthaei ([5], sec. II-D, p. 585).

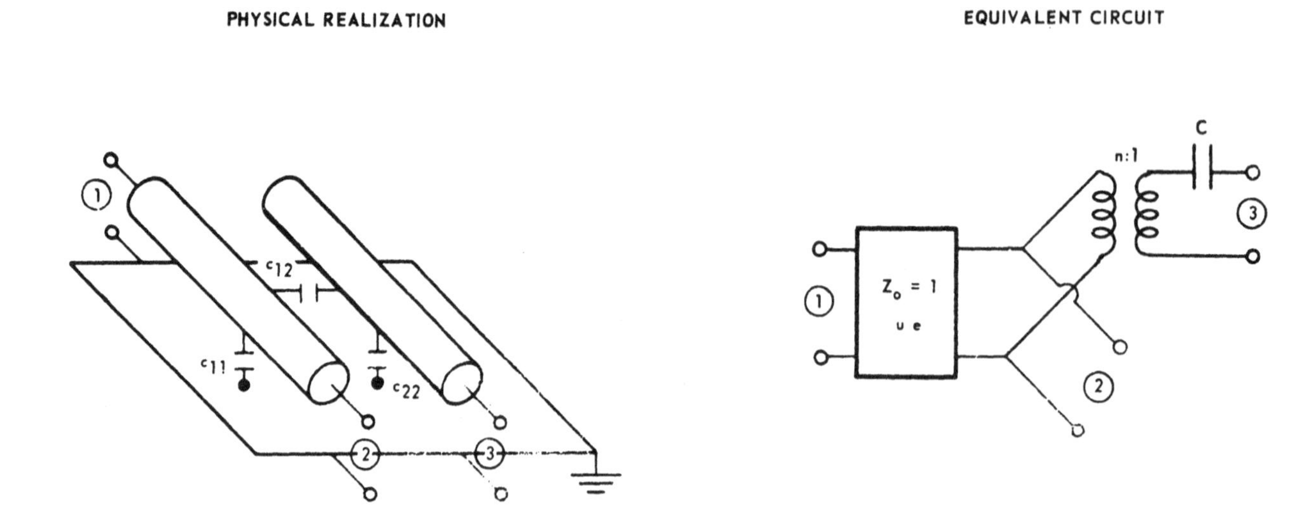

(a) $c_{11} = a^* - a(n-1)C/n^2, \quad c_{12} = aC/n, \quad c_{22} = aC(n-1)/n, \quad n^2 > (n-1)C$

For both connections:
① is the common input,
② is the complementary bandstop port
③ is the bandpass port.

(b) $c_{11} = a(n-1)/n \quad c_{12} = a/n, \quad c_{22} = a/L - a(n-1)/n^2, \quad n^2 > (n-1)L$

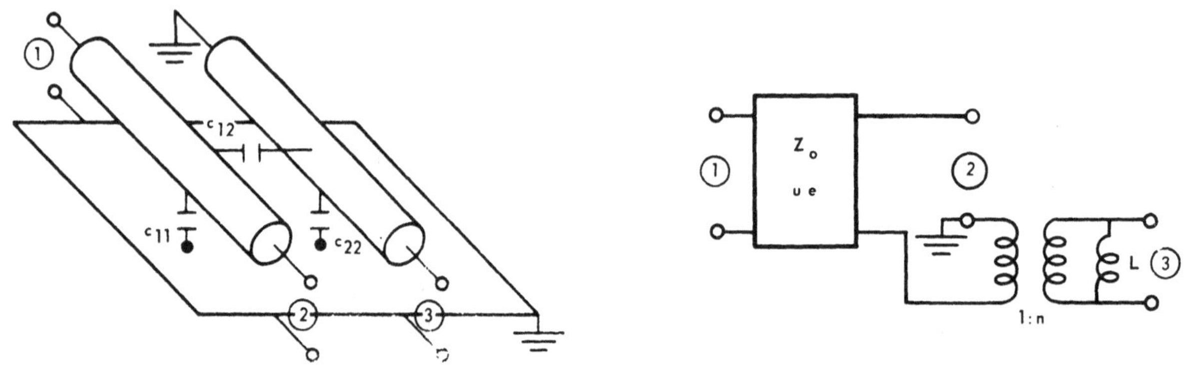

* $a = 376.7/Z_0\sqrt{\epsilon_r}$ for Z_0 ohm terminations.

Fig. 1. Interconnection networks for narrowband complementary filters. (a) Parallel connection. (b) Series connection. (The c_{ik} are the static capacitances per unit length for use with Getsinger's charts.[13])

procedure.[4],[7] However, practical narrowband-bandpass designs requir that at least one redundant unit element be added from the front end of the network and a simple series or parallel connection of component filters cannot be utilized to achieve a complementary filter pair.

The interconnection problem for narrowband diplexers can be solved by use of the parallel-coupled line configurations shown in Fig. 1. These three-port configurations provide a complementary interconnection and allow the arbitrary choice of impedance level for the interior sections of the bandpass filter by suitably choosing the transformer turns ratio n. The physical circuits shown in Fig. 1 are defined in terms of their normalized static capacitance values.[8] The circuit of Fig. 1(a) applies to parallel-connected diplexers and 1(b) to series-connected[3] diplexers. The equivalent circuits are derivable by direct three-port matrix comparison or by use of a capacitance matrix transformation[8] as demonstrated in the Appendix.

In a diplexer design, one synthesizes an appropriate singly terminated narrowband-bandpass filter with input admittance Y_1 (or input impedance Z_1). Convenient dimensions are obtained by adding one unit element of impedance Z_0 to each end of the filter and applying suitable equivalent circuit transformations.[6],[8] A complementary filter pair is then obtained by connecting a narrowband-bandstop filter with input admittance $Y_{1C} = 1 - Y_1$ (or input impedance $Z_{1C} = 1 - Z_1$ for the dual case) at port 2 as indicated in Fig. 1. The bandstop complement is *always* connected at the same physical location for any choice of transformer turns ratio n. In effect, one can design the bandpass filter with one redundant unit element in front of the network without considering the bandstop complement. The bandstop physical-interconnection point is unchanged by equivalent circuit transformations of the first parallel-coupled section of the band-

[3] The possible equivalence of the series connection of Fig. 1(b) was postulated by Matthaei and Cristal ([3], p. 330), and is applicable to networks incorporating line elements with short-circuited ends.

pass filter. For narrowband diplexers, approximate design procedures are quite accurate and might also be used to design the component filters (on a Y_{12} or singly terminated basis).[4] An example of a typical design procedure for the series connection using an approximate design technique can be found in Matthaei and Cristal.[3] An exact design procedure and test results for a printed-circuit stripline diplexer are given in Sections III and IV of this paper. A multichannel filter is obtained by cascading the component diplexing filter pairs as described in Wenzel[2] and Matthaei and Cristal.[3] The parallel connection allows the realization of an N-channel multiplexer using a single printed circuit board without series stubs or short-circuiting straps to the ground plane.

III. Synthesis and Design of Component Filters

A detailed discussion of the network characteristics required in microwave multiplexer design has been given[1]–[4] and only a specific configuration applicable to printed circuit realizations will be described. However, the equivalent circuits of Fig. 1 are quite general and can be applied to the design of multiplexing filters incorporating component filters realizable in many different forms.

The general physical form of the narrowband diplexer, incorporating the component bandpass and bandstop filter configurations to be described, is shown in Fig. 2(a). The bandpass filter can be designed by approximate methods,[5] with suitable modifications, or by exact synthesis using the functions

$$|Y_{12}|_{HP}^2 = \left[1 + \left(\frac{S_c}{S}\right)^2 \left(\frac{1-S_c^2}{1-S^2}\right)^{N-1}\right]^{-1}$$

given in equations (8) and (13) of Horton and Wenzel.[7] The bandstop filter is designed by introducing redundant unit elements from the load end of a complementary $[|Y_{12}|_C^2 = 1 - |Y_{12}|_{HP}^2]$ singly terminated low-pass prototype in the conventional manner.[2],[3],[6] It has been shown that the exact complement of an N-section singly terminated maximally flat parallel-coupled filter, whose S-plane[5] prototype contains one LC element and $N-1$ unit elements, is an N-section LC-ladder filter.[4] The elements of this LC ladder differ in value from those of the conventional singly terminated maximally flat ladder by an amount that decreases as the bandwidth decreases, becoming equal in the limit of zero bandwidth.

The S-plane equivalent circuit for the coupled bandstop configuration utilized in Fig. 2(a) is given in 2(b). The physical realization contains one more parameter (Z_1) than its equivalent circuit and thus allows the arbitrary choice of one physical constant.[6] This choice can be used to advantage in obtaining convenient dimensions. It has been determined experimentally that each bandstop resonator can be tuned over a 5 to 10 percent frequency range by changing the length of the open-circuited stub. This allows simple alignment of the diplexing filter.

To illustrate in detail the exact synthesis approach, consider a two-section filter pair consisting of a parallel-coupled bandpass filter and its complement. The theoretical response and definition of pertinent symbols are shown in Fig. 3(a). For the two-section maximally flat parallel-coupled filter, one obtains from Horton and Wenzel[7]

$$|Y_{12}|_{HP}^2 = \frac{1}{1 + \left(\frac{S_c}{S}\right)^2 \frac{(1-S_c^2)}{(1-S^2)}} = \frac{-S^2(1-S^2)}{-S^2(1-S^2)+k^2} \tag{1}$$

where

$$S = j \tan \theta$$

and

$$k^2 = -S_c^2(1-S_c^2) = \tan^2 \theta_c \sec^2 \theta_c.$$

The transfer admittance $Y_{12_{HP}}$ is obtained by factoring the denominator of (1) and grouping all left-half-plane singularities with $Y_{12_{HP}}$. Therefore,

$$|Y_{12}|_{HP}^2 = Y_{12}(S)Y_{12}(-S) = \left[\frac{S\sqrt{1-S^2}}{S^2+(\sqrt{1+2k})S+k}\right]\left[\frac{-S\sqrt{1-S^2}}{S^2-(\sqrt{1+2k})S+k}\right]. \tag{2}$$

Using the Gewertz procedure ([9], p. 287), one readily obtains $Y_{in_{HP}}$ from $Y_{12_{HP}}$ as

$$Y_{in_{HP}} = \frac{S^2 + \frac{1+k}{\sqrt{1+2k}}S}{S^2+(\sqrt{1+2k})S+k}. \tag{3}$$

Since the network is known to contain one LC element and one unit element,[7] the network can be synthesized by one application of Richards' theorem and a simple pole removal. The synthesized nonredundant network with closed-form element values is given in Fig. 3(b). Note that in nonredundant form the load resistance for the parallel-coupled filter is not unity. Introducing a redundant capacitor by application of Kuroda's identity gives the unit load equivalent circuit of Fig. 3(c).

For the complementary filter,

$$Y_{in_C} = 1 - Y_{in_{HP}} = \frac{\frac{k}{\sqrt{1+2k}}S + k}{S^2+(\sqrt{1+2k})S+k}. \tag{4}$$

Since $|Y_{12}|_C^2$ has all its transmission zeros at $S=j\infty$, the corresponding network is a two-section LC ladder that can be synthesized by two pole removals. The synthesized closed-

[4] In most cases some modification of the approximate design procedures and equivalent circuits is necessary. See, for example, Matthaei and Cristal ([3], p. 331).

[5] A detailed discussion of S-plane elements and nomenclature can be found in Wenzel[6] and in other referenced papers.

[6] This assumes the use of equal width coupled strips. For unequal strip widths, there are two arbitrary parameters. Unequal strip width realizations are readily obtained by application of a capacitance matrix transformation.[8]

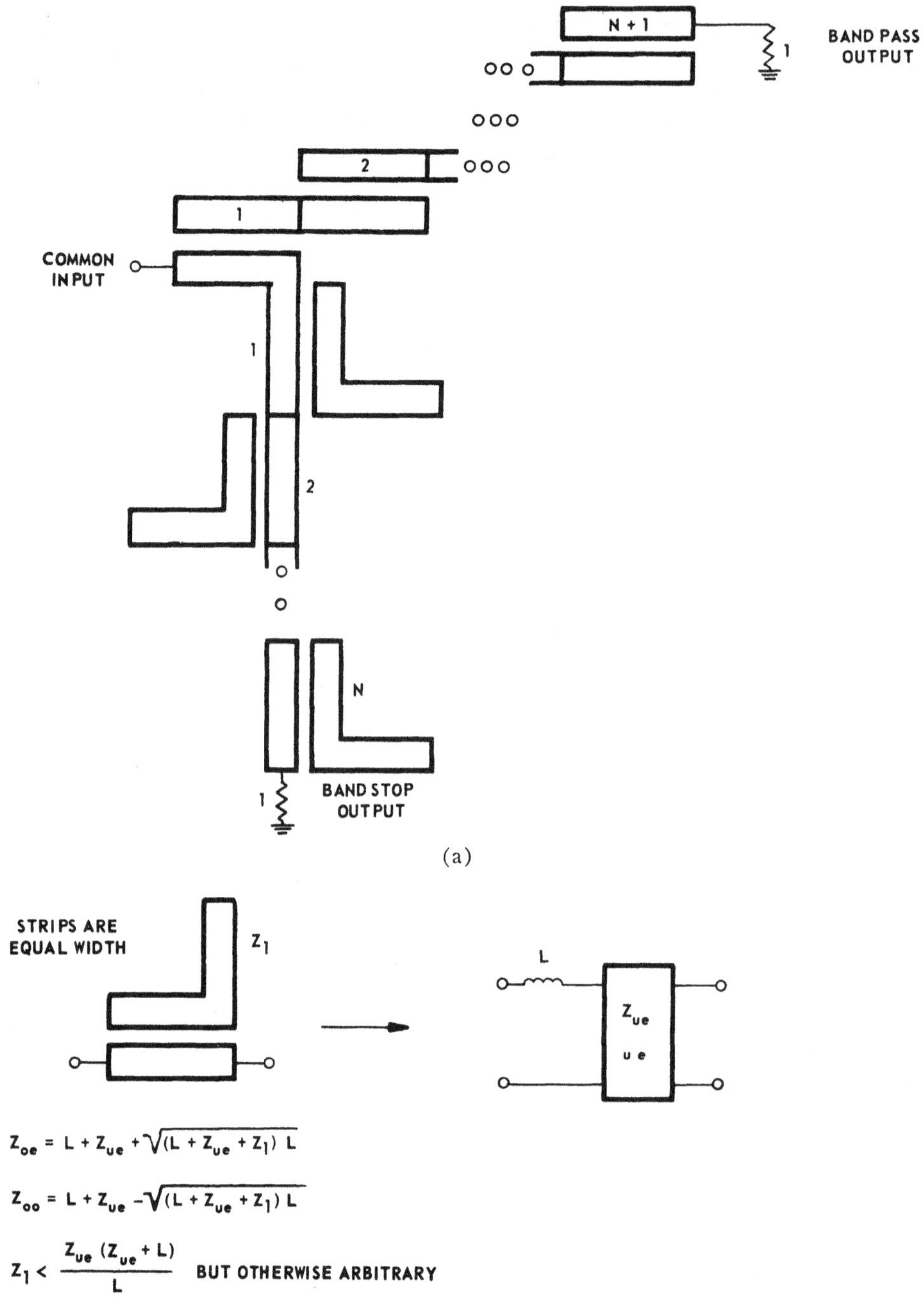

Fig. 2. Networks for stripline complementary filters. (a) Physical form of practical stripline filter pair. (b) Bandstop section equivalent circuit.

form element values for the bandstop prototype are given in Fig. 3(d) and the diplexer equivalent circuit with no redundant unit elements is given in Fig. 3(e). Introduction of redundant unit elements and two ideal transformers with arbitrary turns ratio n gives the final practical equivalent circuit of Fig. 3(f). Convenient dimensions are obtained by properly choosing n.

The equivalent circuit of Fig. 3(f) is the dual of that utilized by Matthaei and Cristal[3] and suitable approximate design procedures are described thoroughly.[3],[5]

The deviation of element values for an exact parallel-coupled complement from those of a conventional singly terminated maximally flat ladder is now readily determined. For the two-section singly terminated conventional ladder, the normalized element values are $\overline{L}=\sqrt{2}$ and $\overline{C}=1/\sqrt{2}$.[10] The bandwidth scaled values are $\overline{L}=\sqrt{2}/\tan\,\theta_c$ and $\overline{C}=1/\sqrt{2}\,\tan\,\theta_c$.[6] Then with suitable simplification

$$\frac{L}{\overline{L}} = \frac{\sqrt{1 + 2\tan\theta_c \sec\theta_c}}{\sqrt{2}\sec\theta_c},$$

$$\frac{C}{\overline{C}} = \frac{\sqrt{2}\tan\theta_c}{\sqrt{1 + 2\tan\theta_c \sec\theta_c}} \tag{5}$$

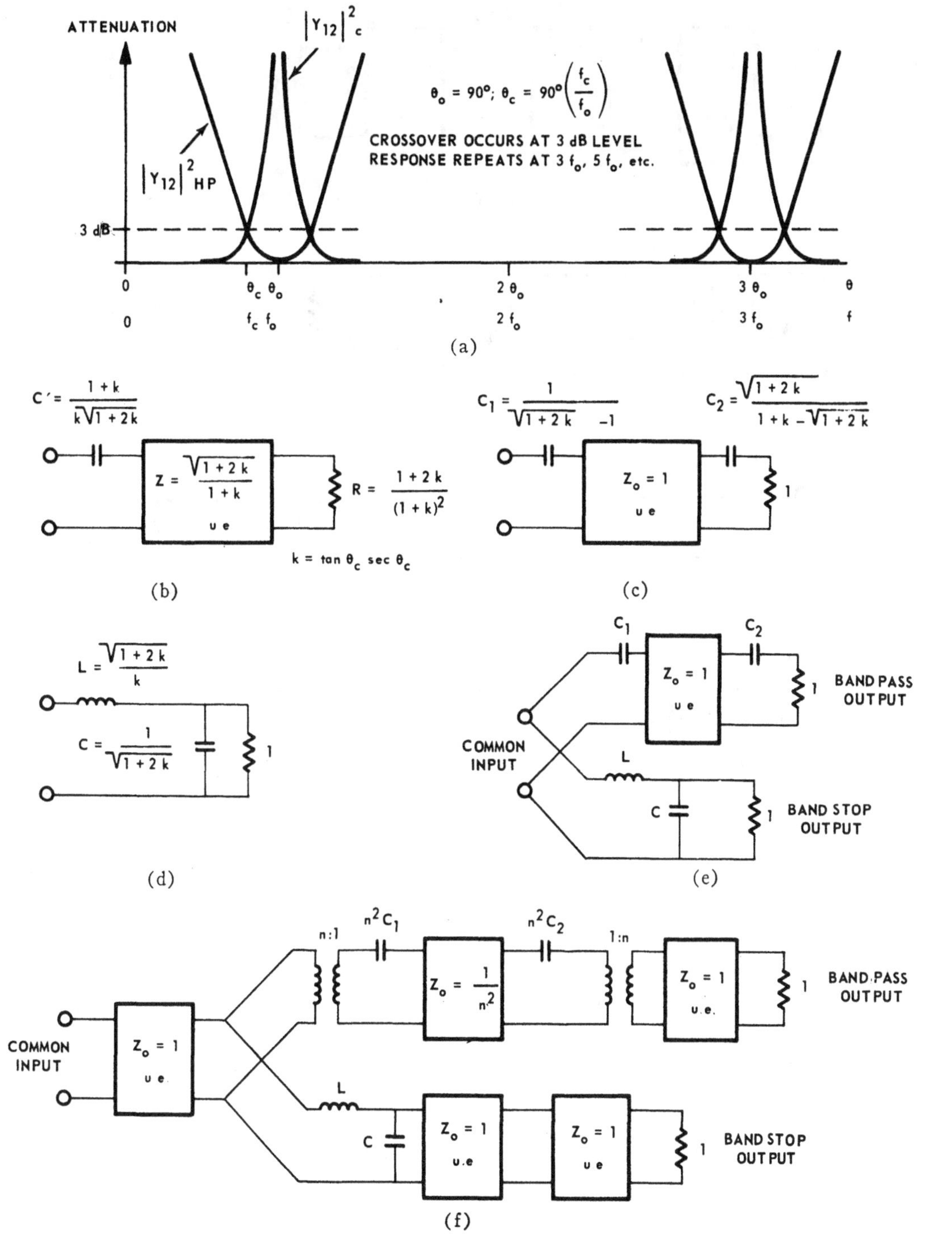

Fig. 3. Exact synthesis and design of stripline complementary filters. (a) Theoretical response of stripline complementary filters. (b) Nonredundant singly terminated bandpass equivalent circuit with closed-form element values. (c) Redundant unity termination equivalent circuit. (d) Exact singly terminated bandstop equivalent circuit. (e) Diplexer equivelant circuit with no redundant unit elements. (f) Practical diplexer equivalent circuit.

where the barred symbols indicate the conventional LC-ladder elements and the unbarred symbols indicate the exact parallel-coupled complement element values.

As the bandwidth (BW) tends to zero, $\theta_c \rightarrow 90$ degrees, or $\tan\theta_c \rightarrow \infty$. Thus

$$\lim_{\mathrm{BW}\rightarrow 0} \frac{\bar{L}}{L}\left(\text{or } \frac{\bar{C}}{C}\right) = 1. \tag{6}$$

An estimate of the error in using the conventional ladder element values rather than the synthesized values is readily obtained from (5).

It is also apparent from Fig. 3(c) that in narrowband diplexers ($k \gg 1$) the series capacitor values in the redundant parallel-coupled filter with unit termination are related to the conventional singly terminated two-section LC-ladder high-pass prototype element values. As before, the two sets of element values become identical in the limit of zero bandwidth.

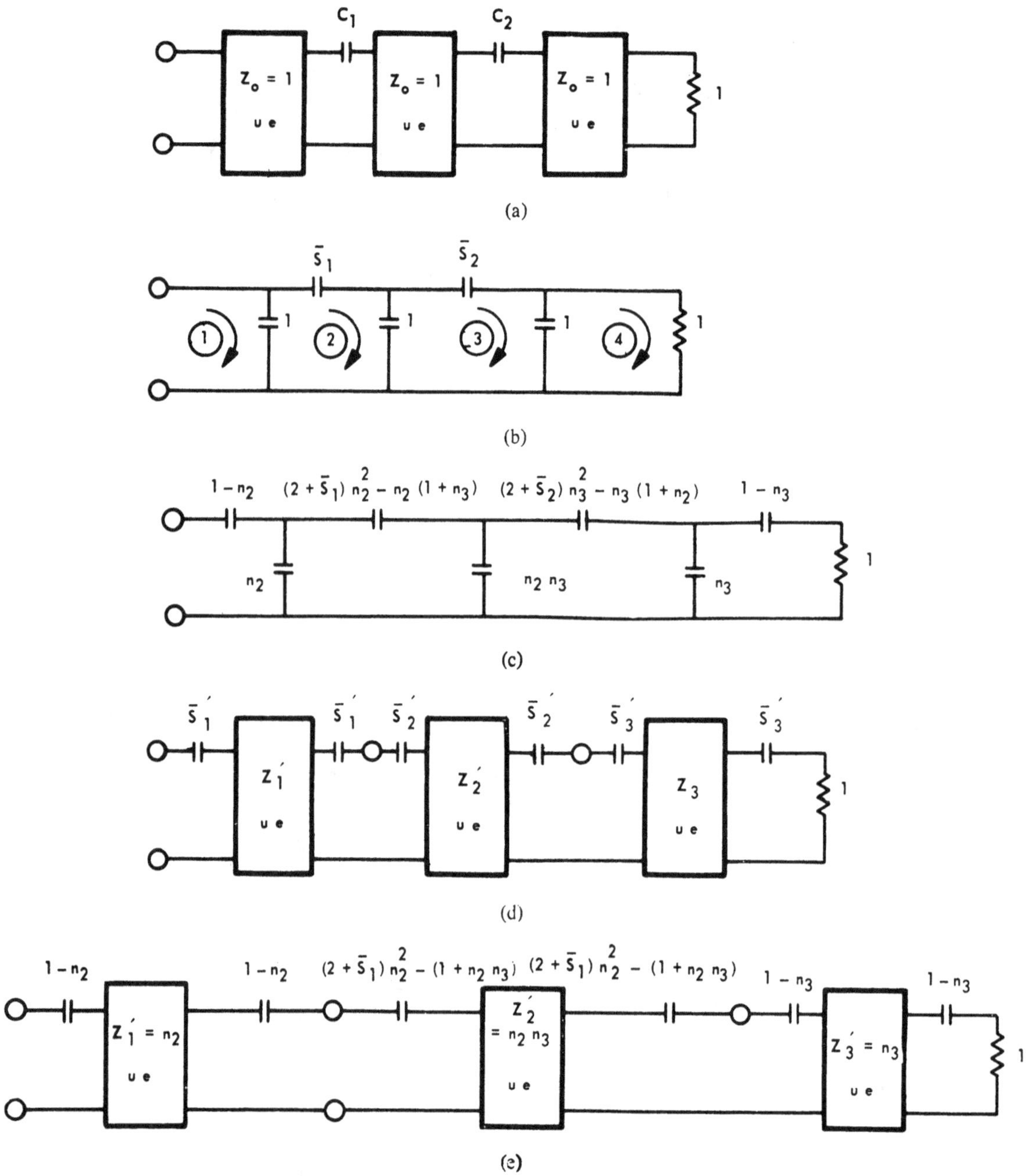

230 Fig. 4. Equal strip width design by use of an elastance matrix transformation. (a) Parallel-coupled equivalent circuit with two redundant unit elements. (b) Elastance network for (a). (c) Elastance network with n_2 and n_3 arbitrary. (d) General S-plane equivalent circuit for equal strip width realization. (e) Equal strip width realization showing specific element values.

It is important to point out that closed-form element values were obtainable for the two-section design because the denominator of $|Y_{12}|^2$ was factorable in closed form. For more complex designs, a digital computer is required to obtain the necessary roots and to maintain numerical accuracy in synthesis. Convenient element value tables can be generated that allow for simple design in an exact manner, but unless these tables are available, the exact design of a multisection filter is quite formidable and one might better consider approximate methods. The exact design approach is only necessary for very precise designs or for filters of wide bandwidths where approximate methods become inaccurate.

Returning to the equivalent circuit of Fig. 3(f), the introduction of the turns ratio n requires unequal strip widths in a parallel-coupled realization of a singly terminated design. While the circuit of Fig. 3(f) provides some flexibility in choosing the interior impedance level, it is still more restrictive than necessary. A design procedure that allows the systematic investigation of *all possible* equivalent circuits will now be given. The design procedure is based on the use of an elastance (reciprocal capacitance) matrix transformation[7] and will be described by means of a specific example.

Consider the parallel-coupled equivalent circuit of Fig.

[7] The transformation procedure can also be carried out by use of the capacitance matrix; however, use of the elastance matrix is simpler in this case. Also, the series capacitors in a nonredundant parallel-coupled narrowband filter are usually quite small and the elastance matrix values are numerically more convenient.

3(c) and the redundant network obtained by adding unit elements as shown in Fig. 4(a). For an elastance matrix transformation, each unit element is considered as a shunt elastance $\overline{S}=Z_{ue}$, and each series capacitor is considered as a series elastance $\overline{S}_i=1/C_i$. The elastance matrix has main diagonal terms given by the sum of all elastances in each circuit loop and off-diagonal terms given by the negative of the elastance mutual to each loop. Multiplication of corresponding rows and columns is then identical to application of the series capacitive Kuroda identity. Detailed properties of elastance matrix transformations are analogous to capacitance matrix transformations and may be reviewed in Wenzel[8] and Guillemin ([9], pp. 141–176).

The elastance network corresponding to the S-plane network of Fig. 4(a) is shown in 4(b) and the corresponding 4×4 elastance matrix is given by

$$\begin{bmatrix} 1 & -1 & 0 & 0 \\ -1 & 2+\overline{S}_1 & -1 & 0 \\ 0 & -1 & 2+\overline{S}_2 & -1 \\ 0 & 0 & -1 & 1 \end{bmatrix} \begin{matrix} \leftarrow n_1 \\ \leftarrow n_2 \\ \leftarrow n_4 \\ \leftarrow n_3 \end{matrix} \qquad (7)$$

$$\begin{matrix} \uparrow & \uparrow & \uparrow & \uparrow \\ n_1 & n_2 & n_3 & n_4 \end{matrix}$$

The elastance network and matrix for networks with more unit elements and series capacitors are obtained by straightforward extension of the preceding procedure.

As with capacitance matrix transformations, equivalent transmission properties and termination impedances are obtained if the input and output loop impedances are unchanged. For the matrix of (7) this requires $n_1=n_4=1$. Changes in termination impedance are readily obtained by choosing n_1 and/or n_4 to be nonunity.[8] The n_i can be chosen arbitrarily subject to the constraint that the transformed elastance matrix represent the elastance matrix of a realizable network.[8],[9]

For the specific case of the matrix of (7), choosing $n_1=n_4=1$ and n_2 and n_3 arbitrarily, gives

$$\begin{bmatrix} 1 & -n_2 & 0 & 0 \\ -n_2 & (2+\overline{S}_1)n_2^2 & -n_2n_3 & 0 \\ 0 & -n_2n_3 & (2+\overline{S}_2)n_3^2 & -n_3 \\ 0 & 0 & -n_3 & 1 \end{bmatrix}. \qquad (8)$$

From a practical viewpoint, a convenient constraint would be to choose n_2 and n_3 such that equal width conductors result for a parallel-coupled realization. The elastance network for arbitrary n_2 and n_3 is shown in Fig. 4(c) and the desired S-plane equivalent circuit is shown in 4(d). For equal strip widths in a parallel-coupled realization, the series capacitors on both sides of each unit element must be identical.[6] Imposing this constraint gives the final equivalent circuit of Fig. 4(e) and the relationship

$$(2+\overline{S}_1)n_2^2 - (n_2n_3+1) = (2+\overline{S}_2)n_3^2 - (n_2n_3+1). \qquad (9)$$

[8] If the load for the parallel-coupled filter is a detector or mixer, nonunity termination levels may be desirable.

Solving for n_2 in terms of n_3 gives

$$n_2^2 = \frac{(2+\overline{S}_2)}{(2+\overline{S}_1)} n_3^2. \qquad (10)$$

Thus, either n_2 or n_3 can be chosen arbitrarily and the other is then determined from (10). By utilizing the full redundancy of the parallel-coupled network an additional degree of freedom is obtained over the approach that led to the equivalent circuit of Fig. 3(f). For filters with more elements, more degrees of freedom are obtained. The elastance matrix transformation can also be used to investigate alternate physical realizations of networks designed by approximate techniques.

IV. Experimental Results

A trial two-section diplexer with a 5 percent bandwidth was constructed using the exact redundant equivalent circuit of 5(a). The element values shown are those obtained by the synthesis procedure described in Section III with $\theta_c=87.75$ degrees. For the bandpass filter, convenient dimensions and equal width strips were obtained by using the technique illustrated in Fig. 4 and choosing $n_2=0.225$ and $n_3=0.314$.[9] Kuroda's identities or the algorithm described by Cristal[11] can be used to obtain the redundant bandstop configuration. The S-plane equivalent circuits of the component filters are shown in Fig. 5(b) and the corresponding physical realization showing even-and odd-mode impedances is given in 5(c). For simplicity, and the fact that practical dimensions resulted, the normalized shunt stub impedances in the bandstop filter were arbitrarily chosen to be 1 ohm. The even- and odd-mode impedances were obtained from Wenzel ([6], Table I) and from Fig. 2(b). Dimensions were determined using the nomographs and length correction factors given by Cohn.[12] For unequal width strips, dimensions are readily determined using Getsinger's charts.[13] For wider bandwidths, the use of unequal strip widths sometimes results in more convenient dimensions. Because of the redundant nature of the parallel-coupled filter and the arbitrary parameters in the bandstop filter, one has considerable freedom in choosing the impedances and coupling coefficients of the final network.

A photograph of an experimental printed circuit board for the two-section diplexer is shown in Fig. 6, and test results for a filter using this board are given in Fig. 7. The response shown was obtained by adjusting slightly the lengths of the open-circuited stubs in the bandstop filter to obtain the proper resonator transmission zeros. No special effort was made to match the connectors, and a time domain reflectometer measurement indicated that the impedance level of the entire structure was slightly below 50 ohms, resulting in the rather constant measured VSWR. Since the maximum VSWR was only 1.25 to 1, no further work was

[9] With the exact design approach, explicit use of the equivalent circuits in Fig. 1 is not required. As described in Section II, the bandstop physical interconnection point is unchanged by equivalent circuit transformations of the first parallel-coupled section of the bandpass filter. The complete diplexer design procedure consists of obtaining a suitable bandpass filter and simply connecting the bandstop complement at the proper point.

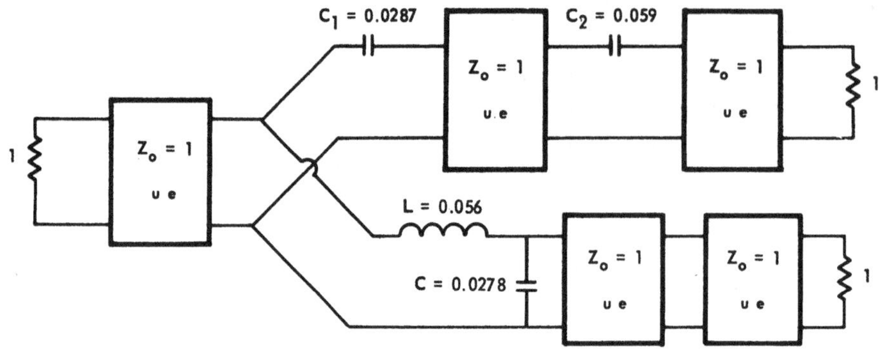

(a)

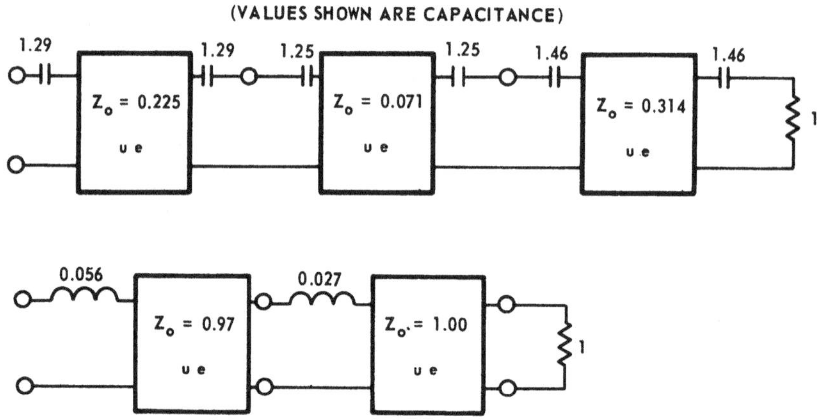

(b)

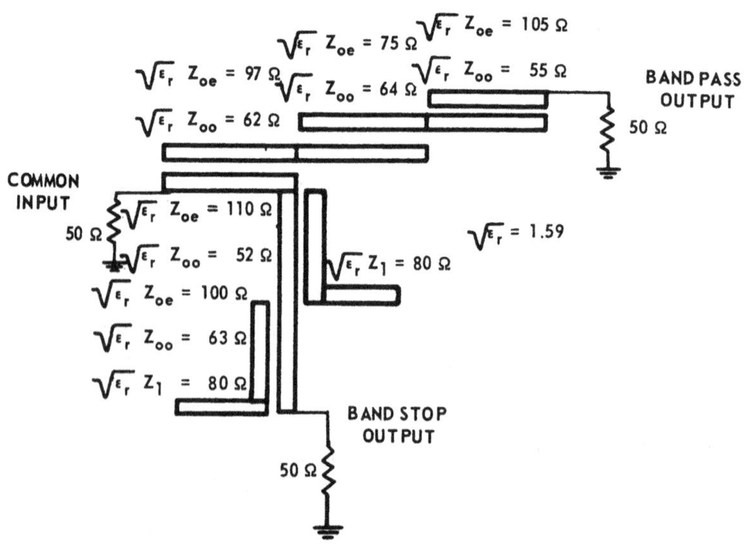

(c)

Fig. 5. Trial design for a two-section complementary filter of 5 percent bandwidth. (a) Exact equivalent circuit for two-section diplexer. (b) S-plane equivalent circuit of redundant component filters. (c) Physical realization showing even- and odd-mode impedances.

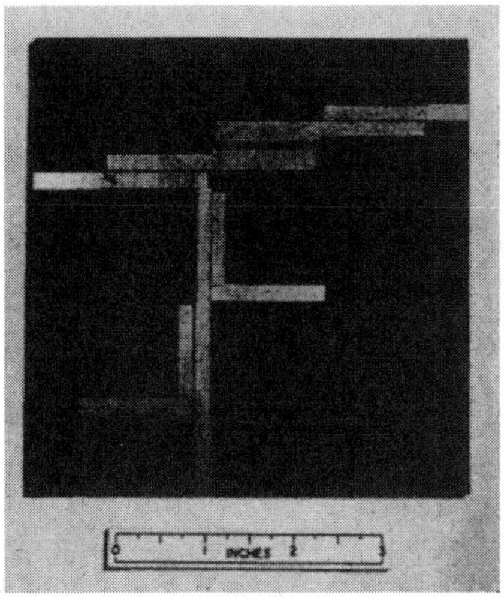

Fig. 6. Experimental diplexer circuit board.

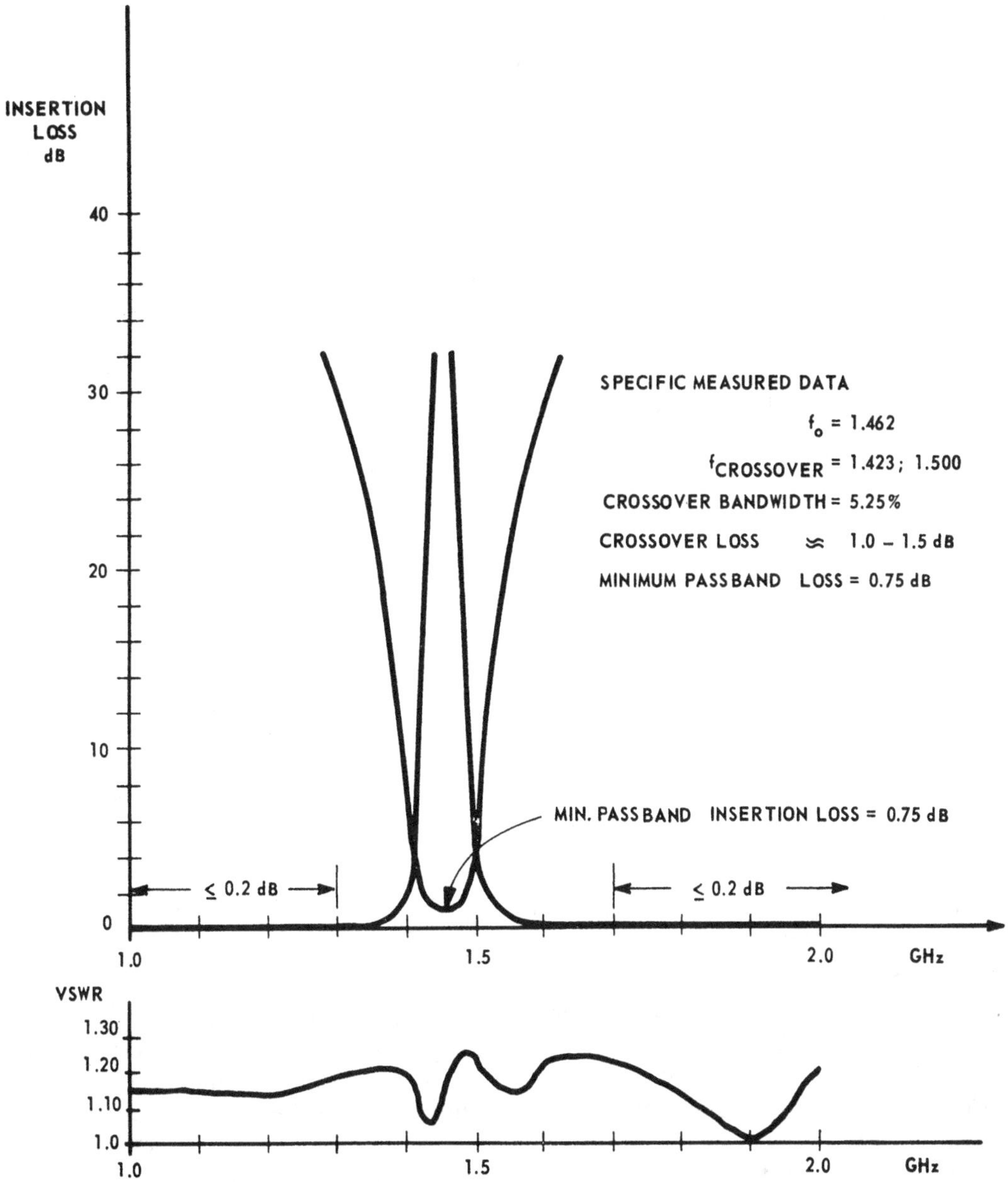

Fig. 7. Measured response of experimental stripline diplexer.

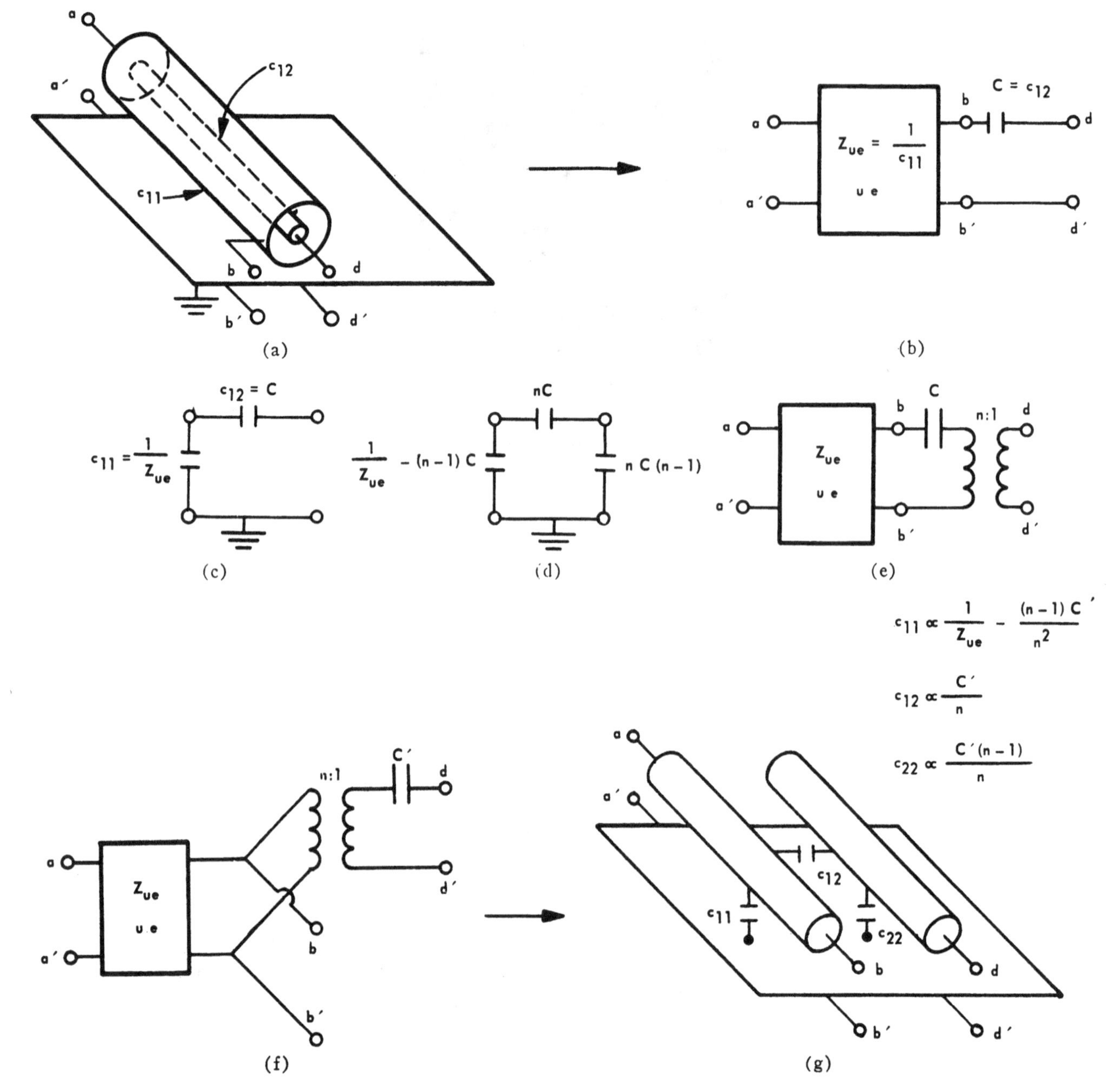

Fig. 8. Derivation of equivalent circuit for parallel interconnection network.

done to reduce this value. The somewhat low impedance level and midband frequency are attributed to the circuit boards having a higher dielectric constant than that used in the calculations. By trimming the ends of the parallel-coupled sections in the bandpass filter and adjusting the shunt stub lengths in the bandstop filter, it was possible to adjust the midband frequency of the diplexer while maintaining a low input VSWR (less than 1.25 to 1). This type of adjustment is important in obtaining optimum performance in a multichannel multiplexer. Teflon-fiberglass circuit boards with a relatively high loss tangent were used in the experimental model. Lower midband loss can be achieved by using higher quality material. Insertion loss at frequencies outside the crossover region was very low due to the loose coupling of the bandstop resonators.

V. Conclusions

Exact equivalent circuits useful in the interconnection of narrowband complementary filters have been given. By using the parallel interconnection and suitable filter forms, a multichannel multiplexer can be designed using a single printed circuit board with no series stubs or short-circuited stubs. The techniques presented are useful in achieving low-cost multiplexers with good response characteristics and can be utilized with filters which can be fabricated in a wide variety of physical forms. Either exact or approximate design techniques can be utilized and the elastance matrix transformation presented allows the systematic investigation of possible equivalent circuits. Test results for an experimental diplexer showed excellent agreement with theory and the structure was easy to construct and align.

Appendix

Derivation of Interconnection Equivalent Circuits

Consider the coaxial circuit of Fig. 8(a) where letters have been added to junction points to help in visualizing the equivalent circuit. Between terminal pairs a–a' and b–b' there exists a unit element of impedance[10] $Z_{ue}=1/c_{11}$, and between terminals b–d there exists a series capacitor $C=c_{12}$. The S-plane equivalent circuit is shown in Fig. 8(b) and the corresponding static capacitance network is shown in 8(c). The capacitance matrix is given by

$$\begin{bmatrix} C+\dfrac{1}{Z_{ue}} & -C \\ -C & C \end{bmatrix} \begin{matrix} \\ \leftarrow n \end{matrix} . \qquad (11)$$
$$\uparrow$$
$$n$$

Transformation of the capacitance matrix by multiplying the second row and column by n gives the matrix

$$\begin{bmatrix} C+\dfrac{1}{Z_{ue}} & -Cn \\ -Cn & Cn^2 \end{bmatrix} \qquad (12)$$

and the capacitance network of Fig. 8(d). As described in Wenzel,[8] the above transformation results in the addition of an ideal transformer of turns ratio n to 1 to both ends of the second line, and consequently gives the S-plane equivalent circuit of Fig. 8(e). Transforming the capacitor C to the other side of the ideal transformer gives $C'=n^2C$ and the equivalent circuit of Fig. 8(f). The transformed network is then realizable[8] as shown in Fig. 8(g) where the static capacitance values are those of 8(d). The equivalent circuit for the series connection is derivable in an analogous manner if the initial coaxial network is chosen to have $c_{11}=0$. Both equivalences can be verified by direct three-port matrix comparison; however, the capacitance matrix approach allows the equivalent circuit to be truly derived whereas the matrix comparison approach requires that one guess the correct equivalent.

[10] For simplicity, the normalization constant between impedance and static capacitance is taken to be unity.

References

[1] R. G. Veltrop and R. B. Wilds, "Modified tables for the design of optimum diplexers," *Microwave J.*, vol. 7, pp. 76–80, June 1964.

[2] R. J. Wenzel, "Application of exact synthesis methods to multichannel filter design," *IEEE Trans. Microwave Theory and Techniques*, vol. MTT-13, pp. 5–15, January 1965.

[3] G. L. Matthaei and E. G. Cristal, "Multiplexer channel-separating units using interdigital and parallel-coupled filters," *IEEE Trans. Microwave Theory and Techniques*, vol. MTT-13, pp. 328–334, May 1965.

[4] R. J. Wenzel, "Some general properties of commensurate line-length complementary and pseudo-complementary microwave filters," *IEEE Trans. Microwave Theory and Techniques*, this issue, pp. 191–192.

[5] G. L. Matthaei, "Design of wide-band (and narrow-band) band-pass microwave filters on the insertion loss basis," *IRE Trans. Microwave Theory and Techniques*, vol. MTT-8, pp. 580–593, November 1960.

[6] R. J. Wenzel, "Exact design of TEM microwave networks using quarter-wave lines," *IEEE Trans. Microwave Theory and Techniques*, vol. MTT-12, pp. 94–111, January 1964.

[7] M. C. Horton and R. J. Wenzel, "General theory and design of optimum quarter-wave TEM filters," *IEEE Trans. Microwave Theory and Techniques*, vol. MTT-13, pp. 316–327, May 1965.

[8] R. J. Wenzel, "Theoretical and practical applications of capacitance matrix transformations to TEM network design," *IEEE Trans. Microwave Theory and Techniques*, vol. MTT-14, pp. 635–647, December 1966.

[9] E. A. Guillemin, *Synthesis of Passive Networks.* New York: Wiley, 1957.

[10] L. Weinberg, *Network Analysis and Synthesis.* New York: McGraw-Hill, 1962.

[11] E. G. Cristal, "Addendum to 'An exact method for synthesis of microwave band-stop filters'," *IEEE Trans. Microwave Theory and Techniques (Correspondence)*, vol. MTT-12, pp. 369–382, May 1964.

[12] S. B. Cohn, "Shielded coupled-strip transmission line," *IRE Trans. Microwave Theory and Techniques*, vol. MTT-3, pp. 29–38, October 1955.

[13] W. J. Getsinger, "Coupled rectangular bars between parallel plates," *IRE Trans. Microwave Theory and Techniques*, vol. MTT-10, pp. 65–72, January 1962.

LINEAR PHASE FILTERS VI

Linear phase filters are becoming increasingly important in modern communications. The reason for this is that non-linear phase-versus-frequency characteristics spell distortion, since the time delay through the filter is then a function of frequency. The distortion is generally most severe near the band edges of the filter. Therefore with conventional filters it is best to allow guard bands between channels, to accommodate the bandedge "ears" (of the delay response) in which distortion is greatest. However, this is wasteful of bandwidth, a natural resource not available in unlimited amounts. Hence the search for filters that have equal ripple amplitude response **without** phase or delay distortion, and so utilize the available spectrum much more efficiently.

The state of the art is still advancing. One of the pioneers is J.D. Rhodes, whose short review paper is reproduced here (p. 239). Other papers by Rhodes are referenced in his review, and still others on the subject of linear-phase filters have been published by him in the June 1970 issue of the **IEEE Transactions on Microwave Theory and Techniques.**

Some brief remarks might be worthwhile on a general principle in the design of equal-delay filters. It has been mentioned that when delay is plotted against frequency for a conventionally coupled filter, "ears" develop near the two band edges, that is, the delay increases near band edge. Picture a conventional filter in which the input is coupled to Resonator 1, which in turn couples to Resonator 2, and so on, until the last (n-th) resonator couples to the output. Note that each resonator couples only to its two neighbors. Now fold the filter about its center so that Resonator 1 is above Resonator n, Resonator 2 is above Resonator n-1, and so on. Next, introduce a little coupling also between 1 and n, between 2 and n-1, and so on. Over most of the pass band these couplings are too small to have any appreciable effect. But as we go to frequencies nearer one band edge or the other, any **portion** of the filter (**not** the **complete** filter) becomes increasingly mismatched. Thus more and more energy begins to "bypass" the rest of the filter and to "leak" out from 1 to n, from 2 to n-1, from 3 to n-2, and so on, thereby reducing the delay through the filter. By proper design, the sum of the outputs via the center resonators and via the "leakage" couplings, i.e. the total output, can be equalized for approximately constant delay.

Of course, filters may also be designed in the first place without regard to phase or delay. Then "equalizers" are added, usually to linearize the phase (that is, giving constant slope on a phase-versus-frequency plot, which results in almost constant delay in the pass band). Such an equalizer is ideally an all-pass structure (passing all frequencies without attenuation, and affecting only the phase). It can be realized in a number of ways. One way is to build another filter having a wider passband than the first and the desired phase characteristic in the band of interest. A more usual way is to combine a circulator with a reactance network or a 3-dB hybrid with two reactance networks. Reactance networks can be obtained in numerous ways, including coupled lines. See Steenart,[20] Cristal[21] and Schiffman.[22] One of the first circuits (still widely used) that made use of coupled lines to obtain constant differential phase is the Schiffman phase shifter.[23]

In connection with delay, it may be stated that delay and dissipation loss in a filter go hand in hand.[24] (In the pass band of a filter they are frequently proportional to within one percent. Hence constant-delay filters are also constant-dissipation-loss filters, to a good approximation. This property is sometimes useful in communication systems.) Formulas for dissipation loss in filters have been given by Cohn[25] and by Young[24] and Wheeler[26] has given formulas that can be used in oonjuction with skin depth to determine the surface resistance of a transmission line and hence the unloaded Q of a resonator. (This useful paper, written nearly 30 years ago, is sometimes forgotten in modern text books.) Delay, dissipation loss and power handling capacity are also related in filters and Young has plotted typical quantitative behavior.[27] See also Ref. 1, Secs. 6.14, 6.15 and 15.03.)

TECHNICAL NOTE

ADVANCES IN NARROWBAND ELLIPTIC AND LINEAR-PHASE FILTERS

DR. J. DAVID RHODES, Microwave Development Laboratories, Inc.
Needham Heights, Massachusetts

INTRODUCTION

It is well known that ordinary microwave filters designed on the Chebyshev equi-ripple basis are non-optimum both in amplitude and phase response characteristics. Many requirements both in military and communications systems can no longer be met by such conventional filters, which require application of more specialized techniques, now becoming available.

Considering first the realization of narrow bandpass filters of high selectivity, the elliptic function filter offers considerably lower insertion loss than the comparable Chebyshev filter, and in a coaxial realization is also smaller in size. The reduced insertion loss is such that where previously in many instances large waveguide filters had to be used, it is now possible to consider their replacement by the new elliptic filters, with enormous savings in size and weight.

The differential group delay characteristics of the elliptic filters are similar to those of Chebyshev filters and can be similarly equalized. However, in many cases it may be preferable to utilize one of a new class of linear phase filters which are designed specifically for the purpose in mind, and which do not require separate equalization. Two of the major drawbacks of separate equalization are the sensitivity of the resulting filter, and the extreme difficulty of producing flat group delay over the *entire* pass band. The new class of self-equalized filters do not suffer from these disadvantages, and compact devices are available in both TEM-line and waveguide forms.

The following sections give more detailed discussions of these new elliptic and linear phase filters.

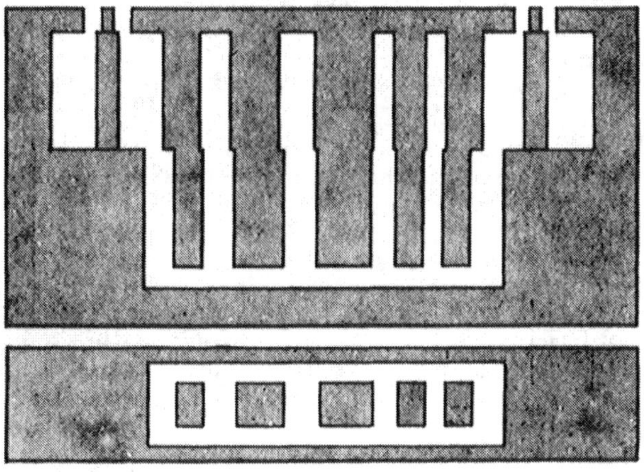

Fig. 1 — Quarter-wave stepped-digital-elliptic filter, (n = 5).

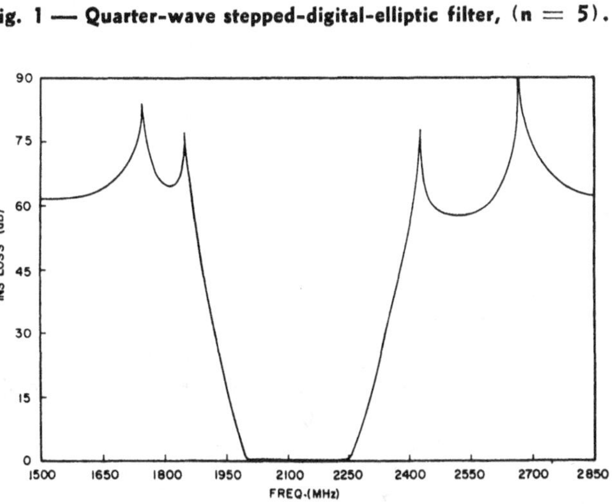

Fig. 2 — Characteristics of a 7-element quarter-wave stepped digital elliptic filter.

THE STEPPED DIGITAL ELLIPTIC FILTER[1]

Apparently this was the first type of low-loss bandpass narrowband elliptic function filter to be constructed in the microwave region. In appearance it is similar to a comb-line filter, consisting of a set of digital lines one quarter-wavelength long at mid-band. Each line is shorted to ground at the same end, and open circuited at the other. Unlike a comb filter, in general each digit is stepped in impedance at its mid point (hence the nomenclature for this class of filters), as shown in Fig. 1. Transformer coupling into the filter is achieved through lines short circuited at the mid-plane. Alternatively direct capacitive couplings can be utilized at the central points of the first and last digits.

Although proportional bandwidths down to 2% may be realized with this structure, it is best suited for the range 5 - 30%, where the impedance values of all the digits are of the order of 70 Ω. Narrow bandwidths are realized using the half-wave stepped-digital-elliptic filter, to be described next. Physically these filters are very compact, and are not difficult to align since the transmission poles and zeros are separately and indepen-

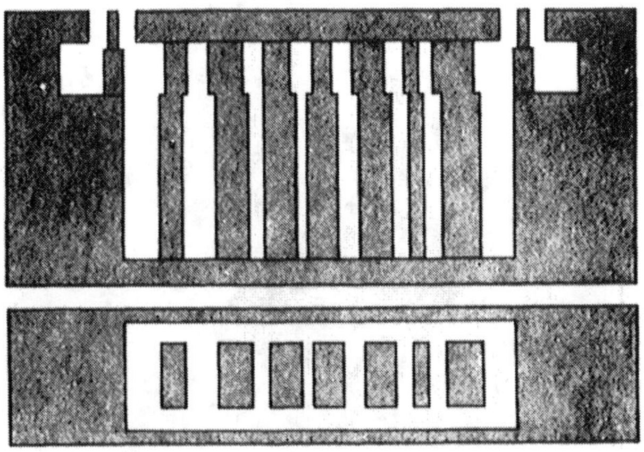

Fig. 3 — Half-wave stepped-digital-elliptic filter, (n = 7).

Fig. 4 — Elliptic and interdigital filters of Fig. 5.

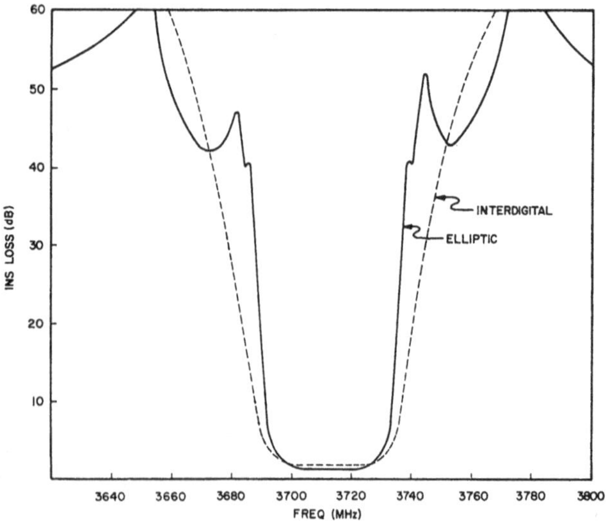

Fig. 5 — Comparison between the performance characteristics of a 7-section elliptic filter and a 7-section interdigital filter.

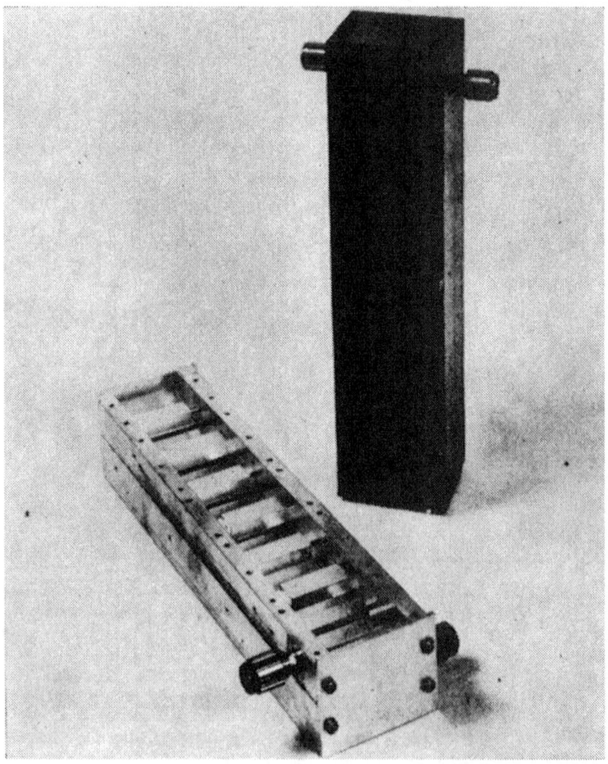

Fig. 6 — L-band linear phase filter.

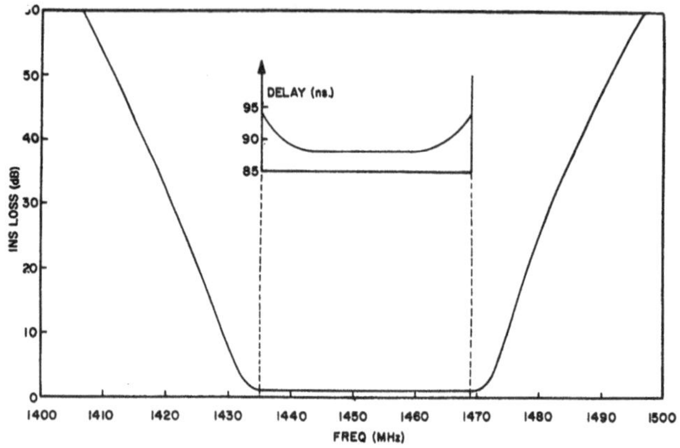

Fig. 7 — Performance characteristics for filter of Fig. 6.

dently tunable. The stop-band is maintained until third harmonic.

The performance characteristics of a typical stepped digital elliptic filter are shown in Fig. 2.

THE HALF-WAVE STEPPED-DIGITAL-ELLIPTIC FILTER[2]

This TEM-line coupled-cavity filter has been designed basically for bandwidths of 5% down to fractional percentage bandwidths and consists of a set of digital lines one half-wavelength long at mid-band, short-circuited to ground at *both* ends. The digits are stepped in impedance along any arbitrary prescribed plane in the structure, and coupling into the filter is again accomplished using interdigital line transformers, as shown in Fig. 3. This half-wave stepped-digital-elliptic filter is rather larger than the quarter-wave version, but is exceptionally rigid in construction since the digits are supported at both ends. In comparison to a 1% bandwidth 7-element interdigital filter, a 7-element half-wave elliptic filter is less than half the size, has 2/3 of the mid-band loss, and has more than twice the skirt selectivity. These comparisons between physical sizes and performances are shown in Figs. 4 and 5 respectively. A further interesting comparison is to that of a conventional waveguide filter designed upon a Chebyshev basis. For the *same* mid-band loss, a half-wave stepped-digital-elliptic filter may frequently be constructed to meet a given selectivity requirement in only a fraction of the weight and size of the waveguide version.

In contrast to the quarter-wave digital elliptic filter, the half-wave version is capable of maintaining the stop-band level only to twice the center frequency. The ground plane spacing may be much larger than that of the quarter-wave filter or of an interdigital filter since there are no fringing capacitances, thus enabling very low insertion loss to be achieved. The limitation on ground plane spacing is set by the onset of a waveguide mode, but an unloaded Q of the order of 4000 may be realized at S-band frequencies.

THE GENERALIZED INTERDIGITAL LINEAR PHASE FILTER[3,4,5]

New approximation theories formulated for microwave TEM-line filters of arbitrary bandwidth give maximally flat group delay and amplitude responses about center frequency.[3,4] More recent developments on a finite band approximation offer considerable improvement, especially in the higher degree cases. The basic, closed form maximally flat solution is in terms of shifted parameter symmetrical Jacobi polynomials,[3,4] whereas the finite band approximation requires the use of a more general class of polynomials.

The required non-minimum phase response is realized by a generalized interdigital filter which consists of two interdigital lines, cross-coupled via slots in a coupling plate. The synthesis of this structure is based upon a recent theory of interdigital lines which have non-adjacent as well as adjacent coupling.[5] Amplitude responses comparable to conventional filters designed upon a Chebyshev basis may be obtained, whilst the group delay variation of typically 1.5:1 or 2:1 is reduced to (typically) less than 2% over the entire passband. Since passband insertion loss and group delay are proportional to each other, a very flat insertion loss characteristic also results. The absolute group delay and mid-band loss are increased by a factor of 2 or 3, which is the case also for conventional equalization of Chebyshev filters. The formal theory gives a group delay which is *too flat* for normal specifications and the selectivity is consequently inferior, but by a simple augmentation of the approximation theory a trade-off can be made between skirt selectivity and group delay flatness.

A photograph of a 14-element generalized interdigital linear phase filter having 38 MHz bandwidth at 1450 MHz is shown in Fig. 6, and its performance characteristics in Fig. 7.

WAVEGUIDE LINEAR PHASE FILTERS

The waveguide version of the linear phase filter is designed in a manner similar to that for the TEM-line version. Physically the filter is in the form of a conventional inductive iris direct-coupled filter folded at its center point through 180°, with side-adjacent cavities cross-coupled through the common narrow wall. This enables the required non-minimum phase response to be achieved.

At the price of increase in physical size, this linear phase filter offers lower insertion loss over its TEM-line counterpart, with a slight increase in delay distortion due to the dispersion of the waveguide. However, for most applications this effect is quite small, and it is hoped to find means for its compensation. Figure 8 shows an X-band waveguide linear phase filter having the performance depicted in Fig. 9.

CONCLUSIONS

It is hoped that these new elliptic and linear phase filters will enable systems designers to prescribe more stringent requirements, particularly in communications systems, where much higher percentage bandwidth utilization may now be feasible. In applications where the requirements are less severe, the small size and low loss of the elliptic filters may be advantageous.

Fig. 8 — X-band waveguide linear phase filter.

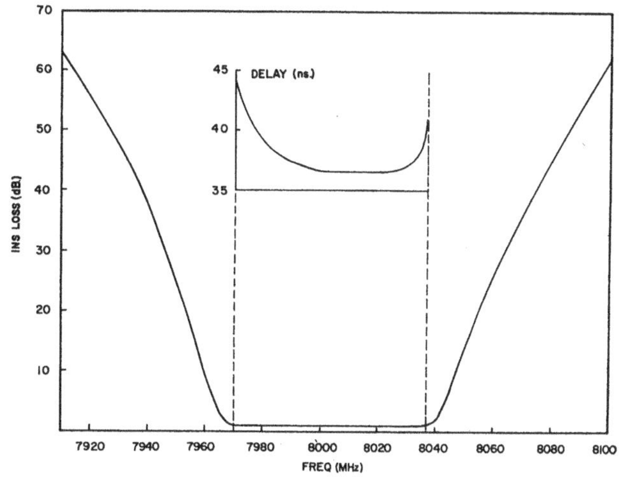

Fig. 9 — Performance characteristics for filter of Fig. 8.

REFERENCES

1. Rhodes, J. D., "The Stepped Digital Elliptic Filter", *IEEE Trans. on Microwave Theory and Techniques,* Vol. MTT-17, April 1969.
2. Rhodes, J. D., "The Stepped Cavity Coupled Elliptic Filter", *1969 G-MTT International Microwave Symposium Digest,* pp. 304-309.
3. Rhodes, J. D., "The Design and Synthesis of a Class of Microwave Linear Phase Filters", *IEEE Trans. on Microwave Theory and Techniques,* Vol. MTT-17, April 1969.
4. Rhodes, J. D., "Generalized Interdigital Linear Phase Networks With Optimum Maximally Flat Amplitude Characteristics", (submitted for publication).
5. Rhodes, J. D., "The Theory of Generalized Interdigital Networks", *IEEE Trans. on Circuit Theory,* Vol. CT-16, August 1969.

J. DAVID RHODES received the B.Sc. and Ph.D. degrees in electrical engineering from the University of Leeds, England in 1964 and 1966, respectively. From 1964 to 1966 he was a Research Assistant in the Department of Electrical and Electronic Engineering, Leeds, where he worked in the areas of microelectronic and microwave network theory. In the academic year 1966-1967, he was a Research Fellow in the same department. In the fall of 1967 he joined Microwave Development Laboratories, Inc., Natick, Massachusetts as a Senior Research Engineer. Since that time he has been engaged in a program to design new classes of microwave filters with particular reference to linear phase and elliptic types.

www.ingramcontent.com/pod-product-compliance
Lightning Source LLC
LaVergne TN
LVHW080042210926
848481LV00011B/3256
9780890060070